CALCIUM, NEURONAL FUNCTION AND TRANSMITTER RELEASE

TOPICS IN THE NEUROSCIENCES

CALCIUM, NEURONAL FUNCTION AND TRANSMITTER RELEASE

Proceedings of the
Symposium on Calcium, Neuronal Function and Transmitter Release
held at the International Congress of Physiology
Jerusalem, Israel—August 28-31, 1984

edited by

Rami Rahamimoff
The Hebrew University
Jerusalem, Israel

Sir Bernard Katz
University College London
London, England

Martinus Nijhoff Publishing
a member of the Kluwer Academic Publishers Group
Boston/Dordrecht/Lancaster

Distributors for North America:
Kluwer Academic Publishers
190 Old Derby Street
Hingham, Massachusetts 02043, USA

Distributors for the UK and Ireland:
Kluwer Academic Publishers
MTP Press Limited
Falcon House, Queen Square
Lancaster LA1 1RN, UNITED KINGDOM

Distributors for all other countries:
Kluwer Academic Publishers Group
Distribution Centre
Post Office Box 322
3300 AH Dordrecht, THE NETHERLANDS

Library of Congress Cataloging-in-Publication Data
Main entry under title:

Calcium, neuronal function and transmitter release.

The symposium was held in Jerusalem during the regional meeting of the International Union of Physiological Sciences.
Includes bibliographies.
1. Neurotransmitters—Congresses. 2. Calcium—Physiological effect—Congresses. 3. Neurons—Congresses. 4. Cellular control mechanisms—Congresses. I. Rahamimoff, Rami. II. Katz, Bernard, 1911- . III. International Congress of Physiological Sciences (1984 : Jerusalem) IV. International Union of Physiological Sciences. [DNLM: 1. Calcium—physiology—congresses. QV 276 C1435 1984]
QP364.7.C33 1986 591.1'88 85-31031
ISBN 0-89838-791-4

Printed in the United States of America

CONTENTS

INVITED LECTURERS

Dr. Daniel L. Alkon
Department of Neurophysiology
National Institutes of Health
National Institute of Neurological Disease & Stroke
Bethesda, Maryland 20014
USA

Prof. M. Blaustein
Department of Physiology
University of Maryland
School of Medicine
Baltimore, MD. 21201
USA

Prof. Arthur M. Brown
Department of Physiology & Biophysics
The University of Texas
Medical Branch
3102 Beluche
Galveston, Texas 77550
USA

Prof. Bruno Ceccarelli
Consiglio Nazionale delle Recherche
Centro per lo Studio delle Farmacologia delle
Infrastrutture Cellulari
20129 Milano
Via Vanvetelli, 32
Italy

Dr. Robert DeLorenzo
Department of Neurology
Yale School of Medicine
333 Cedar Street
New Haven, Connecticut 06510
USA

Prof. Joseph Dudel
Physiologisches Inst.
TUM Biedersteiner STR. 29
8000 Munchen 40
FRG

Dr. Chris J. Duncan
Department of Zoology
University of Liverpool
POB 147
Liverpool L69 3BX
UK

Dr. Simona Ginsburg
Department of Physiology
Hebrew University
Hadassah Medical School
Jerusalem 91010
Israel

Prof. Allan Grinnell
Department of Physiology
University of California
School of Medicine
Los Angeles, CA. 90024
USA

Dr. Bruce Howard
Department of Biological Chemistry
University of California
School of Medicine
Los Angeles, CA. 90024
USA

Dr. W.P. Hurlbut
The Rockefeller University
1230 York Avenue
New York, N.Y. 10021
USA

Dr. B. Kanner
Department of Biochemistry
Hebrew University
Hadassah Medical School
Jerusalem 91010
Israel

Dr. R. Kretz
Department of Anatomy
University of Fribourg
CH-1700 Fribourg
Switzerland

Dr. M.E. Kriebel
Department of Physiology
State University of New York
Upstate Medical Center
766 Irving Avenue
Syracuse, N.Y. 13210
USA

Dr. Rodolfo Llinas
Department of Physiology & Biophysics
New York University Medical Center
550 First Avenue
New York, N.Y. 10016
USA

Prof. H.D. Lux
Max-Planck-Institut fur Psychiatrie
Kraepelinstr 2
D-200 Munchen 40
FRG

Dr. A. Mallart
Centre National de la Recherche Scientifique
Laboratoire de Neurobiologie Cellulaire
92290 Gif-sur-Yvette
France

Dr. Halina Meiri
Department of Physiology
Hebrew University
Hadassah Medical School
Jerusalem 91010
Israel

Prof. J.W. Moore
Department of Physiology
Duke University Medical Center
Durham, North Carolina 27710
USA

Prof. Itzhak Parnas
Dean of Faculty of Sciences
Neurobiology Unit
Institute of Life Sciences
Hebrew University
Givaat Ram, Jerusalem
Israel

Dr. Hannah Parnas
Neurobiology Unit
Institute of Life Sciences
Hebrew University
Givaat Ram, Jerusalem
Israel

Dr. P. Pawson
Department of Physiology
University of California
School of Medicine
Los Angeles, CA. 90024
USA

Prof. Israel Pecht
Department of Chemical Immunology
The Weizmann Institute of Science
Rehovot 76100
Israel

Prof. Jonathan Pincus
Department of Neurology
Yale School of Medicine
33 Cedar Street
New Haven, Connecticut 06510
USA

Prof. David M. Quastel
Department of Pharmacology
University of British Columbia
2176 Health Sciences Mall
Vancouver, B.C. V6T 1W5
Canada

Dr. Hannah Rahamimoff
Department of Biochemistry
The Hebrew University
Hadassah Medical School
Jerusalem, 91010
Israel

Prof. Harald Reuter
Dean
Faculty of Medicine
University of Bern
3008 Bern, Friedbuhlstrasse 49
Switzerland

Dr. Julie Rosenheimer
Department of Physiology
University of Wisconsin
Madison, WI. 53706
USA

Prof. Brian M. Salzberg
Department of Physiology & Pharmacology
University of Pennsylvania
4001 Spruce Street A1
Philadelphia, PA. 19104
USA

Dr. Michel Simonneau
Centre National de la Recherche Scientifique
Laboratoire de Neurobiologie Cellulaire
91190 Gif-sur-Yvette
France

Prof. D.O. Smith
Department of Physiology
University of Wisconsin
Madison, WI. 53706
USA

Dr. A.E. Stuart
Department of Physiology
Medical Research Center
206H University of North Carolina
Chapel Hill, North Carolina 27514
USA

Prof. Stephen Thesleff
Department of Pharmacology
University of Lund
Lund, S-223 62
Sweden

Prof. W.G. Van der Kloot
Department of Physiology & Biophysics
State University of New York
at Stony Brook
Stony Brook, N.Y. 11794
USA

Prof. A. Wernig
Department of Physiology
University of Bonn
Wilhemstr 31
D-5300, Bonn 1
FRG

Prof. Robert S. Zucker
Department of Physiology
University of California
Berkeley, CA. 94720
USA

LIST OF PARTICIPANTS

M. Abdul-Ghani
Dept. of Physiology
Hebrew University-Hadassah
Medical School
P.O.B. 1172
Jerusalem 91010
Israel

M. Abeles
Dept. of Physiology & Computer Ctr.
of the Hebrew University &
Hadassah Medical School
P.O.B. 1172
Jerusalem, 91010
Israel

D. Alkon
Section on Neural Systems
Laboratory of Biophysics, NIH
Marine Biological Laboratory
Woods Hole, MA 02543
U.S.A.

C. Allweis
Dept. of Physiology
Hebrew University
Hadassah Medical School
P.O.B. 1172
Jerusalem 91010
Israel

R.M. Alvarado-Mallart
INSERM U-106
42 Rue Desbassayns de Richemont
Suresnes 92150
France

A. Amsterdam
Dept. of Hormone Research
The Weizmann Institute of Science
P.O.B. 26
Rehovot 76100
Israel

R. Ashkenazi
Biochemistry Dept.
Weizmann Institute of Science
P.O.B. 26
Rehovot 76100
Israel

S. Barash
Dept. of Physiology
Hebrew University
Hadassah Medical School
P.O.B. 1172
Jerusalem, 91010
Israel

Y. Beer
Max Starkman Laboratory
Dept. of Physiology
Hebrew University
Hadassah Medical School
P.O.B. 1172
Jerusalem 91010
Israel

O. Ben-Arzi
The Hebrew University
Jerusalem
Israel

H. Bergman
Rappaport Inst. for Research
P.O.B. 9697
Haifa 31096
Israel

F. Bergmann
Dept. of Pharmacology
The Hebrew University
Hadassah Medical School
P.O.B. 1172
Jerusalem 91010
Israel

O. Binah
Rappaport Inst. for Research
P.O.B. 9697
Haifa 31096
Israel

M.P. Blaustein
Dept. of Physiology
University of Md, Sch of Medicine
655 West Baltimore St.
Baltimore, MD 21201
U.S.A

B. Blum
Sackler School of Medicine
Tel Aviv University
Ramat Aviv, 69978
Israel

N. Browarnik
The Hebrew University
Jerusalem
Israel

A.M. Brown
Dept. of Physiology and Biophysics
University of Texas Medical Branch
Galveston, Texas 77550
U.S.A.

R.M. Buijs
Netherlands Institute for Brain Research
Z. Ijdijk 28,
1095 KN Amsterdam
Netherlands

G. Burnstock
Dept. of Anatomy & Embryology
University College London
Gower Street
London WC1E 6BT
U.K.

J.M. Camhi
Dept. of Zoology
The Hebrew Univesity
Jerusalem
Israel

P.L. Carlen
33 Russell Street
Toronto, Ontario M5S 2S1
Canada

M. Castel
Institute of Life Sciences
The Hebrew University
of Jerusalem
Jerusalem
Israel

B. Ceccarelli
Dept. of Pharmacology
CNR Center & Center of Perispheral
Neuropaties
Via Vanvitelli 32 20129 Milano
Italy

E. Cerasi
Hebrew University
Hadassah Medical School
P.O.B. 1172
Jerusalem 91010
Israel

D. Chernichovsky
Dept. of Physiology
Hebrew University
Hadassah Medical School
P.O.B. 1172
Jerusalem 91010
Israel

S. Constantini
Dept. of Physiology
Hebrew University
Hadassah Medical School
P.O.B. 1172
Jerusalem 91010
Israel

D. Dagan
Rappaport Inst. for Research
P.O.B. 9697
Haifa 31096
Israel

M. Dalith
Dept. of Pharmacology
Hebrew University
Hadassah Medical School
P.O.B. 1172
Jerusalem 91010
Israel

M. Dambach
Zoological Inst. University
Weyertal 119
D-5000 Koeln 41
FRG

G. David
Dept. of Physiology
Hebrew University
Hadassah Medical School
P.O.B. 1172
Jerusalem 91010
Israel

R.D. De Lorenzo
Dept. of Neurology
Yale Medical School
New Haven CT 06510
U.S.A.

D. De Wied
Rudolf Magnus Institute for Pharmacology
Vondellaan 6
3521 GD Utrecht
Netherlands

J.E. Desmedt
Brain Research Unit
University of Brussels
Belgium

M. Devor
Hebrew University,
Jerusalem
Israel

J.J. Dreifuss
Dept. of Physiology
C.M.U. 1, Rue Michel Servet
1211 Geneve 4
Switzerland

C. Duncan
Department of Zoology
University of Liverpool
P.O.Box 147
Liverpool L69 3BX
U.K.

H. Edery
Israel Institute for Biological Research
P.O.B. 19
Ness Ziona
Israel

E. Eilender
Zahal
Israel

D.A. Eisner
Department of Physiology
University College London
Gower Street
London WC1E 6BT
U.K.

I. Engberg
Aarhus University
DK 8000 Aarhus C
Denmark

A. Erdreich
Dept. of Biochemistry
Hebrew University-Hadassah Medical School
P.O.B. 1172
Jersualem 91010
Israel

G. Falk
Biophysics Dept.
University College
London WC1E 6BT
U.K.

S. Feldman
Hadassah Medical School
The Hebrew University
P.O.B. 1172
Jerusalem 91010
Israel

R. Fink
Department of Physiology
The Hebrew University
Hadassah Medical School
P.O.B. 1172
Jerusalem 91010
Israel

I. Friedberg
Tel-Aviv University
Ramat Aviv 69978
Israel

K. Fuxe
Dept. of Histology
Karolinska Institute
Box 60400
104 10 Stockholm
Sweden

H.M. Gerschenfeld
Lab. Neurobiologie
Ecole Normale Superieure
46 rue d'Ulm
75005 Paris
France

S. Ginsburg
Hebrew University
Hadassah Medical School
P.O.B. 1172
Jerusalem 91010
Israel

M. Globus
Dept. of Neurology
Hebrew University
Hadassah Medical School
P.O.B. 1172
Jerusalem 91010
Israel

E. Goren
Rappaport Inst. for Research
P.O.B. 9697
Haifa, 31096
Israel

I. Gozes
Dept. of Hormone Research
The Weizmann Inst. of Science
P.O.B. 26
Rehovot 76100
Israel

A. Grinvald
Weizmann Institute of Science
P.O.B. 26
Rehovot 76100
Israel

Y. Grossman
Unit of Physiology
Faculty of Health Sciences
Ben Gurion University of the Negev
Beer Sheva
Israel

Y. Gutman
Hebrew University
Hadassah Medical School
P.O.B. 1172
Jerusalem 91010
Israel

M. Gutnick
Ben Gurion University of the Negev
Health Sciences Faculty
Beer Sheva
Israel

C. Haimann
32 Via Vanvitelli
Milano 20 129
Italy

E. Heilbronn
University of Stockholm
Unit of Neurochemistry
Enkopingsv 126
S-172 46 Sundyberg
Sweden

U. Heinemann
Max-Planck-Inst. fur Psychiatrie
Am Klopferspitz 18a,
D-8033 Planegg-Martinsried
FRG

P. Hilman
Hebrew University
Jerusalem
Israel

S. Hocherman
Technion Faculty of Medicine
POB 9649
Haifa 31096
Israel

S. Hochstein
Faculty of Life Sciences
The Hebrew University
Givat Ram
Jerusalem
Israel

B.D. Howard
Dept. of Biological Chemistry
University of California
Los Angeles CA
U.S.A.

W.P. Hurlbut
c/o Rockefeller University
1230 York Avenue
New York, NY 10021
U.S.A.

O.F. Hutter
Institute of Physiology
University of Glasgow
Scotland G12 8QQ
U.K.

A. Ilani
Department of Physiology
The Hebrew University
Hadassah Medical School
P.O.B. 1172
Jerusalem 91010
Israel

J. Jansen
Oslo University
Institute of Physiology
Karl Johans Gate 47
Oslo 1
Norway

B. Kanner
Dept. of Biochemistry
Weizmann Institute of Science
P.O.B. 26
Rehovot 76100
Israel

S. Karlish
Dept. of Biochemistry
Weizmann Institute of Science
P.O.B. 26
Rehovot 76100
Israel

B. Katz
Dept. of Biophysics
University College London
Gower St.
London WC1E 6BT
U.K.

Y. Kloog
Dept. of Biochemistry
Tel Aviv University
Ramat Aviv, 69978
Israel

Y. Koch
Weizmann Institute of Science
P.O.B. 26
Rehovot 76100
Israel

M. Konishi
California Tech
California
U.S.A.

R. Kretz
Dept. of Anatomy
University of Fribourg
1 Rue Bockel
CH-1700 Fiebourg
Switzerland

M.E. Kriebel
Dept. of Physiology
Upstate Medical Center
Syracuse, N.Y. 13210
U.S.A.

Y. Lass
Dept. of Physiology
Sackler School of Medicine
Tel Aviv University
Ramat Aviv 69978
Israel

M. Lazdunski
Centre de Biochimie C.N.R.S.
Parc Valrose F 06034
Nice Cedex L
France

A. Lev Tov
Dept. of Anatomy
Hebrew University
Hadassah Medical School
P.O.B. 1172
Jerusalem 91010
Israel

A. Levitzki
Dept. of Biological Chemistry
Hebrew University
Hadassah Medical School
P.O.B. 1172
Jerusalem 91010
Israel

D. Lichtenberg
Dept. of Physiology & Pharmacology
Sackler School of Medicine
Tel Aviv University
Ramat Aviv, 69978
Israel

D. Lichtstein
Dept. of Physiology
The Hebrew University
Hadassah Medical School
P.O.B. 1172
Jerusalem 91010
Israel

H.D. Lux
Dept. of Neurophysiologie
Max-Planck-Institute of Psychiatry
Am Klopferspitz 18a
D-8033 Planegg-Martinsried
FRG

W.J. Malaisse
Laboratory Experimental Medicine
Blvd. de Waterloo
B-1000 Brussels
Belgium

A. Mallart
Laboratoire de Neurobiologie
Cellulaire
C.N.R.S.
91190 Gif sur Yvette
France

H. Meiri
Department of Physiology
Hebrew University
Hadassah Medical School
P.O.B. 1172
Jerusalem 91010
Israel

E. Melamed
Dept. of Neurology
Hebrew University
Hadassah Medical School
P.O.B. 1172
Jerusalem 91010
Israel

L. Mendell
SUNY-Stony Brook College
Dept. of Neurobiology
Stony Brook, NY 11794
U.S.A.

N. Meyerstein
Faculty of Health Sciences
Ben Gurion University of the Negev
Beer Sheva
Israel

D.M. Michaelson
Dept. of Biochemistry
Sackler School of Medicine
Tel Aviv University
Ramat Aviv 69978
Israel

B. Minke
Dept. of Physiology
Hebrew University
Hadassah Medical School
P.O.B. 1172
Jerusalem 91010
Israel

Y. Mizrachi
Dept. of Neurobiology
The Weizmann Institute of Science
P.O.Box 26
Rehovot 76100
Israel

J.W. Moore
Dept. of Physiology
Duke University, Medical Center
Durham NC 27710
U.S.A.

J.G. Nicholls
Dept. of Pharmacology Biocenter
Kingelbergstrasse 70
Bazel 4056
Switzerland

Y. Oron
Sackler School of Medicine
Tel Aviv University
Ramat Aviv, 69978
Israel

J. O'Keefe
Anatomy Dept.
University College London
Gower St.
London WC1E 6BT

Y. Palti
Rappaport Inst. for Research
Technion School of Medicine
P.O.B. 9697
Haifa 31096
Israel

H. Parnas
Dept. of Neurobiology
Life Science Inst.
The Hebrew University
Givat Ram
Jerusalem
Israel

I. Parnas
Inst. of Life Sciences
The Hebrew University
Givat Ram
Jerusalem
Israel

M. Pecot-Dechavassine
91, Rue de Normandie
94700 Maisons-Alfort
France

D. Pelzer
Dept. of Physiology 2
University of the Saarland
6650 Homburg/Saar
FRG

O.H. Petersen
University of St. Andrews
Physiological Laboratory
Fife
Scotland, KY16 9JU
U.K.

J. Pincus
Yale University
New Haven, CT
U.S.A.

S. Publicover
Dept. of Zoology
University of Birmingham
POB 363
Birmingham B15 2TT
U.K.

R. Pumain
Inserm
Inserm U97
2 ter rue d'Alesia, 75014
Paris
France

D.M.J. Quastel
Dept. of Pharmacology & Therapeutics JHILT
The University of British Columbia
2176 Health Sciences Mall
Vancouver, B.C. V6T 1W5
Canada

R. Radian
Dept. of Biochemistry
Hebrew University
Hadassah Medical School
P.O.B. 1172
Jerusalem 91010
Israel

H. Rahamimoff
Dept. of Biochemistry
The Hebrew University
Hadassah Medical School
P.O.B. 1172,
Jerusalem 91010
Israel

R. Rahamimoff
Dept. of Physiology
The Hebrew University
Hadassah Medical School
P.O.B. 1172
Jerusalem 91010
Israel

S. Redman
JCSMR, A.N.U.
Canbera, A.C.T.
Australia

H. Reuter
Dept. of Pharmacology
University of Berne
3010 Berne
Switzerland

S. Ron
Occupational Health &
Rehabilitation Center
Raanana POB 3 43 100
Israel

J. Rosenheimer
University of Wisconsin
Dept. of Physiology
470 North Charter Street
Madison 53706
U.S.A

S. Rotshenker
Dept. of Anatomy
The Hebrew University
Hadassah Medical School
P.O.B. 1172
Jerusalem 91010
Israel

P. Rudomin
Centro de Investigacion del I.P.N.
Apartado Postal 14-740
Mexico 14, D.F.

R. Sagi-Eisenberg
Dept. of Chemical Immunology
The Weizmann Institute of Science
P.O.B. 26
Rehovot 76100
Israel

B. Sakmann
Max-Planck Institute
Biophy, Chemie
P.O.B. 968 AM Fassberg
D-3400 Cottingen
FRG

B.M. Salzberg
4010 Locust St.
Philadelphia PA. 19104
U.S.A.

S. Sampson
Dept. of Life Sciences
Bar Ilan University
Ramat Gan 52100
Israel

D.J. Saphier
Dept.of Neurology
Hebrew University
Hadassah Medical School
P.O.B. 1172
Jerusalem 91010
Israel

M. Scherg
Max Planck Inst. for Psychiatry
Kraepelinstr, 10
D-8000 Munich 40
FRG

Y. Schiller
Dept. of Physiology
Hebrew University
Hadassah Medical School
P.O.B. 1172
Jerusalem 91010
Israel

M. Schramm
Bayer AC Institut fur Pharmakologie
Aprather Weg 18 a POB 10 17 09
5600 Wuppertal 1
FRG

I. Schulz
Max-Planck-Institut fur Biophysik
Kennedyallee 70
6000 Frankfurt/main 70
FRG

M. Schwartz
Dept. of Neurobiology
The Weizmann Institute
P.O.B. 26
Rehovot, 76100
Israel

M. Segal
The Weizmann Institute
P.O.B. 26
Rehovot, 76100
Israel

Z. Selinger
Institute of Life Sciences
The Hebrew University
Givat Ram, Jerusalem
Israel

O.B. Shahrabany
The Hebrew University
Jerusalem
Israel

A. Shainberg
Bar Ilan University
Ramat Gan 52100
Israel

Y. Shimoni
Hebrew University
Hadassah Medical School
P.O.B. 1172
Jerusalem 91010
Israel

S.A. Siegelbaum
Dept. of Pharmacology
Columbia University
New York, N.Y.
U.S.A.

F. Sigworth
Max Planck Institute
POB 968 AM Fassberg
D-3400 Gottingen
FRG

A. Silberberg
Weizmann Institute of Science
P.O.B. 26
Rehovot, 76100
Israel

S. Silberberg
Dept. of Physiology
Hebrew University
Hadassah Medical School
P.O.B. 1172
Jerusalem 91010
Israel

M. Simonneau
Neurobiologie Cellulaire
CNRS
91190 Gif sur Yvette
France

R. Spanier
Hebrew University
Hadassah Medical School
P.O.B. 1172
Jerusalem 91010
Israel

M. Spira
Dept. of Neurobiology
Inst. of Life Sciences
The Hebrew University
Jerusalem,
Israel

I. Steinberg
Weizmann Institute of Science
P.O.B. 26
Rehovot, 76100
Israel

Y. Steinitz
The Hebrew University
Jerusalem
Israel

A. Stuart
Dept. of Physiology 206 H
University of North Carolina
at Chapel Hill
UNC, Chapel Hill N.C. 27514
U.S.A.

A.J. Susswein
Dept. of Life Sciences
Bar Ilan University
Ramat Gan 52100
Israel

S. Thesleff
Dept. of Pharmacology
University of Lund
Solvegatan 10, S-223 62 Lund
Sweden

R. Tirosh
Dept. of Cell Biology
The Weizmann Institute of Science
P.O.B. 26
Rehovot, 76100
Israel

W. Trautwein
Dept. of Physiology
University of the Saarland
6650 Homburg/Saar
FRG

G. Trube
Max-Planck-Institut fur
Biobphys. Chemie,
POB 2841
D-3400 Gottigen
FRG

K.J.M. Ullrich
Max-Planck-Institut fur
Biophysik, Kennedyallee 70
6000 Frankfurt/Main 70
FRG

E. Vaadia
John Hopkins Medical School
Baltimore, Maryland
U.S.A.

W.G. Van der Kloot
Physiology HSC
SUNY at Stony Brook
Stony Brook, N.Y. 11794
U.S.A.

H. Van Wilgenburg
University of Amsterdam
Dept. of Pharmacology
Polderweg 104
Amsterdam 1093 KP
Netherlands

G. Vassort
Inserm
U-241 Bat, 443
Unite de Recherches de Physiologie
Univ. Paris Sud 91405 ORSAY
France

R. Voskuyl
University of Leiden
Leiden
Holland

J. Weidenfeld
Laboratory of Experimental Endocrinology
Dept. of Neurology
Bikur Holim Hospital-
Hebrew University
Jerusalem
Israel

M. Weinstock-Rosin
Dept. of Pharmacology
Hebrew University
Hadassah Medical School
P.O.B. 1172
Jerusalem 91010
Israel

R. Werman
Faculty of Life Sciences
The Hebrew University
Givat Ram, Jerusalem
Israel

J.M.W. Werner
Ruhr-Universitat
AG Electrophysiologie
MA4/59 D 4630
Bochum
FRG

A. Wernig
Dept. Physiol.
University Bonn
31 Wilhelm Str.
D-5300 Bonn
FRG

H. Winkler
Institut f. Physiologie
Fritz-Presel Str. 3
A-6020 Innsbruck
Austria

Z. Wollberg
Dept. of Zoology
Tel Aviv University
Ramat Aviv 69978
Israel

Y. Yaari
Dept. of Physiology
Hebrew University
Hadassah Medical School
P.O.B. 1172
Jerusalem 91010
Israel

Y. Yarom
The Hebrew University
Jerusalem
Israel

M. Yoon
Dept. of Psychology
Dalhousie University
Halifax, N.S. B3H 4J1
Canada

M.B. Youdim
University of Arizona
College of Medicine
Dept. of Pharmacology
Tucson, Arizona 85724
U.S.A.

R.S. Zucker
Physiology-Anatomy Dept.
University of California
Berkeley, Ca 94720
U.S.A.

PREFACE

This symposium entitled Calcium, Neuronal Function and Transmitter Release, was in the framework of the regional meeting of the International Union of Physiological Sciences, that took place in Jerusalem between August 26-31, 1984.

The symposium dealt with the role of calcium ions in regulation of a large number of important processes in modern neurobiology, from molecular and cellular points of view. In this context, we consider heart and most cells as 'honorary neurons'.

The meeting was comprised of lectures and quite intense discussions. We hope that the transcription of the discussions which follow the articles will give the reader a feeling of the intense, but pleasant atmosphere that pervaded during this symposium.

It is our pleasure to thank Mrs. Miriam Silber, the assistant editor of this book, for her hard work in transcribing the discussions, retyping large portions of the book, and getting the approval of the authors. We express our thanks to Dr. Halina Meiri and Dr. Simona Ginsburg for their editorial assistance. The unfailing help of Ms. Rachel Klein and Ms. Shoshana Wineberg is greatly appreciated.

SECTION 1

CALCIUM CHANNELS, TRANSPORTERS & CALCIUM REGULATED CHANNELS

EFFECTS OF TEMPERATURE ON SINGLE CHANNEL AND WHOLE CELL Ca CURRENTS

A. M. BROWN, D. L. KUNZE, H.D. LUX AND D. L. WILSON

INTRODUCTION

Calcium tail currents have a large, fast component which is not detectable when the currents are turned on or activated from the completely rested state (Brown et al, 1983). This is probably because a rate limiting process in the earlier stages of activation dominates the turn-on. There is also evidence that Ca channels are regulated by transmitters (Reuter, 1983) and by Ca ions themselves (Brehm and Eckert, 1978; Tillotson, 1979; Brown et al, 1981), although Lux and Brown (1984) have recently speculated that the latter effect may not be due to a direct action of entering Ca ions. These findings suggest a chemical reaction may be involved in Ca current activation and expectation is that the kinetics of Ca currents might be strongly affected by changes in temperature. Moreover the temperature effects might have specificity for certain kinetic components. Therefore, in these experiments we have examined the effects of varying temperature on single Ca channels and whole cell Ca currents.

METHODS

The experimental set up is shown in Figure 1. Whole cell and patch clamp currents were measured simultaneously (Figure 1B) in neurons from Helix Pomatia using the methods described in Lux and Brown (1984a). A Na-free solution containing (mM) 40 $CaCl_2$, 5 $MgCl_2$, 35 tetraethylammonium (TEA)-Cl, 5 4-aminopyridine (AP) and 20 Tris buffered to pH 7.6 was used in the bath and the patch pipettes. When necessary whole cell Ca currents were blocked by the addition of Ni 10 uM to the bathing solution. The bath temperature was controlled by a thermoelectric device which used a thermistor in a negative feedback circuit. Room or control temperature was 20°C.

In some experiments whole cell Ca currents from neurons of Helix Aspersa were studied using the combined microelectrode-suction pipette method (Figure 1A) as described in Brown et al (1983). The internal dialysis solution contained (mM) 135 Cs Aspartate, 10 TEA-Cl, buffered to pH 7.35 and the bathing solution contained (mM) 35 Tris-Cl, 5 CsCl, 50 TEACl, 5 4-AP, 10 $CaCl_2$, 15 $MgCl_2$ and 5 Glucose and was buffered to pH 7.55.

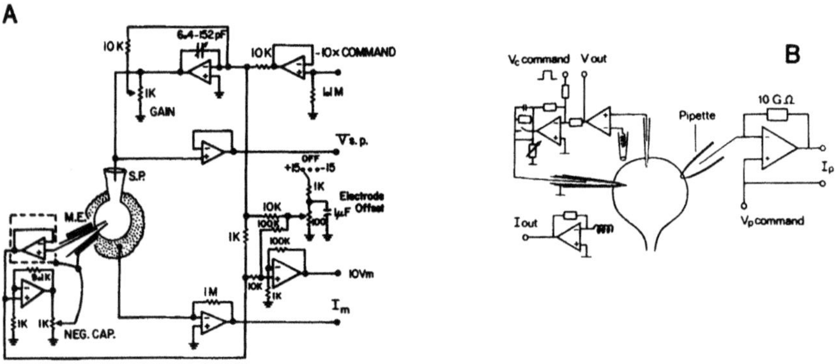

FIGURE 1. A. Diagram of the hybrid suction pipette-microelectrode voltage-clamp system for measuring macroscopic currents. A micro-electrode (M.E.) was used for voltage-sensing and a suction pipette (S.P.) was used for passing current and internal perfusion of the cell; they are shown with a neurone immersed in a patch fluid. The micro-electrode headstage was attached directly to the micro-electrode holder, and it, and the micro-electrode shield were driven from the capacitance neutralization circuit.
B. Experimental arrangement for recording patch and whole cell clamp currents. A two microelectrode system was used for voltage clamp and voltage commands were normally applied to this clamp circuit. Patch currents were recorded with a conventional patch clamp amplifier. Whole cell current was recorded from an I-V converter connected to an electrode in the bath. The reference for the patch clamp was the same as that for the voltage clamp.

Currents were recorded at frequencies from DC to 5kHz and filtered by a 4-pole Bessel filter before digitizing at rates usually well above those required to satisfy the sampling theorem (Nyquist, 1928). Linear components of leakage and capacitance were subtracted. For single channel data the upper frequency limit at -3dB was usually 1 and occasionally 1.5 kHz. This filtering was done using a zero phase digital filter. The single channel data were analyzed using a threshold detector program described in Lux and Brown (1984a) with the threshold set at levels that were usually at least 3 times the standard deviation of the background noise. The true amplitudes were measured and the width of the event at half maximum amplitude was used to set the times of channel opening and closing. Idealized noise-free rectangular events were then constructed from the time and amplitude data and were used for amplitude, open time and closed time histograms and other measurements which will be described at appropriate points in the text. Observable events less than 300 usec in duration were usually excluded from the analysis because they fell below the dead time of our recording system.

RESULTS AND DISCUSSION

Figure 2 (AA) shows that warming produced a large increase in the rate of turn-on or activation of the Ca current from the completely rested state. The latter was achieved by using holding potential of -50mV where the fraction of channels activable, h_{ss}, was 1.0. The linear components of leakage and capacitance have been subtracted and a small asymmetry current precedes the inward Ca current. In contrast to the marked effects on turn-on the rate of turn-off or deactivation of the tail currents was scarcely affected by the change in temperature (Fig 2AB & AC). In addition to the effect on the rate of activation the peak amplitude obtained during the 1.0 msec test pulse was greater. Inactivation also began earlier. As we have already noted (Brown et al, 1983) the tail taus have a ratio of about ten (fast:slow) which excludes a Hodgkin-Huxley type of activation scheme. Nevertheless the turn-on could be conveniently fit by an m^2 H-H formulation as indicated in the figure legend. It is interesting to note (Fig 2B) that tau m changed by a ratio of 4-6 over the potential range plotted, whereas the turn-off deactivation kinetics were little altered.

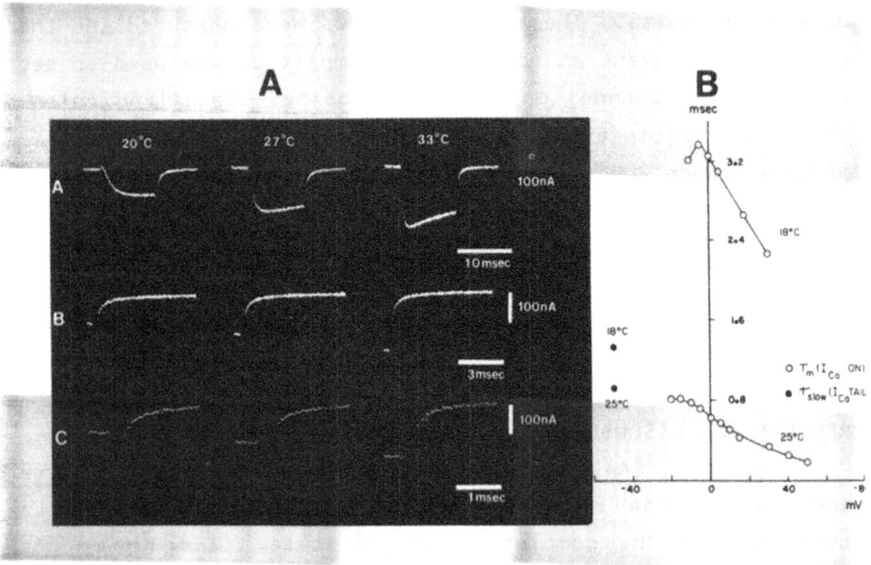

FIGURE 2 A. Effects of temperature changes on Ca currents at +25 mV in A and on tail currents following a repolarizing step to the holding potential, V_H, of -50 mV in B and C. Note the different time scales in A, B and C.
B. In another cell, the effect of temperature on m^2 (see text) parameters was obtained at two temperatures. Turn-on was fitted to the following equation

$$I = G_1(1 - e^{-t/\tau m})^2 + G_2 e^{-t/\tau m}$$

where the second term is used to describe the asymmetry current present. τ_m is plotted as a function of the pulse potential. Also shown are two points representing τ_{slow} (the slower time constant obtained from tail relaxations) at V_H.

Next we analyzed the effects of changes in temperature on single Ca channels (Fig 3). At both temperatures the average patch currents from a single channel and the entire Ca current from the whole cell overlaid each other when scaled, and both showed similar activation and inactivation processes during a sufficiently prolonged voltage step. In both cases the peak current I-V relations were simply scaled by changes in temperature. The single channel records show the bursting discharge which produces two closed time distributions, one due to the gaps within a burst, the other to the intervals between bursts. Inactivation occurs at both temperatures and, as we have pointed out (Lux and Brown, 1984b), the process is not associated with any change in unitary current amplitude, mean open time or the mean value of the faster of the two closed times. The inactivation is due to a reduction in the opening probability, P_o, with time and this in turn is due to a progressive increase in the mean value for the longer of the closed time distributions. Thus we can rule out processes of accumulation and depletion suggested for other Ca channels (Almers et al, 1984) and, as we have recently noted, the Ca entry hypothesis also does not seem to be applicable.

Changing the temperature from 29 to 9°C decreased peak whole cell current to about 1/3 at zero mV potential (Fig 3). This was due to two factors: a fall in elementary current to 2/3 its previous value (Fig 4A), and a reduction by 50 per cent in the peak P_o. The latter may be calculated from

$$\langle I \rangle = P_o \, N \, i$$

where i is the elementary current, $\langle I \rangle$ the averaged single channel current and N the number of channels in the patch. In the patch there was only one single channel (N=1) as judged by the absence of simultaneous openings which should have been observed for the recording times we used, were the binomial theorem to apply. The close correspondence between the changes in whole cell currents predicted from the single channel results and the changes we measured indicates that the number of channels in the entire cell also was not altered with the change in temperature.

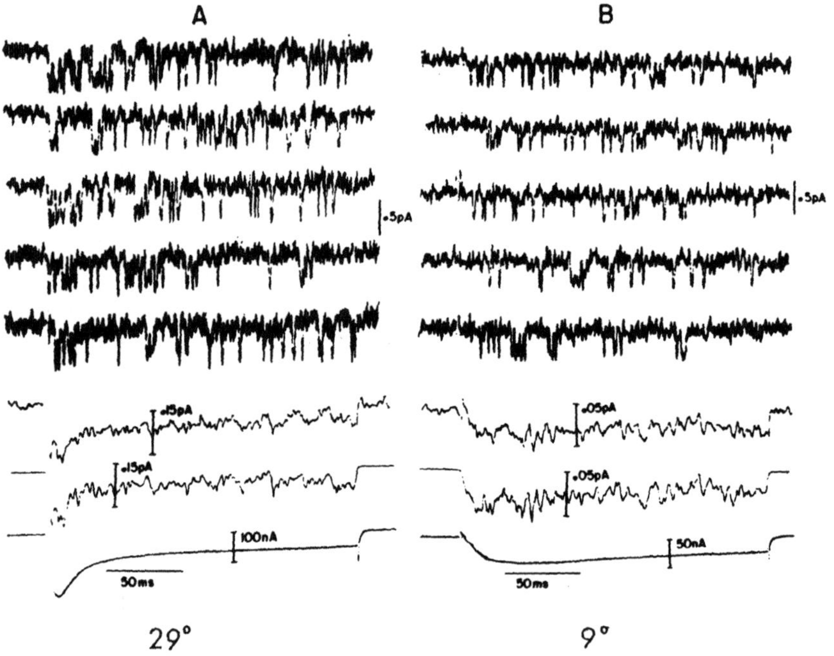

FIGURE 3. Single channel recordings at 29 (A) and 9°C (B). V_H was -50 mV, V_M was zero mV, pulse duration 200 msec. Representative single channel records are shown for each case. The three traces below in each set of single channel records are from top to bottom: the averaged current from the single unit patch records, the averaged current from the idealized single unit traces, and the current from the whole cell recorded simultaneously with the patch currents. Only a single sample of the latter is shown. Currents from the single unit patch recordings are averaged from 31 traces at 29°C and 49 traces at 9°C. Only one failure occurred and that was at the lower temperature.

The open times were not significantly altered by the change in temperature (Fig 4). Bursting was also not altered; neither burst length nor gaps per burst were different. The major changes as the figure legend describes were in the closed times, the longer of the two measured closed time distributions being particularly affected. The other most significant changes were in the time to peak current. This was altered by a factor of 4-6 (Fig 2 B), and at the single channel level a change of similar magnitude was observed in the latencies to first opening or waiting times (Fig 5).

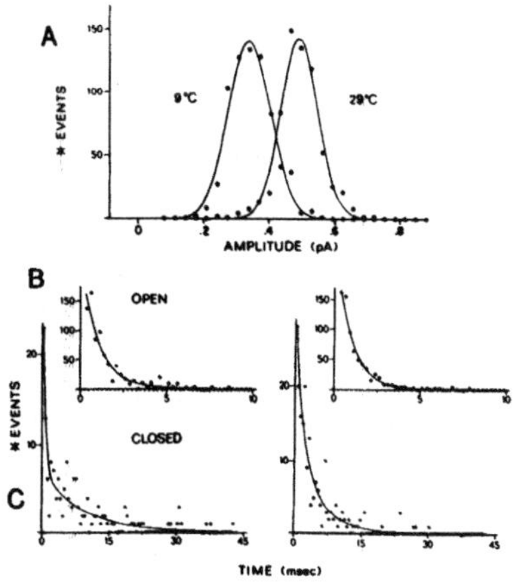

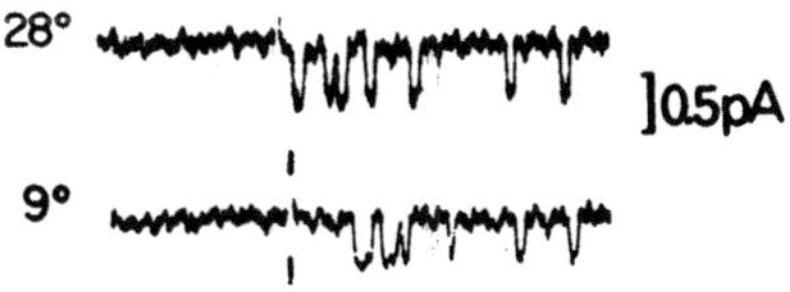

FIGURE 4. A. Amplitude histograms for the single Ca channel of Figure 3 at OmV. Values are .483 ±.52 PA at 29°C and .32 ±.58 PA at 9°C.
B. Open time distributions. Under the same conditions mean open times were 1.05 msec at 29°C and 1.20 msec at 9°C. Average values in msec at the same potential were from 10 cells at 29°C 1.48 ±.52; from 5 cells at 9°C 1.52 ±0.48; and from 20 cells at 18°C 1.60 ±0.56. At least 500 openings were measured for each cell. Thus open times at this bandwidth were little altered by these temperature changes.
C. Closed time distributions. Histograms of closed times were fit with a sum of two exponentials with "fast" and "slow" time constants. The fast time constant was 1.88 msec at 29°C and 1.82 msec at 9°C. The slow time constant had a mean value of 4.9 msec at 29°C and 11 msec at 9°C. Thus the longer of the closed time distributions were changed markedly by temperature.

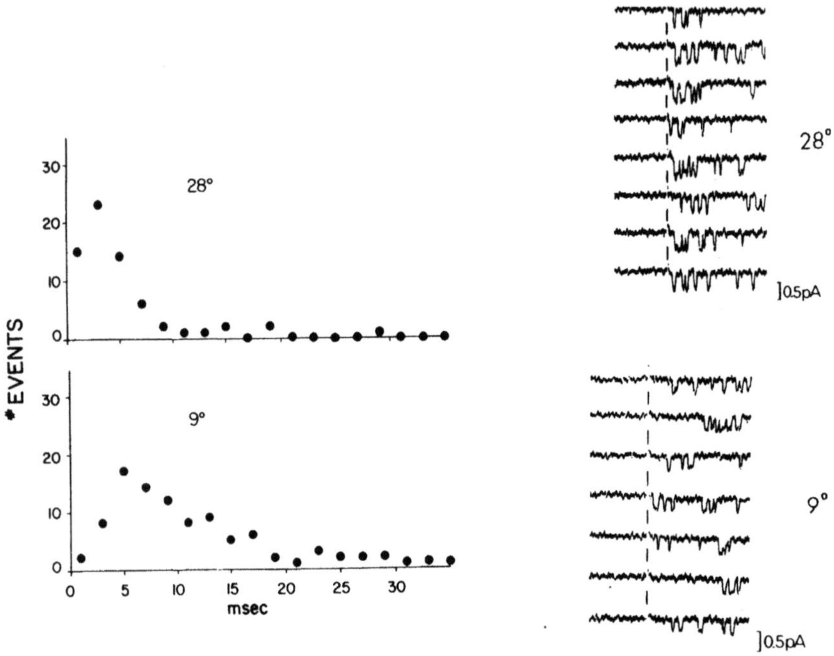

FIGURE 5. Effect of temperature on the waiting time histogram at -5 mV using the single channel data shown. The channels shown consist of records of ∿40 ms duration. The number of measurements was 68 at 28°C and there were zero failures. The number of measurements at 9°C was 72 and there were 15 failures. Note that with the decrease in temperature there is a marked "slowing" and broadening of the waiting time distribution.

The effects of temperature on single channel and whole cell behavior correlated very well. The mean open time which has little voltage dependence is not temperature dependent either. While effects on steady state parameters such as unitary current amplitude and peak opening probability were clear, the most striking effects were on the rates of activation and inactivation, although the latter effect was not analyzed in any detail. The results are consistent with at least one temperature dependent transition near the rested state and this may be related to a metabolically or transmitter regulated step in activation.

SUMMARY

At lower temperatures, turn-on of whole cell Ca currents slowed, but turn-off or tail currents were only slightly affected. Inactivation during a voltage was very much slowed. Average patch and whole cell currents overlayed when scaled, and no change in the size of the unitary Ca events, i_{Ca}, during a pulse was detected. Peak current IV's appeared to be simply scaled by changes in temperature. Changing temperature from 29 to 9°C decreased whole cell Ca current I_{Ca} by ∿1/3 at 0 mV. This was due to decreases in i_{Ca} (Q_{10} = 1.2) and peak Po (Q_{10} = 1.5). Kinetic properties of activation appeared to be much more affected than did steady-state properties; the time-to-peak increased by a factor of 4-6 and the single channel times-to-first-opening were markedly slowed. Open times were not markedly affected (1.05 ms (29°) and 1.2 ms (9°)). Closed times were fit by a sum of two exponentials; the fast time constant was relatively little affected but the slow time constant was doubled. The results are consistent with a temperature-dependent transition near the rest state, which may also be regulated metabolically or by transmitters.

REFERENCES

1. Brown AM, Tsuda Y. Wilson DL: A description of activation and conduction in calcium channels based on tail and turn-on current measurements. J Physiol 344:309-324, 1983.

2. Reuter H: Calcium channel modulation by neurotransmitters, enzymes and drugs. Nature 301:564, 1983.

3. Brehm P, Eckert R: Calcium entry leads to inactivation of calcium channel in paramecium. Science 202:1203-1206, 1978.

4. Tillotson D: Inactivation of Ca conductance dependent on entry of Ca ions in molluscan neurons. Proc Natl Acad Sci USA 3:1497-1500, 1979

5. Brown AM, Morimoto K, Tsuda Y, Wilson D: Calcium current-dependent and voltage-dependent inactivation of calcium channels in Helix aspersa. J Physiol 320:193-218, 1981.

6. Lux HD, Brown AM: Patch and whole cell calcium currents recorded simultaneously in snail neurons. J Gen Physiol 83:727-750, 1984a.

7. Nyquist H: Thermal agitation of electric charge in conductors. Phys Rev 32:110-113, 1928.

8. Lux HD, Brown AM: Single channel studies on Ca inactivation. Science 225:432-434, 1984b.

9. Almers W, McCleskey EW: Sodium and barium current in calcium channels of frog muscle: Calcium selectivity in a single-file pore. J Physiol 353: in press, Aug. 1984.

This work was supported in part by NIH grant NS-11453.

13

DISCUSSION OF THE ARTICLE BY A.M. BROWN (U.S.)

I. Parnas (Israel): If you plot the delay histograms of the mepps or the epps after a nerve impulse in different conditions of calcium or after a train, the delay histograms are exactly the same. You probably cannot account for slowing of the delay in changes of the calcium currents, but for the events that occur after the calcium enters the terminal and triggers the release.

A.M. Brown (U.S.): No comment.

D. Pelzer (Germany): Your four-state kinetic model of Ca channel gating related to Ca channel activation requires that both the distribution of all closed times and the tail currents are composed of the sum of three exponential terms whose time constants have to be well-separated whereas your experimental data and those of others indicate only two. What is the reasoning for your proposed sequence of Ca channel states passage through which accounts for channel activation in the light of the apparent lack of a third time constant? For Ca channels in other excitable cell membranes (e.g. heart and bovine chromaffin cells), a three-state sequential reaction scheme with two closed states preceding the open state has turned out to be sufficient to describe Ca channel activation and to account for the occurrence of Ca channel activity in bursts of individual channel openings.

(Fenwick, Marty, Neher (1882). J. Physiol. 331, 599.
Cavalie, Ochi, Pelzer, Trautwein (1983). Pflugers Arch. 398, 284;
Reuter (1984). Ann. Rev. Physiol. 46, 473.
Tsien (1983). Ann. Rev. Physiol. 45, 341).

A.M. Brown (U.S.): Thank you for this question since I didn't have time to go into this during the talk. The initial minimum model for activation which accounted for the bursting and for two time constants of deactivation has three states. This is shown here but this model would not account for all of our results. In the tail currents, we clearly had two time constants; a fast one and a slow one which differed by an order of magnitude. However, if we took the time constants from these tail currents and applied them to the turn-on process, they could not account for the delay that was observed. The fast process was simply too fast to give us the delay that was necessary for the turn-on process. Therefore, we had to introduce an additional step. Where is the third time constant as you ask? De-activation tails only showed two time constants but it turns out that the slow time constant is actually the average value for the two slower time constants we used to fit the turn-on process. Thus, the tail currents do in fact have three time constants but one of them is the weighted sum of the two slower time constants used for turn-on which differ as it turns out by a factor of two.

A CHANNEL VIEW AT CA INACTIVATION

H.D. LUX and A.M. BROWN

INTRODUCTION

Ca currents rise and fall during a step change in potential (see 1-3) and these changes are termed activation and inactivation. Activation is voltage-dependent but the voltage dependence of inactivation has been recently contested (4-8). The occurrence of inactivation may be exaggerated because outward K currents, particularly Ca-activated currents, and other non-specific outward currents have time courses that coincide with inactivation. The process is greatly reduced when outward currents are minimized. Tail currents measured at return potentials where outward currents should be negligible have envelopes that follow the rise and fall of I_{Ca} and the result is readily taken as evidence of inactivation of I_{Ca} (9, see also 10).

The basic observation pointing to current-dependent inactivation stems from a conditioning-test pulse protocol. It was observed that inactivation decreases as the prepulse potential approaches levels at which Ca influx is minimal. The interpretation is that Ca ions entering the cell produce inactivation and it is strengthened by the observation that buffering intracellular Ca slows inactivation. A principal model for what we shall call Ca-influx dependent inactivation has been proposed (11,12).

Nevertheless voltage-dependent inactivation of Ca currents is clearly demonstrated (13,14) in particular by employing pulse protocol with conditioning potentials below threshold for I_{Ca}. Other evidence addressing this question has been reported. Hence, inactivation of Ca currents may be voltage-dependent, Ca-influx dependent or both.

Recently the behavior of single Ca channels with Ca as the conducting ion has been described (15-17). Openings of single Ca channels provide by themselves instantaneous and quantitative Ca injections and their significance for Ca inactivation can directly be studied on the subsequent behavior of the channels.

The results show that single Ca channels inactivate during a voltage clamp step. There is no correlation between opening activity and subsequent intervals as predicted by a Ca-influx dependent process (18). We conclude that at the single channel level inactivation during a voltage clamp step depends on other factors such as on the pre-existing intracellular Ca activity and on voltage.

METHODS

The experiments were done on neuronal cell bodies of Helix pomatia (60 to 120 μm in diameter) located in the right parietal ganglion of the subesophageal mass (19). During experiments, a Na-free solution having (mM): 40 $CaCl_2$, 5 $MgCl_2$, 35 Tetraethylammonium chloride, 5 4-aminopyridine, 20 Tris (pH 7.4) was used. This was also the filling solution of the patch pipettes. For Ba studies, Ba ion was substituted isosmotically for Ca ion. Electrical recording was done using the scheme in Figure 1A. Whole cell and gigaseal patch clamp currents were registered simultaneously (for details see (19)). Recordings were made at a bandwidth between DC and 3 kHz using a high frequency compensation circuit. The data were filtered at 1 kHz and digitized using a PDP 11/34 minicomputer at rates that ranged from 0.1 to 0.25 ms per point. Control and test records had lengths of 200 ms to 2 sec and intervals of 20 to 30 sec were used between trials to avoid commulative inactivation as checked by the whole cell currents. Linear components of leakage and capacitance were subtracted from the test records. The pipette current was carefully nullified before formation of a seal. Unitary events were analyzed using a zero phase digital filter (20).

The background r.m.s. noise at optimum high frequency compensation and 1 kHz bandwidth was about 0.08 pA. It was measured before depolarization and in channel-free intervals and used to set up two threshold detectors, one for openings with level ϕ and the other for incomplete closings. The latter was set below baseline at about 1.5 times the background rms and ϕ was set at 3-4

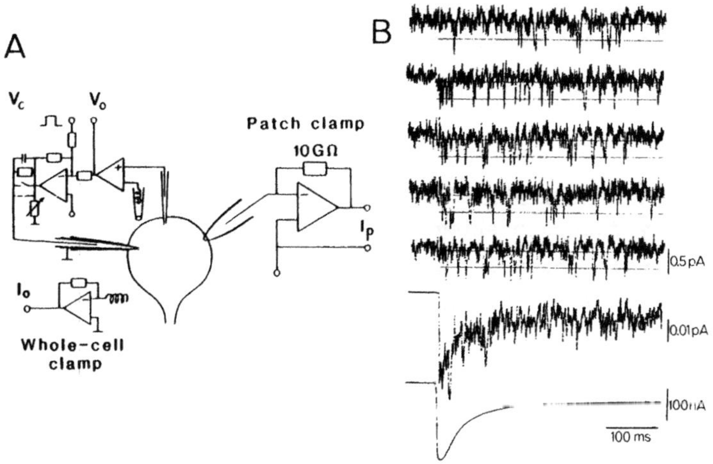

Fig. 1. (A) Experimental arrangement. A two-microelectrode voltage clamp was used to deliver voltage steps, and whole-cell current was measured with the bath current-to-voltage converter. The patch clamp was used for recording patch currents; voltage commands were usually not applied to the patch electrode. (B) Comparison of averaged single-channel and whole-cell currents. Five samples of single channel events are shown. Immediately below is the averaged current obtained from 54 idealized channel records. The bottom record shows a whole-cell current that was recorded simultaneously, showing a similar inactivation time course. Holding potential (V_H) was -50 mV; test potential was 0 mV. Temperature was 29°C to increase rates of activation and inactivation. The effects were also present at 20°C. Bandwidth was 1.0 kHz.

times rms. This was usually about two thirds the amplitude of the openings studied. With an opening probability of 0.1, this choice of ø should keep the average frequency of false events at or below 1/s (22, see 19). An opening transition was identified when the higher level was crossed and a closing transition was identified when the lower level was crossed. Once an event was detected, the program computed the average amplitude and measured the open time at the width corresponding to half-maximum amplitude. Idealized events were construced by setting the points below the lower level to zero and taking the amplitude from the average of the values between the opening and closing transitions. These idealized values were used to construct the histograms. Events of 300 μs duration were rejected from the histogram analysis since they fell below the resolution of the recording system.

RESULTS

Behavior of states during inactivation

The primary data are shown in Fig. 1B. The unitary currents are from a single channel and are due to the flow of Ca ions from the pipette into the cell. The average current from an ensemble of single channel records is shown below along with the whole cell current recorded simultaneously. The similarities of rise and fall of averaged single channel and whole cell currents are obvious. There was no evidence of any outward currents. This removes previous uncertainties about a simulated inactivation by Ca or non-Ca-related outward currents. Raising the bath temperature to near 30°C as done in the present study on Helix neurons speeds and pronounces inactivation. The decay of the whole cell current is well approximated by two exponentials of similar sizes with time constants of about 40 and 300 ms. A similar biphasic time course appears in averages of larger numbers of channel recordings. Of interest is the fact that the Ca channels re-open following closures while inactivation is present in the averaged patch current. The inactivated state is obviously not absorbing in contrast to recent proposals for the Na channel (see 23). This feature of the channel would facilitate the analysis of dwell times of inactivated states.

Mean open times and amplitudes of openings were compared between the rising and falling phases and were found to remain unchanged indicating one open state throughout. The closed times of Ca channels appear distributed in at least a double exponential manner (16-20,24,25). For our purposes, we compared the average values of the closed times during the first 50-75 ms, which encompassed the peak of the current, with the average values during the last 200 msec when the current was inactivated and had attained an almost steady level (Fig. 2).

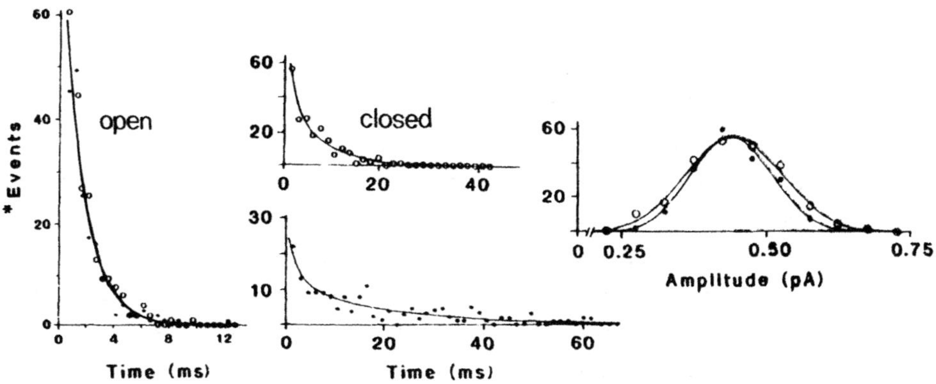

Fig. 2. Distributions of open times, closed times, and amplitudes for the first 50 ms (peak activity), o) and the last 150 ms (inactivation, ●) of an ensemble of 70 samples. V_H, -50 mV; V_{max}, 0 mV. The prolongation of the average closed time was the only significant change.

The average value of the closed time increased about 3-fold between these times which is thus accounts for the reduction of the occupation time of the open state seen in the averaged currents.

Inactivation does not appear to affect the average open time of the channel but the question is whether inactivation is coupled to or independent on activation towards opening. If the open state is to be filled before the channel can transit into a singular inactivated state as recently suggested for the Na channel (22) some predictions of its behavior can be made on the basis of previously proposed activation schemes (16-20,23, 24). We let R, A, O and I stand for "rest", "activated", "open" and "inactivated" states, respectively, and consider the probability of transitions between the open state and both A and I.

$$R \underset{k_{-1}}{\overset{k_1}{\rightleftharpoons}} A \underset{k_{-2}}{\overset{k_2}{\rightleftharpoons}} O \underset{q_{-1}}{\overset{q_1}{\rightleftharpoons}} I$$

The change of the occupancy of the inactivated state is described by

$$dp_I/dt = q_1\ p_O - q_{-1}\ p_I \qquad (1)$$

and similarly

$$dp_O/dt = k_2\ p_A + q_{-1}\ p_I - (k_{-2} + q_1)\ p_O \qquad (2)$$

holds for the open state. $p_I(t)$ can be assumed to follow the averaged decay of the single channel current in line with the observations of the increase of long closed times which is attributable to inactivation. Solving Eqns. (1) and (2) for the rate constants q is possible with previous determinations of k_2 and k_{-2} (20) and observed averages of $p_o(t)$. At low dp_I/dt near steady inactivation and with $k_2 \hat{=} 0.4/ms$, $k_{-2} \hat{=} 0.7/ms$ and $p_o \hat{=} 0.05$, q_1 and q_{-1} attain values of 10^{-2} to $10^{-3}/ms$. The mean open time, which is the reciprocal of the sum of the rates for leaving that state, is not much different from that in the absence of the inactivating pathway since k_{-2} ($\hat{=} 0.7/ms$) $\gg q_1$. Calculations on the basis of such model with these values of $q_{1,-1}$ predict inactivation time courses which are generally longer than those observed. In particular, they fell short of accounting for the early rapid inactivation slope. The open state can be combined with the short lived closed state to define the bursting time (16,17,20, 24,25). If inactivation proceeds from this complex, the same defect is produced because both the occupancy of the bursting 'state' and the rate constants for leaving are low. The conclusions derived from evaluating the above model and the model itself are rather approximate but sufficient to suppose that inactivation in addition or in general affects closed states.

Search for Ca current dependence

A specific deterministic model for Ca channel inactivation utilizes the idea that Ca entering through a channel in turn inactivates that channel (11,12). In this view, the lengths of individual openings or bursts of openings determine the subsequent reaction which may be probabilistic. A simple case is a Ca ion reacting with a site to form a complex that closes the channel. The equilibration with dissociation constant K is assumed to be instantaneous on the time scale of Ca channel activation. The change of intracellular calcium ($\dot{Ca}$) by entry through a channel relates to the probability that a channel having been opened will continue to open i.e.

$$\dot{Ca} = D \cdot \frac{K}{Ca + K} = D \cdot p_o \qquad (3)$$

where the constant D relates to geometric factors and constant Ca buffering (11,12). Hence

$$\dot{Ca}\ Ca + \dot{Ca}\ K = D \cdot K \qquad (4)$$

The solution (18) for this differential equation is

$$Ca = K \pm (2L)^{1/2}\{(t - t_o) + \frac{(Ca_o + K)^2}{2L}\}^{1/2} \qquad (5)$$

where t_o and Ca_o are integration constants defining the initial conditions and L = DK.

dCa/dt is proportional to current $I(t) = ip_o(t)$ for single channels. Since i is constant throughout inactivation both in amplitude and mean open time, the changes are due to changes in $p_o(t)$ and it follows from Eqn. (5) that

$$dP_o(t)/dt = -R(p_o(t))^3 \qquad (6)$$

where R is a constant. The result means that a decrease of the occupancy of the open state is related inversely by the third power to the preceding opening duration.

We investigated this relationship by searching for correlations between durations of open and closed times in the data. Such correlations was not apparent when the open time and its succeeding closed time are compared or vice versa (18). However, comparisons of unitary events, event by event, suffer from the fact that brief openings or closings may be occurring faster than the equilibrium of the reaction proposed in Eqn. (3). Therefore, a method we refer to as a burst catching scattergram was used to compare averaged data occurring over longer time periods corresponding to times of bursts of openings (18).

The burst catchers are established as follows. A time window of 5-15 ms, W_1, is set up to contain openings with an average p_o within two thresholds, T_1 and T_2, $0 \leq T_1 \text{–} T_2 \leq 1$. The window scans the records with the start of openings. If the average in W_1 falls outside the interval T_2-T_1, a second window, W_2 is instituted where W_2 is usually longer than W_1 by a factor of about two. This window length (of 10 to 40 ms) is of the order of both average closed times and initial time constant of inactivation. W_2 samples the openings and an average p_{o2} is determined during its time. The conditional probability of finding an opening with average time τ_o at the end of W_1 ($p_{o1} \cdot \tau_o$) which then spills into W_2 is considered and a similar correction for the W_2 time biasing is being made. For a probabilistic inactivation model then no remembrance of the activity in W_1 should be expected for the samples in W_2. In the deterministic Ca-current dependent model and according to Eqn. (6) high W_1 activity should be followed by a relatively low frequency of openings in W_2. The difference between the samples of W_1 and W_2 should in any case be greater than that of a probabilistic model which results in average or slightly higher than average activity in W_2. The latter should result from time biasing of bursts by the length of W_1. The effect is small since W_1 encompasses well the average burst duration of about 4 ms (20). The bias is further reduced in the recordings with low average occupation time with well separated bursts and by choosing sufficiently long W_2 sampling times.

The results from averaging after-effects of activity periods of fixed length over long pulse duration (200 to 400 ms) show that the ratio of Δp_o (activity drop between W_1 and W_2) and p_{o1} is nearly constant for varied p_{o1} if the latter is beyond the average of the recordings p_o (18). However, Eqn. (6) predicts a strong monotonic rise for increased average open times in W_1. Thus Ca-entry dependent inactivation as related to single openings or bursts of openings is not obvious in the records.

If this method is applied to study after-effects within time sections of the sample recordings, an interesting observation is made (Fig. 3) when samples with either high or low initial opening rates are compared. This initial selection period comprised the peak of the average current of the entire sample. Periods of high initial opening rates are in the average but regularly followed by periods of activity much beyond average of the sample. We found no example in which the larger initial Ca currents depressed the subsequent current average as it is expected from Ca-entry dependent inactivation. A final period with low occupation of the open state by inactivation is reached regardless of whether the preceding opening activity and thus the Ca entry is high or low. It was surprising to find that sample averages with low initial activity failed to produce a delayed peak, which should

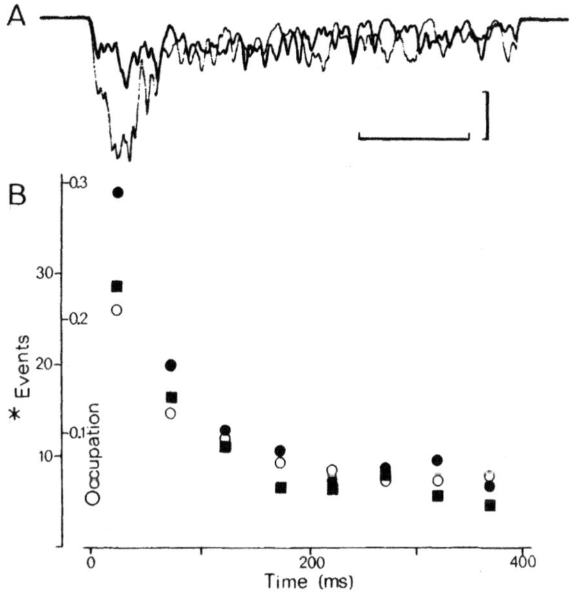

Fig. 3. (A) Comparison of the effects of high and low initial single channel activity on the steady single channel activity. V_H, -50 mV; V_{max}, 0 mV. Bath solution contained Mg and Ni to block whole cell calcium current. Seventeen records with low initial activity ($P_o \leq 0.085$) and 21 records with high initial activity ($P_o \geq 0.16$) were sampled and the averages filtered at 100 Hz. The complete ensemble contained 67 records. Despite considerable differences between initial activities, the steady levels were identical. Calibrated at 100 ms, 0.1 pA. (B) Pulse steps to 0 mV of 450 ms (44 samples) were sectioned into periods of 49 ms and the occupation time of openings was averaged over these periods (O). A time window W_1 of 7 ms selected openings and burst of openings to obtain a window occupation time in the limits of 0.7 and 1. The distribution of these W_1 values is given by (●). The opening activity following each of these W_1 samples is collected and averaged in the subsequent window W_2 of 16 ms duration. The W_2 data (■) compare with the averages (O) of the sectioned inactivation slope except for in initial section where a surplus of activity in W_2 is obtained. The data are corrected for fixed-time biasing on the basis of the average opening duration of 1.44 ms.

occur if the sampling was selective for randomly delayed activation processes. Thus it appears that inactivation starts early during the step change of voltage.

If inactivation was due to a cytoplasmic process mediated by Ca entry, this process must be localized to the vicinity of single channels. Evidence against a role for generalized Ca entry was provided by experiments in which Ca was omitted from the bath solution (18). Ca was replaced by Mg and 10 mM Ni was added so that the whole cell Ca current was completely blocked. Single channel patch currents similar to those shown in Fig. 1 were obtained, confirming that for a Ca entry model only currents through single channels need to be considered. Still another way to examine this possibility is to record from a depolarized patch. The averaged patch currents obtained in this manner were similar to whole cell currents recorded subsequently.

DISCUSSION

The results show clearly that single Ca channels inactivate. There is no contamination by outward currents, nor is there a change in unitary amplitude. The average open time is unchanged and the channels inactivate because the rate at which they open decreases with time. Furthermore, runs of initial high activity do not correlate with subsequent intervals of low activity. A Ca influx-dependent model of inactivation predicts a strong correlation of the change of the occupancy of the open state with its pre-existing occupation. On the other hand, a Markov chain including an inactivation state with voltage-dependent transition rates could fit the results.

We conclude that inactivation of single Ca channels is independent of actual Ca current flow. Our results do not rule out an indirect modulatory effect that internal Ca may have on transitions among states of the Ca channel. If this applies exclusively to resting states, the channel, once activated, would be free to pass through the remaining states including returns to the open state.

How then can the apparent Ca influx dependence of whole cell currents, where it occurs, be explained? It is necessary to restate the two strongest arguments for Ca influx dependence - the rising phase of the curve relating a test Ca current with the potential of a conditioning pulse and the slowing of inactivation by intracellular Ca buffer (EGTA). As suggested before (26), the former can be due to reactivation of Ca channels in much the same way as described for Na channels (27). The $EGTA_i$ effect might then be explained by removal of the inactivation gate. This may receive support from the fact that $EGTA_i$, when combined with low divalent action concentrations, can alter Ca channels selectively (1).

As for whole cell currents, single channel Ba currents inactivate more slowly than single channel Ca currents. If Ca-dependent inactivation is excluded, then the remaining possibilities are that Ba in some way inhibits voltage-dependent inactivation or that Ca and Ba act differently upon the inactivation particle.

A distinction between Ca-dependent inactivation and Ca-influx dependent inactivation should be made. It is possible that Ca can interact directly with a voltage-sensitive inactivation particle at sufficiently high levels of internal Ca. This could explain previous observations (28) that intracellular injection of EGTA can unmask Ca currents and the findings (26,2) that high levels of internal free Ca ($> 10^{-6}$ M) reduce Ca currents. At intermediate levels of free internal Ca (0.5 to 1 x 10^{-7} M) both strength and time course of inacivation were similar to those of untreated cells (M.J. Gutnick and H.D. Lux, in preparation). Influx-related changes of Ca should have been largely prevented in these experiments in which quantitative injections were done to achieve intracellular buffer concentrations between 30 and 80 mM.

The results reported here are by no means restricted to Ca channels in Helix neurons. Inactivation of single Ca channels in adult and neonatal rat hearts occurs under a similar set of conditions (D.L. Kunze & A.M. Brown, in preparation).

A strongly voltage-sensitive inactivation is displayed by second type of Ca channel recently described in a vertebrate preparation (30). This channel which is not found in Helix can co-exist wlith the formerly described type of channel.

The decision about current-dependent inactivation of this second Ca channel can be made by inspection. A significant part of the increase of inactivation with voltage occurs in conjunction with an average decrease of both channel activity and opening amplitude.

REFERENCES

1. Kostyuk PG, Krishtal OA: Effects of calcium and calcium-chelating agents on the inward and outward current in the membrane of mollusc neurones. J. Physiol. (Lond.) 270: 569-580, 1977.
2. Akaike N, Lee KS, Brown AM: The calcium currents of Helix neuron. J. Gen. Physiol. 71: 509-532, 1978.
3. Byerly L, Hagiwara S: Calcium currents in internally perfused nerve cell bodies of Limnea Stagnalis. J. Physiol. (Lond.) 322: 503-528, 1982.
4. Brehm P, Eckert R: Calcium entry leads to inactivation of calcium channel in Paramecium. Science 202: 1203-1206, 1978.
5. Tillotson D: Inactivation of Ca conductance dependent on entry of Ca ions in molluscan neurons. Proc. Natl. Acad. Sci. USA 77: 1497-1500, 1979.
6. Brehm P, Eckert R, Tillotson D: Calcium-mediated inactivation of calcium current in Paramecium. J. Physiol. (Lond.) 306: 193-203, 1980.
7. Eckert R, Ewald D: Residual calcium ions depress activation of calcium-dependent current. Science 216: 730-733, 1982.
8. Ashcroft FM, Stanfield PR: Calcium dependence of the inactivation of calcium currents in skeletal muscles fibers of an insect. Science 213: 224-225, 1981.
9. Eckert R, Ewald D: Inactivation of calcium conductance characterized by tail current measurements in neurones of Aplysia Californica. J. Physiol. (Lond.) 345: 549-565, 1983.
10. Brown AM, Tsuda Y, Wilson DL: A description of activation and conduction in calcium channels based on tail and turn-on current measurements. J. Physiol. (Lond.) 344: 549-584, 1983.
11. Standen NB, Stanfield PR: A binding-site model for calcium channel inactivation that depends on calcium entry. Proc. R. Soc. (Lond.) B217, 101-110, 1982.
12. Chad J, Eckert R, Ewald D: Kinetics of calcium-dependent inactivation of calcium current in voltage-clamped neurones of Aplysia Californica. J. Physiol. (Lond.) 347: 279-300, 1984.
13. Fox AP: Voltage-dependent inactivation of a calcium channel. Proc. Natl. Acad. Sci. USA 78: 953-956, 1981.
14. Fukushima Y, Hagiwara S: Voltage-gated Ca^{2+} channel in mouse myeloma cells. Proc. Natl. Acad. Sci. USA 80: 2240-2242, 1983.
15. Lux HD, Nagy K: Single channel Ca^{2+} currents in Helix pomatia neurons. Pflügers Arch. 391: 252-254, 1981.
16. Brown AM, Camerer H, Kunze DL, Lux HD: Similarity of unitary Ca currents in three different species. Nature 299: 156-158, 1982.
17. Cavalié A, Ochi R, Pelzer D, Trautwein W: Elementary currents through Ca^{2+} channels in guinea pig myocytes. Pflügers Arch. 398: 284-297, 1983.

18. Lux HD, Brown AM: Single channel studies on inactivation of calcium currents. Science 225: 432-434, 1984.
19. Lux HD, Brown AM: Patch and whole cell calcium currents recorded simultaneously in snail neurons. J. Gen. Physiol. 83: 727-750, 1984.
20. Brown AM, Lux HD, Wilson DL: Activation and inactivation of single calcium channels in snail neurons. J. Gen. Physiol. 83: 751-769, 1984.
21. Lux HD: Observations on single Ca channels. In: B Sakmann and E Neher (eds) Single-channel recording. Plenum Publ. Corp., New York, 1983, pp 437-449.
22. Colquhoun D, Sigworth FJ: Fitting and statistical analysis of single-channel recordings. In: B Sakmann and E Neher (eds) Single-channel recording. Plenum Publ. Corp., New York, 1983, pp 191-263.
23. Aldrich RW, Corey DP, Stevens CF: A reinterpretation of mammalian sodium channel gating based on single channel recording. Nature 306: 436-441, 1983.
24. Fenwick EM, Marty A, Neher E: Sodium and calcium channels in bovine chromaffin cells. J. Physiol. (Lond.) 331: 599-635, 1982.
25. Hagiwara S, Ohmori H: Studies of single calcium channel currents in rat clonal pituitary cells. J. Physiol. (Lond.) 336: 649-661, 1983.
26. Brown AM, Morimoto K, Tsuda Y, Wilson D: Calcium current-dependent and voltage-dependent inactivation of calcium channels in Helix aspersa. J. Physiol. (Lond.) 320: 193-218, 1981.
27. Chandler, WK, Meves H: Sodium and potassium currents in squid axons perfused with fluoride solutions. J. Physiol. (Lond.) 211: 623-652, 1970.
28. Hagiwara S, Naka K: The initiation of spike potential in barnacle muscle fibers under low intracellular Ca^{2+}. J. Gen. Physiol. 48: 141-162, 1964.
29. Kostyuk PG, Krishtal OA, Sakhovalov Yu: Separation of sodium and calcium currents in the somatic membrane of mollusc neurones. J. Physiol. (Lond.) 270: 545-568, 1977.
30. Carbone E, Lux HD: A low voltage-activated, fully inactivating Ca channel in vertebrate sensory neurones. Nature 310: 501-502, 1984.

DISCUSSION OF THE ARTICLE BY H.D. LUX (GERMANY)

Comment: Probably you would agree that occupancy of your inactivated state is several times longer than any occupancy on the activation sequence. But this model would then predict that on some depolarizations you should see no activity at all. Did you see those blanks or empty sweeps or something like that?

H.D. Lux (Germany): Yes. Mean open times do not decrease during the inactivation time course seen in the sample average. During inactivation, there are returns to the open state from prolonged inter-burst intervals. Initial long lasting pauses occur more frequently at low temperatures, say 10°C than at about 30°C where we recorded not more than 1% of 'nulls' in trials of 100 to 400 msec duration at a membrane potential of 0 to +10 mV.

Comment: Could you then speculate a little bit about your correlation in your single channel recordings to steady-state inactivation (if you modify the holding potential and keep the time potential constant), did you test that?

H.D. Lux (Germany): In Helix, activation and inactivation appears not to be sensitive to changes in holding potential negative to -40mV. This does not apply to the other type of Ca channel discovered in dorsal root ganglionic cells. The latter channel is characterized by showing almost complete steady-state inactivation at membrane potentials positive to -60mV.

G. Trube (Germany): In the last part of your talk, you showed us another type of calcium channel. You mentioned that you see it in outside-out patches. Do you only see it in outside-out patches or do you also see it in cell-attached patches? Most people have problems in seeing calcium channels in isolated patches because of a very fast run-down (Cavalie et al (1983). Pflugers Arch. 398, 284-297).

H.D. Lux (Germany): A systemic study of this channel necessitates outside-out patches because in cell-attached patches, the membrane potential is not sufficiently well known and probably too low for a significant activation of this channel. Fortunately, this second channel shows very little run-down, much in contrast to the usual Ca channel. It appears as stable as a Na channel.

G. Trube (Germany): Indeed, one may have the impression that this is a modified Na channel which carries Ca ions to a higher proportion.

H.D. Lux (Germany): In whole cell recordings, no significant changes of the inward current were observed after replacing choline or Tris by Na in the presence of 20mM Ca. However, we know that the channel can carry Na if the external Ca concentration is much reduced.

MODULATION OF CALCIUM CHANNELS IN CARDIAC CELLS BY NEUROTRANSMITTERS

H. Reuter

Voltage dependent Ca channels are widely distributed in excitable cells, and important cellular functions are regulated by Ca ions moving through such channels. Adrenergic neurotransmitter effects on Ca currents, I_{Ca}, in cardiac cells have served as prime examples of modulation of voltagegated ion channels. Recently it has become possible to analyze molecular events involved in Ca channel modulation. Ca channels are activated by changes in membrane potential. The probability, p_o, steeply with depolarization. Analysis of the gating kinetics of Ca channels in the presence of catecholamines or 8-bromo-cAMP showed (a) an increase in the mean open time of Ca channels, (b) a decrease in closed times between bursts of openings, and (c) a reduction in failures of a channel to open upon repetitivo depolarizations. These kinetic changes result in an overall increase in p_o. The number of functional channels, N_f, may also be increased, while the single channel current, i, is not affected. These results explain the increase of $I_{Ca} = N_f \cdot p_o \cdot i$. cAMP-dependent phosphorylation of the channels is a likely intermediate step in this modulation.

DISCUSSION OF THE ARTICLE BY R. REUTER (SWITZERLAND)

W. Trautwein (Germany): There is evidence that the cascade you showed in your second slide plays also a role in the absence of the beta stimulation for the control of the calcium permeability. If you inject regulatory subunit, which is as the biochemists tell us are supposed to bind the catalytic subunit in the absence of beta stimulation, you reduce the calcium current.
(Trautwein E., Brum G. and Ostereicher W. (1985). Proc. 16th Meet. Fed. Europ. Biochem. Soc. VNU Science Press BV).
Have you any evidence along this line?

H. Reuter (Switzerland): We don't. Injecting regulatory subunit probably is the best proof so far. We have not done this sort of experiment. However, we get a very rapid run-down of calcium channels if we isolate the patches. Within seconds to minutes, the calcium channel activity disappears. We can, however, prolong the run-down by applying the catalytic subunit of cAMP-dependent protein kinase to such isolated inside-out patches.

W. Trautwein (Germany): When you find a high opening probability and you average your ensemble, you see a high amplitude of the average current. What about epinephrine? You can see a four or five-fold increase in average current amplitude (Bean B.P., Nowycky M.C. and Tsien R.W. (1984). Nature 307, 371). This would mean that the probability would have to be increased to a very high extent. However, sometimes you already find a high opening probability under control conditions which in this case, the average current could not increase considerably anymore in the presence of epinephrine.

H. Reuter (Switzerland): This depends a little upon the definition of p_o, the opening probability. It is a question of semantics, to some extent, whether you include in the calculations of the opening probabilities the 'null' traces or whether you don't. We have, for example, in our calculations of p_o, always included 'nulls' as well but I knew that not all people are doing this and for that reason they start off with a rather high opening probability. The 'nulls' are considered to be a channel that is not available to open. We have looked for some evidence whether it is really N that can increase and we have found no evidence for that. For example, we have used patches where there was no calcium channel activity and looked if we could evoke channel activity after application of isoproterenol, but we have never found channel activity in a patch where there was no calcium channel to begin with.

A.M. Brown (U.S.): Can you explain the discrepancy in which noise analysis leads to the conclusion that catecholamines increase N - the # of channels - and your results in which it is p_o that is responsible?

H. Reuter (Switzerland): Actually, the noise analysis does not distinguish between N and 'nulls'.

H.D. Lux (Germany): In DRG cells, norepinephrine reduces the Ca current in a weakly voltage-dependent manner. Could your reaction schemes account for this effect?

H. Reuter (Switzerland): It is known that alpha-2 agonists decrease the cAMP level in cells, and, for that particular reason, I would expect that a reduction in the calcium current would fit into this scheme. Actually, my very last slide which I did not show you because of lack of time shows exactly this effect. Paul Adam's group showed a reduction in the calcium current after application of norepinephrine (Brain Res. (1982). 244, 135-144).

D. Pelzer (Germany): The p_o value has been used to calculate actual numbers for kinetic transition rates (especially the rate - limiting rate constant k_1) in a C-C-O sequence which seems to be related to Ca channel inactivation and does not take into account an inactivated state (e.g. Reuter (1984). Ann. Rev. Physiol. 46, 473). Don't you think you might introduce a bias if you include the traces with no channel openings ('nulls') in the calculation of p_o, since there is evidence that these blank sweeps might to some extent represent the sojourn of the Ca channel in the inactivated state.

(Cavalie, Ochi, Pelzer and Trautwein (1983). Pflugers Arch. 398, 284; Trautwein and Pelzer (1985). In Zipes, Cardiac Electrophysiology and Arrhythmias, Grune and Stratton, N.Y., in press; Trautwein and Pelzer (1985). Proc. Satellite Symp. 9th. Intern. Cong. Pharmacol. Intracell. Ca Regulat., Karger, Basle, in press)

H. Reuter (Switzerland): Again, this is a question of semantics, whether it is inactivation or whether it is something else. You can also call it a dephosphorylated state of the channel which is not available for activation. I think that this sort of kinetic scheme has to be worked out in greater detail before we can draw conclusions with regard to the regulatory mechanisms involved in each of the respective steps.

D. Pelzer (Germany): But anyway, your p_o is a mixture of several processes, some of them taking place on a millisecond time scale, whereas the others occur over a time scale of at the least hundreds of milliseconds.

H. Reuter (Switzerland): Yes.

A.M. Brown (U.S.): I would like to know if there is an argument on how opening probability is defined. I define opening probability for a single channel as the number of times a 1 state occurs as opposed to a 0 state at any given point in time for an ensemble. This includes ensemble samples that are 'nulls'. This is consistent with usage of macroscopic currents when we refer to channel availability. We use the 'H-infinity V curve' to tell us how many channels are available at a given potential. In the case of a single channel, the corresponding value is the peak opening probability as I have defined it.

H. Reuter (Switzerland): Absolutely. One way of doing this is by adding a subscript to the p, namely p_f.

P. Hess (U.S.): I agree and I think that if you do it that way, you do of course include the 'nulls' but if you want the actual numbers for the kinetic transition on the C-C-O scheme then you cannot include them because then you are dealing with a state which clearly has a lifetime which is a lot longer than anything on that scale.

H. Reuter (Switzerland): Of course.

SOME PROPERTIES OF CALCIUM CURRENT IN MOUSE MOTOR ENDINGS

A. Mallart

It is now well established that the chain of events leading to phasic transmitter release by nerve endings is triggered by Ca^{2+} influx through channels activated by depolarization of the presynaptic membrane. Direct evidence for this has been obtained by Llinas et al (1), in the giant synapse of the squid. Little is known, however, about the relationship between membrane currents and transmitter release in vertebrate endings whose small size prevents a direct approach to presynaptic events. For this reason, many of the basic concepts used to describe excitation-secretion coupling in vertebrate synapses come from the study of postsynaptic responses in neuromuscular junctions. Here, acetylcholine (Ach) release is a brief event that occurs 0.5 - 1 ms (at room temperature) after arrival of nerve impulses to motor endings (2,3). Since both Ach diffusion in the cleft and interaction with postsynaptic receptors are very fast processes, most of the synaptic delay would appear to be caused by Ca channel opening and by the release mechanism itself. Little is known about the kinetics of opening and closing of Ca channels in vertebrate synapses, although an extrapolation from the data obtained by Llinas et al (4) in the giant synapse of the squid suggests that Ca channels start conducting a little before the peak of the action potential and begin closing down near the halfway point of its repolarizing phase. Indeed, experiments reported by Katz & Miledi (2) showed that a hyperpolarizing pulse delivered by a focal electrode to frog nerve terminals during the decay of a nerve-evoked action potential suppresses transmitter release, presumably by Ca channel closure upon membrane repolarization. If the hyperpolarizing pulse was delivered a little later, no effect on transmitter release was observed. Thus, it is likely that Ca^{2+} enters presynaptic membrane for a very short period of time. Unfortunately, because of the small size of vertebrate endings, direct evidence for the relationship between Ca^{2+} influx and transmitter release has been missing until now.

The first attempt to relate electric signals recorded from motor endings by external electrodes and specific membrane currents was that of Brigant &

Mallart (5). Their research was made possible by the use of a new neuromuscular preparation, the triangularis sterni muscle of the mouse (6). This muscle consists of two to three layers of fibers whose motor endplates can be visualized for electrode placement and recording at a x 500 magnification by using an immersion lens of 1.6 mm working distance. Later, the same method was successfully applied to motor endings of frog cutaneus pectoris muscle (7), although this preparation proved to be less amenable to study than that of the mouse. In this latter preparation, membrane channels show a particularly convenient distribution: unmyelinated terminal branches are practically devoid of Na channels but contain K and Ca channels which, in turn, are absent from the nodes of Ranvier (8,9). The endings, thus, are passively invaded by nerve impulses which generate depolarizations of slower rise time, smaller amplitude and faster decay time as compared to action potentials generated in preterminal nodes of Ranvier. As a result, terminal membrane becomes less depolarized than the parent axon which allows local circuit current to flow between preterminal nodes of Ranvier (including the heminode) and terminal branches.

Consequently, the current recorded at the terminals flows outwards and shows two positive peaks (Fig 5A). The first is passive current which is the inverted replica of that carried by Na^+ at the nodes; the second peak is K^+ current. Corresponding currents can be recorded with inverted polarity either by focal electrodes from the heminode or by microelectrodes inserted into the perineurium of preterminal nerve bundles.

The experiments presented in this paper were performed at room temperature and centered on the stimulation of the motor nerve in fully curarized mouse triangularis sterni nerve-muscle preparations. Most of the recordings are from the perineurium, where K and Ca currents generated at the terminals become reversed and the 'Na spike' regains its true polarity. Despite changes in polarity with electrode emplacement, currents will be referred to as inward or outward according to the direction in which they traverse the membrane at the site of the permeability change.

Furthermore, recorded currents will be described in terms of potential change in the terminals relative to that in the axon. Since the flow of current is proportional to potential differences between preterminal nodes and nerve endings, it is easy to reconstruct the potential change at the latter by using as a reference the action potential generated at the nodes of Ranvier obtained from data in the literature (see 9).

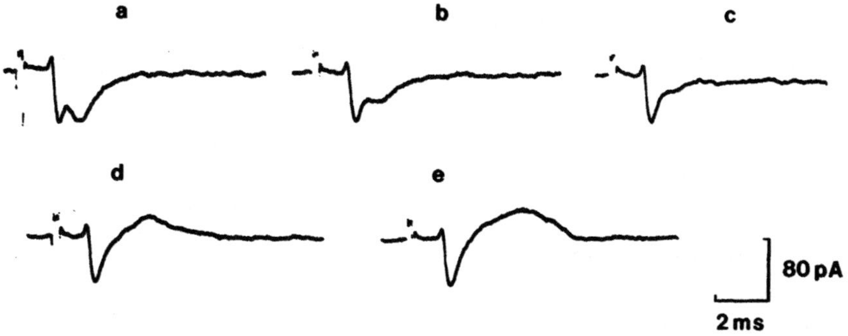

FIGURE 1. Changes in presynaptic responses induced by bath perfusion of 1mM TEA and 0.1mM 3,4-DAP. a, control. b to e, shows progressive suppression of the K-dependent component and its replacement by "Ca current". Microelectrode recordings from the preterminal perineurium. Ten sweeps have been averaged.

Calcium current in motor endings

Calcium currents cannot be distinguished as a separate entity in recordings effected in standard Krebs solution because its intensity is too small with regard to that of potassium. It was necessary, therefore, to block first IK by tetraethylammonium (TEA) and/or 3,4-diaminopyridine (3,4-DAP) to show ICa. Fig 1 shows changes in presynaptic currents induced by bath perfusion of 0.1mM 3,4-DAP and 1.0mM TEA. IK disappears and is replaced by a component of inward current which has been shown to correspond to ICa (5). The configuration of presynaptic signals suggests that under these circumstances, the action potential in the terminals bears an important Ca component which crosses over the action potential generated at the nodes of Ranvier. In other words, with TEA + 3,4-DAP in the bath, action potentials are generated in the terminals by a brief flow of current which discharges membrane capacity and a prolonged Ca^{2+} current which, if K conductance is completely blocked, may become regenerative and give way to plateau responses (see below). In standard Krebs solution, however, few Ca channels have the opportunity to open because of the potent action of the delayed rectifier which rapidly brings the membrane potential back to its resting level.

Calcium current generated at mouse motor endings was readily blocked by 30mM Mg^{2+}, 10mM Co^{2+}, 5mM Mn^{2+} or 0.2mM Cd^{2+}, but unexpectedly, it was unaffected by 100µM D 600, which is in agreement with the lack of effect of D 600 on transmitter release by motor endings (10).

Calcium plateau responses

In the presence of K channel blocking agents, Ca current does not fall immediately to zero but presents a prolonged phase of slow decay before terminating in a rather abrupt manner. This was observed in all experiments provided: (1) sufficient amounts of TEA and 3,4-DAP had been added to the bath; (2) the external calcium had been increased to 4 - 6mM; and (3) the nerve had not been too frequently stimulated. Under these conditions, plateau duration was about 300 ms but shortened during successive pulses delivered at 1 - 2 sec intervals (Fig 2A). A similar effect was observed with paired pulse stimulation where the shorter the pulse interval, the shorter the response to the second pulse (Fig 2B). Also, for a given interval, the longer the duration of the first response, the shorter the response to the second pulse.

In other neural systems, two mechanisms have been proposed to explain plateau shortening with repetitive stimulation (11,12). One is progressive calcium accumulation inside the membrane, the other is the activation of a Ca-dependent K current (IK(Ca)).

Ca accumulation and binding to a site at the internal membrane surface is generally believed to cause inactivation of Ca channels (13). Thus, during long plateau responses, the amount of Ca^{2+} entering the membrane may be large enough to saturate the cellular mechanisms for Ca^{2+} buffering and extrusion and accumulate inside the nerve terminal. Above a critical internal Ca^{2+} concentration, Ca channels are turned off and cannot be activated again until internal calcium concentration decreases.

The other possibility is that not all the K conductance has been blocked by the doses of TEA and 3,4-DAP used in these experiments. Indeed, when the minimal dose of TEA and 3,4-DAP required to elicit a Ca plateau response was used (i.e., 1mM and 0.1mM, respectively), a notch was frequently observed at the top of the plateau (Fig 2 C,D,E). This notch appeared near the beginning of the Ca response and its presence caused plateau depression in an almost all-or-none manner when repetitive stimulation was used (Fig 2E). Thus, the possibility exists that a K conductance is triggered shortly after the onset of the Ca response which repolarizes the membrane and prevents plateau development. An interesting observation was that plateau responses obtained in a medium in which Ba^{2+} or Sr^{2+} had been substituted for Ca^{2+} never presented this particular outward component. Similarly, when 1mM Ba^{2+} was added to the standard Krebs solution, plateau amplitude increased and the initial notch disappeared (Fig 2F). It is highly probable that the action of

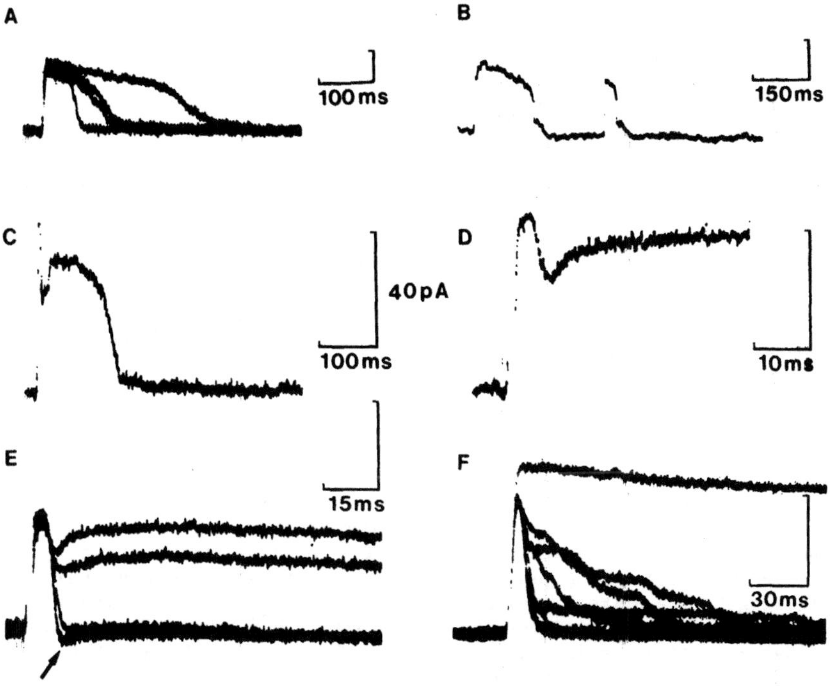

FIGURE 2. Calcium plateau responses obtained from the perineurium in 0.1mM DAP + 1mM TEA and 3mM (A,B) or 6mM (C,D,E,F) external calcium. A shows plateau shortening with repeated stimuli. B, plateau duration is reduced when elicited shortly after a conditioning stimulus. C,D, a transient outward current is visible at the beginning of the plateau. E, the plateau is cut off with repeated stimuli at the start of transient outward current. Net outward current is revealed when the plateau is compeletely depressed (arrow). F, addition of 1mM Ba^{2+} suppresses transient outward current and makes the plateau decrease gradually on repeated stimulation.

Ba^{2+} on the configuration of plateau responses is due to its ability to block K channels (14). The ability of Ba^{2+} to restore calcium plateau currents is shown, furthermore, by the experiment illustrated in Fig 3. Plateau responses were lost after the preparation was exposed for about 20 min to a temperature of 12 - 14°C and then returned to room temperature (Fig 3B). Addition of 1mM Ba^{2+} promoted prompt plateau recovery (Fig 3 C,D). The mechanism by which temporary exposure to cold causes plateau suppression is not clear. Perhaps metabolic pumps that remove intracellular calcium are depressed and Ca^{2+} accumulates inside the cell (15), thus inactivating calcium channels, as

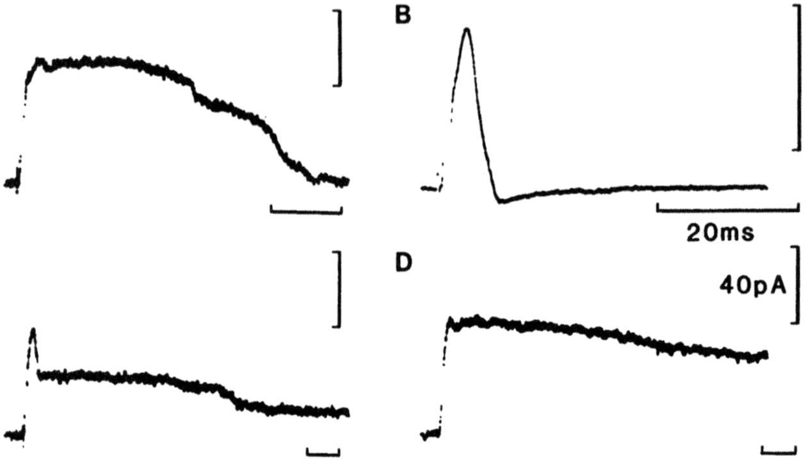

FIGURE 3. Action of transient cold exposure and Ba^{2+} on plateau responses. A, control plateau Ca current. B, the preparation was exposed for several minutes to 12°C and then returned to 21°C. Plateau responses could not be elicited even after long periods of rest. Note transient outward current following the Ca spike. Average of 10 sweeps. C,D, Ba^{2+} (1mM) was added, restoring the plateau. Recordings were taken 3 and 5 min after admission of the Ba^{2+}-containing solution. 0.1mM 3,4-DAP, 1.0mM TEA and 6mM Ca^{2+} were present throughout the experiment.

described above. Another possibility is that increased intracellular calcium activates a K conductance which prevents plateau development. As a matter of fact, a small undershoot follows early calcium current in recordings taken after cold exposure (Fig 3B). The timing of this undershoot corresponds to that of the notch observed frequently on top of the plateau (Fig 2). Moreover, as shown above, both events were suppressed by Ba^{2+}, strongly suggesting that both are caused by IK(Ca). Further characterization of IK(Ca) in mouse motor endings has been carried out in a series of experiments whose data are reported in a preliminary communication (16). A full account of these will be published elsewhere. Suffice it to say here that the evidence for IK(Ca) is based on the following observations: (1) this outward current is unaffected by 3,4-DAP; (2) it is blocked by 2mM TEA or 1mM Ba^{2+}; (3) its amplitude is reduced in low external Ca^{2+} solutions; (4) it is suppressed by Co^{2+}, Mn^{2+} or Cd^{2+}.

The presence of IK(Ca) in motor endings does not necessarily imply that it is the sole cause for plateau depression. Indeed, when IK(Ca) has been blocked by TEA or Ba^{2+}, calcium plateau responses become shorter and shorter with repetitive stimulation, indicating that intracellular Ca^{2+} accumulation progressively inactivates Ca channels. The main difference between both mechanisms is that IK(Ca) acts ont he early phase of plateau generation, whereas activity-dependent Ca accumulation primarily affects plateau duration.

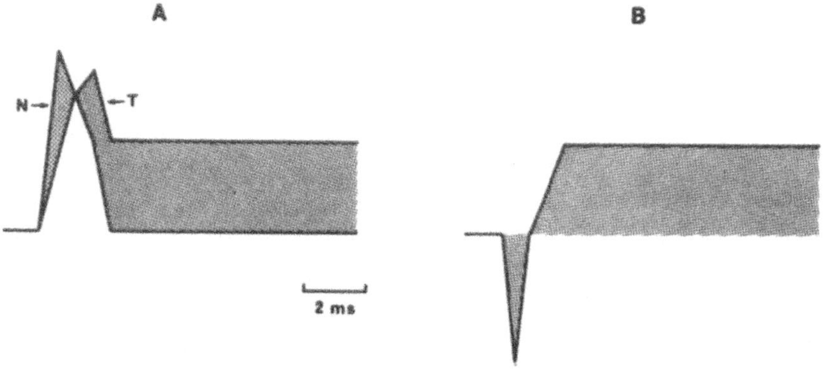

FIGURE 4. A, reconstruction of Ca plateau action potentials at the endings, in the presence of 3,4-DAP + TEA, as inferred from the shape of the nodal action potential (see Ref 9) and the configuration of external currents. The shaded area corresponds to potential differences between the endings and the axon which promote external current. B shows this current, which is proportional to the shaded area in A.

Reconstruction of potential changes at the endings caused by K-channel blockers may be helpful to understand changes in the shape of external currents. As stated above, the externally recorded currents are caused by potential differences between the endings and the last few nodes of Ranvier which, in turn, depend on specific K^+ and Ca^{2+} conductance changes promoted only at the former (since nodes are devoid of K or Ca channels) by the surge of capacitive current. It will thus be possible to reconstruct the probable shape of the potential at the endings by taking the nodal action potential as a constant and by regarding the external current as a variable proportional to the amount of current required to bring the potential of the endings to the same level as that of the nodes. Such a reconstruction is shown in Fig 4.

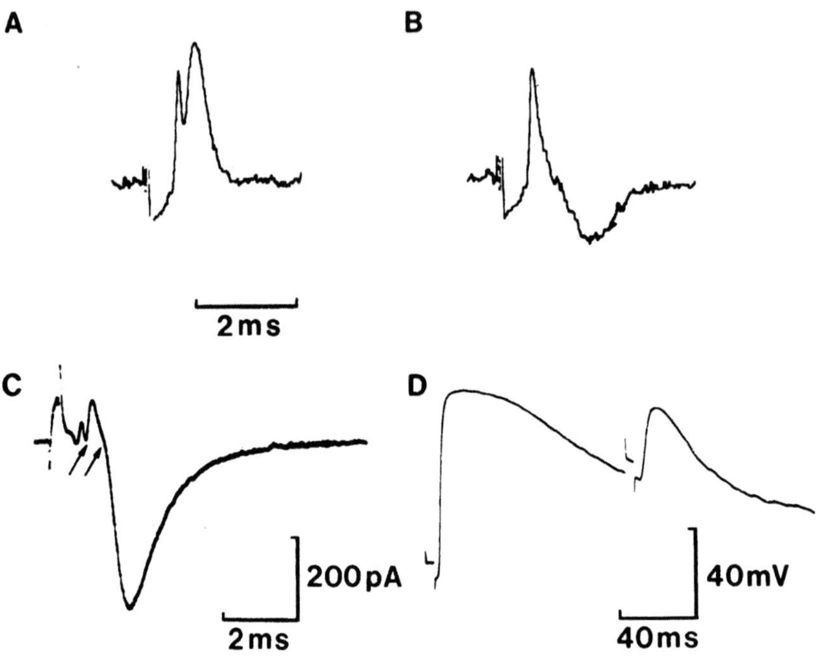

FIGURE 5. A,B, presynaptic currents recorded by fire-polished electrodes from terminal branches before (A) and during (B) perfusion with 1.0mM TEA and 0.1mM 3,4-DAP. Reversal of membrane current indicates crossing over of potential changes at terminal and nodal membranes induced by calcium current. Thus, it can be said that calcium current starts at the end of capacitive current. C, simultaneous recording of pre- and postsynaptic currents in the presence of 2μM curare to show synaptic delay (arrows). D, intracellularly recorded endplate potentials evoked in the presence of 1μM TTX and 5mM TEA by electrotonic stimulation of the preterminal axon (courtesy of J. Molgo). Note the flat top of the first one as membrane potential approaches the reversal potential for Ach response. Otherwise, its amplitude would be larger and its decay faster. Despite large release levels, a second pulse is still able to evoke a response.

Calcium current and transmitter release

Fig 5 shows simultaneous recordings of presynaptic currents from the terminal part of the endings and postsynaptic current. It appears that transmitter release is delayed by approximately 0.4 ms (at 21°C) from the end of the first positive spike, which responds to the peak of presynaptic depolarization. That this delay is not due to a retard of calcium current is shown in Fig 5, where it can be seen tha Ca current starts at the peak of presynaptic action potential, in agreement with findings in the giant synapse of the squid by Llinas et al (4). It follows that most of the synaptic delay is consumed by the release process itself, which may include vesicle movement and interaction with the presynaptic membrane (see 17).

An intriguing question is that of the transmitter release during Ca plateau currents. Calcium regenerative responses elicited in the presence of 0.2mM 3,4-DAP + 2.0mM TEA may last as long as 300 - 400 ms with a flat plateau phase of about 200 ms. The time course of the postsynaptic response is, however, considerably shorter and decays at a much faster rate in comparable experimental situations (Fig 5; see also 18). In other words, terminals cannot maintain high levels of transmitter output despite sustained calcium influx. At first glance, high release levels could cause depletion of the fraction of transmitter which is readily available. Another possibility is that excessive accumulation of calcium on the inside of the membrane could inactivate the release mechanism (19,12). That this is not the case is shown by the experiment of Fig 5D (kindly communicated by Dr. J. Molgo), where a second pulse is able to promote further release of transmitter during the decay phase of the endplate potential. It is worth reporting in this context that a second impulse arriving at the terminals during a plateau response evokes only a brief upsurge of calcium current (A. Mallart, unpublished).

A possible explanation for the observed discrepancy between sustained Ca^{2+} influx and decaying Ach release is that the release mechanism depends not only on the entry of Ca^{2+} inside presynaptic membrane but also on membrane depolarization (20). Since it is likely that synaptic vesicles bear net negative surface charge (21), a voltage jump which brings the inside of the membrane to positive potential levels will promote synchronous vesicle displacement and apposition to the presynaptic membrane. A further step in the release process is Ca^{2+} entry close to the active zones, stimulating, by a not well-understood mechanism, Ach permeation through vesicular and plasmalemmal membranes. Steady depolarization during the plateau phase will cause random vesicle displacement and apposition which would produce only asynchronous release.

REFERENCES

1. Llinas R., Steinberg I.Z. and Walton K. (1981). Presynaptic calcium currents in squid giant synapse. Biophys. J. 33, 289-322.
2. Katz B. and Miledi R. (1965). The measurement of synaptic delay and the time course of acetylcholine release at the neuromuscular junction. Proc. R. Soc. Lond. B. 161, 483-495.
3. Datyner N.B. and Gage P.W. (1980). Phasic secretion of acetylcholine at a mammalian neuromuscular junction. J. Physiol. (Lond.) 303, 299-314.
4. Llinas R., Sugimori M. and Simon S.M. (1982). Transmission by presynaptic spike like depolarization in the squid giant synapse. Proc. Natl. Acad. Sci. USA 79, 2415-2419.
5. Brigant J.L. and Mallart A. (1982). Presynaptic currents in mouse motor endings. J. Physiol. (Lond.) 333, 619-636.
6. McArdle J.J., Angaut-Petit D., Mallart A., Bournaud R., Faille L. and Brigant J.L. (1981). Advantages of the triangularis sterni muscle of the mouse for investigations of synaptic phenomena. J. Neurosci. Meth. 4, 109-115.
7. Mallart A. (1984). Presynaptic currents in frog motor endings. Pflugers Arch. 400, 8-13.
8. Brismar T. (1980). Potential clamp analysis of membrane currents in rat myelinated nerve fibres. J. Physiol. (Lond.) 298, 171-184.
9. Chiu S.Y., Ritchie J.M., Rogart R.B. and Stagg D. (1979). A quantitative description of membrane currents in rabbit myelinated nerve. J. Physiol. (Lond.) 292, 149-166.
10. Bregestowski P.D., Miledi R. and Parker I. (1980). Blocking of endplate channels by the organic calcium antagonist D 600. Proc. R. Soc. Lond. B. 211, 15-24.
11. Alvarez-Leefmans F.J. and Miledi R. (1980). Voltage sensitive calcium entry in frog motoneurones. J. Physiol. (Lond.) 308, 241-257.
12. Katz B. and Miledi R. (1969). Tetrodotoxin resistant electric activity in presynaptic terminals. J. Physiol. (Lond.) 203, 454-487.
13. Standen N.B. and Stanfield P.R. (1982). A binding site model for calcium channel inactivation that depends on calcium entry. Proc. R. Soc. Lond. B. 217, 101-110.
14. Eaton D.C. and Brodwick M.S. (1980). Effects of barium on the potassium conductance of squid axon. J. Gen. Physiol. 75, 727-750.
15. Dietmer J.W. and Schlue W.R. (1983). Intracellular Na^+ and Ca^{2+} in leech Retzius neurones during inhibition of the Na^+ - K^+ pump. Pflugers Arch. 397, 201.
16. Mallart A. (1984). Calcium activated potassium current in presynaptic terminals. Biomed. Res. 5, 287-290.
17. Kelly R.B., Deutch J.W., Carlson S.S. and Wagner J.A. (1979). Biochemistry of neurotransmitter release. Ann. Rev. Neurosci. 2, 399-466.
18. Katz B. and Miledi R. (1979). Estimates of quantal content during "chemical potentiation" of transmitter release. Proc. R. Soc. Lond. B. 205, 369-378.
19. Adams D.J., Takeda K. and Umbach J.A. (1983). Metabolic inhibitors and lithium depress synaptic transmission at the squid giant synapse. J. Physiol. (Lond.) 345, 147P.
20. Dudel J. (1983). Transmitter release triggered by a local depolarization in motor nerve terminals of the frog: role of calcium entry and of depolarization. Neurosci. Lett. 41, 133-138.
21. Van der Kloot W. and Kita H. (1973). The possible role of fixed membrane surface charges in acetylcholine release at the frog neuromuscular junction. J. Membr. Biol. 14, 365-382.

DISCUSSION OF THE ARTICLE BY A. MALLART (FRANCE)

S. Thesleff (Sweden): Do you know if the calcium activated potassium current is sensitive to membrane potential. Will small changes in presynaptic membrane potential affect your current?

A. Mallart (France): It is difficult to say with this technique but I don't think so because at least the time course is not affected as is seen when it appears on top of the plateau. The time course is similar to that obtained near the base line. We can reconstruct potential changes at the terminal from these currents and we can conclude that there is a post hyperpolarization of the endings with this current.

Y. Yaari (Israel): We have made a very similar observation in the frog nerve-muscle preparation, which I will present this afternoon. We found that this presumably calcium current is not abolished by perfusing the preparation with EGTA. Therefore, we wonder whether this is indeed a pure calcium current or whether other ions participate in this response. Did you try to remove the calcium in these conditions?

A. Mallart (France): No. The only thing that we put was 0.1mM calcium and magnesium to abolish these currents.

Y. Yaari (Israel): Could it be that in the absence of extracellular calcium, sodium ions penetrate through Ca^{2+} channels?

A. Mallart (France): Sodium current is not expected to be blocked by cobalt except if it uses the calcium channel, but I don't know whether these sodium ions would activate this potassium current.

H.D. Lux (Germany): Potassium channels can be blocked by the entry of sodium ions (Bezanilla F. and Armstrong A.M. (1972). Negative conductance caused by the entry of sodium and cesium ions into the potassium channels of squid axons. J. Gen. Physiol. 60, 588-608). But this occurs under the condition of strong depolarizations. Did you see effects of elevated intracellular sodium?

A. Mallart (France): I have no experience with this.

MOLECULAR PROPERTIES OF ISOLATED Ca^{2+} TRANSPORT SYSTEMS FROM NERVE TERMINALS

Hannah Rahamimoff, Ari Barzilai, Anat Erdreich and Rivka Spanier
Department of Biochemistry, Hebrew University-Hadassah Medical School, Jerusalem 91010, Israel

1. INTRODUCTION

Most of our knowledge on the subcellular events taking place from the moment that the action potential invades the nerve terminal until response is recorded in the postsynaptic cell comes from experiments performed on the neuromuscular synapse. This preparation which is most suitable for electrophysiological studies is unfortunately not adequate for biochemical experiments. Therefore, we are using synaptosomes from brain (intact or following osmotic disruption) as the alternative biochemical model for studying calcium transport systems in nerve terminals. Synaptosomes are the closest biochemical model structures to intact synapses (1,2). They maintain membrane potentials as measured by voltage sensitive fluorescent probes (3), have osmotic properties similar to an intact cell, are more permeable to K than to Na (1) and release neurotransmitters in media containing calcium - provided depolarizing potassium concentrations or other depolarizing agents are present (1,2,3).

Calcium ions play a major role in the regulation of neurotransmitter liberation (4,5) and thus in the entire process of information transfer across chemical synapses. Several processes are known to participate in the regulation of intracellular calcium concentration. At least three processes are associated with the presynaptic membrane: Calcium entry through voltage sensitive channels (6), extrusion of calcium ions through the nerve membrane in an exchange process with sodium ions and/or calcium ions (7,8,9) and an ATP dependent calcium pump for extrusion purposes (10,11). Inside the cell, the involvement of mitochondrial calcium uptake in the regulation of $[Ca]^{in}$ has been described in detail (12). Coated synaptic vesicles were shown (13) to take up calcium in an ATP dependent fashion. ATP dependent calcium uptake has been detected in disrupted synaptosomes after their exposure to mitochondrial blockers (14). Fixation of frog neuromuscular junction preparations in high calcium revealed calcium binding sites on synaptic vesicles (15) and electron probe analysis showed calcium binding to endoplasmic reticulum in nerve terminals (16).

The contribution of the different Ca^{2+} transporting systems to the maintenance of the intracellular Ca^{2+} homeostasis has been estimated by various approaches mainly in the squid giant axon (17). Studying the regulation of intracellular Ca^{2+} ion concentration in the chemical synapse can be approached from different angles: One can measure the overall functional response of the entire system following a well-defined change in one parameter or one can choose the approach that in order to understand the coordinated response of the entire system, one has to study each individual Ca^{2+} transporting component separately. Following the latter approach, we started several years ago to isolate the different synaptic membranes, study their Ca^{2+} transport properties in the native preparation, then try to isolate and purify the individual Ca^{2+} transport activities, study their molecular organization and transport activity after suitable reconstitution into phospholipid membranes.

In all following experiments, rat or calf SPM vesicles prepared by differential centrifugation from lysed brain synaptosomes were used. SPM vesicles, like other membrane vesicle preparations, permit the formation of well defined intravesicular and extravesicular ionic environments by choice and the transport across the vesicles' membrane can be measured free of cytosolic or metabolic perturbations.

Besides these general advantages, the preparation of SPM vesicles offers also the specific advantage that both types of Ca^{2+} transport systems, the ATP dependent one and the ATP independent ones - the Na^+-Ca^{2+} exchanger, and the negative inside membrane potential driven Ca^{2+} uptake system are present in it (18) and each one of them can be demonstrated separately. The Na^+-Ca^{2+} antiport system is investigated by measuring Ca^{2+} influx under conditions of preformed outward oriented Na^+ gradient in the absence of ATP. The negative inside membrane potential driven Ca^{2+} uptake is measured in the absence of both Na^+ gradient and ATP. The ATP dependent Ca^{2+} pump is studied in the absence of a Na^+ gradient and negative inside membrane potential and in the presence of added ATP. ATP dependent Ca^{2+} influx in SPM vesicle is exhibited by that fraction of the vesicles that were resealed after hypoosmotic treatment of the synaptosomes in an inverted fashion - with the "cytosolic" face of the synaptic plasma membrane on the outside ("inside-out" vesicles). The Na^+ gradient dependent Ca^{2+} transporter can transport Ca^{2+} across the synaptic membrane in both directions, depending on the direction of the Na^+ gradient (19). The direction of the Na^+ gradient dependent Ca^{2+} flux respective to the orientation of the membrane can be distinguished by selectively preloading with Ca^{2+} only the "inside-out" vesicles in the absence of a Na^+ gradient and in the presence of ATP. Ca^{2+} efflux from these vesicles can be initiated via the Na^+-Ca^{2+} exchanger by increasing the Na^+ concentration in their external medium (19,20).

In this presentation, we introduce some of our results describing properties of native SPM vesicles, demonstrate the isolation, purification and reconstitution of the Na^+-Ca^{2+} exchanger into liposomes, and show some of the advantages of using the purified carrier for mechanistic studies.

2. METHODS

2.1. Preparation and characterization of synaptic plasma membrane (SPM) vesicles

SPM vesicles are prepared from 14 days old rat brains. The procedure consists of gentle homogenization of the rat brains in isoosmotic mannitol solution containing 1mM EDTA followed by differential centrifugation and step ficoll gradient (2-20% in 5 steps) to obtain a preparation enriched in synaptosomes. Following exposure to hypoosmotic 5mM tris HCl, pH 7.4-1mM EDTA solution, the synaptic membranes are collected by centrifugation. Characterization of this preparation by specific enzyme marker activities shows that it contains less than 10% mitochondrial contamination (18) and is highly enriched relative to other fractions in the following activities: Na^+/K^+ ATPase, Na^+ coupled GABA uptake, Na^+-Ca^{2+} antiport, and 5'AMP nucleotidase activity. The membranes are frozen in liquid N_2 and kept in small aliquots at -90°. Their Na^+-Ca^{2+} and ATP dependent Ca^{2+} transport activity is maintained for several months.

2.2. Purification and reconstitution of the SPM Na^+-Ca^{2+} exchanger

The detailed procedure for isolated purification and reconstitution of the Na^+-Ca^{2+} exchanger has been described in detail (21) recently. In principle, the procedure involves solubilization of the native membranes in cholate in the presence of a very large excess (40-80 fold) of added phospholipids. Following dialysis and vesicle reformation, the native proteins become separated among a large number of liposomes. Specific Na^+ gradient dependent Ca^{2+} loading renders the Ca^{2+} containing vesicles denser than the rest of the liposome population and permits their separation by density gradient centrifugation. Figure 1 is a schematic presentation of the reconstitution procedure. It should be noted that while successful purification and reconstitution of the SV and SPM Ca^{2+} ATPase (22,23,24) could be achieved employing asolectin (soybean phospholipids), successful reconstitution of the Na^{+}-Ca^{2+} exchanger required purified brain phospholipids.

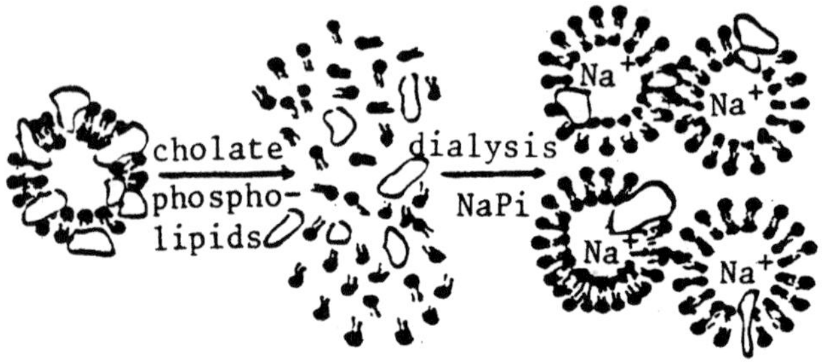

FIGURE 1. Schematic presentation of the reconstitution process of the Na^{+}-Ca^{2+} exchanger.
Native SPM vesicles were added to a solution containing 30-40 mg/ml purified brain phospholipids, 2% Na-cholate, 0.2M NaPi buffer pH 7.4, 5mM β-mercaptoethanol to yield a final concentration of 1 mg/ml protein. Dialysis was performed against 0.2M NaPi buffer pH 7.4.

2.3. Transport assays

Transport assays are performed in two ways depending on the experimental conditions: 1) Filtration technique (employing Schleicher and Schull BA 85 cellulose nitrate filters); 2) Mini ion exchange columns - Dowex 50 (25); Ca^{2+} transport assays with native SPM vesicles are usually performed by millipore filtration, while $^{22}Na^{+}$ transport with native membranes and any assay with phospholipid fortified reconstituted membrane is carried out by using the Dowex mini columns. $[^{3}H]TPP^{+}$ (tetraphenylphosphonium) uptake was measured by the filtration method except, that cellulose acetate filters were used (27).

2.4. Preparation of purified brain phospholipids

Successful reconstitution and subsequent purification of the SPM Na^{+}-Ca^{2+} exchanger depend on the presence of purified brain phospholipids. Freshly slaughtered calf brains are obtained from a local slaughterhouse. Brain phospholipids are prepared by a modification of the Bligh and Dyer extraction of lipids. 20 g pieces of calf brain are extracted with a mixture of chloroform (20ml):methanol (40ml) (26). Lipids are purified further by chromatography on a K-oxalate washed column of silicic acid to separate neutral lipids from phospholipids. Total chloroform:methanol (20:80) extractable phospholipids are used for reconstitution. The composition of the phospholipid mixture is determined by thin layer chromatography. It contained 31.4% phosphatidylcholine, 35.8% phosphatidylethanolamine, 17.7% phosphatidylserine, 15.1% sphingomyelin in the reconstitution experiments described here. The phospholipids are kept frozen in the cholate containing dialysis solution (in small portions) at -20°, and defrosted prior to usage.

3. RESULTS AND DISCUSSION

3.1. Properties of ATP independent Ca^{2+} transport in SPM vesicles

In the absence of added ATP, two types of Ca^{2+} transport can be demonstrated in SPM vesicles: 1) Na^+ gradient dependent Ca^{2+} transport; 2) Na^+ gradient independent Ca^{2+} transport. The major properties of these Ca^{2+} transport systems are demonstrated in Figure 2.

In Figure 2B (closed triangles), synaptic plasma membrane vesicles were preloaded with 0.15M Na phosphate (NaPi) pH 7.4, and diluted 83-fold into a medium containing 0.15M buffered KCl (at pH 7.4). A large calcium uptake into the vesicles results, reaching values of about 19 nMole Ca^{2+}/mg of protein in 10 minutes. This calcium uptake depends largely on the sodium gradient across the vesicle's membrane, since if one abolishes it by equilizing the sodium concentration inside the vesicles and in the medium, calcium transport is reduced to less than 15% of the control (Figure 2D, open circles) or if one reverses the direction of the sodium and the potassium gradients (Figure 2D, filled triangles) the net uptake of calcium is very small - as also when KPi preloaded vesicles are diluted into a buffered isoosmotic KCl solution containing Ca^{2+} (Figure 2B, open circles).

Figure 2A demonstrates that the Na^+ gradient is not the sole driving force for calcium uptake into synaptic plasma vesicles, since there is a substantial calcium entry in the virtual absence of sodium ions, when the vesicles are loaded with K phosphate (KPi) and diluted into isoosmotic sucrose medium (Figure 2A, open circles). The Ca^{2+} uptake in KPi preloaded vesicles diluted into an isoosmotic sucrose solution probably represents Ca^{2+} entry due to formation of a membrane potential (negative inside) since K^+ can diffuse out through the synaptic plasma membrane. When NaPi preloaded SPM vesicles are diluted into the buffered isoosmotic sucrose solution (Figure 2A, closed triangles), the Ca^{2+} uptake into the vesicles reaches about 32 nMoles/mg protein in 10 minutes. This value represents the sum of Na^+ gradient

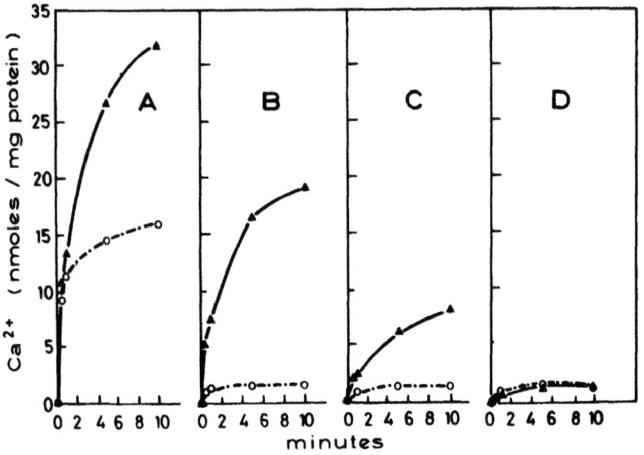

FIGURE 2. Calcium transport in SPM vesicles in the absence of ATP. 3 µl of NaPi (▲-▲-▲) or KPi (o-o-o) preloaded SPM vesicles (about 40 µg of protein) were diluted into 250 µl of: A. 0.3M sucrose - 10mM tris HCl pH 7.4; B. 0.15M KCl - 10mM Tris HCl pH 7.4; C. 0.15 M LiCl - 10mM tris HCl pH 7.4 and; D. 0.15M NaCl - 10mM tris HCl pH 7.4. The external solution contained also 50 µM $^{45}CaCl_2$ (0.1 µCi).

dependent (17 nMoles/mg protein) and the Na^+ gradient independent Ca^{2+} uptake. Substituting the external sucrose solution in a solution containing permeant cations (NaCl or KCl) leads to considerable decrease or even to total disappearance of the Ca^{2+} uptake into KPi preloaded vesicles (Figure 2B or 2D, open circles for example) probably due to a decrease in the negative inside membrane potential relative to the outside - and absence of any other driving force. The requirement for Na^+ gradient in the countertransport (in the absence of ATP or membrane potential) is unequivocal and no other ion can replace it. Partial substitution of the internal Na^+ solution with either K^+ or Li^+ leads to similar decrease in the Ca^{2+} influx (not shown), reflecting probably the reduction in the driving Na^+ gradient.

Figure 2C demonstrates that not all permeant ions are equal for supporting Na^+ gradient dependent Ca^{2+} uptake. Substitution of external KCl with LiCl causes a considerable reduction in the Ca^{2+} uptake although the driving force - the Na^+ gradient - was not reduced. There are several possible ways to explain this finding. For example, Ca^{2+} and Li^+ ions could compete for binding sites present on the carrier's external face; or, - both

Ca^{2+} and Li^+ could compete for the available Na^+ gradient by the existence of two apparently unrelated transport systems one exchanges Na^+ for Ca^{2+} and the other Na^+ for Li^+. Another explanation could be that the hydrated Li^+ ion is much less permeable than K^+ and consequently there is a lesser positive polarization on the membranes' internal face. This, would in turn lead to a decrease in Na^+ gradient dependent Ca^{2+} influx due to the electrogenic nature of the carrier.

In order to demonstrate directly that indeed in the absence of a Na^+ gradient when KPi preloaded vesicles are diluted into a buffered isoosmotic sucrose solution a negative inside membrane potential is developed, we have measured TPP^+ (tetraphenylphosphonium)$^+$ uptake into SPM vesicles. TPP^+ is a lipophilic cation used widely (27) to measure membrane potential in small vesicles - where direct measurements employing microelectrodes are not possible.

Figure 3 shows the $[^3H]TPP^+$ uptake obtained when KPi preloaded vesicles are diluted into buffered isoosmotic sucrose solution (closed circles) or buffered isoosmotic KCl solution with valinomycin (open circles) or without valinomycin (closed triangles).

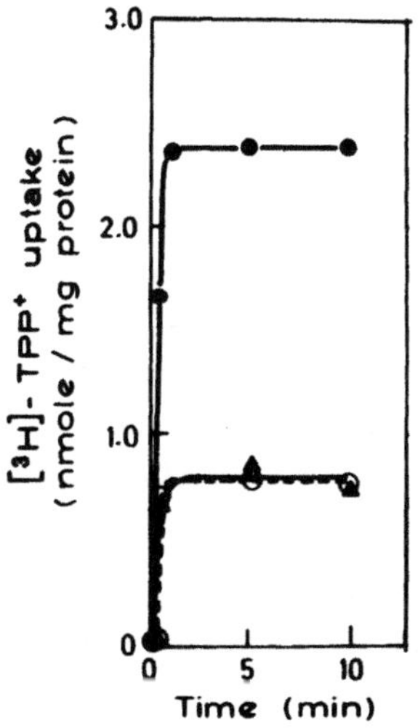

FIGURE 3. $[^3H]TPP^+$-uptake into SPM vesicles.
3 µl (about 30 µg protein) of 0.15M KPi buffer pH 7.4 preloaded SPM vesicles were diluted into 0.3M sucrose, 0.01M tris buffer pH 7.4, 50 µM $CaCl_2$ and 4.48 µM $[^3H]TPP^+$ (about 4 nCi) (●-●-●) or 0.15M KCl, 0.01M tris buffer pH 7.4, 50 µM $CaCl_2$ and 4.48 µM $[^3H]TPP^+$ (about 4 nCi): with valinomycin (o-o-o) or without (▲-▲-▲).

In the presence of valinomycin, the KPi preloaded SPM vesicles' membrane diluted into KCl is presumably freely permeable to K^+ and consequently there should be no membrane potential difference across the vesicle's membrane. The amount of TPP^+ taken up by these vesicles represents the equilibrium distribution of TPP^+ in addition to the amount of TPP^+ solubilized in the membrane's lipid phase. In any attempt, to calculate the membrane potential difference created across the vesicles membrane, this value should be subtracted. It can be seen that KPi preloaded vesicles in the presence of external KCl do not take up additional TPP^+ (Figure 3, closed triangles) beside the equilibrium value. Additional amounts of TPP^+ are taken up by the KPi preloaded vesicles when the external KCl solution is replaced by buffered isoosmotic sucrose.

The intravesicular volume of the vesicles has been determined from the distribution of ^{14}C-inulin and ^3H_2O - this value corresponds to 7 ul/mg protein. Therefore, we can calculate the internal TPP^+ concentration and hence the negative inside membrane potential created when KPi preloaded vesicles are diluted into the buffered sucrose solution, according to the Nernst equation. This value corresponds to about -100mV. This potential difference should lead to a ratio of:

$$\frac{[Ca^{2+}]in}{[Ca^{2+}]out} \approx 1000.$$

Since this ratio was not reached in the time scale of these experiments, the data presented are under non-equilibrium conditions for Ca^{2+}. This finding can be substantiated also by adding the Ca^{2+} ionophore A23187 (Figure 4A, closed triangles) to NaPi preloaded vesicles diluted into an isoosmotic sucrose solution. The addition of the Ca^{2+} ionophore enhances in this case the rate of Ca^{2+} entry, as expected in a non-equilibrium state. Addition of the Ca^{2+} ionophore to NaPi preloaded vesicles diluted into a KCl solution - (Figure 4B, open triangles) - in the absence of additional driving force beside the Na^+ gradient, leads to loss of Ca^{2+} from the vesicles.

It should be noted that addition of nigericin, an ionophore inducing electroneutral exchange of $H^+/Na^+/K^+$, to the NaPi preloaded vesicles diluted into buffered isoosmotic KCl solutions (Figure 4B, open circles) when the Na^+ gradient is the sole driving force for Ca^{2+} uptake - prevents accumulation of Ca^{2+}, while addition of nigericin to the same vesicles diluted into the buffered isoosmotic sucrose solution (Figure 4A, closed circles) has only a very small effect. The reason for this is probably the relative contribution of the Na^+ gradient respective to the membrane potential as a driving force for Ca^{2+} uptake.

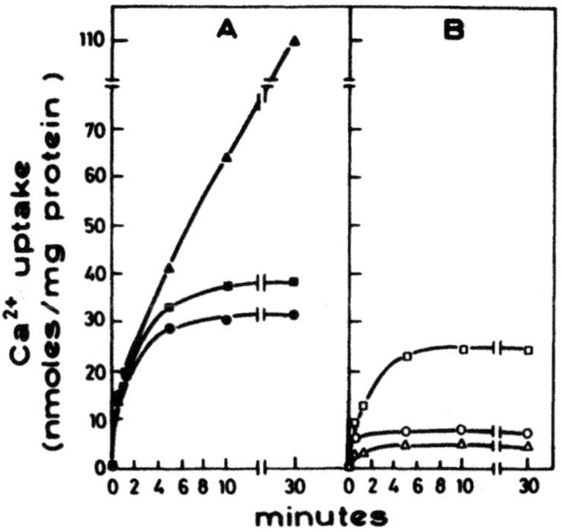

FIGURE 4. The effect of the Ca^{2+} ionophore A23187 and nigericin on Ca^{2+} uptake in SPM vesicles.
A. 3 µl (about 30 µg protein) 0.15M NaPi preloaded SPM vesicles were diluted into 250 µl of 0.3M sucrose, 10mM tris HCl pH 7.4, 50 µM $^{45}CaCl_2$ and with 10 µM A23187 (▲-▲-▲) or with 10 µM nigericin (●-●-●) or without the ionophores (■-■-■); B. 3 µl (about 30 µg protein) 0.15M NaPi preloaded SPM vesicles were diluted into 250 µl of 0.15M KCl, 10mM tris HCl pH 7.4, 50 µM $^{45}CaCl_2$ with 10 µM A23187 (△-△-△) or 10 µM nigericin (o-o-o) or without the ionophores (□-□-□).

In the heart, both in the intact preparation and in isolated cardiac membrane vesicles, a 6-3 Na^+ to 1 Ca^{2+} stoichiometry was proposed (28,29,30). We have performed measurements in which the ratio of the apparent initial rate of Ca^{2+} dependent $^{22}Na^+$ efflux from SPM vesicles has been compared with Na^+ gradient dependent apparent initial rate of Ca^{2+} influx in the same preparation. The data obtained in five different experiments are shown in Figure 5. It can be seen that a ratio of 5 Na^+ ions exchanged for each Ca^{2+} is obtained, suggesting that if no other other ions are involved in the exchange process - the antiport is electrogenic. It should be remembered however, that the exact stiochiometry of the process might be somewhat different due to several reasons: the initial rates measured are apparent initial rates and not true initial rates due to technical limitations of flux measurements. In addition, in native SPM vesicles as also in other multi ion channel systems, unrelated conductances might affect the measurements attributed to Na^{+}-Ca^{2+} antiport solely. To overcome these problems, steady-state measurements under conditions of fixed membrane potentials should be done employing the purified and reconstituted Na^{+}-Ca^{2+} exchanger preparation.

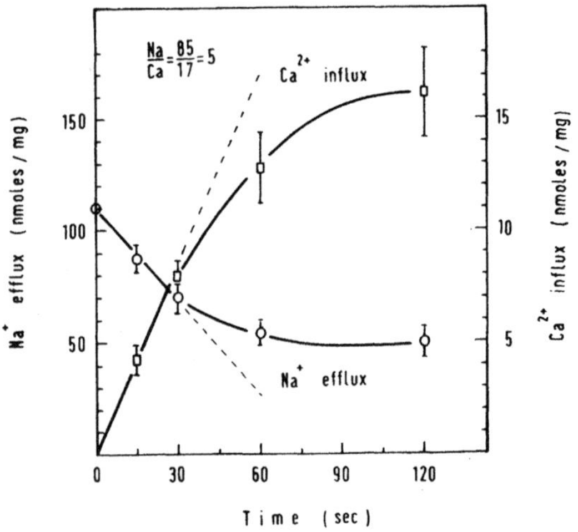

FIGURE 5. Stoichiometry of the SPM Na^{+}-Ca^{2+} antiporter.
3 µl (30 µg protein) SPM vesicles preloaded in buffered 0.15M ^{22}NaCl were diluted into 100 µl of buffered 0.15M KCl with or without 50 µM $CaCl_2$. $^{22}Na^{+}$ efflux has been determined at time points specified. Ca^{2+} independent $^{22}Na^{+}$ efflux has been subtracted (o-o-o). The same experiment was repeated except that the NaCl solution was unlabelled and the Ca^{2+} added to the reaction mixture was labelled (□-□-□) (the bars represent standard deviation).

Several lines of evidence support the idea that the Na^{+}-Ca^{2+} antiporter exhibits asymmetrical properties when embedded in its native membrane: Previous experiments we have performed (19) in which the inhibitory effect of lanthanides on Na^{+}-Ca^{2+} antiport has been studied showed, that the inhibition by lanthanides on the Na^{+} gradient dependent Ca^{2+} influx or efflux was manifested only when the lanthanides were added from the external (extracellular surface) side of "right side-out" vesicles. Studying the effect of changes in the pH of the external solution on the Na^{+}-Ca^{2+} antiport process reveals also, that Na^{+} gradient dependent Ca^{2+} influx (Figure 6, open

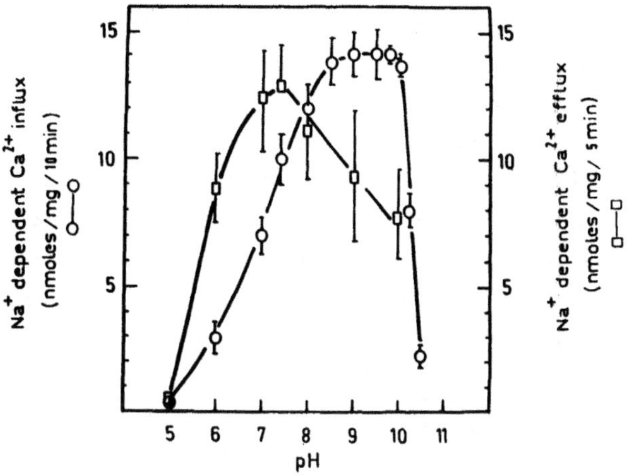

FIGURE 6. pH Dependence of [Na^+] gradient dependent Ca^{2+} influx and efflux. Influx: 3 µl (30 µg protein) 0.15M NaCl preloaded SPM vesicles were diluted 83 fold into 0.15M KCl containing 50 µM $^{45}CaCl_2$. Both internal and external solutions contained also a 10mM buffer system: pH 5-6 acetate pH 6-7 imidazole; 7-8.5 tris HCl, pH 8-9 glycine, pH 9-10.5 CAPS. Ca^{2+} uptake was measured for 10 min (o-o-o).
Efflux: Ca^{2+} preloaded SPM vesicles in a Na^+ gradient dependent manner at pH 7.4 as in the influx experiment were separated by centrifugation from the influx medium. Then, the vesicles were diluted 83 fold into 0.15M NaCl buffered with the same buffer systems as used for the influx experiment. Na^+ gradient independent Ca^{2+} influx has been subtracted.

circles) exhibits a different pH dependancy than Na^+ gradient dependent Ca^{2+} efflux (Figure 6, open squares), suggesting that Na^+-Ca^{2+} antiport is manifested differently at the two sites of the synaptic plasma membrane. The steep increase in Na^+ gradient dependent Ca^{2+} influx between pH 6 to pH 8.5, followed by a plateau till pH 10 above which rapid decline in the transport is obtained, is not observed when Na^+ gradient dependent Ca^{2+} efflux is measured. The efflux process is gradually activated as the pH increases between pH 5 and pH 7.5 above which a gradual decrease in the process is observed. No considerable difference in Na^+-Ca^{2+} antiport by either addition of H^+ ionophores such as FCCP (experiments not shown) or creating a pH gradient across the vesicles' membrane has been obtained, suggesting that Na^+-H^+ or Ca^{2+}-H^+ antiport was not an intermediate step in Na^+-Ca^{2+} antiport and the pH dependency of the process reflected may be the nature of the protein groups involved in binding Ca^{2+} or Na^{+} ions. pH regulation of transport selectivity has been observed (31) by studying Ca^{2+} and K^{+} symport coupled to H^{+} antiport by the lipophilic dicarboxylic acid - dicarboxamide [18] - O_6 macrocycle 1. Ca^{2+} transport increased as the pH increased between 2 and 9 while K^{+} transport decreased and was almost abolished above pH 6.5

In order to study the molecular properties of the Na^{+}-Ca^{2+} exchanger, we have isolated, purified and reconstituted it into phospholipid membranes (21). At the end of the reconstitution, NaPi loaded vesicles were exposed to $^{45}Ca^{2+}$ in the presence (external KCl) or in the absence (external NaCl) of Na^{+} gradient. Following Ca^{2+} loading, the vesicles were separated by sucrose density gradient centrifugation. Ca^{2+} loaded vesicles were displaced to a denser region of the sucrose gradient than the majority of the liposomes.

Figure 7 shows the distribution of the $^{45}Ca^{2+}$ and phospholipid phosphate contents of the isopycnic sucrose density gradient. It can be seen (Figure 7A) that considerable separation between the vesicles containing $^{45}Ca^{2+}$ and the major phosholipid peak occured when the $^{45}Ca^{2+}$ loading was performed in the presence of Na^{+} gradient. In its absence (Figure 7B), no appreciable $^{45}Ca^{2+}$ loading was obtained - and none of the vesicles were displaced to a denser region of the sucrose gradient - below the phospholipid peak. Figure 8

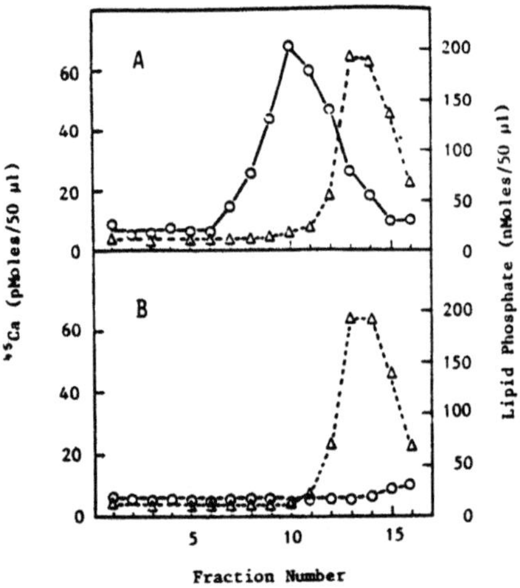

FIGURE 7. Sucrose gradient profile of reconstituted SPM vesicles following $^{45}Ca^{2+}$ loading.
Reconstituted NaPi preloaded vesicles were exposed to $^{45}Ca^{2+}$ in the presence of buffered 0.2M KCl (A) or buffered 0.2M NaCl (B) for 20 minutes and subjected to centrifugation. Fractions of 0.5ml were collected from the bottom. $^{45}Ca^{2+}$ (o-o-o) and phospholipid phosphate (Δ-Δ-Δ) were determined in each fraction.

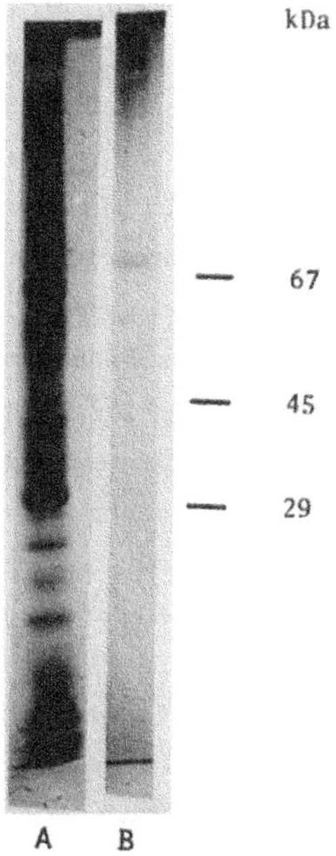

FIGURE 8. $NaDodSO_4$ gel electrophoresis pattern of the purified and reconstituted Na^+-Ca^{2+} exchanger.
A. Native SPM vesicles - 20 µg protein; B. Pooled fractions containing the $^{45}Ca^{2+}$ peak from a similar sucrose gradient to that shown in Figure 7A. 1 µg protein; gel is stained by silver stain.

shows the gel electrophoresis pattern of the protein profile obtained when the fractions below the major phospholipid peak containing most of the $^{45}Ca^{2+}$ from the experiment shown in Figure 7A (fractions 7-11) were pooled and analysed by $NaDodSO_4$ (Na dodecyl sulphate) polyacrylamide gel electrophoresis. One major protein band corresponding to a molecular weight of 70,000 daltons is consistently revealed by ultrasensitive silver staining (32). No specific protein band was revealed when the corresponding fractions from the control experiment shown in Figure 7B were analysed in an identical fashion. In addition to the 70,000 dalton molecular weight protein which consistently appeared in all our experiments in a highly enriched form, other minor bands could be inconsistently seen. Direct identification of the Na^+-Ca^{2+} exchanger protein will be possible only after specific antibodies that either immunoprecipitate the activity or impair it are obtained. The reason for this is since no specific inhibitor of the Na^+-Ca^{2+} exchanger is available that could be used as an additional direct tool for its identification.

The purification obtained in these experiments has been considerable: based on protein or phospholipid content of the Ca^{2+} loaded sucrose gradient purified and reconstituted vesicles an increase of about 130 fold in Na^+ gradient dependent Ca^{2+} transport could be measured (21).

The reconstituted phospholipid fortified Na^+-Ca^{2+} exchanger, beside its analytical value in studying the protein itself, is also extremely useful for studies designed to understand its mechanism of action. For example, the effect of changes in membrane potential on the Na^+-Ca^{2+} antiport can be measured. This has been done, by adding valinomycin to K^+ containing medium and measuring the extent of Na^+ dependent Ca^{2+} uptake respective to an identical system except that valinomycin was not added. Figure 9 shows such an experiment. Native SPM vesicles were preloaded with NaPi and diluted into a KCl solution containing Ca^{2+} (Figure 9A). The open circles present the control experiment and the closed circles represent identical conditions except that valinomycin was included in the external medium as well. It can be seen that except in the very early time points, no significant stimulation of the Na^+ gradient dependent Ca^{2+} influx is obtained by valiomycin.

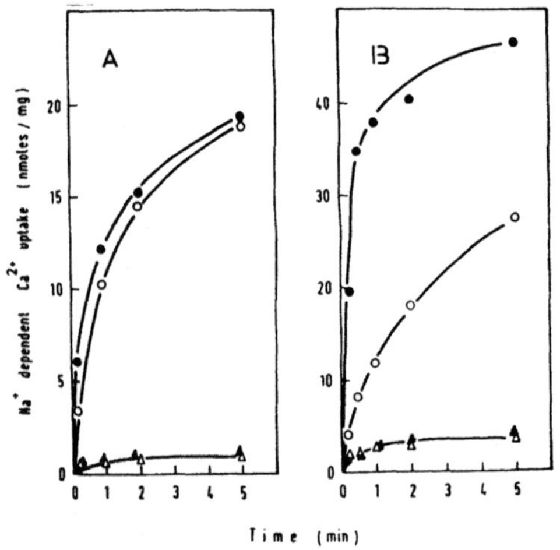

FIGURE 9. The effect of valinomycin on Na^+-Ca^{2+} antiport.
A. 3 ul native SPM vesicles (30 μg protein) were preloaded in 0.15M NaPi buffer pH 7.4 and diluted into 100 μl of buffered 0.15M KCl, 50 μM $^{45}CaCl_2$ with (●-●-●) or without (o-o-o) valinomycin; B. The same experiment as in A except that reconstituted SPM vesicles were used. The effect of valinomycin on Na^+ gradient independent Ca^{2+} uptake is also shown (▲-▲-▲ with valinomycin; Δ-Δ-Δ without valinomycin).

Figure 9B shows an experiment which is identical to the one shown in Figure 9A except that reconstituted SPM vesicles were used. The stimulation obtained by adding valinomycin (closed circles) in this case is quite impressive, as expected for an electrogenic process. No stimulation by valinomycin is obtained in either case for the Na^+ gradient independent Ca^{2+} uptake. The reconstituted phospholipid enriched preparation has two main advantages in these mechanistic studies: 1) The passive membrane permeability to ions is much smaller than in the native vesicles. Therefore, the stimulation of Na^+ dependent Ca^{2+} influx obtained by valinomycin induced K^+ entry could be separated from passive K^+ diffusion and; 2) In this reconstitution process a relative high phospholipid to protein ratio has been maintained. Thus, ion conductances unrelated to the Na^+-Ca^{2+} exchanger protein itself were separated into other liposomes following solubilization, dialysis and reformation of vesicles. Therefore, their indirect effects on Na^+-Ca^{2+} antiport are eliminated.

ACKNOWLEDGEMENTS

The results presented here were supported in part by a research grant from the Muscular Dystrophy Association, New York, and the Israel Academy of Sciences. A.E. is a recipient of the Foulkes Fellowship.

REFERENCES

1. Bradford HF: Isolated nerve terminals as an in vitro preparation for the study of dynamic aspects of transmitter metabolism and release. In: Iverson, Iverson and Snyder (Eds), The Handbook of Psychopharmacology, Plenum Press, 1975, 1: 191-252.
2. Kelly RB, Deutsch JW, Carlson SC, Wagner JA: Biochemistry of neurotransmitter release. Ann. Rev. Neurosci. 2: 399-446, 1979.
3. Blaustein MP, Goldring JM: Membrane potentials in pinched off presynaptic nerve terminals, monitored with a fluorescent probe. Evidence that synaptosomes have potassium difusion potential. J. Physiol. 247: 589-615, 1975.
4. Rahamimoff R: Principles of chemical signaling in the nervous system. In: JG Nicholls (Ed), The Role of Intercellular Signals: Navigation, Outcome and Consequences. Berlin: Dahlem.
5. Katz B: Nerve, Muscle and Synapse. New York, MacGraw-Hill, 1966.
6. Hagiwara S, Byerly L: Calcium channel. Ann. Rev. Neurosci. 4: 69-125, 1981.
7. Baker PF, Blaustein MP, Hodgkin AL, Steinhardt RA: The influence of calcium on sodium efflux in squid axon. J. Physiol. 200: 431-458, 1969.
8. Blaustein MP, Ector AC: Carrier mediated sodium dependent and calcium dependent calcium efflux from pinched off presynaptic nerve terminals in vitro. Biochem. Biophys. Acta 419: 295-308, 1976.
9. Rahamimoff H, Spanier R: Sodium dependent calcium uptake in membrane vesicles derived from rat brain synaptosome. FEBS Lett. 104: 111-114, 1979.
10. DiPolo R: Ca pump driven by ATP in squid axon. Nature 274: 390-392, 1978.
11. Sorensen RG, Mahler HR: Calcium stimulated adenosine triphosphatase in synaptic membrane. J. Neurochem. 37: 1407-1418, 1981.
12. Alnaes E, Rahamimoff R: On the role of mitochondria in transmitter release from motor nerve terminals. J. Physiol. 248: 285-306, 1975.
13. Blitz AL, Fine RE, Tosselli PA: Evidence that coated vesicles isolated from brain are calcium sequestering organelles resembling sarcoplasmic reticulum. J. Cell. Biol. 75: 135-147, 1977.
14. Kendrick NC, Blaustein MP, Fried RC, Ratzlaff RW: ATP dependent calcium storage in presynaptic nerve terminals. Nature 265: 246-248, 1977.
15. Politoff AL, Rose S, Pappas GD: Calcium binding sites of synaptic vesicles of frog sartorius neuromuscular junction. J. Cell. Biol. 68: 818-823, 1978.
16. Henkart MP, Reese TS, Brinley FJ: Endoplasmic reticulum sequesters calcium in the squid giant axon. Science 202: 1300-1302, 1978.
17. Requena J: Calcium transport and regulation in nerve fibers. Ann. Rev. Biophys. Bioeng. 12: 237-257, 1983.
18. Erdreich A, Spanier R, Rahamimoff H: The inhibition of Na-dependent Ca uptake by verapamil in synaptic plasma membrane vesicles. Europ. J. Pharmacol. 90: 193-202, 1983.
19. Rahamimoff H, Spanier R: The asymmetric effect of lanthanides on Na^+-gradient dependent Ca^{2+} uptake in synaptic plasma membrane vesicles. Biochim. Biophys. Acta 773: 279-289, 1984.
20. Schellenberg GD, Swanson PD: Sodium dependent and calcium dependent calcium transport by rat brain microsomes. Biochim. Biophys. Acta 648: 13-27, 1981.
21. Barzilai A, Spanier R, Rahamimoff H: Isolation, purification and reconstitution of the Na^+ gradient dependent Ca transporter (Na^+-Ca^{2+})

exchanger from brain synaptic plasma membrane. Proc. Natl. Acad. Sci. 81: 6521-6525, 1984.
22. Papazian D, Rahamimoff H, Goldin SM: Reconstitution and purification by transport specifity fractions of an ATP dependent calcium transport component from synaptosomes. Proc. Natl. Acad. Sci. 76: 3708-3712, 1979.
23. Papazian D, Rahamimoff H, Goldin SM: Partial purification and functional reconstitution of a calmodulin-activated, adenosine-5'-phosphate dependent calcium pump from synaptic plasma membrane. J. Neurosci. 4: 1933-1943, 1984.
24. Chan SY, Hess EJ, Rahamimoff H, Goldin SM: Purification and immunological characterization of a calcium pump from bovine brain synaptosomal vesicles. J. Neurosci. 4: 1468-1478, 1984.
25. Gasko DD, Knowles AF, Shertzer HG, Suolinna EH, Racker E: The use of ion exchange resin for studying ion transport in biological systems. Anal. Biochem. 72: 57-65, 1976.
26. Sheltway A, Dawson RMC: Chromatographic and Electrophoretic Techniques. In: I Smith (Ed), Vol 1, pp.450-493, 1969.
27. Schuldiner S, Kaback R: Membrane potential and active transport in membrane vesicles from E. Coli. Biochem. 24: 5451-5460, 1975.
28. Mullins LJ: The generation of electric current in cardiac fibers by Na-Ca exchange. Ann. J. Physiol. 236: C103-C110, 1979.
29. Pitts BJR: Stoichiometry of sodium-calcium exchange in cardiac sarcolemmal vesicles. J. Biol. Chem. 254: 6232-6235, 1979.
30. Horackova M, Vassort G: Sodium-calcium exchange in regulation of cardiac contractility. J. Gen. Physiol. 73: 403-424, 1979.
31. Hriciga A, Lehn JM: pH regulation of divalent/monovalent Ca/K cation transport selectivity by a macrocyclic carrier molecule. Proc. Natl. Acad. Sci. U.S.A. 80: 6426-6428, 1980.
32. Oakley BR, Kirsch DR, Morris NR: A simplified ultrasensitive silver stain for detecting proteins in polyacrylamide gel. Anal. Biochem. 105: 361-363, 1980.

DISCUSSION OF THE ARTICLE BY H. RAHAMIMOFF (ISRAEL)

A.M. Brown (U.S.): You showed a slide which was supposed to show electrogenicity. This showed that the uptake was increased in the presence of K-valinomycin gradient but what was not clear was could not a component of this uptake be the passive influx component that you described previously so that it is not electrogenic.

H. Rahamimoff (Israel): The increased influx was in the reconstituted preparation where you have almost no passive permeability of other ions.

M.P. Blaustein (U.S.): You tantalized us with a very nice data on Na/Ca exchange reconstituted in liposomes but you did not tell us the stochiometry of the exchange. Have you measured it?

H. Rahamimoff (Israel): No. We are trying to do it. The problem with the stochiometry there is that you are preparing and purifing the reconstituted vesicles by calcium loading so they are always calcium loaded and then we have to have a double reconstitution to get rid of the calcium that was inside for the purfication and then to try to use them. We are trying to do this.

CALCIUM-ACTIVATED POTASSIUM CHANNELS IN PRESYNAPTIC NERVE TERMINALS

D.K. BARTSCHAT and M.P. BLAUSTEIN
Department of Physiology, University of Maryland School of Medicine, Baltimore, Maryland 21201

INTRODUCTION

In their classical study of excitability in squid axons, Hodgkin and Huxley (cf. Hodgkin, 1967) showed that the nerve action potential could be explained by selective, transient sodium (Na) and potassium (K) conductance changes that were explicitly described by sets of voltage- and time-dependent parameters. Subsequent work on a wide variety of excitable cells has revealed that, in contrast to Na conductances, a number of K conductances do not conform to the simple voltage- and time-dependent mechanism that occurs in squid axons. For example, in many types of cells, some K conductances are activated by a rise in the calcium (Ca) activity in the cytoplasm (cf. Meech, 1978). Also, in some cells, K conductances may be modulated by intracellular cyclic nucleotides, for example, as a consequence of neurotransmitter action (Castelluci et al., 1980; De Peyer et al., 1982). Klein and Kandel (1980) have demonstrated that inactivation of a specific K conductance, due to transmitter action on presynaptic neurons in *Aplysia*, may have profound behavioral effects. In addition, it has been suggested that the complex schizophrenia-like syndrome elicited by phencyclidine intoxication in man and animals may involve block of a specific class of presynaptic K channels in the central nervous system (CNS) (Albuquerque et al., 1981; Bartschat and Blaustein, 1984c).

Most of our information regarding K channel function has been obtained from the neurons of lower animals: much less is known about K channels in the mammalian CNS. Information on the types and functions of K channels in presynaptic nerve terminals is particularly scarce, even though alterations in K channel function here would have a major impact on synaptic transmission and integrated neuronal function.

Presynaptic nerve terminals from mammalian CNS are too small for microelectrode impalement; we have therefore developed methods, based on radiotracer efflux from isolated nerve terminals ("synaptosomes") from rat brain, that enable us to assess nerve terminal K permeability (P_K) changes on a time scale approaching physiologically relevent events. We show here that nerve terminals from mammalian CNS possess Ca-activated K channels; these K channels are physiologically and pharmacologically similar to Ca-activated K channels observed in other excitable tissues.

METHODS

Synaptosomes were prepared from the forebrains of female albino rats by a one-step sucrose gradient method (Krueger et al., 1979). The 0.8M sucrose fraction containing the synaptosomes was equilibrated slowly with three volumes of ice-cold physiological saline containing (in mM): NaCl, 145; KCl, 5; $MgCl_2$, 2.0; RbCl, 0.1; glucose, 10; and HEPES buffer, 10, titrated to pH 7.4 with NaOH. The equilibrated synaptosomes were centrifuged and resuspended to a protein concentration of 30-40 mg/ml; tracer quantities of ^{86}Rb (20 μCi/ml) were added, and the synaptosomes were allowed to accumulate the tracer for 30-40 min at 30^oC. Control experiments revealed that >75% of the ^{86}Rb was accumulated by the terminals via a ouabain-sensitive Na-K pump.

In order to measure ^{86}Rb efflux, aliquots (50 μliters) of the ^{86}Rb-loaded synaptosome suspension were placed on glass fiber filters, and were then washed five times with 2 ml aliquots of tracer-free equilibration medium. Efflux was initiated by the rapid addition of "efflux media" (see Results for details on the composition of these solutions), followed by the rapid addition of quench medium to terminate tracer efflux. Vacuum was then applied, and the effluent was collected and analyzed for released radioactivity by liquid scintillation techniques. The efflux of ^{86}Rb was expressed as the percentage of total accumulated Rb released into the effluent. Thus, for each time point:

$$\%^{86}Rb\ Efflux = \frac{^{86}Rb_{Effluent}}{^{86}Rb_{Effluent} + {}^{86}Rb_{filter}} \times 100$$

The quench medium contained (mM): tetraethylammonium chloride (TEA), 145; tetrabutylammonium chloride (TBA), 5; $MgCl_2$, 5; $NiCl_2$, 10; and HEPES buffer, pH 7.4, 20. Control experiments indicated that high K-stimulated ^{86}Rb efflux was reduced to negligible rates immediately (less than one sec) after the addition of the quench medium.

RESULTS

Previous experiments (Bartschat and Blaustein, 1983 and

1984a-c) have shown that voltage-regulated K channels in synaptosomes can be studied by measuring Rb efflux from ^{86}Rb-loaded terminals with a manual quench technique. An aliquot of ^{86}Rb-loaded synaptosomes was placed on a glass fiber filter, was washed free of the bulk of extracellular ^{86}Rb, and was then "pulsed" with an efflux solution of desired composition. Efflux was rapidly terminated by the addition of quench solution containing high concentrations of the K channel blockers TEA and TBA and the Ca channel blocker Ni^{+2}, followed immediately by the application of suction to the filters. By collecting the filtrate, and subjecting both the filter and the filtrate to liquid scintillation counting, it was possible to determine the fraction (or the percentage) of the previously-accumulated ^{86}Rb that was released. This method can resolve tracer effluxes as small as 0.2-0.3% of the total tracer accumulated by the synaptosomes with an error of 3% or less.

^{86}Rb was used as a K analog because it has a much longer half-life than ^{42}K. Furthermore, Rb is actively accumulated into excitable cells via the ouabain-sensitive Na-K pump (Baker et al., 1969; Bartschat and Blaustein, 1984a), and Rb passes through various neuronal K channels nearly as well as K itself (cf. Hille, 1975; Reuter and Stevens, 1980). In preliminary experiments Rb and K effluxes were measured simultaneously after the synaptosomes had been allowed to accumulate both ^{86}Rb and ^{42}K. These results showed that basal and depolarization-evoked Rb and K effluxes were qualitatively quite similar, although Rb permeates K channels in synaptosomes about only 0.6-0.8 times as well as K itself (Bartschat and Blaustein, 1984a).

When experiments of this type were carried out, the pattern of release illustrated in Fig. 1 was seen. Efflux promoted by media containing 5 mM K was about 0.3-0.4%/sec; this component of efflux ("R", Fig. 1) was unchanged whether or not 1 mM Ca was present in the medium. Component R probably represents ^{86}Rb efflux through K channels responsible for the resting K conductance of the terminals (Blaustein and Goldring, 1975).

Extrapolation of component R to the ordinate ("zero-time") revealed the presence of another component of efflux (component "B", Fig. 1). Control experiments (data not shown) indicated that component B is not a physiological K conductance, since: i) reversing the order of addition of efflux media and quench media (i.e., adding quench medium first) yielded a value of efflux equal to the extrapolated "zero-

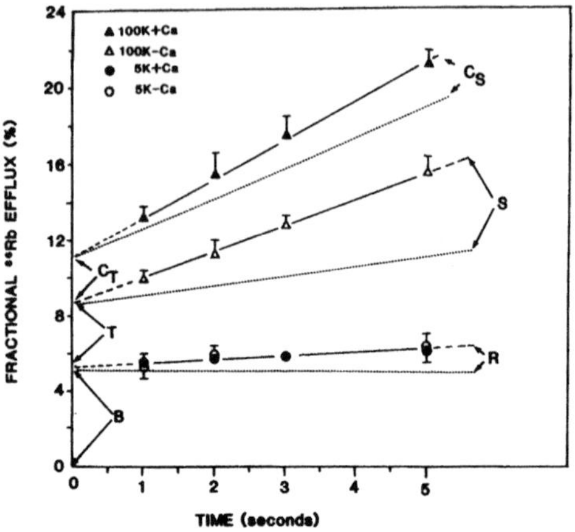

FIGURE 1. ^{86}Rb efflux from synaptosomes: the effects of depolarization and Ca. Rb efflux from ^{86}Rb-loaded synaptosomes was perfomed as described (see Methods). In those experiments employing elevated $[K]_o$ ± 1 mM Ca, these ions replaced Na or Mg, respectively, mole-for-mole. The values are the means ± SEM of four determinations.

time" efflux in 5 mM K medium; ii) altering the $[Mg]_o$ concentration in the efflux solutions, which modifies the release of Rb evoked by K-rich solutions, had no effect on component B; and iii) known K channel blockers, which inhibit K-evoked release (see below), had no effect on B.

Since component B does not represent a physiological K conductance, but probably reflects release of extracellularly bound tracer that was incompletely removed during the wash procedure, the ordinate intercept obtained by extrapolation of ^{86}Rb efflux in 5 mM K medium was assumed to be the true "zero-time" efflux. With this assumption, and assuming $[K]_i$ = 110 mM (Bartschat and Blaustein, 1984a), and the "resting" membrane potential (V_R) in 5 mM K = -60 mV (Blaustein and Goldring, 1975), and the synaptosomal surface area-to-volume ratio is 8 X 10^4 cm^{-1} (Blaustein, 1975), the resting Rb efflux was 5.5 pmol/cm^2-sec, corresponding to a resting Rb permeability of 1.9 X 10^{-7} cm/sec. The analogous values for ^{42}K efflux and P_K were 7.2 pmol/cm^2-sec and 2.4 X 10^{-7} cm/sec, respectively, in good agreement with other excitable tissues (eg., Carmeliet et al, 1976).

Experiments with voltage-sensitive fluorescent dyes (Blaustein and Goldring, 1975) indicate that synaptosomes have membrane potentials that behave like K diffusion potentials, so we examined the effect of depolarization of the synaptosomes (by elevating $[K]_o$) on Rb efflux. Depolarization of the synaptosomes with nominally Ca-free medium containing 100 mM K markedly stimulated Rb efflux; the stimulated release occured in two phases. A fast, transient K conductance that apparently inactivated in less than 1 sec (component "T", Fig. 1) was seen, as well as a slower, sustained K conductance that did not inactivate in less than 5 secs (component "S", Fig. 1). These K conductances could be distinguished pharmacologically as well as kinetically. Component T was selectively blocked by low (<1 mM) concentrations of TEA, TBA, and 4-aminopyridine, while component S was much less sensitive to these agents. On the other hand, component S was selectively blocked by nano- to micromolar concentrations of the psychotomimetic phencyclidine, while component T was blocked only by 100-fold greater concentrations of this agent (Bartschat and Blaustein, 1984a-c).

Addition of 1 mM Ca to media containing 5 mM K had no effect on Rb efflux. However, when Ca was added to K-rich media, a greater ^{86}Rb efflux was seen than with elevated K and no added Ca (Fig. 1). This Ca-dependent increment in Rb efflux in K-rich media could also be divided into two components: a small, somewhat variable slow component, corresponding to the Ca-dependent increment in Rb efflux between 1 and 5 secs ("C_S" in Fig 1), and a fast, transient component, corresponding to the Ca-dependent increase in Rb efflux seen when the efflux was extrapolated to "zero-time" ("C_T" in Fig. 1). Since Ca is ineffective in 5 mM K, but requires depolarization to act, Ca probably exerts its action by entering the synaptosomes through voltage-activated Ca channels (Nachshen and Blaustein, 1980). This leads to an increase in the cytoplasmic Ca activity, thereby activating ^{86}Rb efflux. These observations are consistent with the presence of Ca-activated K channels in mammalian central presynaptic nerve terminals.

The data in Fig. 2A illustrate the effect of varying $[Ca]_o$ on the time course of Rb efflux into media containing 100 mM K. The magnitudes of both Ca-dependent components (C_T and C_S) are graded with increasing $[Ca]_o$ although, at saturating $[Ca]_o$, C_T is larger than C_S (Fig. 2B). In addition, C_S appears to be more sensitive to $[Ca]_o$ than is C_T, but its relatively small magnitude and its variation (in magnitude) from preparation to preparation has made this difficult to evaluate.

The transient nature of C_T may be a result of spon-

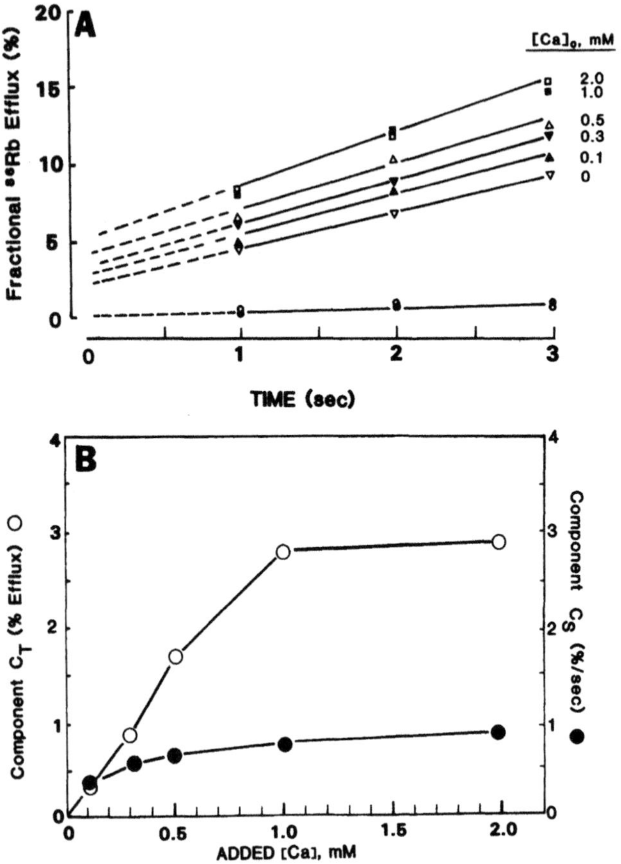

FIGURE 2. a). The effect of Ca on ^{86}Rb efflux. Rb efflux was measured in media containing either 5 mM K + 2 mM Ca (●) or no added Ca (0), or was performed in 100 mM K plus the $[Ca]_o$ indicated in the figure. In these experiments, Ca replaced Mg mole-for-mole, such that Ca + Mg = 2 mM. b). The dependence of component C_T (0) or C_S (●) on $[Ca]_o$. These components were quantitated as described in Methods.

taneous inactivation of this conductance, although other explanations exist. About 80% of the depolarization-elicited Ca entry in synaptosomes is inactivated within one sec; thus it is likely that the apparent inactivation of a large fraction of the Ca-dependent ^{86}Rb efflux may reflect the decline in Ca entry during a prolonged incubation in K-rich solutions (cf., Nachshen and Blaustein, 1980; Nachshen, 1982). Because

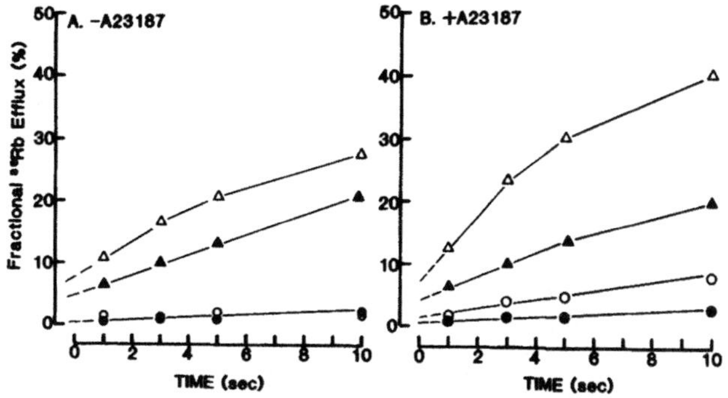

FIGURE 3. The effect of A23187 on ^{86}Rb efflux. a). Control. b). In the presence of 1 M A23187. (●)= 5 mM K, 0 Ca. (O)= 5 mM K + 1 mM Ca. (▲)= 75 mM K, 0 Ca. (△)= 75 mM K + 1 mM Ca.

of this dilemma, as well as the fact that $[Ca]_i$ may change with time during prolonged depolarizations as a result of buffering and extrusion (McGraw et al., 1982), we are unable to resolve the kinetics of the Ca-dependent ^{86}Rb efflux with the present techniques. Nevertheless, these manual rapid quench methods do enable us to study many other properties of this conductance.

The Ca-selective ionophore, A23187, was used to determine whether Ca entry into synaptosomes in the absence of membrane depolarization can stimulate ^{86}Rb efflux. Fig. 3A shows the typical pattern of Rb efflux into 5 or 75 mM K media ± 1 mM Ca in the absence of A23187; comparable fluxes, in the presence of 1 μM A23187, are illustrated in Fig. 3B. Comparison of these data indicates that A23187 had no effect on the efflux of Rb in either 5 mM or K-rich media in the absence of external Ca. However, with Ca present, the ionophore increased ^{86}Rb efflux into both 5 and 75 mM K media. Thus, Ca entry into either polarized or depolarized synaptosomes can activate Rb efflux. Note that the Rb efflux promoted by Ca+A23187 lasts for several seconds. This may indicate that the early decline ("inactivation") in the Ca-dependent Rb efflux, observed after 1 sec (see Figs. 1 and 2), is due largely to inactivation of the voltage-activated Ca channels and reduction in Ca entry, and not to inactivation of the Ca-activated Rb (K) conductance.

To determine whether or not the Ca+A23187-induced

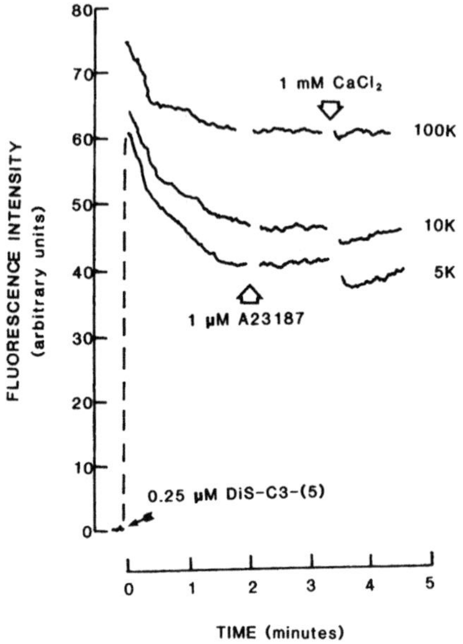

FIGURE 4. Effect of Ca+A23187 on membrane potentials in synaptosomes. The voltage-sensitive fluorescent dye, di-S-C3-(5), was added to a cuvette containing synaptosomes suspended in media containing the indicated $[K]_o$ and 50 μM EGTA. After a stable fluorescence signal was obtained, A23187 was added, followed by the addition of $CaCl_2$.

increase in Rb efflux reflects a selective increase in P_K, synaptosome membrane potentials were monitored with a voltage-sensitive fluorescent dye, DiS-C3-(5) (Fig. 4). The fluorescence intensity (FI) of the dye + synaptosomes increased with increasing $[K]_o$, consistent with membrane depolarization (Blaustein and Goldring, 1975). Addition of A23187 to nominally Ca-free media did not alter FI at any $[K]_o$; subsequent addition of 1 mM Ca decreased FI in media containing 5 or 10 mM K, but not in 100 mM K medium. This is to be expected if P_K is selectively increased by this maneuver: in 100 mM K, the membrane potential is essentially equal to the K equilibrium potential, E_K, whereas in 5 or 10 mM K, the membrane potential is normally some value more positive than E_K, owing to the significant Na permeability of the plasma membrane (cf., Blaustein and Goldring, 1975). Thus an increase in P_K should hyperpolarize the membrane towards E_K in 5 or 10 mM K. Therefore, the Ca-activated increase in Rb efflux most likely reflects a specific increase in P_K, and not the activation of a non-specific cation channel (e.g., Yellen, 1982), which would have caused the membrane to <u>depolarize</u> in 5 or 10 mM K media.

The data presented above indicate that the Ca-activated Rb efflux is an <u>in vitro</u> manifestation of Ca-activated K channels. Further evidence supporting this view is provided by experiments on the pharmacolgical sensitivity of this

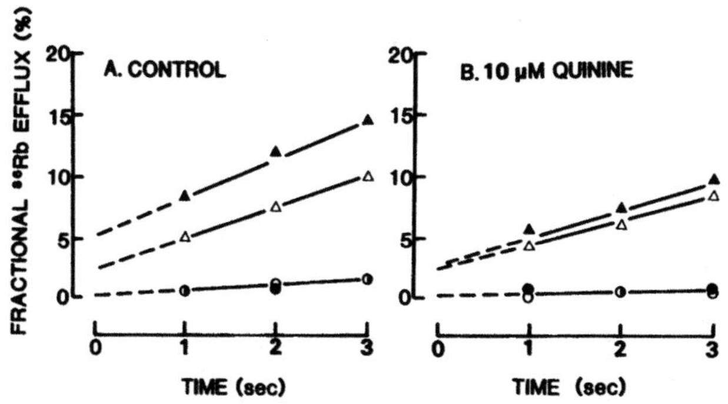

FIGURE 5. The effect of quinine sulfate on ^{86}Rb efflux. a). Control. b). 10 μM quinine sulfate was added to all solutions. (O)= 5 mM K, 0 Ca. (●)= 5 mM K + 1 mM Ca. (△)= 100 mM K, 0 Ca. (▲)= 100 mM K + 1 mM Ca. The values are the means of 4 determinations.

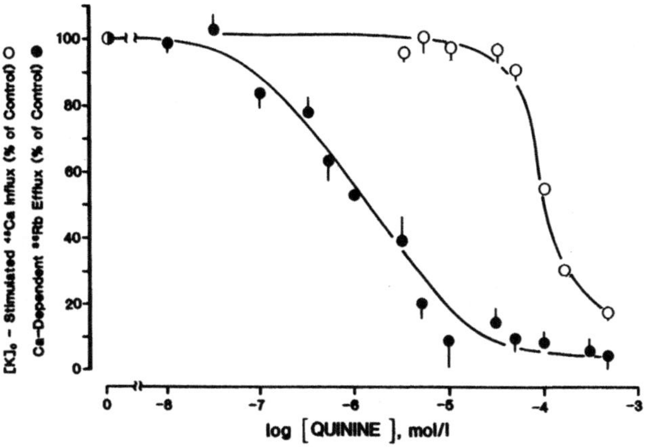

FIGURE 6. Dose-response curves for the inhibition of ^{86}Rb efflux and ^{45}Ca uptake by quinine sulfate. Ca-dependent Rb efflux was defined as the increase in Rb efflux during a 1 sec incubation promoted by the addition of 1 mM Ca to 100 mM K media. K-dependent Ca uptake was defined as the increase in ^{45}Ca uptake (1 mM $[Ca]_o$) at 1 sec induced by raising $[K]_o$ from 5 to 100 mM (cf., Nachshen and Blaustein, 1980).

conductance. The antimalarial, quinine, blocks Ca-activated K channels in a number of cells (Schwarz and Passow, 1983). Figs. 5A (control) and 5B illustrate an experiment in which the effect of 10 micromolar quinine sulfate was tested on the

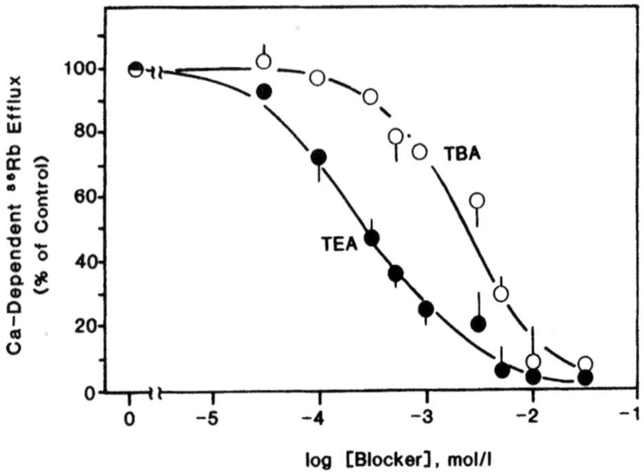

FIGURE 7. The effect of tetraalkylamines on Ca-dependent ^{86}Rb efflux. The experiment was performed essentially as decribed in Figure 6, except that the blockers employed were tetraethylammonium (TEA) or tetrabutylammonium (TBA). The values are the means of six determinations ± SEM.

various components of Rb efflux. At this quinine concentration, the predominant effect was a nearly complete block of Ca-activated Rb efflux, while the other conductances were virtually unaltered. The dose-response curve for the effects of quinine sulfate on the Ca-activated Rb efflux (measured at 1 sec; Fig 5C, closed circles) shows that the half maximal inhibitory concentration ("K_I") for quinine's block of this conductance was approximately 1 - 2 μM. The effect of quinine was not mediated by inhibition of Ca influx, since voltage-activated ^{45}Ca uptake, also measured at 1 sec (see Nachshen and Blaustein, 1980), was inhibited only by much higher concentrations (Fig. 5C, open circles).

Quinine sulfate was also tested for its ability to inhibit the Ca+A23187 - induced increase in ^{86}Rb efflux (data not shown); quinine blocked this Rb efflux as effectively as it blocked the K-stimulated, Ca-dependent Rb efflux. These results support the view that the properties of the Ca-activated Rb efflux is similar to those of Ca-activated K channels in other tissues, and that quinine can be used to identify and selectively block this conductance in nerve terminals.

Several tetraalkylammonium ions, including TEA and TBA, block K channels in many excitable cells (Armstrong, 1975). For example, Hermann and Gorman (1981b) reported that TEA blocked Ca-activated K currents in Aplysia (K_I = 0.4 mM), but TBA was a much less active blocker. Fig. 7 illustrates the dose-dependence of block of the Ca-activated Rb efflux by TEA and TBA. As in Aplysia (Hermann and Gorman, 1981b), TEA was a potent blocker of the K-stimulated, Ca-dependent conductance in synaptosomes (K_I = 0.2-0.4 mM), but TBA was much less potent. TEA is not selective, however: this drug also blocks the voltage-regulated, Ca-independent component T (K_I = 0.6 mM; data not shown).

Several aminopyridines, including 4-aminopyridine (4-AP), are potent blockers of some voltage-regulated K channels (cf. Thompson, 1977). However, 4-AP has been found to have either no effect (Hermann and Hartung, 1982) or, at high concentrations, a stimulatory effect on Ca-activated K currents (Hermann and Gorman, 1981a). Figs. 8A (control) and 8B show data on the effect of 1 mM 4-AP on the various components of Rb efflux. Ca-independent component T was totally abolished by 1 mM 4-AP, but the Ca-dependent Rb efflux was not inhibited. The concentration dependence of 4-AP's effect on Ca-activated Rb efflux, measured at 1 sec (Fig. 9), indicates that 4-AP produces a small (about 20%) inhibition of this conductance at low doses; however at concentrations greater than 1 mM, Ca-activated Rb efflux is stimulated. It is apparent that the Ca-activated ^{86}Rb efflux in synaptosomes shows a characteristic pharmacological sensitivity similar to that observed for Ca-activated K channels in other excitable cells.

DISCUSSION

The small size of presynaptic nerve terminals and the complexity of mammalian brain precludes the use of standard electrophysiological techniques to assess K channel activity in this tissue. A large variety of K conductance mechanisms, with varying physiological and pharmacological properties exist: it is clear that the ionic conductance properties of cells (or parts of cells) must be studied in the preparation of interest if we hope to learn how these specific cells function. Furthermore, modulation of presynaptic K channels may have profound behavioral consequences, and these channels are therefore of considerable interest. We have developed methods that enable us to examine K conductances in presynaptic nerve terminals isolated from mammalian brain. Although the tracer efflux technique detailed in this report is limited by a relatively slow time resolution and limited membrane voltage control (as

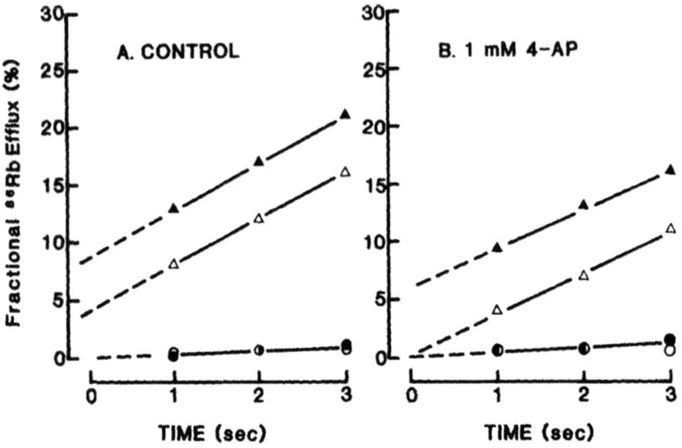

FIGURE 8. The effect of 4-aminopyridine (4-AP) on ^{86}Rb efflux. See Fig. 5 legend for the details of the experimental protocol. a). Control. b). In the presence of 1 mM 4-AP. The values are the means of 12 determinations.

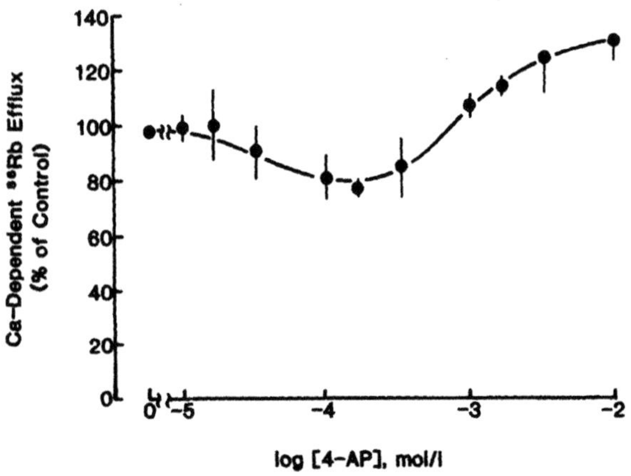

FIGURE 9. Dose-response curve for the effect of 4-AP on ^{86}Rb efflux. See Fig. 6 legend for details of the experimental protocol. The values are the means of 12 determinations ± SEM.

compared to voltage clamp methods), these shortcomings are partially compensated for by the fact that we can measure the unidirectional flux of a single ionic species, rather than net ionic current. This technique should also enable us to determine the relative permeabilities of the specific ionic species of interest. By applying automated rapid quench techniques (cf., Nachshen, 1982), it should also be possible to measure tracer fluxes on a time scale of tens of milliseconds.

Rb efflux from ^{86}Rb-loaded synaptosomes can be separated into at least four physiologically and pharmacologically distinct components (Table I; also see Bartschat and Blaustein, 1983 and 1984a-c). i) There is a slow, sustained efflux of Rb from synaptosomes incubated in media containing 5 mM K (component R). This most likely represents Rb efflux through "resting" K channels. ii) Under depolarizing conditions in Ca-free media, two voltage-regulated components of Rb efflux are observed: Component T appears to inactivate within 1 sec, and is selectively blocked by low concentrations of 4-AP; Component S inactivates slowly, if at all. Furthermore, while it is relatively insensitive to "classical" K channel blockers such as the aminopyridines and tetraalkylammonium ions, component S is selectively inhibited by nano- to micromolar concentrations of phencyclidine (Bartschat and Blaustein, 1984c). iii) There are also two components of Rb efflux (C_T and C_S) that are activated by increasing $[Ca]_i$; the rise in cytoplasmic Ca activity can be mediated either by activation of depolarization-regulated Ca channels or by the introduction of the Ca-selective ionophore A23187.

Several lines of evidence lead us to conclude that Ca-dependent Rb efflux represents Ca-activated K channels: i) The Ca-dependent ^{86}Rb efflux closely parallels the Ca-dependent ^{42}K efflux (data not shown; see Bartschat and Blaustein, 1984a), although ^{42}K efflux is about 20% larger than ^{86}Rb efflux in double label experiments. This is consistent with reports demonstrating that Rb passes through Ca-activated K channels nearly as well as K itself (eg., Gorman et al., 1982). ii) The Ca-dependent Rb efflux is activated by conditions that promote Ca entry into the terminals. iii) The activation of Rb efflux by Ca+A23187 is associated with a specific increase in P_K, which is reflected in membrane hyperpolarization when the membrane potential is more positive than E_K. iv) The pharmacological properties of the Ca-activated Rb efflux are also consistent with the view that this flux is mediated by Ca-activated K channels.

The foregoing discussion was based on the assumption that the two components of Ca-dependent Rb efflux, C_T and C_S, are manifestations of the same conductance. However, the apparent difference in the sensitivity to $[Ca]_o$ of these components (Fig. 2B), and their apparently different $P_{Rb}:P_K$

ratios (data not shown) imply that these may be two distinct classes of Ca-activated K channels. In this regard, Marty (1983) has argued that more than type of Ca-activated K channel may exist, and Romey and Lazdunski (1984) have suggested that embryonic rat muscle may contain two classes of Ca-activated K channels, one of which is TEA sensitive, and one of which is not blocked by high concentrations of TEA. We are currently studying the two components of Ca-activated Rb efflux in synaptosomes in more detail to determine whether C_S and C_T might also be pharmacologically distinct Ca-activated K channels, since each channel type may be involved differently in the regulation of action potential duration and neuronal excitability (Adams et al., 1982).

TABLE I

PHARMACOLOGICAL PROPERTIES OF SYNAPTOSOME POTASSIUM CHANNELS

Potassium Channel	^{86}Rb Efflux Component	Sensitivity to K Channel Blockers			
		Q.S.	TEA	4-AP	PCP
1. Resting	R	0	+	+	0
2. Ca-activated	$C_T + C_S$	+++	++	stim	0
3. V_M-regulated inactivating	T	+	++	++	++
4. V_M-regulated non-inactivating	S	+	++	+	+++

Q.S. = quinine sulfate; PCP = phencyclidine.
0 = no effect; + = K_I > 1000 μM; ++ = K_I 50-1000 μM;
+++ = K_I <5 μM; Stim = stimulation.
(Data from Bartschat and Blaustein, 1983; 1984a-c)

CONCLUSIONS

Our data from synaptosomes provide the first detailed physiological and pharmacological profiles of several classes of K channels in mammalian presynaptic nerve terminals; this report has focused primarily on Ca-activated K channels. In addition to the basic characterization of these conductances, the methods employed here should also enable us to test whether these K channels are modulated, for example as a consequence of neurotransmitter action. Indeed, our experiments have already led to a new hypothesis concerning the mechanism of action of the psychotomimetic drug phencyclidine (Albuquerque et al., 1981; Blaustein and Ickowicz, 1983; Bartschat and Blaustein, 1984c). Synaptosome preparations from specific brain regions should also be useful for examining the regional distribution of these K channels. Further, the pharmacological "fingerprints" of the various K channels described in this and other reports (Bartschat and Blaustein, 1983; 1984a-c) may be used to help elucidate the influence of presynaptic K channels on central synaptic transmission. Finally, it is anticipated that the pharmacological data may also be useful for biochemical identification and purification of specific K channel proteins, and for the identification of unitary K conductances when plasma membrane fragments are incorporated into planar lipid bilayers (eg., Nelson et al., 1983).

REFERENCES

Adams, PR, Constanti, A, Brown, DA, and Clark, RB (1982) Intracellular Ca activates a fast voltage-sensitive K current in vertebrate sympathetic neurones. Nature 296:746-749.

Albuquerue, EX, Aguayo, LG, Warnick, JE, Weinstein, H, Glick, SD, Maayani, S, Ickowicz, RK, and Blaustein, MP (1981) The behavioral effects of phencyclidines may be due to their blockade of potassium channels. PNAS USA 78:7792-7796.

Armstrong, CM (1977) Potassium pores of nerve and muscle membranes. In, Membranes- A Series of Advances. G. Eisenman, ed. Vol. 3. Marcel Dekker, Inc., New York. pp 325-358.

Bartschat, DK, and Blaustein, MP (1983) Ca-activated K channels in rat brain synaptosomes. Soc. Neurosci. Abstr. 9:21.

Bartschat, DK, and Blaustein, MP (1984a) Potassium channels in isolated presynaptic nerve terminals from rat brain. Submitted.

Bartschat, DK, and Blaustein, MP (1984b) Calcium-activated potassium channels in isolated presynaptic nerve terminals. Submitted.

Bartschat, DK, and Blaustein, MP (1984c) Phencyclidine (PCP), in low doses, selectively blocks a specific presynaptic K channel in mammalian CNS. Soc. Neurosci. Abstr., in press.

Blaustein, MP (1975) Effects of potassium, veratridine, and scorpion venom on calcium accumulation and neurotransmitter release by nerve terminals *in vitro*. J. Physiol. 247:617-655.

Baker, PF, Blaustein, MP, Keynes, RD, Manil, J, Shaw, TI, and Steinhardt, RA (1969) The ouabain-sensitive fluxes of sodium and potassium in squid giant axons. J. Physiol. 200:459-496.

Blaustein, MP, and Goldring, JM (1975) Membrane potentials in presynaptic nerve terminals monitored with a fluorescent probe. Evidence that synaptosomes have potassium diffusion potentials. J. Physiol. 247:589-615.

Blaustein, MP, and Ickowicz, RI (1983) Phencyclidine in nanomolar concentrations binds to synaptosomes and blocks certain potassium channels. PNAS USA 80:3855-3859.

Carmeliet, E, Horres, CR, Lieberman, M, and Vereeke, JS (1976) Developmental aspects of potassium flux and permeability of the embryonic chick heart. J. Physiol. 254:673-692.

Castellucci, VF, Kandel, ER, Schwartz, JH, Wilson, FD, Nairn, AC, and Greengard, P (1980) Intracellular injection of the catalytic subunit of cyclic AMP-dependent protein kinase stimulates facilitation of transmitter release underlying behavioral sensitization in *Aplysia*. PNAS USA 77:7492-7496.

De Peyer, JE, Cachelin, AB, Levitan, IB, and Reuter, H (1982) Ca^{+2}-activated K^{+} conductance in internally perfused snail axons is enhanced by protein phosphorylation. PNAS USA 77:5013-5017.

Gorman, ALF, Woolum, JC, and Cornwall, MC (1982) Selectivity of the Ca-activated and light-dependent K channels for monovalent cations. Biophys. J. 38:319-322.

Hermann, A, and Gorman, ALF (1981a) Effects of 4-aminopyridine on potassium currents in molluscan neurons. J. Gen. Physiol. 78:63-86.

Hermann, A, and Gorman, ALF (1981b) Effects of tetraethylammonium on potassium currents in a molluscan neuron. J. Gen. Physiol. 78:87-110.

Hermann, A, and Hartung, K (1982) Properties of a Ca-activated K conductance in *Helix* neurones investigated by intracellular Ca iontophoresis. Pflug. Arch. Eur. J. Physiol. 393:248-253.

Hille, B (1975) Ion selectivity of Na and K channels of nerve membranes. In, *Membranes- A Series of Advances*. G. Eisenman, ed. Vol. 3. Marcel Dekker, Inc., New York. pp. 255-323.

Hodgkin, AL (1967) The Conduction of the Nerve Impulse. Liverpool University Press, Liverpool, UK. 108 pp.

Klein, M, and Kandel, ER (1980) Mechanism of calcium current modulation underlying presynaptic facilitation and behavioral sensitization in Aplysia. PNAS USA 77:6912-6916.

Krueger, BK, Ratzlaff, RW, Strichartz, GR, and Blaustein, MP (1979) Saxitoxin binding to synaptosome membranes and solubilized binding sites from rat brain. J. Membr. Biol. 50:287-310.

Marty, A. (1983) Ca-dependent K channels with large unitary conductance. Trends Neurosci. 6:262-265.

McGraw, CF, Nachshen, DA, and Blaustein, MP (1982) Calcium movement and regulation in presynaptic nerve terminals. In, Calcium and Cell Function. Vol. 2. Cheung, CY, ed. Academic Press, New York. pp. 81-100.

Meech, RW (1978) Calcium-dependent potassium activation in nervous tissue. Ann. Rev. Biophys. Bioeng. 7:1-18.

Nachshen, DA, and Blaustein, MP (1980) Some properties of potassium-stimulated calcium influx in presynaptic nerve endings. J. Gen. Physiol. 76:709-728.

Nachshen, DA (1982) The initial time-course of Ca influx in synaptsomes. Biophys. J. 37(No. 2, Pt. 2): 314a.

Nachshen, DA, and Blaustein, MP (1982) The influx of calcium, strontium, and barium in presynaptic nerve endings. J. Gen. Physiol. 79:1065-1087.

Nelson, MT, Roudna, M, and Bamberg, E (1983) Single K-channel current measurements from brain synaptosomes in lipid bilayers. Am. J. Physiol. 245:C151-C156.

Reuter, H, and Stevens, CF (1980) Ion conductance and ion selectivity of potassium channels in snail neurons. J. Membr. Biol. 57:103-118.

Romey, G, and Lazdunski, M (1984) The coexistence in rat muscle cells of two distinct classes of Ca-dependent K channels with different pharmacological properties and different physiological functions. Biochem. Biophys. Res. Comm. 118:669-674.

Schwarz, W, and Passow, H (1983) Ca-activated K channels in erythrocytes and excitable cells. Ann. Rev. Physiol. 45:359-374.

Thompson, SH (1977) Three pharmacologically distinct potassium channels in molluscan neurons. J. Physiol. 265:465-488.

Yellen, G (1982) Single Ca-activated non-selective cation channels in neuroblastoma. Nature 296:357-359.

DISCUSSION OF THE ARTICLE BY M.P. BLAUSTEIN (U.S.)

O.F. Hutter (U.K.): In using ^{86}Rb as an isotope, you presumably made the presumption that Rb behaves like potassium. This is probably true if the mole ratio is low enough, but not necessarily so when it is higher. Could you tell us what was the mole ratio of Rb to potassium, i.e., the highest mole ratio used in experiments with low potassium concentrations?

M.P. Blaustein (U.S.): We actually only used Rb as a tracer for potassium so that the mole ratio is very low for Rb. We have done exactly the same experiment in parallel samples ^{42}K and ^{86}Rb and we get qualitatively the same result. We find that the terminals are about 80% as permeable to Rb as they are to potassium for most of the channels that I have described.

O.F. Hutter (U.K.): Can you put a figure to the mole ratio, e.g., 1 in 1000, 1 in 500?

M.P. Blaustein (U.S.): The ratio for the 5mM K solution was 50:1 (K:Rb) and up to 750-1000:1 for the K-rich solutions.

I. Steinberg (Israel): Is it possible to get an idea about the topographic distribution of these channels? I understand that you have tried it on the terminals but maybe you have some idea about other places too?

M.P. Blaustein (U.S.): We haven't tried it in other parts of the neuron. An important question is whether or not these channels are uniformly distributed in the terminals that we look at or whether they are present in only some terminals and whether there is a non-uniform distribution. That is something very interesting which we haven't addressed but which needs to be addressed. What I tried to show you is that we think there are at least 4, or perhaps 5 components of Rb efflux which are all pharmacologically distinct and kinetically distinct. I think they may not be distributed uniformly so it is going to turn out to be very interesting when we actually begin to test this. We haven't done these experiments yet.

HISTAMINE AS A PUTATIVE TRANSMITTER OF PRESYNAPTIC INHIBITION: EFFECT ONTO THE CA^{++} CHANNEL IN APLYSIA NEURONS

Kretz, R., Shapiro, E.*, Bailey, C.H.* and Kandel, E.R.*, University of Fribourg, Dept. of Anatomy, CH-1700 Fribourg, Switzerland, and *Columbia University, Center for Neurobiology and Behavior and Dept. of Anatomy and Cell Biology, New York, N.Y. 10032, USA

Utilizing voltage-clamp techniques, we examined the synaptic conductance mechanism underlying presynaptic inhibition in Aplysia californica, in a circuit in which all the neural elements are identified cells. L10 is an interneuron in the abdominal ganglion, whose output to RB cells (EPSP excitatory postsynaptic potential) and LUQ cells (IPSP inhibitory postsynaptic potential) is presynaptically inhibited by stimulating cells of the L32 cluster. Stimulation of L32 cells causes a slow inhibitory synaptic potential onto L10 which is associated with an apparent increased membrane conductance. Among various putative transmitters tested, histamine was most effective in mimicking the postsynaptic effects of L32 cells onto L10: (1) decrease of the voltage-dependent Ca^{++} current, and (2) increase of a voltage-dependent K^{+} current. Both effects inhibit transmitter release from L10 either directly by decreasing the amount of Ca^{++} entering during spike depolarization or indirectly by hyperpolarizing L10 and decreasing the amplitude and duration of spikes in L10. The response of L10 to histamine and the L32-L10 IPSP could be reversibly blocked by cimetidine, a histamine antagonist in Aplysia. In addition, the synaptic vesicles in identified L32 cells (using horseradish peroxidase as a marker) are morphologically strikingly similar to vesicles in cell C2, an identified histaminergic neuron in Aplysia. The vesicle distribution in L32 cell varicosities also is identical to the vesicle distribution in C2. These results support our model of presynaptic inhibition which states that presynaptic inhibition is caused by a direct transmitter-mediated reduction in presynaptic Ca^{++} current.

DISCUSSION OF THE ARTICLE BY R. KRETZ (SWITZERLAND)

H.M. Gershenfeld (France): Did you try to find out if you can pharmacologically disassociate the effect on the hyperpolarization of L10 from the effect on the calcium current. Is cimetidine also blocking presynaptic inhibition?

R. Kretz (Switzerland): Do you mean the PSP?

H.M. Gershenfeld (France): No. You have shown that histamine has two effects on L10. What you pretend to be a presynaptic effect would be a change of the calcium current, which is clear, but you have a very powerful effect on the rate of firing this cell too. We have seen in some of your slides all epsps disappearing and you have shown also a slide of the histamine effect on L10, also causing a hyperpolarization and the rate of firing being zero. In a pharmacological way you have shown us that cimetidine decreases but does not block the effect of the hyperpolarization. Do you have a pharmacological effect with another blocker to distinguish between the hyperpolarizing effect of L10 and the effect on the calcium current?

R. Kretz (Switzerland): No, we have not and we did not try. We have used cimetidine only to show that it can block the hyperpolarization of L10 caused either by L32 stimulation or by the local application of histamine. Because cimetidine is known to be a specific blocking agent of the slow potassium mediated response (hyperpolarization) elicited by the application of histamine onto Aplysia neurons (Gruol D.L. and Weinreich D. (1979). Two pharmacologically distinct histamine receptors mediating membrane hyperpolarization on identified neurons of Aplysia californica. Brain Res. 162, 281-301), we got another indication that the transmitter involved in presynaptic inhibition might be histamine. I did not go into the analysis of the hyperpolarization of L10, but as far as we have looked at the underlying mechanism that causes this hyperpolarization, it seems to be an increased potassium conductance. Today, I concentrated my talk only onto the calcium current and its role in presynaptic inhibition.

PERSISTENT INACTIVATION OF POTASSIUM CURRENTS BY ELEVATION OF INTRACELLULAR CALCIUM

Daniel L. Alkon

ABSTRACT

An early voltage-dependent K^+ current, I_A, and a late Ca^{2+}-dependent K^+ current, $I_{Ca^{2+}-K^+}$, across the soma membrane of an identified cell, the Type B photoreceptor, remain reduced for days after associatively conditioning the nudibranch mollusc Hermissenda crassicornis. Because intracellular calcium, $Ca^{2+}{}_i$ (as measured by differential absorption spectrophotometry of calcium complexed with the dye arsenazo III), undergoes prolonged elevation during the acquisition of the conditioned behavior we hypothesized that $Ca^{2+}{}_i$ elevation might reduce these K^+ currents on a time scale of many minutes and ultimately, with repeated elevation as occurs during training, on a time scale of many days. Intracellular iontophoresis of Ca^{2+} (for 30 seconds) through one microelectrode while voltage clamping the isolated Type B photoreceptor soma with two other electrodes, caused reduction of I_A and $I_{Ca^{2+}-K^+}$ for several minutes. Iontophoresis of EGTA greatly reduced prolonged inactivation of either I_A or $I_{Ca^{2+}-K^+}$. Manipulations which prevented elevation of $Ca^{2+}{}_i$ via activation of a sustained voltage-dependent Ca^{2+} current, including substitution of Ba^{2+} for external Ca^{2+} and the use of blockers such as Co^{2+} and Cd^{2+} reduced I_A and $I_{Ca^{2+}-K^+}$ inactivation lasting 2-3 minutes following a 10-second positive command to 0mV (which substantially activates the Ca^{2+} current). Inactivation of $I_{Ca^{2+}-K^+}$ could be separated from activation by graded intracellular release of Ca^{2+} with photic stimulation of a visual pigment during steady-state activation of $I_{Ca^{2+}-K^+}$ by command depolarizations to -30 to 0mV from a holding potential of -60mV. Evidence is presented which suggests that transitions from inactivation of I_A and $I_{Ca^{2+}-K^+}$ lasting minutes to that lasting days is mediated by Ca^{2+}-dependent phosphorylation of low molecular weight proteins which either regulate or are a part of the relevant membrane channels.

Calcium has been known for some time to activate K^+ current across neuronal (1-8) and non-neuronal (8) membranes. Elevation of intracellular calcium due to either release from intracellular stores or to voltage-dependent flux across the cellular membrane (9-11) opens K^+ channels to generate a maximum outward current within tens to hundreds of milliseconds. Presumably, calcium binds to the surface of the internal membrane to activate K^+ channels (see 7).

Elevation of intracellular calcium has also been shown to inactivate membrane channels. Brown and his colleagues observed that increased intracellular calcium could inactivate ligh-induced Na^+ current in the Limulus photoreceptor and thereby contribute to the regulation of the photoreceptor's state of light adaptation (12,13). Accumulated evidence also indicates increased intracellular calcium inactivates voltage-dependent calcium current (14-20). Calcium-inactivation of calcium current has the indirect effect of reducing the calcium-dependent K^+ current which is a consequence of voltage-dependent flux of calcium across the cell membrane.

In recent years, my colleagues and I have observed that elevation of intracellular calcium not only activates K^+ channels, it also causes inactivation of K^+ channels (21-24). Calcium inactivation of K^+ channels may be of particular interest to students of long-term regulation and transformation of neuronal function during such phenomena as learning and development. We found that transient elevation of intracellular calcium can cause inactivation of K^+ channels for minutes and, with repetition of calcium-elevating conditions, K^+ current inactivation can persist for days (22,25-28). If the magnitude of the K^+ current inactivation represents a significant fraction of the total membrane current as it does in the photoreceptors of the nudibranch mollusc Hermissenda crassicornis, the excitability of a neuronal membrane, and thus its responsiveness to input, may be transformed by such inactivation (29).

The Type B photoreceptor has an approximately spherical (20-30μm in diameter) soma which is joined at its anterior end by a narrower process (~10μm in diameter) to an elaborately infolding rhadbomeric membrane. An axon (~1μm in diameter) exits from the posterior (or caudal) end of the photoreceptor soma and courses for 60-70μm before ending in a spray of fine terminal branches on which are pre- and postsynaptic sites. At these sites, the Type B photoreceptor has mutual inhibitory synaptic interaction with each of the two other Type B cells in the eye - and the lateral Type A photoreceptor. The medial Type A photoreceptor receives unidirectional synaptic inhibition from the medial and intermediate but not the lateral Type B cells. The Type B photoreceptor also causes synaptic inhibition of second order visual interneurons within the optic ganglion and in the cerebropleural ganglion and synaptic excitation of other interneurons. The Type B photoreceptor receives as well synaptic excitation from the S/E optic ganglion cell and synaptic inhibition from hair cells caudally located within the ipsilateral statocyst.

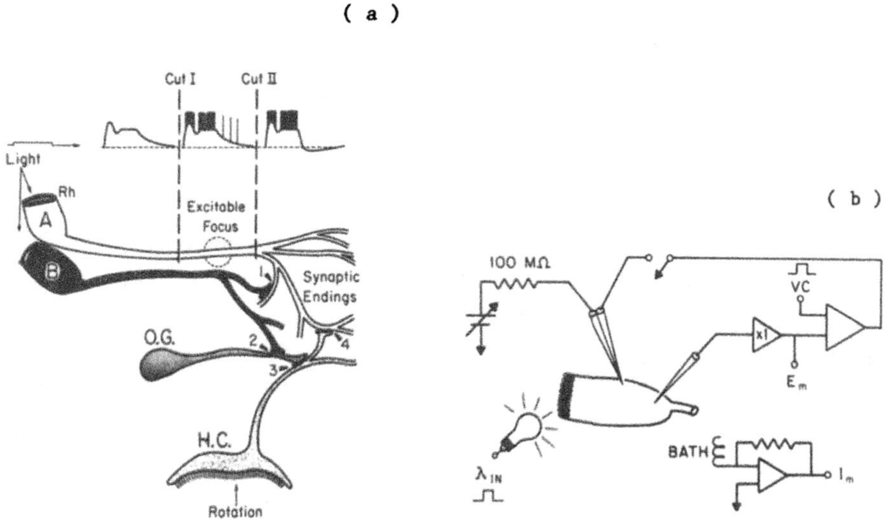

FIGURE 1. a) Inhibition of Type A photoreceptor. Diagram of Type A photoreceptor soma with rhabdome, axon, excitable focus and terminal branches. Light depolarizes photoreceptors at rhabdomes (Rh). Type B photoreceptors (of which there are three) inhibit, at synaptic endings, lateral Type A, directly (1) and indirectly by inhibiting (2) optic ganglion (OG) cells (of which there are 13). Ipsilateral hair cells (HC) receive less inhibition from optic ganglion cells (3) when the Type B fires and thus increases its inhibition of the lateral Type A (4). Hair cells also inhibit Type A when they depolarize in response to rotation. In response to light, the Type A cell depolarizes without impulses or afterhyperpolarization with cut I lesion. It depolarizes with impulses but without afterhyperpolarization with cut II lesion. The response of an intact Type A cell is represented at upper right (see ref. 49). b) Schematic of the voltage-clamp apparatus. The circuit is a conventional two microelectrode voltage clamp with current measured by a virtual ground amplifier. For iontophoresis, a double-barreled micropipette supplied clamp current from one barrel and iontophoretic current out of the other barrel via the DC voltage source and 100 MΩ dropping resistor. The light source is controlled by a negative feedback system to improve rise time (21).

These and other (29,30-36) synaptic interactions can be eliminated (Fig 1) by cutting the Type B axon just before its terminal branches (i.e., about 60um from the soma). There are no synaptic interactions at or in proximity to the soma. All impulse generating membrane is located distal to the first half of the axon as it exits from the soma. Axotomy within 10-20μm of the soma therefore eliminates all impulse activity as well as all synaptic interactions (37). The severed axon tends to seal as indicated by the increased input resistance measured in the soma following axotomy. The axotomized soma with attached rhabdomeric membrane can then be voltage-clamped with standard two or three microelectrode techniques or patch-clamp methods.

Ionic currents of Type B photoreceptors

In darkness there are three major voltage-dependent outward K^+ currents:

1. An early, rapidly activating and inactivating current, I_A, (4,38-40) which is blocked (22,41) by 1-3mM external 4-aminopyridine (Fig 2). Following activation by a depolarizing command to 0mV from a holding potential -60mV, 20-30 seconds are necessary for complete recovery from inactivation. Furthermore, although most I_A activation requires significant depolarization above the resting membrane potential, there is clear evidence that some small but significant steady-state activation occurs at membrane potentials below and just above the resting membrane potential. External application of 3mM 4-aminopyridine, which has little or no effect on the other voltage-dependent K^+ currents (but blocks I_A), can cause a substantial increase of input resistance. Three mM 4-AP also substantially increases the steady-state depolarization of the Type B cell during a prolonged light step. This suggests that although the greatest fraction I_A (e.g. 90-95%) is inactivated during prolonged depolarization (e.g. to -20mV) enough remains to significantly contribute to the total membrane current and the potential across the Type B cell membrane. Thus, I_A is both transiently and tonically activated from -60 to 0mV which defines the effective limits of the Type B cell membrane potential during the range of its response to light stimuli of varying intensity in its natural environment.

2. A delayed, slowly activating and inactivation current, I_K, which is blocked by 100mM external TEA but is unaffected by 1-3mM 4-AP. This K^+ current, commonly known as the delayed rectifying K^+ current (1), is minimally activated at potentials $\leq$0mV and thus contributes little to Type B cell responses to a natural range of stimulus intensities.

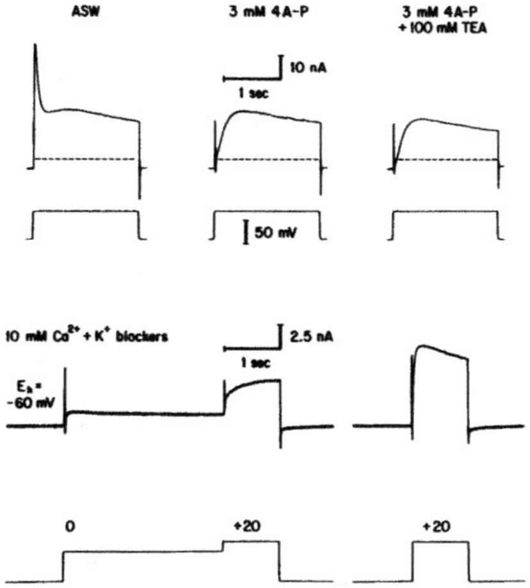

FIGURE 2. a) Voltage-dependent outward currents across the membrane of the isolated Type B cell soma. From left to right, ASW, 3mM 4-aminopyridine (4-AP) added to ASW, 4-AP and 100mM tetraethylammonium (TEA) ion added to ASW. Note that addition of 4-AP and TEA remove only a small portion of the late outward current elicited by command to 0mV from a holding potential of -60mV. The dashed lines indicate the level of the non-voltage-dependent or 'leak' current (22). b) Voltage-dependent outward current in presence of 4-AP and TEA. Voltage-dependent activation of outward calcium-dependent K^+ current ($I_{Ca^{2+}-K^+}$) in Type B photoreceptor, and reduction by pre-pulse depolarization. Current (top) and voltage (bottom) records from voltage-clamp experiment illustrate that a 3 sec depolarization to 0mV elicits a small net outward current (top left), followed by a larger outward current when the membrane is stepped to +20mV (top middle). The pre-pulse depolarization reduced by ~40% the outward current normally evoked by an 80mV step from -60 to +20mV (top right) and slowed the rise time as well. Bathing solutions included 10mM 4-AP and 100mM TEA to block the fast (I_A) and delayed (I_K) K^+ currents (22).

3. A delayed slowly activating and inactivating current, $I_{Ca^{2+}-K^+}$, which is dependent on the level of free intracellular Ca^{2+} and which is blocked by calcium current blockers such as Co^{2+}, Cd^{2+} and Mn^{2+} ions (Fig 2). Substitution of Ba^{2+} for Ca^{2+} in the external bathing medium also greatly reduces $I_{Ca^{2+}-K^+}$ as does reduction of free intracellular Ca^{2+} ions by iontophoretic injection of the calcium-chelator, ethylene glycol tetracetic acid (EGTA). Following activation by a depolarizing command to 0mV from a holding potential of -60mV, 30-60 seconds are necessary for complete recovery from inactivation (Fig 3). As for I_A, although most $I_{Ca^{2+}-K^+}$ activation requires significant depolarization above the resting membrane potential, there is clear evidence that some small but significant steady-state $I_{Ca^{2+}-K^+}$ activation occurs at membrane potentials immediately below and above the

resting membrane potential. Substitution of 10mM Ba^{2+} for Ca^{2+} in the external bathing medium has little effect on the other voltage-dependent K^+ currents (actually slightly increasing I_A) but does cause a substantial increase of input resistance, presumably by reducing a resting $I_{Ca^{2+}-K^+}$ current.

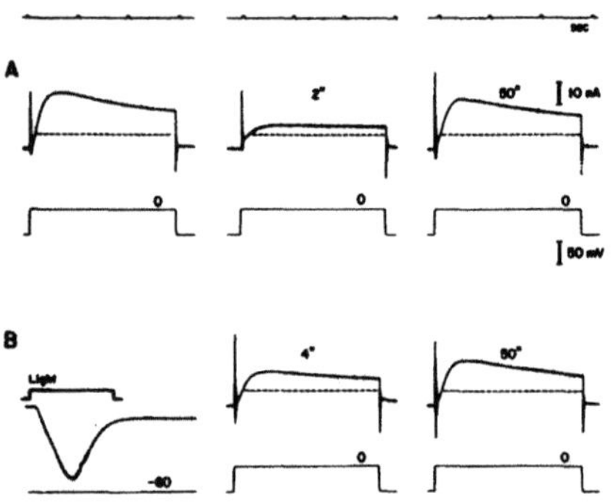

FIGURE 3. Long-lasting inactivation of $I_{Ca^{2+}-K^+}$. Depolarizing command to 0mV (from holding potential of -60mV) elicits an outward current in 3mM 4-AP and 100mM TEA. a) The maximum current, $I_{Ca^{2+}-K^+}$, is not elicited by subsequent command steps until the interval between commands is at least one minute. b) Following a light step, a delay of one minute or more is also necessary before a command to 0mV elicits the maximum $I_{Ca^{2+}-K^+}$. Dashed lines indicate the level of the non-voltage-dependent or 'leak' current (22).

In addition to the aformentioned voltage dependent K^+ currents, there is in darkness a voltage-dependent inward calcium current. The calcium current is most apparent under conditions which minimize the activation of $I_{Ca^{2+}-K^+}$. Substitution of Ba^{2+} for Ca^{2+} in the external medium for example reveals a voltage-dependent inward Ba^{2+} current (Figs 4,5) which is accompanied by a markedly reduced late K^+ current (which is eliminated by further washing with Ba^{2+}-ASW). Elevation of external K^+ from 10 to 300mM brings the reversal potential for K^+ flux to ~0mV where the calcium current is close to maximally activated. Command steps to 0mV (from a holding potential of -60mV) in 300mM K^+_o therefore allows observation of the calcium current (which is unaffected by removal of Na^+ from the external bathing medium) in the absence of voltage-dependent K^+ currents. This calcium current has a very rapid onset (<1 msec) and reaches a steady-state level within 100-200 msec. It shows no appreciable inactivation with 10mM Ca^{2+} in the external bathing medium (Fig

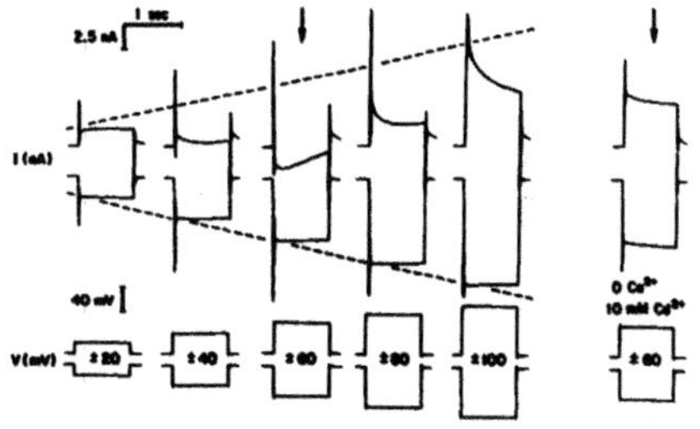

FIGURE 4. Records of voltage-dependent inward current present in Type B photoreceptors under conditions of 100mM external Ba^{2+} and K^{+} current blockers (10mM 4-AP, 100mM TEA). From a holding potential of -60mV, successive command steps in multiples of 20mV reveal activation of an inward current at potentials more positive than -40mV. This current increases non-linearly, reaching its peak value at 0mV (left arrow) and thereafter diminishing. From a holding potential of -60mV, successive commands to potentials $<$-60mV elicit a non-voltage-dependent 'leak current', indicated by lower dashed lines. The upper dashed line is extrapolated from potentials $\leq$-50mV. Note that when 10mM Cd^{2+} is substituted for Ca^{2+} in the external bathing medium, the current elicited by the command to 0mV absolute begins to approach the extrapolated leak level (indicated by the dashed line) (22).

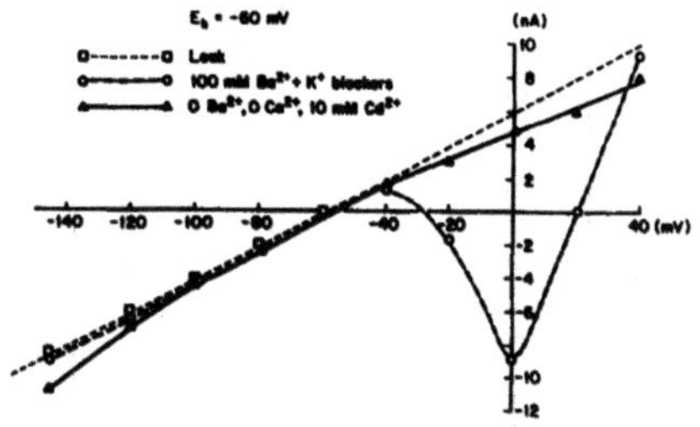

FIGURE 5. Steady-state current-voltage plot of voltage-dependent inward current present in Type B photoreceptor (same cell as in Fig 4). Under conditions of high external Ba^{2+} (100mM) and blockade of the voltage-dependent K^{+} currents, the current is inward over the range of -40 to +40mV (absolute), reaching its peak at $\sim$0mV. Removal of Ba^{2+} and Ca^{2+} from the bath, and addition of the calcium-channel blocker Cd (10mM) substantially reduced after five minutes the inward current at all levels of membrane potential (triangles). The residual deviations from the extrapolated leak current values may be due to incomplete block of the inward current (22).

6). With elevated external calcium (e.g. 100mM) the current increases and shows clear inactivation (Fig 7). The calcium current is blocked by divalent cations such as Co^{2+}, Cd^{+}, and Mn^{2+} (cf. Fig 4).

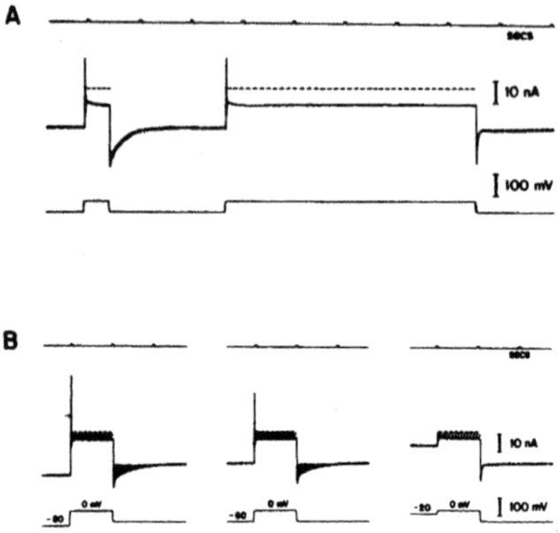

FIGURE 6. a) Inward current and tail currents with brief and prolonged depolarization. An inward current (leak current level is indicated by dashed line) elicited by command to 0mV, reaches its maximum level within ~300 msec in the presence of 4-AP, TEA and 300mM K^+. During a prolonged depolarizing command, the inward current shows little or no inactivation. The inward tail current following a brief command to 0mV is largely eliminated when the depolarization is prolonged. The reversal potential for the inward tail current was determined to be ~0mV. With settling time of the voltage clamp used here, tail current magnitudes can only be compared 15 msec or more following the offset of the test depolarization (22). b) Effects of conditioning comands on currents elicited by depolarization to 0mV. 10-second steps to -80mV and -20mV preceded a depolarizing step to 0mV in the presence of 4-AP, TEA and 300mM K^+. Note that the inward current (represented by the hashed lines below the 'leak' level, indicated by the dashed lines) is largely unchanged under the three different conditions. The inward tail current, which reversed at ~0mV, was slightly enhanced by the conditioning step to -80mV and almost entirely eliminated by the conditioning step to -20mV. With the settling time of the voltage clamp used here, tail current magnitudes can only be compared 10 msec or more following the offset of the test depolarization (22).

In response to light, there are three currents elicited:

1. Light induces a rapidly activating and inactivating inward Na^+ current (Fig 3). The light-induced inward Na^+ current is eliminated by thorough washing in 0 Na^+-ASW. It becomes progressively smaller at increasingly positive holding potentials (Fig 8A), reversing at approximately +30mV.

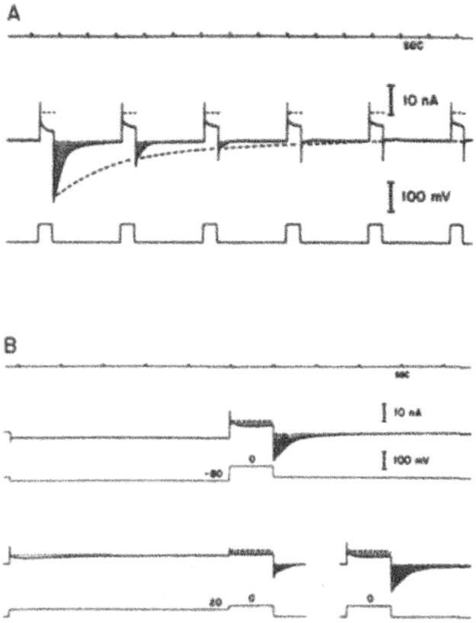

FIGURE 7. a) Inactivation of inward current and inward tail currents with repetitive depolarizations. Depolarizing commands occur at 0.5Hz in upper record. Inward current (beneath leak current indicated by upper dashed line) inactivates as does the inward tail current (indicated by shaded areas) with repetition of steps to 0mV in the presence of 100mM Ca^{2+}, 4-AP, TEA, and 300mM K^+ (22). b) Effects of conditioning commands on currents elicited by depolarization to 0mV. 5-second steps to −80mV and −20mV preceded a depolarizing step to 0mV in the presence of 100mM Ca^{2+}, 4-AP, TEA, and 300mM K^+. The inward current (beneath leak current indicated by dashed line) shows clear inactivation during and following the conditioning step to −20mV. This was also true for inward tail current following the step to 0mV which had been preceded by the conditioning step to −20mV. The inward tail current as well as the inward current during a step to 0mV were largely unaffected by the conditioning step to −80 mV. A command step to 0mV without a conditioning step (i.e., held at −40mV) is shown in the lower right. The reversal potential for the inward tail current was determined to be ~0mV (22).

2. In 0 Na^+-ASW, i.e., in the absence of the light-induced inward Na^+ current, light elicits a slowly activating and inactivating outward K^+ current (Fig 9). This current, whose magnitude varies as a function of external K^+, is not blocked by calcium current blockers (e.g., Co^{2+}, Cd^{2+}, and Mn^{2+}) but is diminished by intracellular iontophoresis of EGTA. This light-induced outward K^+ current is therefore considered to be an $I_{Ca^{2+}-K^+}$ resulting from a light-induced rise of intracellular calcium.

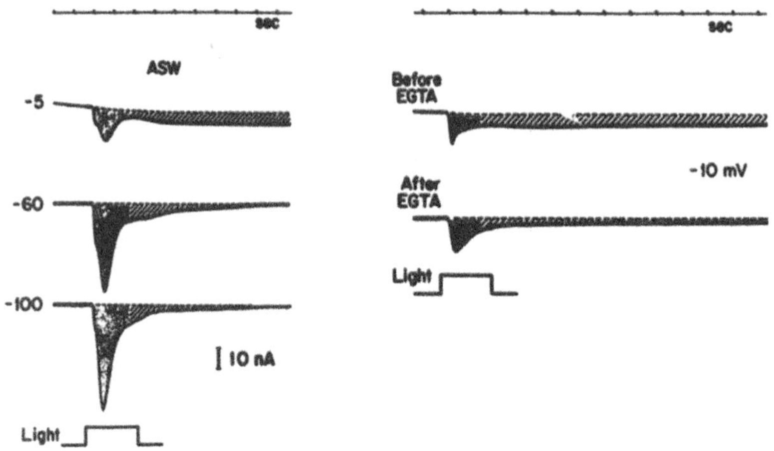

FIGURE 8. a) (on left): Light-induced inward sodium current. Light steps (10^4 ergs/cm^2 x sec), presented at two minute intervals after 10 minutes dark adaptation, elicited an inward current of constant amplitude at holding potential of -60mV. When the light occurred 5 seconds after the onset of a 20-second depolarizing command step (e.g., to -5mV), the peak amplitude of the large initial phase of the inward current became progressively smaller as the depolarizing command became more positive. A small delayed phase of the inward current, however, became larger with more depolarization (24). b) (on right): EGTA injection reduces the delayed light-induced current. Light-induced current at -10mV before (upper record) and after (lower record) intracellular injection of EGTA (-3.0 nA, 1 minute under isopotential conditions). Note that the light-induced inward Na^+ current is slightly increased and prolonged while the delayed inward component is reduced. Light steps (10^4 ergs/cm^2 x sec) occurred at 2 minute intervals after 10 minutes' dark adaptation. Light steps occurred at 5.0 seconds after the onset of command depolarization to -10mV. Dashed lines indicate the current elicited by the command steps in the absence of light stimuli (24).

3. In ASW, light elicits a slow 'apparently' inward current (Fig 8). This current has a very delayed onset (1-2 seconds depending on light intensity) and an even more prolonged relaxation (20-60 seconds, again depending on light intensity). The delayed 'apparent' inward current becomes greater at holding potentials progressively more positive than -60mV and its magnitude varies with K^+ concentration in the external bathing medium. The delayed light-induced inward current also is apparent in 0 Na^+-ASW (Fig 9). It persists in the presence of 3mM 4-AP and 100mM TEA but not when Ba^{2+} is substituted for Ca^{2+} in the external bathing medium. Taken together, these results indicate that light elicits a delayed and persistent decrease of $I_{Ca^{2+}-K^+}$ as well as a rapidly activating and inactivating inward Na^+ current. More will be said below of how light might cause this decrease of $I_{Ca^{2+}-K^+}$.

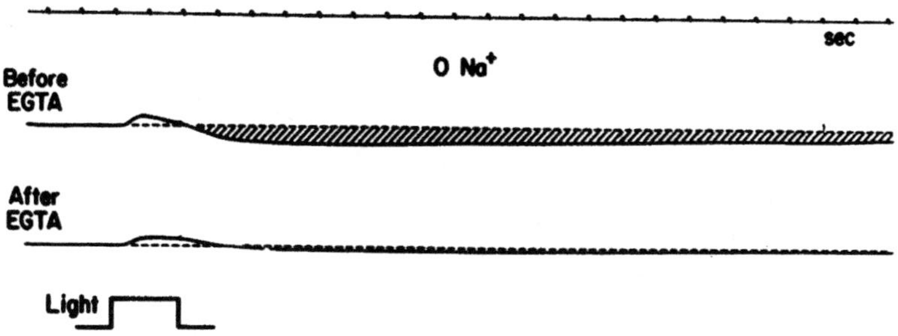

FIGURE 9. Light-induced potassium currents. In 0 Na^+-ASW light steps (10^4 ergs/cm^2 x sec), presented at two minute intervals after 10 minutes' dark adaptation, elicited little or no current at holding potential of -60mV (not illustrated here). When the light occurred 5 seconds after the onset of a 20-second depolarizing command step (here to 0mV), an initial outward current and a delayed inward current were elicited. With progressively more positive commands, both the outward and inward currents became more apparent. The dashed line indicates the current elicited by the command steps in the absence of light stimuli (24). Note that after EGTA injection (-3.0 nA for 3 minutes under isopotential conditions) the delayed inward current is markedly reduced while the initial light-induced outward current is slightly affected.

Ca^{2+}-activation of ionic currents

Iontophoretic injection (~0.5nA for 30 seconds) of Ca^{2+} through one microelectrode into a Type B photoreceptor which is voltage-clamped with two additional microelectrodes causes prolonged (lasting several minutes) reduction of at least three ionic currents which flow across the Type B membrane (Fig 10): the light-induced inward Na^+ current, I_A, and a slowly activating voltage-dependent K^+ current (largely $I_{Ca^{2+}-K^+}$). That the mechanism for this calcium-initiated inactivation may vary in specific ways with the channel inactivated was suggested by the observation that iontophoretic injection of Mg^{2+} instead of Ca^{2+} caused a profound reduction of light-induced I_{Na^+} but had no effect on the voltage-dependent K^+ currents. Because I_A and $I_{Ca^{2+}-K^+}$ remain reduced for days (Fig 11) after conditioning Hermissenda with paired stimulation of the visual and statocyst pathways (which converge on the Type B photoreceptor) whereas I_{Na^+} was not significantly changed, we will focus here on calcium inactivation of K^+ currents.

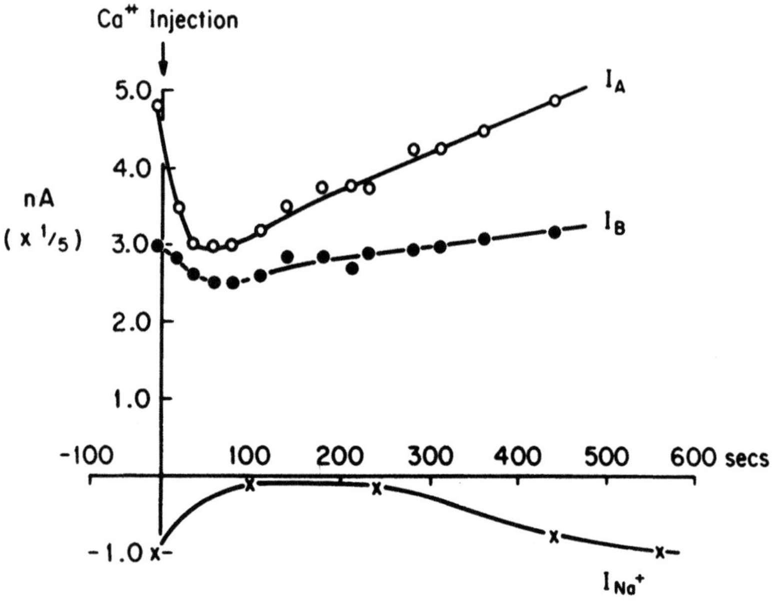

FIGURE 10. Response of B photoreceptor to potential steps to -5mV before and after Ca^{2+} injection. First measurement of post-injection membrane current occurred at 80 sec. I_A and I_{NA} (the light-induced inward current) are determined from peak values while I_B (largely $I_{Ca^{2+}-K^+}$) is determined from current amplitude after 250 ms from step onset. All three currents are reduced for many seconds following Ca^{2+} injection (21).

1. I_A. Measurement of intracellular calcium concentration by spectrophotometric determination of the quantity of arsenazo III (iontophoretically injected) complexed with free calcium showed that it is the combination of light stimulation and depolarization which causes the greatest and most prolonged elevation of intracellular Ca^{2+} (Fig 12). This combination of light and depolarization, when repeated can cause reduction of I_A lasting for several minutes (Fig 13). Such reduction is eliminated when Ba^{2+} is substituted for Ca^{2+} in the external bathing medium or when Ca^{2+} is simply removed from the medium.

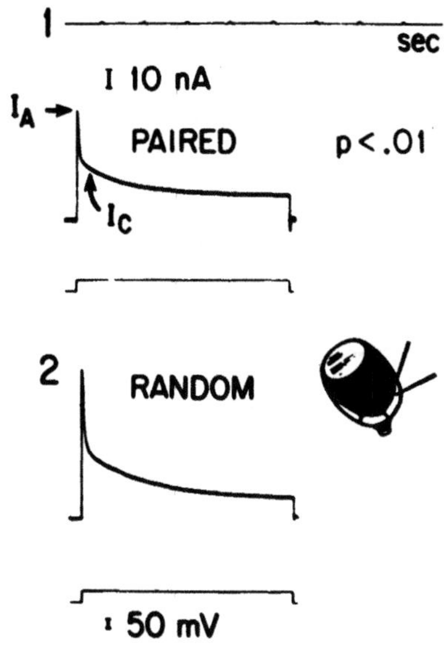

FIGURE 11. Membrane current changes with learning. Type B cell is isolated by axotomy after behavioral measurements are made preceding and following training. Cartoon depicts placement of two microelectrodes for voltage clamp. Ionic currents from holding potential of -60mV: (1) Paired I_A (early peak outward K^+ current) and $I_{Ca^{2+}-K^+}$ (late current, maximal at 300 msec after onset of command) are similar than random (2) or naive values (not shown) (29).

A calcium load can also be effected by a depolarizing command when large identifiable central neurons are voltage-clamped and monitored with the arsenazo technique. A 15 second command to +10mV from a holding potential of -40mV for the giant pedal ganglion neuron, Pedal 1, for example, will cause a very substantial rise of intracellular calcium which declines to the previous level over 2-3 minutes. Reduction of I_A closely follows the measured elevation of $Ca^{2+}{}_i$ and its gradual decline. Treatments which prevent elevation of intracellular calcium (e.g., substitution of Ba^{2+} for external calcium, Ca^{2+} current blockers as Co^{2+} or Cd^{2+}, and prior iontophoresis of EGTA) all similarly prevent or greatly diminish I_A reduction (42,43).

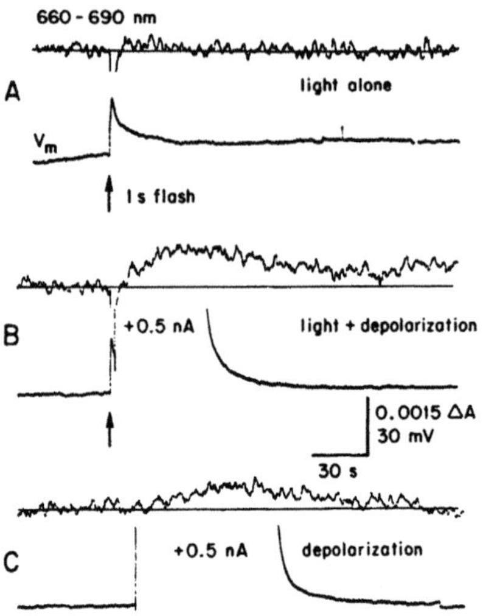

FIGURE 12. Differential absorption spectrophotometric signals (at 660-690) from Type B cell injected with arsenazo III. Light flash (~10^4 ergs/cm^2 x sec) occurs at arrow. Note that depolarizing current (which depolarizes cell by ~20mV) injection occurring immediately after light causes larger Ca^{2+} signal than light or current injection alone (44).

2. $I_{Ca^{2+}-K^+}$. As mentioned above, light elicits a prolonged decrease of $I_{Ca^{2+}-K^+}$. Iontophoretic injection of EGTA greatly reduces this light-induced 'apparent' inward current (Fig 8B,9). This observation suggests that light-induced elevation of intracellular calcium not only activates K^+ current but also inactivates the same current. Substitution of Ba^{2+} for external Ca^{2+} eliminates the delayed inward current but not some early light-induced outward current. This is consistent with the inference that light releases intracellular calcium (light has been shown to cause intracellular release of calcium in other invertebrate photoreceptors; cf. 12,13) which then activates a calcium-dependent K^+ current. This inference is given additional support by the finding that light together with depolarization of the Type B cell causes substantially greater elevation of intracellular calcium than depolarization alone. Presumably, light-induced release of intracellular calcium adds in a non-linear manner to the elevation of intracellular calcium which results from flux across the Type B membrane through voltage-dependent

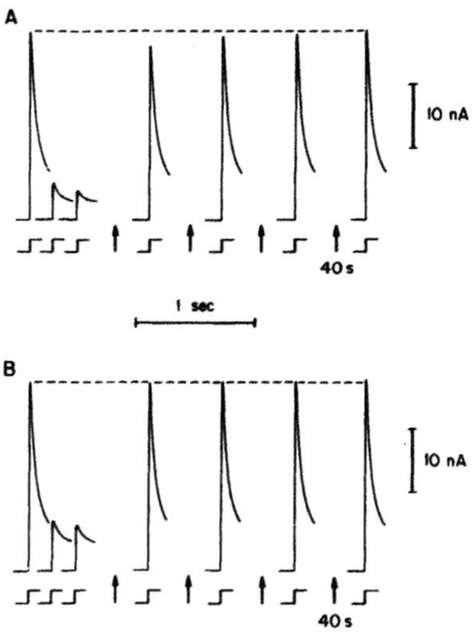

FIGURE 13. Rates of I_A decrease and recovery during and following repetitive command depolarizations. Current responses to only three of the first five depolarizing steps used are shown. The steps following arrows were given at 40 s intervals after the five depolarizations (2.2 s) to 0mV occurred with a cycle time of 4.0 s. Each 2.2 s step followed by a second command (800 ms) to +10mV. A) Command depolarizations paired with light. A light step (2.0 s) was presented 150 ms after the onset of each command depolarization. Light intensity ($10^3 \times 10^5$ ergs/cm^{-2}sec). B) Command depolarizations alone. Arrows indicate 40 s intervals. Lower rectangular traces under A and B indicate onset of 60mV command steps. Note that I_A (peak-inward currents) decreases to lower values in A than in B, and peak I_A takes minutes to return to original values in A but not B. The first three currents included in A and B are the first, second, and fifth currents elicited by the five successive command steps.

Ca^{2+} channels. That there is no light-induced activation of soma membrane Ca^{2+} channels was demonstrated by the absence of any light-induced current at 0mV, the reversal potential for K^+ flux, in 300mM $K^+{}_o$. That light had no effect on the voltage-dependent Ca^{2+} current (measurable at E_K in 300mM $K^+{}_o$) rules out an indirect inactivation of $I_{Ca^{2+}-K^+}$ by increased [$Ca^{2+}{}_i$] - namely, due to inactivation of the calcium current by elevation of Ca $^{2+}{}_i$ as shown for other neurons (14-20). Direct inactivation of $I_{Ca^{2+}-K^+}$ by a light-induced rise of $Ca^{2+}{}_i$ without inactivation of $I_{Ca^{2+}}$ is consistent with the aforementioned demonstration of $I_{Ca^{2+}-K^+}$ inactivation in the absence of $I_{Ca^{2+}}$ inactivation. In normal external Ca^{2+} (10mM), $I_{Ca^{2+}}$ does not inactivate while $I_{Ca^{2+}-K^+}$ does inactivate. In elevated $Ca^{2+}{}_o$, both currents inactivate (Fig 7). $I_{Ca^{2+}-K^+}$ then could have a lower threshold for Ca^{2+}-mediated inactivation than $I_{Ca^{2+}}$.

It is worth noting that in response to dim light, $I_{Ca^{2+}-K^+}$ decreases in the absence of any initial increase of $I_{Ca^{2+}-K^+}$. At holding potentials where $I_{Ca^{2+}-K^+}$ remains significantly activated during steady-state conditions, small increments of intracellular calcium apparently cause more inactivation than activation $I_{Ca^{2+}-K^+}$. It is also worth noting that in the presence of calcium current blockers, such as Co^{2+}, light-induced decrease (albeit diminished) of $I_{Ca^{2+}-K^+}$ still occurs. This is consistent with the interpretation that there is a steady-state $I_{Ca^{2+}-K^+}$ which is due to the level of intracellular free calcium at the resting potential (~-60mV) of the Type B cell, a level of $Ca^{2+}{}_i$ which is independent of the Type B membrane potential.

Light- and voltage-dependent soma channels

Two microelectrode voltage clamp analysis as discussed, has shown that for the voltage range -60 to 0mV, there are only two major voltage-dependent outward potassium currents in the dark, I_A and $I_{Ca^{2+}-K^+}$. $I_{Ca^{2+}-K^+}$ is activated by elevation of $Ca^{2+}{}_i$ resulting from (1) light-induced release from intracellular stores and (2) voltage-dependent flux across the Type B soma membrane. Patch-clamp of this soma membrane (Weiner & Alkon, unpublished observations; and 47) demonstrated light and voltage-dependent K^+ flux at the level of single ionic channels. Voltage commands in the dark to -20 and -10mV ($V_H \cong$ -60mV) elicited outward channels of 2-3 pA (Fig 14). The same voltage commands given immediately after the onset of a light step (~10^5 ergs/$cm^{-2}sec^{-1}$) elicited a much greater frequency of outward channels (even with 0 Na^+ in the external bathing medium to eliminate light-induced depolarization) which also showed what was apparently greater incidence of summation (Fig 14). By thirty seconds after the light onset, the frequency of the channels and the incidence of apparent summation had decreased markedly. These ionic channels, which are not blocked by 3mM 4-aminopyridine of 100mM tetraethylammonium ions, and which have a reversal potential consistent with that for K^+ flux, then are most likely $Ca^{2+}-K^+$ channels which are at first activated by elevation of $Ca^{2+}{}_i$ and then, as demonstrated for the macroscopic soma currents, show marked inactivation. No light-induced inward Na^+ currents were recorded from soma patch recordings. Furthermore, the action spectra of the macroscopic light-induced Na^+ current and the light-induced outward $Ca^{2+}-K^+$ current were both that of rhodopsin. These results together indicate that a light-induced rise of $Ca^{2+}{}_i$ is initiated in the rhadbome. Most likely $Ca^{2+}{}_i$ then diffuses to the inner surface of the soma membrane to open $Ca^{2+}-K^+$ channels.

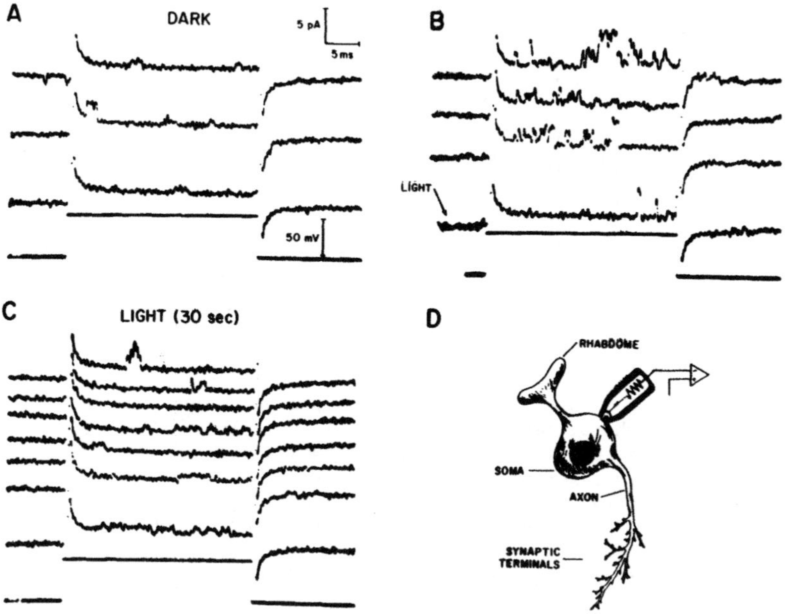

FIGURE 14. Patch-clamp of Type B soma membrane currents. Under gigaseal conditions, voltage-dependent outward channel currents were apparent in darkness (A). Within one second of the onset of steady illumination (10^6 ergs/cm^2 x sec) the frequency of unitary and summated currents markedly increased (B). After 30 seconds of steady illulmination, the frequency of summated currents was substantially reduced (C). These channel currents, which had a reversal potential of ~-70mV and persisted in the presence of 3mM 4-aminopyridine and 100mM tetraethylammonium ion, are unitary components of the macroscopic $I_{Ca^{2+}-K^+}$ and show a similar dependence on $Ca^{2+}{}_i$ elevation, whether via light-induced intracellular release or via voltage-dependent flux of Ca^{2+} across the membrane. The initial increase and then decrease of channel currents during illumination is consistent with transient activation followed by prolonged inactivation by $Ca^{2+}{}_i$. In each panel depolarizing commands occur at 0.5 sec intervals beginning at the bottom of each panel. (D). Diagram of patch electrode on the Type B cell soma and its relationship to other structures. Note that the rhadbomeric membrane which contains light-activation Na^+ channels is remote from the patch electrode (Weiner & Alkon, unpublished observations; see also 47).

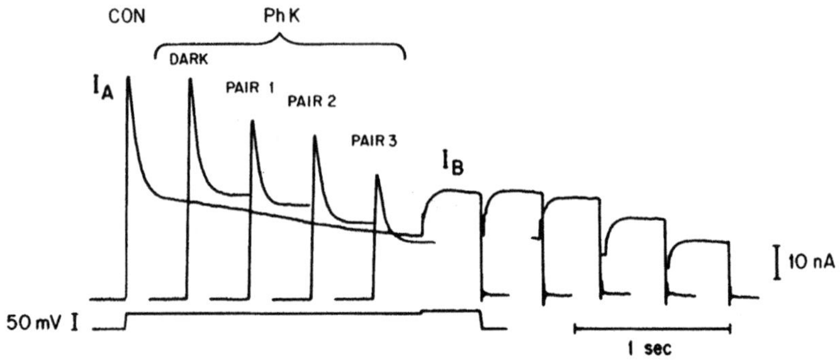

FIGURE 15. Effects of a single injection of phosphorylase kinase (40 nC) on axotomized Type B photoreceptor K^+ currents in normal Ca^{2+}-ASW. From a V_H of -60mV, I_A (left) was elicited by the first command step to -10mV and decayed within approximately 1 second to the steady state. A second superimposed step to 0mV, applied 1.9 seconds after the first step, elicited I_B. The family of K^+ currents (I_A and corresponding I_B) show that prior phosphorylase kinase had no effect on I_A or I_B in darkness, but reduced I_A after a 20-second pairing of light with depolarization to give a Ca^{2+} load (pair 1, to -10mV). Further reduction of I_A and also I_B was evident after two additional pairings associated with increasing command steps (pair 2, to 0mV; pair 3, to 15mV). A full record is shown only for the control condition before the phosphorylase kinase injection (45).

CONCLUSION

Our data indicate that intracellular calcium can at different levels inactivate at least four major currents across Type B photoreceptor membrane; the light-induced I_{Na^+}, I_A, $I_{Ca^{2+}-K^+}$, and $I_{Ca^{2+}}$. $I_{Ca^{2+}}$ seems to be relatively insensitive to elevation of $Ca^{2+}{}_i$ and probably is not regulated by $Ca^{2+}{}_i$ during responses of the cell to its naturally encountered range of stimuli. The net effect of $Ca^{2+}{}_i$ elevation on Type B responses will depend on the magnitude and persistence of inactivation of I_{Na^+}, I_A and $I_{Ca^{2+}}$. Under conditions of prior light adaptation with bright light, the Type B cell hyperpolarizes in response to subsequent bright light, the opposite of its depolarizing response when in a dark adapted state. Light adaptation, in addition to bleaching rhodopsin (responsible for the light-induced I_{Ca^+} and depolarization), will be accompanied by elevated $Ca^{2+}{}_i$ and thus inactivation of I_{Na^+}, I_A, and $I_{Ca^{2+}-K^+}$. Light of moderately bright intensity is then sufficient to cause further release of $Ca^{2+}{}_i$ and an initial activation of $I_{Ca^{2+}-K^+}$, causing hyperpolarization. This is followed by additional inactivation of $I_{Ca^{2+}-K^+}$ and often some delayed depolarization.

As mentioned above, during conditioning with paired light and rotation, mean Ca^{2+}_i undergoes repeated prolonged elevation as a consequence of prolonged and cumulative depolarization of the Type B photoreceptor. Elevation of Ca^{2+}_i and cumulative depolarization is an effect on the Type B cell of the integrated response of the visual statocyst neural system during training of Hermissenda. On days after training, I_A and $I_{Ca^{2+}-K^+}$ are substantially reduced, making the Type B cell more excitable. A significant conditioning-induced effect on I_{Na^+} was not found. This new balance of current inactivation, therefore, most likely cannot be explained simply as a consequence of Ca^{2+}_i remaining elevated on the days following training. A difference in the biochemical sequences responsible for inactivation of I_{Na}, I_A, and $I_{Ca^{2+}-K^+}$ may be critical for determining inactivation lasting for days (as with conditioning) as opposed to many minutes (as with light adaptation).

A number of experiments suggest that the relevant biochemical specificity could involve calcium-calmodulin-dependent phosphorylation. Iontophoretic injection of a Ca^{2+}-calmodulin-dependent protein kinase (phosphorylase kinase) markedly enhances Ca^{2+}-mediated inactivation of I_A and a delayed voltage-dependent K^+ current, at least in part $I_{Ca^{2+}-K^+}$ (45; Fig 14). As already described, a light step paired with a command depolarization is associated with a substantial rise of Ca^{2+}_i and with repetition causes prolonged reduction of I_A and $I_{Ca^{2+}-K^+}$. A single pairing of light and depolarization has relatively small and short lived effect. Following iontophoresis of the Ca^{2+}-dependent protein kinase, the effect of these pairings is enhanced and with two repetitions the two voltage-dependent K^+ currents (I_A and the late current) often remain reduced for the duration of the recording (20-45 minutes) while I_{Na^+} is essentially unaffected.

One and two dimensional gel electrophoresis on eyes isolated from conditioned as well as control Hermissenda revealed changes of phosphorylation of particular low molecular weight proteins (46) and more recently of a 55,000 molecular weight protein (47) specific to the conditioned animals. Phosphorylation of these proteins, furthermore, was found to be dependent on Ca^{2+} concentration. Inhibitors of Ca^{2+}-calmodulin-dependent phosphorylation by trifluoroperazine or calmidozolium not only reduced phosphorylation of the 55,000 MW protein - it also very specifically prevented Ca^{2+}-mediated inactivation of $I_{Ca^{2+}-K^+}$ (as induced by light during a command depolarization; 47). Finally, subjecting the isolated eyes to prolonged depolarization (by bathing in 100mM K^+_o) to stimulate the effect of training also produced changes of phosphorylation of the low molecular weight proteins (48).

It remains to be determined what role Ca^{2+}-dependent phosphorylation actually plays in the regulation of the K^+ channels. This may not be phosphorylation of a primary constituent of a membrane channel but only one of many steps in a long sequence of biochemical reactions involving pre- and/or post-translational modification of the channels beginning with prolonged elevation of $Ca^{2+}{}_i$. Nevertheless, characterization of those specific proteins which undergo conditioning-induced changes in Ca^{2+}-calmodulin-dependent phosphorylation and establishing the relationship of those proteins to the channels which have been shown to remain inactivated for days after conditioning offers a promising starting point for localizing molecular mechanisms by which neural systems undergo long-lasting transformation.

REFERENCES

1. Neher E. and Lux H.D. (1972). Differential action of TEA^+ on two K^+ current components of a molluscan neuron. Pfleug. Arch. 336, 87-100.
2. Meech R.W. (1974). The sensitivity of Helix aspersa neurones to injected calcium ions. J. Physiol. (Lond.) 237, 259-277.
3. Heyer C.B. and Lux H.D. (1976). Control of the delayed outward potassium currents in bursting pacemaker neurones of the snail Helix pomatia. J. Physiol. (Lond). 262, 349-382.
4. Thompson S.H. (1977). Three pharmacologically distinct potassium channels in molluscan neurons. J. Physiol. (Lond.) 265, 465-488.
5. Eckert R. and Tillotson D. (1981). Calcium mediated inactivation of the calcium conductance in caesium-loaded giant neurones of Aplysia californica. J. Physiol. (Lond.) 314, 265-280.
6. Meech R.W. (1978). Calcium-dependent potassium activation in nervous tissues. Ann. Rev. Biophys. Bioeng. 7, 1-18.
7. Hermann A. and Hartung K. (1983). Ca^{2+}-activated K^+ conductance in molluscan neurones. Cell Calcium 4, 387-405.
8. Atwater I., Rosario L. and Rojas E. (1983). Properties of the Ca-activated K^+ channel in pancreatic β-cells. Cell Calcium 4, 451-461.
9. Adams D.J. and Gage P.W. (1979). Ionic currents in response to membrane depolarization in an Aplysia neurone. J. Physiol. (Lond.) 289, 115-142.
10. Connor J.A. (1979). Calcium current in molluscan neurones: measurement under conditions which maximize its visibility. J. Physiol. (Lond.) 286, 41-60.
11. Hagiwara S. and Byerly L. (1981). Calcium channel. Ann. Rev. Neurosci. 4, 69-125.
12. Brown J.E. and Blinks J.R. (1974). Changes in intracellular free calcium during illumination of invertebrate photoreceptors: detection with aequorin. J. Gen. Physiol. 64, 643-655.
13. Brown J.E., Brown P.K. and Pinto L.K. (1977). Detection of light induced change of intracellular ionized calcium concentration in Limulus ventral photoreceptors using arsenazo III. J. Physiol. (Lond.) 267, 299-230.
14. Tillotson D. and Horn R. (1978). Inactivation without facilitation of calcium conductance in caesium-loaded neurons of Aplysia. Nature 273, 312-314.
15. Tillotson D. (1979). Inactivation of Ca conductance dependent on entry of Ca ions in molluscan neurons. Proc. Natl. Acad. Sci. USA 76, 1497-1500.
16. Eckert R. and Tillotson D. (1978). Potassium activation associated with intraneuronal free calcium. Science 200, 437-439.
17. Eckert R. and Ewald D. Calcium mediated calcium channel inactivation determined from tail current measurements. Biophys. J. 33, 145a.
18. Eckert R., Tillotson D. and Brehm P. (1981). Calcium mediated control of calcium and potassium currents. Fed. Proc. 40, 2226-2232.
19. Eckert R. and Ewald D. (1983). Calcium tail currents in voltage-clamped intact nerve cell bodies of Aplysia californica. J. Physiol. (Lond.) 345, 533-548.
20. Eckert R. and Ewald D. (1983). Inactivation of calcium conductance characterized by tail current measurements in neurones of Aplysia californica. J. Physiol. (Lond.) 345, 549-565.
21. Alkon D.L., Shoukimas J. and Heldman E. (1982). Calcium-mediated decrease of a voltage-dependent potassium current. Biophys. J. 40, 245-250.
22. Alkon D.L., Farley J., Sakakibara M. and Hay B. (1984).

Voltage-dependent calcium and calcium-activated potassium currents of a molluscan photoreceptor. Biophys. J. (in press).

23. Alkon D.L. and Sakakibara M. (1984). Prolonged inactivation of a Ca^{2+}-dependent K^+ current but not Ca^{2+}-current by light induced elevation of intracellular calcium. Soc. Neurosci. (in press)
24. Alkon D.L. and Sakakibara M. (1984). Ca^{2+}-activated and inactivated K^+ flux with illumination of Hermissenda photoreceptors. (in preparation)
25. West A., Barnes E.S. and Alkon D.L. (1982). Primary changes of voltage responses during retention of associative learning. J. Neurophysiol. 48, 1243-1255.
26. Alkon D.L., Lederhendler I. and Shoukimas J.J. (1982). Primary changes of membrane currents during retention of associative learning. Science 215, 693-695.
27. Farley J., Sakakibara M. and Alkon D.L. (1984). Associative-training correlated changes in I_{Ca-K} in Hermissenda Type B photoreceptors. Soc. Neurosci. (in press)
28. Forman R., Alkon D.L., Sakakibara M., Harrigan J., Lederhendler I. and Farley J. (1984). Changes in I_A and I_C but not I_{Na} accompany retention of conditioned behavior in Hermissenda. Soc. Neurosci. (in press).
29. Alkon D.L. (1984). Calcium-inactivated potassium currents: A biophysical memory trace. Science (in press)
30. Alkon D.L. and Fuortes M.G.F. (1972). Responses of photoreceptors in Hermissenda. J. Gen. Physiol. 60, 631-649.
31. Alkon D.L. (1973). Intersensory interactions in Hermissenda. J. Gen. Physiol. 62, 185-202.
32. Alkon D.L. (1974). Sensory interactions in the nudibranch mollusc Hermissenda crassicornis. Fec. Proc. 33, 1083-1090.
33. Alkon D.L., Akaike T. and Harrigan J.F. (1978). Interaction of chemosensory, visual and statocyst pathways in Hermissenda. J. Gen. Physiol. 71, 177-194.
34. Alkon D.L. (1980). Membrane depolarization accumulates during acquisition of an associative behavioral change. Science 210, 1375-1376.
35. Alkon D.L. (1983). Learning in a marine snail. Sci. Am. 249, 70-84.
36. Alkon D.L. (1984). Persistent changes of identified membrane currents as a cause of associative learning. In: D.L. Alkon and J. Farley (Eds.) Primary Neural Substrates of Learning and Behavioral Changes. Cambridge University Press, New York and London. (in press)
37. Alkon D.L. (1979). Voltage-dependent calcium and potassium ion conductances: A contingency mechanism for an associative learning model. Science 205, 810-816.
38. Hagiwara S. and Saito N. (1959). Voltage-current relations in nerve cell membrane of Onchidium verruculatum. J. Physiol. (Lond.) 148, 161-179.
39. Connor J.A. and Stevens C.F. (1971). Voltage clamp studies of a transient outward current in gastropod neural somata. J. Physiol (Lond.) 213, 21-30.
40. Neher E. (1971). Two fast transient current components during voltage clamp on snail neurons. J. Gen. Physiol. 58, 36-53.
41. Shoukimas J.J. and Alkon D.L. (1980). Voltage-dependent, early outward current in a photoreceptor of Hermissenda crassicornis. Soc. Neurosci. 6, 17.
42. Acosta-Uriquidi J., Alkon D.L., Connor J.A. and Neary J.T. (1983). Intracellular injection of a Ca^{2+}-dependent protein kinase amplifies Ca^{2+}-mediated inactivation of transient K^+ current (I_A) in Hermissenda giant neurons. Soc. Neurosci. 9, 501.
43. Connor J.A. and Alkon D.L. (1982). Light-induced changes of

intracellular Ca^{2+} in Hermissenda photoreceptors measured with arsenazo III. Soc. Neurosci. 8, 944.

44. Connor J.A. and Alkon D.L. (1984). Light- and voltage-dependent increases of calcium ion concentration in molluscan photoreceptors. J. Neurophysiol. 51, 745-752.

45. Acosta-Urquidi J. Alkon D.L. and Neary J.T. (1984). Ca^{2+}-dependent protein kinase injection in a photoreceptor mimics biophysical effects of associative learning. Science 224, 1254-1257.

46. Neary J.T., Crow T.J. and Alkon D.L. (1981). Change in a specific phosphoprotein band following associative learning in Hermissenda. Nature 293, 658-660.

47. Alkon D.L., Naito S., Neary J., Harrigan J., Sakakibara S., Weiner A. and Heldman E. (1984). Inactivation of voltage-dependent K^+ currents is related to Ca^{2+}-calmodulin-dependent phosphorylation of specific proteins. Biol. Bull. (in press)

48. Neary J.T. and Alkon D.L. (1983). Protein phosphorylation/dephosphorylation and the transient voltage-dependent potassium conductance in Hermissenda crassicornis. J. Biol. Chem. 258, 8979-8983.

49. Alkon D.L. and Grossman Y. (1978). Long-lasting depolarization and hyperpolarization in eye of Hermissenda. J. Neurophysiol. 41, 1328-1342.

DISCUSSION OF THE ARTICLE BY D.L. ALKON (U.S.)

B. Minke (Israel): Can you estimate the increase in intracellular calcium concentration following light and following a voltage step and can you also estimate what percentage of this increase is from intracellular stores and what percentage from extracellular?

D.L. Alkon (U.S.): We cannot at this time, but we are in the process of making the necessary measurements to arrive at such estimates.

D.O. Smith (U.S.): You proposed that your long-term changes are due to phosphorylation which is dependent on calmodulin. Another mechanism which has been hypothesized in other systems for these long-term changes involves calcium-activated proteases. Do you have any evidence for or against the role of these proteases in this system?

D.L. Alkon (U.S.): I don't have any evidence for that. However, the potential complexity of the biochemical basis at learning-induced channel transformations shouldn't be underestimated. For instance, the cAMP-dependent protein kinase when injected iontophoretically does not produce exactly the same changes of ionic currents that the training does, as does the calcium calmodulin-dependent protein kinase, when activated by calcium elevation. Nevertheless, there is no reason to discount the possibility that cAMP can effect the amount of free calcium inside the cell and vice versa, that the concentration free calcium can influence the levels of cAMP.

R.S. Zucker (U.S.): You have two ways of increasing intracellular calcium. One is with light which releases calcium from intracellular stores. The other is by depolarization which causes influx through channels. Why don't these two effects summate? It seems that when you do them simultaneously you get a response much bigger then the sum of the twc individually.

D.L. Alkon (U.S.): I don't have the answer but you pointed out a particularly important feature. Namely, there is a non-linearity of the effects of the two sources of elevation of intracellular calcium. I don't know why this is true but I could imagine that the buffering or the ways of disposing of the calcium, such as via pumping mechanisms, have non-linear aspects to them and once a certain level of intracellular calcium is reached, re-equilibration to previous resting levels may be a much slower process.

R.S. Zucker (U.S.): I guess we all knew that learning was a non-linearity. The other question is, I think in the past you thought that it was primarily the block of I_A that was responsible for the long lasting depolarization and modulation of learning. Now it looks like it is primarily the block of calcium-activated potassium channels. Is that true?

D.L. Alkon (U.S.): No. First of all we have to break down the question. The A current is very important in determining the long lasting depolarization. Bob referred to a long lasting depolarizing tail that follows the light response. This depolarizing tail is very much dominated by inactivation of the A current. Therefore, if you block the A current with 4-AP, you do a lot of damage to the tail. If you block only the A current, you will account for a significant enhancement of the input resistance of the cell body which we also measured in other experiments. In fact, it doesn't account for all of it. If you block the calcium-dependent potassium current, you do the same thing. Reduction of both of these currents as a result of learning can account for enhanced excitability at the Type B cell. We think that there is a resting or steady-state A current as well as a resting calcium-dependent potassium current. If you block the A current with 4-AP, you will see an enhancement of the input resistance by 50 to 100 percent. If you substitute barium ror calcium, reducing the calcium-dependent potassium current at rest, you also elevate the input resistance. They are both contributing. As with many features of cells' responses, the kinetics of those currents determine when they are most important. The A current may be more important during rapidly depolarizing and repolarizing potential changes. The calcium-dependent potassium current could have more profound influence on steady-state membrane potential.

References

Alkon, D.L. and Sakakibara, M. (1984). Prolonged inactivation of a Ca^{++}-dependent K^{+} current but not Ca^{++} current by light-induced elevation of intracellular calcium. Soc. Neurosci. Abstr. 10.

Alkon, D.L., Farley, M., Sakakibara, M. and Hay, B. (1984). Voltage-dependent calcium and calcium-activated potassium currents of a molluscan photoreceptor. Biophys. J. (in press)

SECTION 2

CALCIUM & TRANSMITTER RELEASE

SOME CONSEQUENCES OF INTRACELLULAR CALCIUM BINDING ON PHASIC SYNAPTIC TRANSMITTER RELEASE

JOHN W. MOORE and MICHAEL HINES

1. INTRODUCTION

Until recently, no mechanism has been offered to account simultaneously for phasic release of transmitter and its facilitation at the frog neuromuscular junction. Katz & Miledi (1) showed that external calcium was necessary at the time of the first nerve impulse and suggested that calcium entered the terminal and, by binding to some receptor site at the presynaptic terminal, released transmitter. The "residual calcium" hypothesis for facilitation evolved from their suggestion (2) that a residuum of "active calcium" remains within the terminal following an action potential for several milliseconds and is able to enhance the amount of transmitter released by succeeding stimuli.

Previously, most models proposed for facilitation were merely mathematical descriptions of the envelope of peak postsynaptic potentials and did not include any mechanisms. Most have avoided the crucial question: If transmitter release is proportional to the internal calcium concentration, how can it be reduced rapidly for phasic release and yet be high enough to show facilitation for many milliseconds? Katz & Miledi (3) raised this question and concluded that a full answer was not provided by the assumption that transmitter release is proportional to the 4th power of the calcium concentration.

In contrast, our laboratory has chosen to carry out computer simulations of physiological processes, using known mechanisms to provide insight into the functioning of a phasic synapse. We have improved on our original description (4) of the mechanisms for the control of calcium which are known to be present at nerve terminals -- influx, efflux, intracellular binding, and diffusion -- by including the interaction of calcium with a finite number of binding sites within the cytoplasm. In the present paper we examine the effects of local saturation of these binding sites and explore how diffusion limits the efflux and thus the apparent pump rate.

We have been able to account for:

1) The rapid transient rise and decay of phasic transmitter release

There is as yet no direct measure of transmitter release on a millisecond time scale. The endplate current gives the best available (but still inaccurate) measure of release. Endplate currents at the neuromuscular junction of the frog begin after a delay of 0.5 ms (at room temperature) following the arrival of a nerve impulse. The rising phase then lasts 0.5 ms, and the falling phase has a time constant of about 1 millisecond (5,6).

2) The dynamics of facilitation

Following single stimuli or trains of stimuli, there is a time during which succeeding stimuli will evoke the release of increased amounts of transmitter. This characteristic has been described as four processes, distinguished by their time constants for decay. The faster pair of processes, referred to together as facilitation, are evident following single stimuli. As described by Mallart and Martin (7), facilitation appears at its peak immediately after a conditioning pulse and decays with time constants of about 35-50 ms and 200-300 ms. The slower pair are referred to as augmentation and potentiation, and arise and decay over periods of seconds, times long compared with the processes in this model. Slow processes will not be dealt with in the current work.

2. MODEL

2.1. Calcium's role in transmitter release

Different aspects of the role of calcium in transmitter release have been demonstrated in a number of ways.

2.1.1. Internal calcium results in release. Miledi (8) showed that transmitter release could be stimulated at the squid giant synapse by intracellular iontophoresis of calcium, but not other divalent cations. Rahamimoff et al. (9) exposed frog neuromuscular junctions (nmj) to calcium-containing liposomes and found that they increased release while liposomes with other ions do not. Charlton et al. (10) showed that the internal calcium concentration (measured with arsenazo III signals) increased with each impulse arriving at the squid giant presynaptic terminal.

2.1.2. Transmitter release is proportional to the fourth power of the bath concentration of calcium. The kinetics of the postsynaptic response in the frog nmj can be described as a first order reaction raised to the third or fourth power (11, 12). Katz and Miledi (1) showed that transmitter release occurred only when external calcium was present during the presynaptic action potential invasion of the nerve terminal.

To summarize, 1) direct injection of calcium into the terminal results in transmitter release, 2) external calcium ions are necessary to support transmitter release by nerve impulses, 3) calcium entry into the terminal with each impulse has been demonstrated, and 4) there is a fourth power relationship between external calcium and transmitter release.

2.2. Model components

The model we examined is illustrated diagramatically in Fig 1 and consists of 1) calcium influx, 2) slow diffusion of calcium due to rapid equilibrium binding of most calcium ions to saturable binding sites assumed to be uniformly distributed, and 3) calcium efflux, linearly related to the internal calcium concentration just inside the cell membrane.

In the model, we take the fourth power of the internal calcium concentration as the release function, the time course of which is compared with the time course of transmitter release as measured by endplate currents. We have made this choice for two reasons. Firstly, under no conditions have we found similarity between the kinetics of transmitter release and the first power of the internal calcium concentration. Secondly, we think that the increase of calcium concentration just under the excitable membrane following an impulse should be proportional to the external concentration. Because the gradient for calcium across the cell membrane is so high, we argue that the inward flux is largely determined by diffusion,

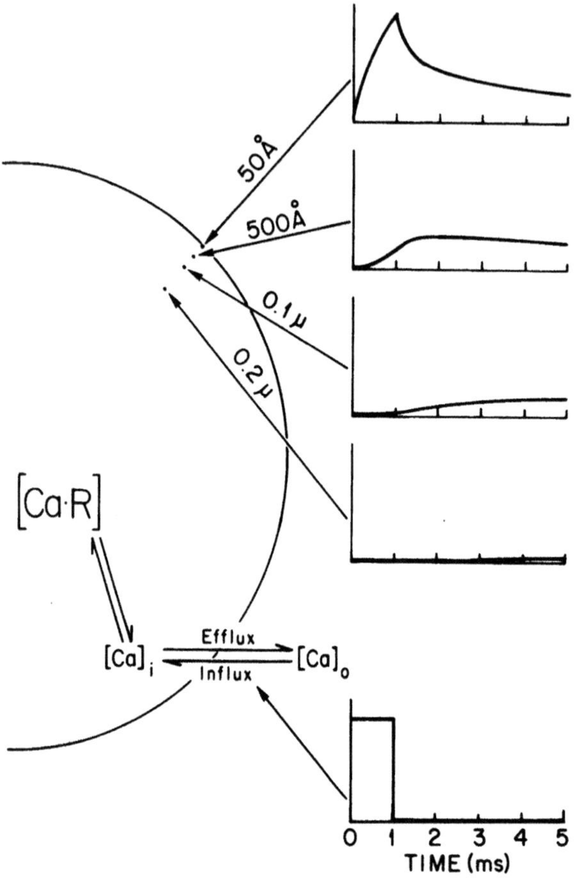

FIGURE 1. Diagram of the model of the various mechanisms which affect free intracellular calcium concentration on a millisecond time scale. The components are calcium influx (by voltage-dependent calcium channels), efflux (by an energy-dependent pump), and, within the terminal, diffusion limited by the rapid binding of a large fraction of the total intracellular calcium to cytoplasmic sites. In response to the calcium influx pattern shown in the lower right diagram, the rise and fall of the free calcium concentration is shown at various distances from the cell surface membrane. Note the large transient increase in free calcium just under the membrane.

and that the fourth power relation observed for external calcium and transmitter release probably reflects a similar strong dependence of release on internal calcium.

2.2.1. Calcium influx. Because Katz & Miledi (1) showed that, for transmitter to be released, external calcium was necessary at the time of arrival of the nerve impulse, voltage-dependent calcium channels probably exist at the frog's nerve terminal. At the squid giant synapse, influx of calcium into the presynaptic terminal has been demonstrated (by arsenazo III absorption changes) following action potentials (10) and under voltage clamp depolarizations (13).

Calcium influx through ion specific channels is modeled as a square pulse having an amplitude of 1000 $pmol/cm^2/sec$ and a duration of 1 msec. A more realistic shape close to that measured by Llinas et. al. (14) was examined in our previous simulations (4) and shown to have no significant effect on the variables which we are examining.

2.2.2 Calcium binding. Axoplasm is the source of a number of components which bind calcium and would be expected to result in slowing the rate of diffusion of intracellular calcium. Baker (15) described mitochondrial and non-mitochondrial components of high affinity (0.3-0.5 µM) and moderate capacity (20-40 µM).

Calcium diffusion in cytoplasm has been found to be much slower than in free solution (6×10^{-6} cm^2/sec). Hodgkin and Keynes (16) found the radial diffusion constant to be 6×10^{-7} cm^2/sec, (implying a bound-to-free ratio of 10) while Blaustein and Hodgkin (17) estimated the rate at 3×10^{-7} cm^2/sec (implying a bound-to-free ratio of 20). In both cases the calcium loads were large compared to physiological loads. One should regard these values for the effective diffusion constant as an upper limit.

Gorman & Thomas (18) estimated that the ratio of bound-to-free calcium was on the order of 50 in an Aplysia neuron. Brinley et. al. (19) have shown that the ratio of bound-to-free calcium can be as high as 2000 in squid axon.

We have explored the effect of the binding ratio over this range of 20 to 2000.

2.2.3. Pump. Any calcium which enters the cell during excitation must be eventually extruded. The types of mechanism by which this is accomplished --in red cells, liver, snail neurons, and squid axon-- are 1) sodium-calcium exchange (20, 21, 22) and 2) an energy-dependent and sodium-independent pump (23).

Blaustein (24) has described extrusion in the squid giant synapse as a saturable process with a maximal rate of 1.7 pmol/cm^{2}s, half maximal at a calcium concentration of 8 uM. At low calcium concentrations, the relationship is nearly linear with a slope of 1.4×10^{-4} cm/s. Mullins (22) reports a linear relation in low calcium with a slope of 10^{-3} cm/s. Our simulations include this range of pumping rates and extend it another 10-fold.

The measurement of the rate constant for extrusion from the efflux depends upon knowledge of the calcium concentration near the membrane since that is the calcium that the extrusion mechanism sees. As calcium is extruded, if there arises a diffusion gradient from the interior of the cell toward the periphery, then an erroneous estimation of the calcium concentration near the membrane will lead to erroneously low values for the calcium pump rate constant. The extent to which this may occur is simulated and discussed.

2.3. Factors omitted

While the role of various intracellular organelles in the control of intracellular calcium has been the subject of a number of investigations, the case remains to be made whether such organelles as mitochondria and smooth endoplasmic reticulum participate in the rapid time course of processes such as transmitter release. Because our initial calculations showed that diffusion is sufficiently slow so that the free calcium ions modulating rapid events (such as transmitter release) never encounter such organelles, their contribution to calcium homeostasis has been excluded from the model we present. No attempt is made here to force transmitter release into discrete quanta. Nor was there explicit inclusion of 1) diffusion of transmitter across the synaptic cleft, 2) the possible generation of concentration gradients in the synaptic cleft, or 3) the reaction with postsynaptic receptors and subsequent opening of channels in the sarcolemma.

These simplifications seemed reasonable because these steps contribute little to the time course of the endplate response (25).

3. METHODS

In this model, the free calcium concentration, [Ca], at a point is affected by [Ca] at adjacent points (diffusion) and by binding of calcium to fixed binding sites at concentration [B] to form [CaB].

The equation expressing the conservation of calcium in the interior of the terminal is then described by a generalization of the diffusion equation,

$$\frac{\partial}{\partial t}[\mathrm{Ca}] = -\vec{\nabla}\cdot(-D\vec{\nabla}[\mathrm{Ca}]) - \frac{\partial}{\partial t}[\mathrm{CaB}] \quad , \qquad 1)$$

where $-\vec{\nabla}\cdot(-D\vec{\nabla}[\mathrm{Ca}])$ is the inward divergence of the flux due to the free calcium gradient and D is the free diffusion constant for calcium (cm^2/sec). In this equation concentration units are moles/cm^3 (10^3M) but for plots and discussion we have converted the numerical results to the normal units of moles/liter. The last term in Eq. 1 expresses the fact that, all other things being equal, an increase in bound calcium implies a corresponding decrease in free calcium. The nerve terminal was taken to be a cylinder, 1µm in diameter, in which circumferential symmetry and uniform properties in the axial direction were assumed. We thus effectively have a one dimensional system in which $-\vec{\nabla}\cdot(-D\vec{\nabla}[\mathrm{Ca}]) = \frac{D\partial}{r\partial r}(r\frac{\partial}{\partial r}[\mathrm{Ca}])$ where r is the radial distance from the center of the terminal. The radially symmetric flux out of the center of the terminal is automatically zero in this coordinate system so that it is not necessary to specify a boundary condition at r=0. We model the flux through the membrane by an inward square wave of current 1 msec in duration, approximating the influx during an action potential. In addition we incorporate a linear pump in which efflux is directly proportional to [Ca] next to the membrane. Flux continuity across the membrane provides the boundary condition at the terminal radius r_o so that the radial component of the calcium flux at the membrane is,

$$-D\frac{\partial}{\partial r}[Ca]\ \Big|_{r=r_o} = -\mathrm{Influx} + k_e[Ca]\ \Big|_{r=r_o} \quad ,$$

where positive influx measured in moles/cm^2 generates a negative flux (positive gradient) at the membrane and a linear pump with flux rate constant k_e in cm/sec causes a positive flux. Note that positive flux is defined here as flow toward the periphery which is the direction of increasing r. The pump rate in this model acts only on free calcium and differs from the pump rate of Stockbridge and Moore (4) which acted on the total calcium next to the membrane.

To complete the model, we write the simplest reaction of free calcium with immobile binding sites,

$$\frac{\partial}{\partial t}[CaB] = -k_1[CaB] + k_2[B][Ca] \quad , \qquad 2)$$

where k_1 and k_2 are the first and second order reaction rate constants. Equations 1 and 2, along with the boundary condition at the membrane and the initial values of the concentrations are in a form which can be easily simulated. However it is conceptually helpful to disregard the detailed time dependence of [CaB] in favor of emphasizing the primary effect of free calcium binding on the effective rate of diffusion. To this end, we assume that the reaction rates in Eq. 2 are extremely large so that [CaB] and [Ca] are always in quasi-equilibrium. We have then, neglecting the time dependent term in Eq. 2, introducing the constant total concentration of binding sites, B_T, using the relation $[B] = B_T - [CaB]$, and defining the constant $\beta = \frac{k_2}{k_1}B_T$,

$$[CaB] = \frac{\beta}{1 + \frac{[Ca]}{B_T}\beta}[Ca] \quad . \qquad 3)$$

When the ratio of total number of binding sites to free calcium is much greater than β, the denominator in Eq. 3 can be neglected and β is the ratio of bound to free calcium. Substituting this into Eq. 1 then shows that D/(1+β) can be thought of as the effective diffusion constant of [Ca].

Equations 1 and 3 were simulated by dividing the terminal into 25 annular compartments of equal width. The resulting set of ordinary differential equations were solved with a first order correct pure implicit method using time steps of 50μsec. No iterations were used within a time step since decreasing the time step did not change the solution by more than the line width of the plots. The "50% efflux at 100 msec" simulations which did not involve a fast influx pulse were carried out with time steps of .2 and 1 msec and differed by less than 1 percent. Numerical errors due to first order simulation of the boundary condition were more serious. Simulations using even 100 segments seriously underestimate the calcium concentration at the membrane when that concentration is changing rapidly (when the spatial profile is exponential). For this reason we extrapolated the concentration values at the three compartments closest to the membrane to obtain the concentration at the membrane. This allowed simulation with 25 segments to yield more accurate results than 100 segments without extrapolation.

4. RESULTS

4.1. Diffusion

We consider first the effect of radial diffusion alone (with no calcium extrusion mechanism) in a cylindrical cell in which the resting free calcium concentration is 10^{-8} M.

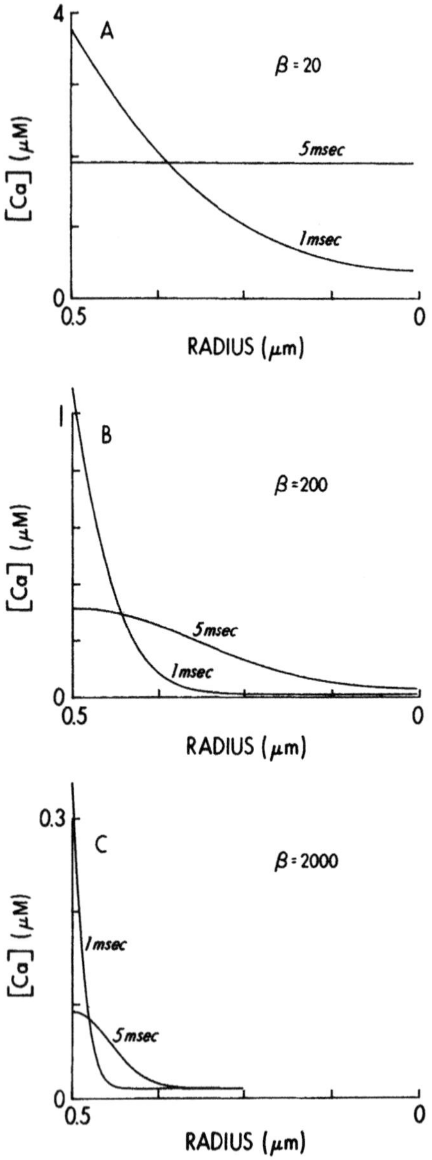

Figure 2

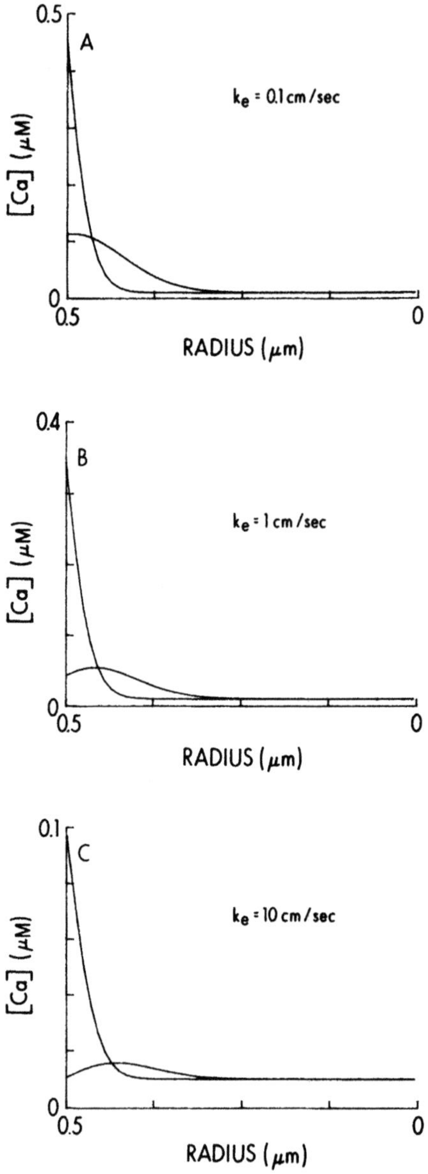

Figure 3

FIGURE 2. The result of influx (but no efflux) and binding on intracellular calcium transients for β equal to 20, 200, and 2000 respectively. Spatial distributions are shown from the outer margin (r = .5 μm) to the center of the model cell (r = 0 μm) at the end of the stimulus (1 msec) and at 5 msec. In each case the total calcium influx was a 1 msec square wave of $1 nM/cm^2/sec$. The simulations were started with an initial uniform CaFree of 0.01μM. The free diffusion constant was 6×10^{-6} cm^2/sec, k_e = 0, and [B] = infinity.

FIGURE 3. Effect of pump rate on the calcium transient in the model cell for β=1000. Shown is the spatial distribution at 1 msec and at 5 msec in response to an influx of CaFree as in Fig 2. The pump rates are as shown and other parameters are the same as in Fig. 2.

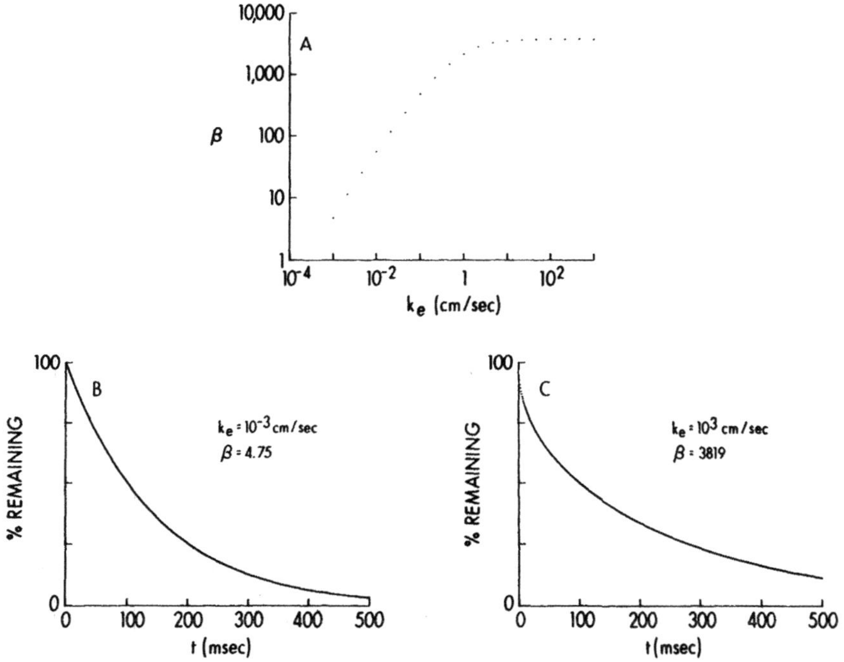

FIGURE 4. Effect of β on the apparent pump rate. The cell was loaded with a uniform initial concentration of calcium. A) For different values of the pump rate, k_e, the value of β was found which resulted in removal of 50% of the total initial calcium in 100 msec. B) and C) show the time course of % calcium remaining within the cell for values of k_e and β corresponding to two different points plotted in panel A. Other parameters are as in Fig. 2.

Furthermore, we initially consider the case of such large concentration of binding sites that saturation does not occur. Figure 2 illustrates the spatial distribution of the intracellular calcium concentration immediately following, and 4 msec after, a constant influx lasting 1 msec. The total influx was 1 pmol/cm^2. The largest peak concentration of free calcium occurs with the highest effective diffusion rate (smallest β). For β = 20, 200 and 2000 respectively, the free calcium peak concentrations reached at the end of the stimulus are about 4., 1.0, and 0.3 µM.

As the equilibrium ratio, β, increases, not only does the peak concentration of free calcium decline but as the diffusion is slowed, most of the calcium remains near the membrane. Note particularly that, at 5msec, the calcium concentration has barely begun to rise at the center if β is above 200.

4.2. Diffusion and pump

The effect of the addition of a pump is shown in Fig 3, where β was held constant at a value of 1000, a value thought to be reasonable on the basis of considerations raised in the MODEL section. A linear pump with a rate constant of.1 cm/sec (or, since β = 1000, .0001 cm/sec acting on total calcium-- 24) has negligible effects on this time scale, but a ten-fold increase in the pump rate to 1 cm/sec (22), causes a small change in the calcium distribution (Fig 3B). The peak calcium concentration near the membrane is reduced somewhat, but the reduction at 5 msec is marked and a reverse gradient is seen. A further ten-fold increase in the pump rate (to 10 cm/sec, Fig 3C), has a more dramatic effect; the peak calcium concentration near the membrane is reduced about 1/3 and the calcium concentration just under the membrane is pulled down to the resting level. Further increases in the rate of extrusion have a diminishing effect because the concentration of calcium at the membrane available for extrusion becomes determined by the rate of diffusion rather than the pump rate. Thus, after the first few milliseconds following an influx, the efflux was rather insensitive to the pump rate constant.

That the diffusion process limits the pumping rate can be seen perhaps more directly by loading the cell with a large initial concentration and comparing the percent total remaining calcium as a function of time for different values of β and k_e (Fig 4B and 4C). Figure 4A shows β as a function of k_e which results in 50% efflux of the total initial calcium in 100 msec. The efflux is entirely diffusion limited for β greater than 2000. Figures 4B and 4C are pump rate limited and diffusion limited respectively. The former has a characteristic exponential time course, and the latter shows an initial drop in internal calcium followed by a slower decay as [Ca] diffuses toward the membrane.

4.3. Transmitter release kinetics

As noted above, the best available measure of transmitter release is the endplate current. Experimental records from neuromuscular junctions of the frog, are characterized by: 1) a delay of at least 0.5 msec (at 20° C and increasing rapidly with cooling, 25), 2) a rapid rise to a peak within another 0.5 msec and 3) a slower exponential decay with a time constant of about 1 msec (5, 6).

In response to a 1 msec square pulse injection of calcium into the terminal, the internal calcium concentration rises without a delay, but the fourth power of the calcium concentration (which we take as proportional to transmitter release, and hereafter refer to as our "release function") of course shows a delay. However this delay (a fraction of a millisecond) is always small compared to the experimentally observed synaptic delay. The rate of fall in submembrane calcium concentration at the end of such a square pulse influx is greater than the rate of rise; the release function, of course, falls even faster. Thus the rounded peak and relatively slow decay of the endplate current are not fitted with this release function. The substitution of a calcium current of more realistic shape (14) does not alleviate these two problems.

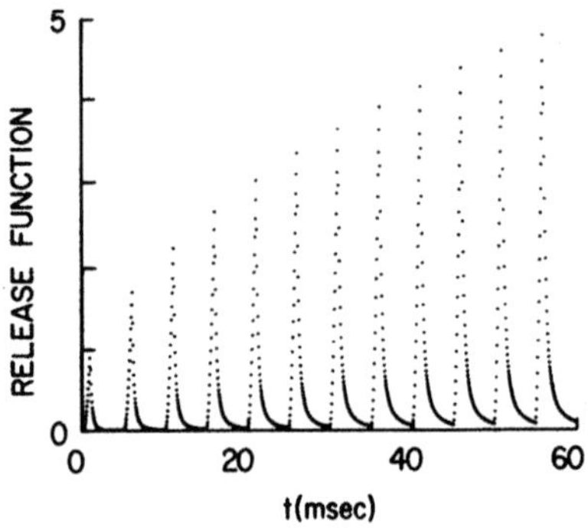

FIGURE 5. Release function for a train of impulses 5 msec apart. $\beta = 1000$, $k_e = .5$ cm/sec, initial [Ca] = 0, and other parameters as in Fig 2. The release function is normalized so that the first release transient has a value of 1.

Clearly some other mechanism is necessary to account for synaptic delay. Presumably calcium participates in some reaction to produce release of transmitter and the additional synaptic delay probably occurs at this stage.

4.4. Facilitation

For values of facilitation below 10 the model predicts a facilitation buildup similar to that seen experimentally. Figure 5 shows the characteristic first order lag response of the normalized release function buildup for $\beta = 1000$ and $k_e =$ .5 cm/sec.

Cessation of transmitter release between impulses on the one hand, and the increased transient response to successive impulses on the other, rule out a linear model based on residual calcium. Figure 6A illustrates that in a nonsaturable model the increased [Ca] to a train of impulses is merely a uniform transient riding on a steadily increasing baseline. The invocation of the highly nonlinear 4th power release function alleviates this dilemma by amplifying the effect of subsequent [Ca] transients (the first transient

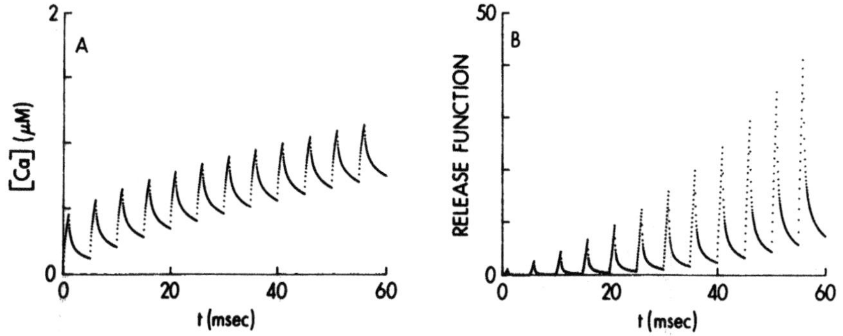

FIGURE 6. Facilitation with non-saturable sites. Panel A shows the intracellular calcium transient at the membrane and panel B shows the release function for a train of impulses 5 msec apart. The figure was designed to illustrate the problem associated with linearity when a condition of high facilitation occurs. The first release transient in panel B has a peak value of 1. Parameters are as in Fig. 2. except β = 1000.

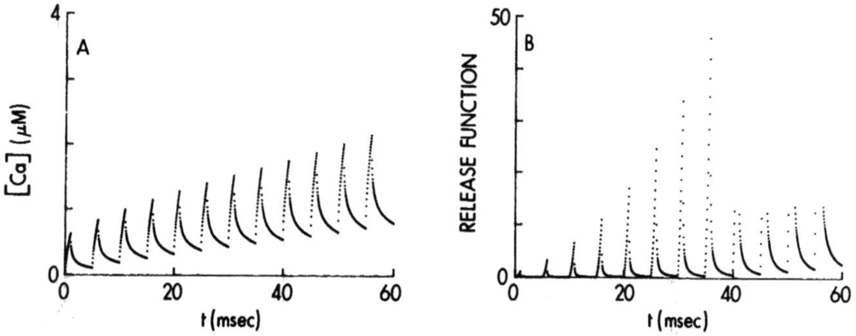

FIGURE 7. Facilitation with saturable sites. As in Fig. 6, panel A shows the intracellular calcium transient at the membrane and panel B shows the release function for a train of impulses 5 msec apart. Much greater facilitation is possible before CaFree fails to become low between impulses. The last few release transients are off scale with the final transient having a peak value of about 120. B_T = 500μM, β = 1000, and k_e = .1cm/sec.

produces unit release by definition) and attenuating the response to [Ca] which falls below the level producing unit release (Fig 6B). However, there are two severe problems associated with such a simple realization of the release function. First, when facilitation begins to approach a value of 15, or [Ca] is doubled, the problem of significant release between impulses reappears. Second, the time course of buildup of facilitation no longer follows a first order lag.

It was unexpected to us that a natural extension of the model to incorporate saturable binding sites potentially allows a maximum facilitation of several hundred before the residual [Ca] level between impulses rises high enough to evoke significant release. Figure 7A shows [Ca] at the membrane in response to a train of impulses as in Fig. 6 but in the presence of 500 µM concentrations of binding sites and strong linear pump. During an impulse entry, Ca quickly saturates the sites close to the membrane so that the latter part of a pulse of entering Ca is free diffusion limited in its ability to increase [Ca] at the membrane. For early impulses the flux of free calcium is very fast because of the steep concentration gradient. For later impulses there is less flux because the free calcium concentration gradient to the "sink" of unsaturated sites is less. Thus the transient riding on the slowly increasing residual [Ca] becomes successively greater. It is clear that a saturable pump can maximize this effect since operation in its velocity limited region would not be as effective as a linear pump in reducing the transient peak but at the low residual [Ca] levels is equivalent to a linear pump.

5. DISCUSSION

This work shows the very important role of diffusion and pumping of calcium in determining the amplitude and time course of transmitter release and facilitation at a synapse; and further defines in considerable detail how "residual calcium" might cause facilitation. Even a moderate concentration of cytoplasmic calcium binding sites so reduces diffusion that many tens of milliseconds are required to achieve a uniform distribution of calcium even within a nerve terminal as small as 1 µm in diameter. A brief (1 msec) influx produces a calcium concentration increase just under the cell membrane which is larger than in the cell interior. We considered the possibility that this localization of calcium near the membrane would enhance its removal by the pump and ran simulations to address this point. Because the pump moves free calcium only, we multiplied its rate by the value of β chosen so as to give an efflux proportional to the total calcium present. We found that, with increasing values of β (bound to free ratio) the time for removal of 50% of the pulsative influx of calcium indeed decreases markedly.

This slowing of diffusion (by binding of calcium to cytoplasmic elements) has important consequences for any intracellular calcium-dependent process. It means that diffusion cannot be ignored if the time course of the process studied is on the order of milliseconds. For example, the marked decay of inward calcium (but not barium) currents in neurons during maintained depolarization has been ascribed to the calcium-dependent inactivation of the channels. Perhaps, a simpler explanation would be that the influx of calcium significantly alters E_{Ca} during the voltage clamp pulse. Such an explanation has in fact been offered several times (26, 18). If that were true, a measure of the reversal potential after a long pulse might be expected to demonstrate the reduction. Unfortunately, difficulties measuring the reversal potential and departures of the calcium current from a linear dependence on E_{Ca} (26) make such a test impractical.

5.1. Our model

The present model is similar to the preceding one (4) in that it uses machinery known to exist in the nerve terminal for the influx of calcium, its binding and diffusion within the cell, the 4th power relationship between free internal calcium and transmitter release, and finally the extrusion of calcium from the terminal. We exclude the contribution of mitochondria or smooth endoplasmic reticulum for the same reasons as Stockbridge & Moore (4). However the present model employs explicit (rather than implicit) treatment of the binding process, allows for its saturation and shows its important consequences.

5.2. Pump rates

The results of our simulations raise the question of accuracy of published values of pump rate constants obtained from experimental measurements of calcium extrusion. The calculated value of the rate constant of a pump is directly proportional to the value of calcium concentration assumed to exist just under the membrane. The simulations of Fig 4 show that as the pump rate is increased, the rate of calcium efflux becomes limited by the rate of diffusion rather than the pump rate.

A second major problem is that the measurement of calcium efflux has very poor temporal resolution because pooled samples are collected over seconds to minutes. The value derived for the maximum flux may be very much underestimated. If at some time a high internal calcium concentration were established and then the pump turned on, the calcium concentration near the cell membrane may be rapidly reduced by a powerful pump. This will induce a gradient, with the concentration higher near the center of the cell and low near the pump at the perimeter, due to slow axial diffusion. But the initial rate measured probably will be in error because both the amount of calcium extruded in the tenth-second time scale is still small, and because of the finite time required to change solutions.

Thus it appears likely that the magnitude of the published values for pump rates are underestimated in cases where there is significant cytoplasmic binding of calcium.

5.3. Facilitation

Previously, two mechanisms have been proposed but have failed to account for facilitation. Broadening of the second nerve terminal spike might result in increases in calcium influx and therefore transmitter release. However, Katz & Miledi (3) were able to produce facilitation at the frog neuromuscular junction when action potentials in the nerve terminal were blocked with TTX and release was stimulated by identical electrical depolarizations. Likewise, this hypothesis was not consistent with the observations of uniformity of the nerve terminal potential at the crayfish neuromuscular junction (27) and the squid giant synapse (28). Furthermore, the idea of successive calcium currents increasing, rather than inactivating, is contrary to the findings in the many preparations in which the calcium current has been studied.

The other proposal to account for facilitation has been the residual calcium hypothesis. Though there is no direct evidence for this idea either, Charlton et al. (12) describe a step-wise increase of the arsenazo III signal in the presynaptic terminal of the squid giant synapse in response to a train of stimuli. This stepwise increase in total intracellular calcium is consistent with both phasic release and facilitation as the calcium load is redistributed toward the center of the terminal (29).

Rather than considering component mechanisms, many previously proposed "models" for facilitation have simply employed mathematical functions to describe the rise and fall of the envelope of junction potential amplitudes in response to a train. Few have attempted to include the constraints of phasic release in conjunction with facilitation.

The plausibility of the residual calcium hypothesis was promoted by Younkin (30) who found that facilitation of endplate potentials could be quite well predicted for a complicated train of impulses based on the fourth power of some intermediate to which successive impulses added equal contributions. The decay of the intermediate, however, was fit with a function which bears no apparent relationship to any physical process or even Mallart & Martin's (7) estimates of decay time constants. Balnave & Gage (31, 32) have proposed, in a less quantitative way, a kinetic model for the release mechanism, but speculate little on the nature of the individual steps.

Rahamimoff (33) has presented a qualitative model for phasic release and facilitation in which a calcium load binds four ions per site in a random manner. A fully loaded site then results in the release of one quantum of transmitter. The model as presented contained no description of the ultimate fate of internal calcium-- no diffusion and no efflux.

Parnas and Segel (34) have similarly modelled facilitation, but not phasic release, by a scheme in which the prime determinant is the nature of efflux and in which diffusion plays no part.

Llinas et al. (14) provide an analysis of phasic transmitter release, but not facilitation, at the squid synapse based on their voltage clamp data of the calcium current in the presynaptic fiber. The model specifically discounts the contribution of diffusion to the time course of transmitter release, based on a diffusion constant for calcium of 10^{-6} cm^2/sec. The current work illustrates that use of a value one to two orders of magnitude lower, a more realistic estimate of the effective diffusion constant for calcium, means that diffusion cannot be dismissed as an important contributor to the time course of transmitter release.

6. SUMMARY

Although a much oversimplified representation of the nerve terminal, this model shows that diffusion plays an important role in the determining the time course of phasic release and facilitation.

The model includes a square-wave influx of calcium, a linear pump for the efflux of calcium, and slow diffusion of calcium within the terminal due to binding of the major portion of the influx to fixed, saturable sites in the cytoplasm. Transmitter release is taken as proportional to the fourth power of the calcium concentration at the surface membrane. The model predicts facilitation, the increased transmitter release in response to a second stimulus for many milliseconds after the initial transmitter release has ceased. This facilitation is similar to that reported at the frog neuromuscular junction in its initial amplitude and time course. The cessation of transmitter release between pulses implies a maximum facilitation of less than 15 with nonsaturable binding sites. Saturable binding permits much greater maximum facilitation. The most serious problem with the model is that the time course of the phasic portion of transmitter release is much faster than observed currents.

ACKNOWLEDGMENT

We are grateful for the support of this work by NIH grants NS 03437, NS11613. Simulations were performed using the facilities of the National Biomedical Simulation Resource.

REFERENCES

1. Katz, B. & Miledi, R. (1967) The timing of calcium action during neuromuscular transmission. J. Physiol. 189: 535-544.
2. Katz, B. & Miledi, R. (1965) The effect of calcium on acetylcholine release from motor nerve terminals. Proc. R. Soc. B. 161: 491-.
3. Katz, B. & Miledi, R. (1968) The role of calcium in neuromuscular facilitation. J. Physiol. 195: 481-492.
4. Stockbridge, N. & Moore, J.W. (1984) Dynamics of intracellular calcium and its possible relationship to phasic transmitter release and facilitation at the frog neuromuscular junction. J. Neurosci. 4: 803-811.
5. Takeuchi, A. & Takeuchi, N. (1959) Active phase of frog's endplate potential. J. Neurophysiol. 22: 395-411.
6. Magleby, K. L. & Stevens, C.F. (1972) The effect of voltage on the time course of end-plate currents. J. Physiol. (Lond.) 223:151-171
7. Mallart, A. & Martin, A.R. (1967) An analysis of facilitation of transmitter release at the neuromuscular junction of the frog. J. Physiol. 193: 679-694.
8. Miledi, R. (1973) Transmitter release induced by injection of calcium ions into nerve terminals. Proc. R. Soc. B. 183: 421-425.
9. Rahamimoff, R., Meiri, H., Erulkar, S.D., & Barenholz, Y. (1978) Changes in transmitter release induced by ion-containing liposomes. Proc. Natl. Acad. Sci. U.S.A. 75: 5214-5216.

10. Charlton, M.P., Smith, S.J. & Zucker, R.S. (1982) Role of presynaptic calcium ions and channels in synaptic facilitation and depression at the squid giant synapse. J. Physiol. 323: 173-193.
11. Dodge, F.A. Jr. & Rahamimoff, R. (1967) Cooperative action of calcium ions in transmission release at the neuromuscular junction. J. Physiol. 193: 419-432.
12. Andreau, R. & Barrett, E.F. (1980) Calcium dependence of evoked transmitter release at very low quantal contents at the frog neuromuscular junction. J. Physiol. 308: 79-97.
13. Llinas, R., Steinberg, I.Z. & Walton, K. (1981) Presynaptic calcium currents in squid giant synapse. Biophys. J. 33: 289-322.
14. Llinas, R., Steinberg, I.Z. & Walton, K. (1981) Relationship between presynaptic calcium current and postsynaptic potential in squid giant synapse. Biophys. J. 33: 323-352.
15. Baker, P.F. (1976) Regulation of intracellular calcium and magnesium in squid axons. Fedn. Proc. 35: 2589-2595.
16. Hodgkin, A.L. & Keynes, R.D. (1957) Movements of labelled calcium in squid giant axon. J. Physiol. 138: 253-281.
17. Blaustein, M.P. & Hodgkin, A.L. (1969) The effect of cyanide on the efflux of calcium from squid axons. J. Physiol. 200: 497-527.
18. Gorman, A.L.F. & Thomas, M.V. (1980) Intracellular calcium accumulation during depolarization in a molluscan neurone. J. Physiol. 308: 259-285.
19. Brinley, F.J., Jr., Tiffert, T., Scarpa, A., & Mullins, L.J. (1977) Intracellular calcium buffering capacity in isolated squid axons. J. Gen. Physiol. 70: 355-384.
20. Baker, P.F., Hodgkin, A.L., & Ridgeway, E.B. (1970) Two phases of calcium entry during the action potential in giant axons of Loligo. J. Physiol. 208: 80P.
21. Baker, P.F. (1972) Transport and metabolism of calcium ions in nerve. Prog. Biophys. Molec. Biol. 24: 177-223.
22. Mullins, L.J. (1976) Steady-state calcium fluxes: membrane versus mitochondrial control of ionized calcium in axoplasm. Fedn. Proc. 35: 2583-2588.
23. DiPolo, R. & Beaugé, L. (1979) Physiological role of ATP-driven calcium pump in squid axon. Nature, (Lond.) 278: 271-273.
24. Blaustein, M.P. (1976) The ins and outs of calcium transport in squid axons: internal and external ion activation of calcium efflux. Fedn. Proc. 35: 2574-2578.
25. Katz, B. & Miledi, R. (1965) The effect of temperature on the synaptic delay at the neuromuscular junction. J. Physiol. 181: 656-670.
26. Hagiwara, S. & Byerly, L. (1981) Calcium channel. Ann. Rev. Neurosci. 4: 69-125.
27. Zucker, R.S. & Lara-Estrella, L.O. (1979) Is synaptic facilitation caused by presynaptic spike broadening? Nature, Lond. 278: 57-59.

28. Charlton, M.P. & Bittner, G.D. (1978) Presynaptic potentials and facilitation of transmitter release in the squid giant synapse. J. Gen. Physiol. 72: 487-511.
29. Zucker, R.S. & Stockbridge, N. (1983) Presynaptic calcium diffusion and the time courses of transmitter release and synaptic facilitation at the squid giant synapse. J. Neurosci. 3: 1263-1269.
30. Younkin, S.G. (1974) An analysis of the role of calcium on facilitation at the frog neuromuscular junction. J. Physiol. 237: 1-14.
31. Balnave, R.J. & Gage, P.W. (1974) On facilitation of transmitter release at the toad neuromuscular junction. J. Physiol. 239: 657-675.
32. Balnave, R.J. & Gage, P.W. (1977) Facilitation of transmitter secretion from toad motor nerve terminals during brief trains of action potentials. J. Physiol. 266: 435-451.
33. Rahamimoff, R. (1974) Modulation of transmitter release at the neuromuscular junction. The Neurosciences Third Study Program, F.O. Schmidt & F.G. Worden, eds., MIT Press, pp 943-952.
34. Parnas, H. & Segel, L.A. (1980) A theoretical explanation for some effects of calcium on the facilitation of neurotransmitter release. J. Theor. Biol. 84: 3-29.

DISCUSSION OF THE ARTICLE BY J.W. MOORE (U.S.)

R.S. Zucker (U.S): I have done similar kinds of models for facilitation at the squid giant synapse and I have gotten many results similar to yours, reported at the last IUPS and Biophysical Society Meetings (Zucker, R.D. (1984). A calcium diffusion model predicts facilitation, but not the time course of transmitter release, during tetanic stimulation. Biophys. J. 45, 164a). For example, If you use 100 spikes, even for saturable binding the accumulation of calcium was too high. I noticed that at the very end of your tetanus also there was too much calcium. How can you account for that failure of the model. The way I account for it is by going to a different model which is a series of punctate entries of calcium through descrete sites and looking near calcium channels. At descrete sites, the peak calcium is much higher and calcium diffuses much faster away from the single channel because diffusion is initially three dimensional. Do you know of any other way to get around this problem that you have in your tetanus?

J.W. Moore (U.S.): No. I tried to point out that on one hand, it may sound like I am interested in developing a particular model to describe in particular a phenomenon. However, here I am simply trying to look at what different things this simple model can do.

H. Parnas (Israel): If you suggest that the residual calcium is somewhat away from the membrane, then it has to affect the diffusion of a second or third pulse which is given. Did you take this into account? If the diffusion is slower in the second pulse because of some residual calcium away from the membrane, it should prolong the time course of this second or third pulse. Was this considered in your simulations.

J.W. Moore (U.S.): My intuition is very poor, even more so than my mathematical ability. Because I cannot keep all these things in my head I handed it to the computer for simulation. I simply write down the equations (simple differential equations), I put these into a computer program, ask the computer to tell the answer to the question to which I know the answer, a special case. Then and only then, I'll accept the program as giving me answers that might be reasonable otherwise. To your specific question, no the rules have not changed for the second or later pulses.

H. Parnas (Israel): If indeed diffusion affects so strongly the pump, or for example, the sodium-calcium exchanger, then it seems to me very difficult how we and Rami Rahamimoff obtained such a huge increase in the duration of facilitation by reducing extracellular sodium or modifying extracellular calcium or magnesium?

J.W. Moore (U.S.): This is certainly another test that it fails. Because this simplistic model works as well as it does, improvements seem to be worthwhile. We are planning to incorporate a sodium-calcium exchange pump in the next version.

R.S. Zucker (U.S.): Actually, this model shows that the late phase of facilitation is sensitive to pump rate, so the model succeeds in simulating these results.

I. Parnas (Israel): The validity of a model is not to simulate one condition. Once you find experimental results which contradict the model, you have to modify the model in order to account for the results or you disregard the model. Did you try to simulate with the computer, with the same equations and same parameters (unless you want to change parameters for every condition), for example, can you account for increase in duration of facilitation with an increase in calcium concentration? Since you measured release, can you account for the lack of change of synaptic current or kinetic of release in different calcium concentrations? Can you predict the delay histograms of release and show they do not depend on changes in extracellular calcium or the pulse of calcium that you've given and so forth? If you can't account for these results, you certainly can adjust a line to the facilitation curve of Martin & Mallart curve, but it will not account for the other results.

J.W. Moore (U.S.): There is an ambivalence in trying to talk about this because on one hand, it looks like one is trying to fit certain experimental data. On the other hand, one of my major goals is to try and gain insight into how the system works. The thing that is impressive to me is that with what I consider the most stripped-down of mechanisms, one cando as much as I have presented. I noted one or two omissions of known mechanisms from the model. There are others which I present in the paper and there are other tests that need to be performed. To reiterate, I don't want to defend the model but only point out how much the stripped-down model can predict.

M.P. Blaustein (U.S.): I found your calculations are very interesting. In terms of dimensions: is what you are really talking about, the squid synapse?

J.W. Moore (U.S.): No. The dimension is 1-2 micrometers diameter, appropriate for the frog neuromuscular junction. On the scale that you saw, it took milliseconds for the concentration to move over that dimension. If one is talking about the squid axon, it would take "years" to achieve a uniform distribution.

M.P. Blaustein (U.S.): There are a number of problems that come up with pump rates. One of the things with particular regard to the sodium-calcium exchange is that I think the rates are going to change dramatically. For example, I would expect to see some reduction in the concentration gradients of Na and Ca across the membrane during activity; the altered or reversed gradients would alter the transport rates.

J.W. Moore (U.S.): We didn't get to do it yet but plan to include the sodium-calcium exchange in our next version of the model.

R. Rahamimoff (Israel): That is very nice that in addition to the two parameters that were considered mainly up to now - concentrations and time, you introduced the parameter of space. I think that is extremely important. I wouldn't be very bothered at the moment by the fact that it doesn't predict properly potentiation, because we know that in potentiation we have some additional cellular processes that may be affected and they will have to be eventually incorporated in the model.

J.W. Moore (U.S.): Thank you.

CALCIUM COOPERATIVITY IN CALCIUM ENTRY AND CALCIUM ACTION, AND ITS IMPLICATIONS WITH REGARD TO FACILITATION, AT THE MOUSE MOTOR NERVE TERMINAL

D. M. J. QUASTEL and D. A. SAINT

1. INTRODUCTION

It is now well established that the coupling of neurotransmitter release to depolarization of nerve terminals depends upon influx of Ca^{2+} from the external medium via voltage-regulated cation-selective channels (reviews, 1, 2, 3). At the vertebrate neuromuscular junction this dependence is manifest in essentially the same way for release provoked by nerve impulses (i.e., end-plate potentials, EPPs) or release manifest as a raised frequency of miniature end-plate potentials (fmepp) induced by intermittent and graded depolarization of nerve terminals by extrinsically-applied currents or by steady depolarization of nerve terminals using raised K^+ (4, 5, 6). The observation that EPP height can be proportional to the fourth power of the external Ca^{2+} (7, 8) has led to the proposition that four atoms of Ca^{2+} are required for the release of each package or "quantum" of neurotransmitter (8). In the first part of this paper there will be presented three independent lines of evidence indicating that this high apparent cooperativity is compounded of two elements: (a) a cooperativity of 2 for entry of Ca^{2+} into the nerve terminal and (b) a cooperativity of about 2 (at most 3) for transmitter release by internal Ca^{2+}.

In addition to the main effect of a nerve impulse, whereby transmitter release is raised transiently, for about 1 msec, to a rate about 10^5 times the spontaneous, there exists a process generally termed facilitation or potentiation, whereby a nerve impulse causes an increase in release by subsequent nerve impulses (9, 10, 11); the facilitation is paralleled in time course and intensity by the frequency of MEPPs (e.g. 12, 13, 14). Since the demonstration by Katz and Miledi that the presence of external Ca^{2+} is required for facilitation (15), it has become usual to attribute facilitation to persistence of "active Ca^{2+}" within the nerve terminal, a view strengthened by the observation that at the squid giant synapse Ca^{2+}

does indeed persist within the nerve terminal for several seconds (16, 17). Data will be presented indicating that the "residual-Ca^{2+}" model for facilitation makes false predictions with regard to the rate of transmitter release between and subsequent to facilitated EPPs, and with regard to the gradation of facilitation with the intensity and/or duration of brief extrinsic depolarizations of the nerve terminal.

2. THEORY

2.1. Relation between transmitter release rate and external Ca^{2+}

A rather flexible and general statement of the Ca^{2+} hypothesis for transmitter release is that of Parnas and Segel (18), which, with minor changes of nomenclature, can be written as follows:

$$L = \lambda/(1 + K_\lambda/C^n) \qquad (1)$$

$$C = C_0 + E \; ; \qquad E = \varepsilon/(1 + K_\varepsilon/C_e^r) \qquad (2)$$

where L is transmitter release rate, C is internal Ca^{2+} concentration at certain critical sites in the nerve terminal (i.e. "active Ca^{2+}"), and C_e is extracellular $[Ca^{2+}]$. C has a resting level, C_0, in the absence of external Ca^{2+} (and in the absence of nerve terminal depolarization), and increases to $C_0 + E$ when Ca^{2+} enters. The term E is governed by ε (which rises with depolarization, i. e. , incorporates Ca^{2+} channel gating) and by the amount of external Ca^{2+} bound to certain membrane sites, with cooperativity r. Maximum possible L is λ; this occurs when internal C becomes large relative to K_λ, while the maximum L obtainable by raising C_e is given by equation (1) with $C = C_0 + \varepsilon$. The terms n and r are Hill coefficients for cooperativity in action of internal Ca^{2+} and in Ca^{2+} entry respectively.

These equations give rise to the close approximation (18):

$$Sl_{max} = n\ r\ (S-1)/(S+1), \text{ where } S = (1 + \varepsilon/C_0)^{1/2} = K_\varepsilon/\gamma^r \qquad (3)$$

Here Sl_{max} denotes the maximum slope of log L vs. log C_e and γ is the concentration of external Ca^{2+} where this maximum occurs. This model successfully accounts for the finding that with increasing depolarization there is a progressive increase in the maximum slope of log release rate (L) vs. log C_e (as ε, and therefore S, increases) with a progressive reduction of γ, as observed by Cooke, Okamoto and Quastel (4). It is important to note that the maximum possible Sl_{max} is n.r, the product of the two cooperativity terms, and the observation that the maximum slope of log EPP vs. log C_e is about 4, and no more than 5 (4, 8) constrains n, if $r > 1$. If $r = 2$, n is no more than 3. However, if $r = 1$, n might be large, since if release saturates because of limited ϵ, $S = (\text{maximum } L/\text{minimum } L)^{1/2n}$; Sl_{max} is limited by the ratio of maximum to minimum L, rather than by n.

2.2. Facilitation

Quantitatively the residual Ca^{2+} hypothesis can be phrased as follows. A nerve impulse injects a set amount of Ca^{2+} into the nerve terminal and transiently increases C from C_o to $C_o + E$, where E now denotes the appropriately weighted average of the increment of "active Ca^{2+}" at the time of heightened release. C then declines towards C_o, but at any time after one or a series of impulses there is a residuum, say G. A test nerve impulse now increases C to $C_o + G + E$. Now we may rewrite equation (1) as:

$$R = \Delta t\ \lambda' / [1 + K'_\lambda / (C_o + E + G)^n]$$

$$L = \lambda / [1 + K_\lambda / (C_o + G)^n]$$

where R is release by the nerve impulse, λ' and K'_λ are the values of λ and K_λ appropriate to a nerve impulse (i.e., either might be transiently altered by the nerve impulse) and Δt is the duration of the high release rate produced by the nerve impulse. At the same time, the frequency of MEPPs is L.

By differentiating the above expressions it may easily be shown that for any G the multiplication of R is always less than the multiplication of L and the difference between the multiplying factors is minimized when C is small relative to K_λ or K'_λ. In view of this result, it is possible to pursue the mathematics considering only cases where $C \ll K'_\lambda$, i.e., release is far less than maximal. This corresponds to facilitation observed when EPPs are made small by low C_e and/or high external Mg^{2+}, which is the usual experimental situation at the vertebrate end-plate. The resulting equations are much the same as those considered by others (e.g. 13, 19):

'spontaneous' (resting) fmepp $= L_o = k_L\, C_o^n$

release (EPP) by 1st impulse $= R_1 = k_R\, (C_o + E)^n\, \Delta t$

fmepp before test impulse $= L_t = k_L\, (C_o + G)^n$

release (EPP) by test impulse $= R_t = k_R\, (C_o + E + G)^n\, \Delta t$

where k_L and k_R, both constants, are λ/K_λ and λ'/K'_λ, respectively. Now,

$$L_t/L_o = (1 + G/C_o)^n > (1 + G/E)^n > R_t/R_1 = [1 + G/(C_o+E)]^n$$

With high n, of course, the E giving rise to EPPs may be of the same order as C_o, and R_t/R_1 (which may be close to L_t/L_1) will grow with C_e. However, if n is not high then for EPPs of quantal content of about 1 or more, $E \gg C_o$ and R_t/R_1 is dependent only on the ratio of G to E and not on the absolute magnitude of either; this agrees with the experimental observation that facilitation is essentially independent of external Ca^{2+}, provided R is far from maximal and the same C_e is present for conditioning and test nerve impulses (12, 15, 20). Moreover, since G/E depends only on the time course of disposal of Ca^{2+} within the nerve terminal, it predicts that for direct EPPs produced by focal nerve terminal depolarization, in the presence of tetrodotoxin to block nerve impulse generation, facilitation should be independent of the intensity and duration of the depolarizing pulse, provided only that the direct EPP is of the same order of magnitude as EPPs evoked by nerve stimulation in low Ca^{2+}/raised Mg^{2+}.

Another testable feature of this model is that the residual Ca^{2+} underlying a facilitation of R should also manifest itself by causing transmitter release between EPPs, to an extent that depends upon the value of n. From the above equations,

$$G/(C_o + E) = (R_t/R_1)^{1/n} - 1 = A$$

$$(C_o + E)/C_o = [(R_1/L_o).(k_L/k_R)/\Delta t]^{1/n} = B$$

and, hence, $L_t/L_o = (1 + G/C_o)^n = (1 + A.B)^n$

When external Ca^{2+} (and therefore E) is zero $k_R/k_L = (R/\Delta t)/L$. Experimentally, in very low external Ca^{2+} there is no discernible increase in release rate at the time an EPP is normally observed (e.g. 4), and therefore k_R/k_L must be near unity and is certainly no more than 10. Moreover, for EPPs (and "direct" EPPs evoked by brief nerve terminal depolarizations), Δt can be taken to be 1 ms (19). Thus, for any n, a predicted L_t can be obtained using $k_R/k_L = 1$; for a larger k_R/k_L dividing this estimate by k_R/k_L gives a fair approximation for L_t. The equations, of course, give L_t only at the time of a test (facilitated) EPP, but in general G will be further increased by the test EPP, and the predicted L_t will represent a minimum for the time period just following the test EPP.

It should be emphasized that the above equations are quite independent of the time course of residual Ca^{2+}, and of whether or not the removal of Ca^{2+} from the active area by diffusion and binding is a linear process.

However, it is of some importance that when facilitation is analysed on the basis of time course into several distinct multiplicative components (13, 14, 21, 22), both L (MEPP frequency) and R (EPP amplitude) show the same components, with similar time course (12, 13). This observation effectively excludes the argument that spontaneous release of normal quanta is not governed by the same mechanisms as release evoked by nerve impulses. However, there is no difficulty in obtaining predicted L_t on the assumption that true C_0 is much lower than appears from the observed spontaneous fmepp (L_0); one merely uses the above equations with any arbitrarily low L_0 and adds true L_0 to the predicted L_t.

3. METHODS

All experiments were performed using conventional recording with intracellular microelectrodes, from mouse diaphragm in vitro, at 25-29°C, using a system for continuous local superperfusion that allowed altering the bathing solution at superficial end-plates within a few seconds. This method, the composition of bathing solutions, and the method for focal polarization of nerve terminals have been described in full elsewhere (23). Transmitter release was evaluated in terms of frequency of miniature end-plate potentials (MEPPs) or amplitude of EPPs, recorded using a Mingograf ink-jet recorder, or digitally, using a PDP-12 computer.

4. RESULTS

4.1. Evaluation of "r" and "n"

4.1.1. Method 1. This method rests on obtaining r from the observed modulation of transmitter release by external Ca^{2+}, at constant nerve terminal depolarization and/or with nerve impulses. That this might be possible arises as follows.

Rearrangement of equation (1) gives:

$$C/K_\lambda^{1/n} = [L/(\lambda-L)]^{1/n}$$

That is, for any arbitrarily chosen λ and n, one can obtain a value for $C/K_\lambda^{1/n}$ for each observed L (transmitter release rate). $C_0/K_\lambda^{1/n}$, of course, corresponds to L_0, the value of L when external Ca^{2+} is zero. At the same time, from equation (2),

$$d \ln E \;/\; d \ln C_e = r/(1 + C_e^r/K_\varepsilon)$$

Therefore, the limiting slope, at low C_e, of ln E vs ln C_e, which is the slope of $\ln [(C - C_0)/K_\lambda^{1/n}]$ vs ln C_e, gives r. In Fig. 1 are

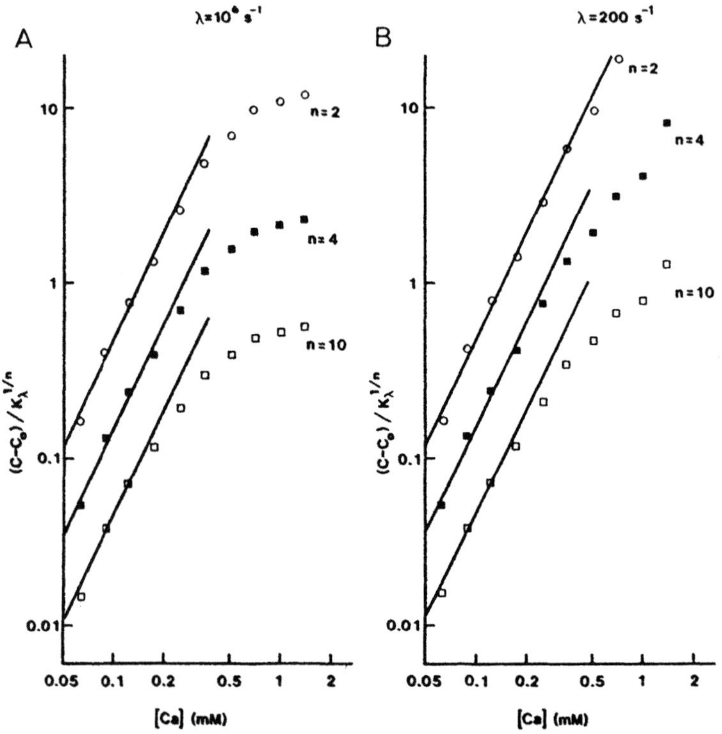

FIGURE 1. Log - log plots of calculated $(C - C_0)/K_\lambda^{1/n}$ vs C_e, from a junction in 20 mM K^+. The original data are shown in ref (4), fig. 4. The lines are drawn with a slope of 2. Note that all plots show a limiting slope ≅ 2 at low C_e, whatever the choice of n or λ.

shown graphs of $\log (C - C_0) - (\log K_\lambda)/n$ vs $\log C_e$, with n = 2, 4, or 10, and with $\lambda >> L_{max}$ or λ close to L_{max}, for an experiment in which fmepp (L) was observed at various C_e at an end-plate equilibrated in 20 mM K^+. It is evident that the limiting slope and therefore r, is at least 2, whatever values are assigned to λ and n, provided only that $\lambda > L_{max}$ and n is positive. A similar result is obtained for each and all of the sets of data obtained by Cooke, et al (4), including those relating EPP height and C_e. These data could generally be fitted closely to the empirical formula

$$\ln (L/L_0) = \beta/[\,1 + \gamma^2(C_e^{-2} + \delta C_e^{-1})]$$

where β, γ, and δ are fitting parameters, with the term δC_e^{-1} contributing little at $C_e < 0.5$ mM. It can also be shown analytically that for any set of points generated by equation (1) and (2), a fit to $\ln(L/L_0) = \beta/[1 + (\gamma/C_e)^\theta]$ implies that $r \geq \theta$. The empirical observation that $\theta = 2$ implies $r \geq 2$.

4.1.2. Method 2. An independent indication of Ca^{2+} cooperativity for entry into the nerve terminal is provided by study of the inhibitory action on transmitter release of neomycin (24), an aminoglycoside antibiotic containing a large sugar moiety. In experiments using raised K^+ to depolarize nerve terminals, we have found that with neomycin there is no alteration in the maximal fmepp obtained by raising external Ca^{2+} (C_e), and that curves of fmepp vs. C_e are shifted in a typically competitive manner. With neomycin (N) as a competitor for Ca^{2+}, one would expect, from elementary receptor kinetics, equation (2) to become

$$E = \varepsilon/[1 + (K_\varepsilon/C_e^r)(1 + [N]/K_N)]$$

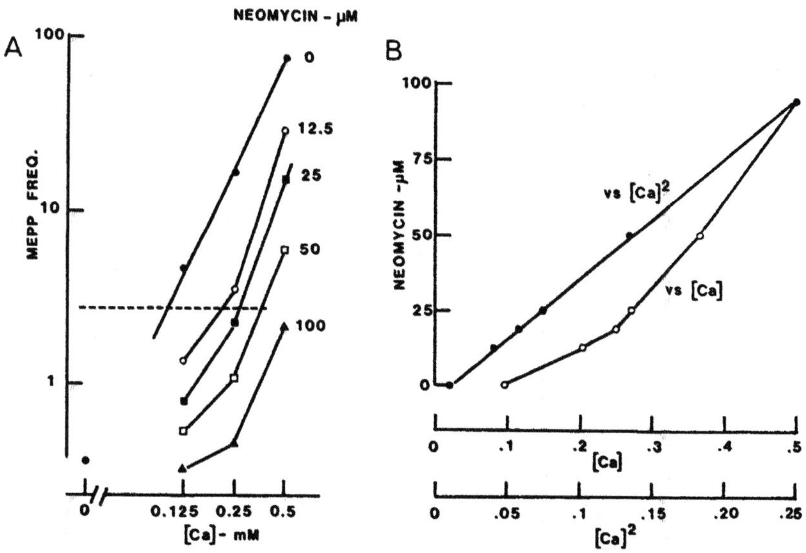

FIGURE 2. (A) Graphs of fmepp vs external $[Ca^{2+}]$ in the presence of 15 mM K^+, at various concentrations of neomycin,. Each point is the geometric mean of fmepp at about 20 junctions, with s.e.m. about 20 per cent. (B) Graphs of [neomycin] vs $[Ca^{2+}]$ and vs $[Ca^{2+}]^2$ for equal fmepp (dashed line in (A)). Some points in these graphs were obtained from plots of fmepp vs [neomycin] at various $[Ca^{2+}]$.

if only one molecule of neomycin can combine with the receptor, with dissociation constant K_N. Rearranging this equation gives

$$[N] = K_N\ [C_e^r\ (\varepsilon/E-1)/K_\varepsilon - 1]$$

That is, for values of C_e and [N] giving the same transmitter release rate (same E) a graph of [N] vs. C_e^r should be linear. As shown in Fig. 2, data show that such linearity is obtained with r set at 2, but not with r = 1.

4.1.3. Method 3. With this method we seek more directly to determine n, the cooperativity for Ca^{2+} at the internal sites governing transmitter release. Our first indication that this might be possible came from the observation that the irreversible increase in fmepp produced by La^{3+} develops at a rate that depends upon nerve terminal depolarization (raised K^+, trains of nerve impulses, or focally applied current), suggesting that La^{3+} acts after entering the nerve terminal via voltage-gated channels. The rate of growth of fmepp should then provide a measure of channel opening. However the irreversibility of the increase in fmepp with La^{3+} precluded use of this ion for quantitative studies. Evidently what was needed was an agent which would enter the nerve terminal via Ca^{2+} channels, there induce transmitter release (either directly or via local release of Ca^{2+}), and persist only long enough that the fmepp just after a brief series of nerve impulses or a short stint of nerve terminal depolarization would provide an index of total entry, and therefore of the average Ca^{2+} channel opening during the test period. All these requirements appeared to be met by Ba^{2+}, which produces after a series of nerve impulses an after-discharge or "tail" of raised fmepp that is graded in amplitude with the number of nerve impulses and declines in a matter of seconds (25). We have been able to show that the tail depends upon the presence of extracellular Ba^{2+} only during the period of nerve terminal depolarization or tetanic stimulation, and the tail is unaffected by the frequency at which the nerve is stimulated, if a fixed number of stimuli are given within a time period that is short compared to the apparent time constant of Ba^{2+} disposal (about 5 sec).

That Ba^{2+} should enter the terminal via Ca^{2+} channels was to be expected (3, 26), and in keeping with this we have found that the high fmepp found with raised K^+, in the presence of Ba^{2+}, is just as sensitive to neomycin or Mg^{2+} as is the high fmepp produced by Ca^{2+}.

In Fig. 3 are shown examples of the response of fmepp to focal depolarization of the end-plate in the presence of Ba^{2+} and absence of Ca^{2+}; fmepp grows slowly with a time course similar to that of the decay of fmepp after the pulse, and the fmepp just after termination of the pulse is essentially the same as that just before. This behaviour contrasts with the rapid rise and fall of fmepp, and brief, attenuated tails, observed with such depolarizations in the presence of Ca^{2+} and absence of Ba^{2+} (see fig 3 and refs 23 and 27). Thus the behaviour of fmepp is consistent with Ba^{2+} entry via voltage-gated Ca^{2+} channels, accumulation and disposal

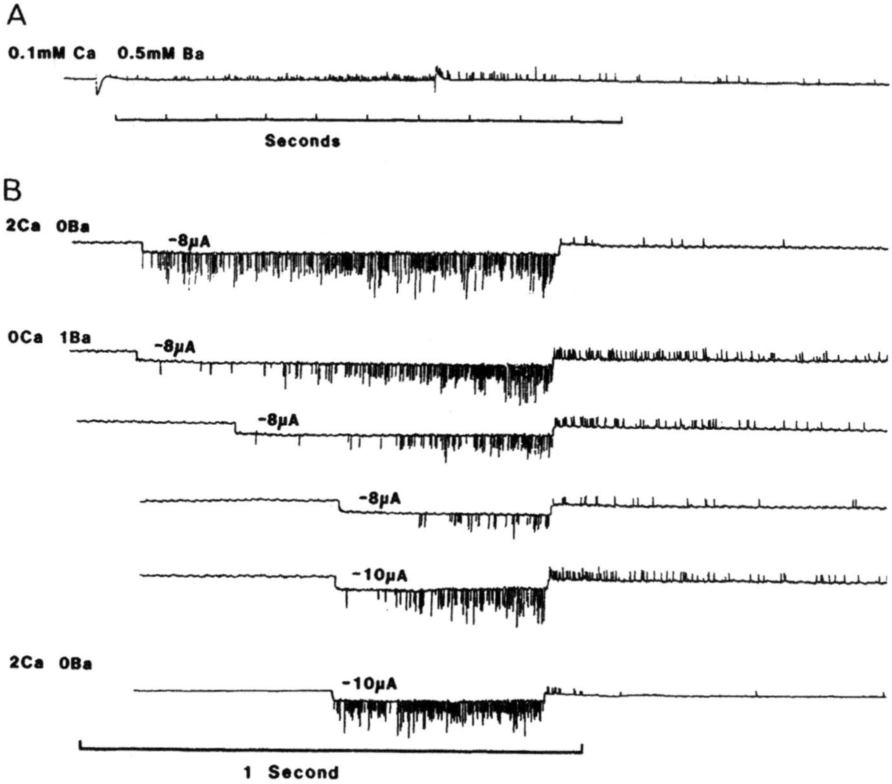

FIGURE 3. Focal depolarization of end-plates in the presence of tetrodotoxin and Ca^{2+} or Ba^{2+} (5 mM K^+). (A) Note slow growth of fmepp during pulse (between artifacts) and subsequent slow decline, characteristic of Ba^{2+}. (B) Modulation of fmepp with pulse intensity and duration. All records from one end-plate. About one minute was allowed for equilibration after each switch of the bathing solution. MEPPs were inverted during pulses because of local field at the end-plate. Most of the 'stimulus artifact' has been subtracted.

of Ba^{2+} from its site of action within the nerve terminal having a time constant of several seconds.

The one assumption that we must now make is that the rate of Ba^{2+} entry is constant over a period of depolarization from about 100 msec to 1 sec. Support for this (not illustrated) comes from our observation that the relation between tail amplitude and duration of steady nerve terminal depolarization is the same as that between tail amplitude and duration of tetanic stimulation, for periods less than about 1 sec, and is the same as that between tail amplitude and number of nerve impulses in a set time period.

Now it can be seen in Fig. 3 that in 1 mM Ba^{2+}, a 420 msec 10 μA depolarizing pulse caused an after-discharge that was much the same as that produced by an 8 μA pulse applied for 820 msec. Thus, in this example average Ba^{2+} entry rate for the 10 μA pulse must have been about twice as much as for the 8 μA pulse.

In Fig. 4 are shown data from a more complete experiment of this type. To begin, focal polarizations were applied in the presence of Ca^{2+} and absence of Ba^{2+} to obtain the slope of ln fmepp vs applied current. Then, the bathing solution was changed to one containing Ba^{2+} and low Ca^{2+}, to obtain the relation between afterdischarge (estimated simply as number of MEPPs in the period 0.2-2.2 sec after the pulse), pulse duration, and pulse amplitude. The graph of pulse duration vs pulse amplitude, for the same after-discharge, shows a linear relation for ln duration vs. depolarizing current, with a slope of -0.37/μA. That is, entry rate for Ba^{2+}, and presumably Ca^{2+} channel opening, was increased exponentially by focal depolarization, with an e-fold increase per 2.8 μA. With the previously determined slope of ln fmepp vs.depolarizing current of 0.72/μA, this gives a slope of d ln fmepp /d ln (channel opening) of 0.72/0.37 = 1.9. This is close to n (actually it is n $(1-C_0/C)$), if we assume that steady state "active Ca^{2+}" grows linearly with the rate of Ca^{2+} entry; n is less if Ca^{2+} disposal mechanisms in the nerve terminal appreciably approach saturation. The experiment was concluded by switching to 0 Ba^{2+}, 0.5 mM Ca^{2+}, and verifying that polarizing currents caused the same increases in fmepp as previously.

To circumvent the problem of the polarizing electrode drifting away from the nerve terminal, which often occurred with this kind of experiment, we have also done experiments with enough Ca^{2+} and Ba^{2+} present simultaneously to allow the slope of ln release rate vs. depolarization for Ca^{2+} to be obtained from the fmepp achieved within a few msec of depolarization, or from "direct" EPPs produced by brief (about 1 msec) depolarizations, early in a train, the after-discharge again serving as an index of accumulated Ba^{2+} and therefore of average channel opening. In all cases the derived values of n $(1-C_0/C)$ have been between 1.5 and 2.3, giving n probably close to 2, and at most 3.

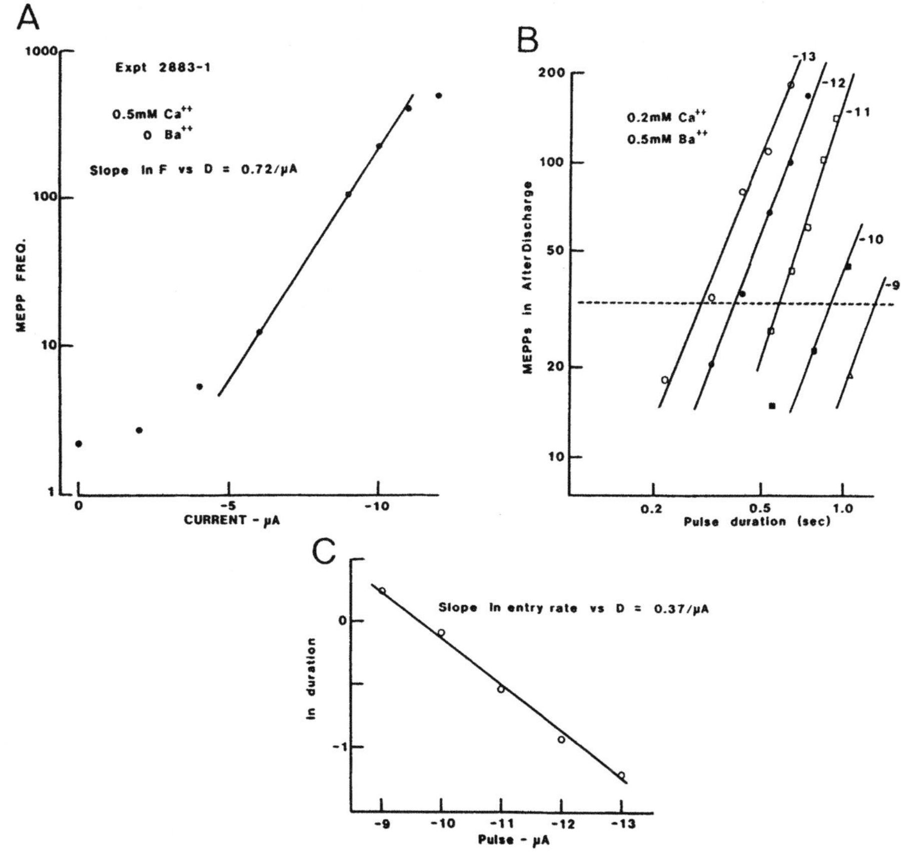

FIGURE 4. Data all from one junction. (A) Semi-log plot of fmepp vs focal polarizing current in the presence of Ca^{2+} and no Ba^{2+}. (B) in the presence of Ba^{2+}, graphs of numbers of MEPPs at 0.2-2.2 sec after pulse, vs pulse duration, at various current intensities. (C). Semi-log plot of pulse duration vs pulse intensity for constant after-discharge, obtained from graphs in B, yields the slope of ln channel opening vs pulse intensity.

4.2. Facilitation

As described under THEORY the residual Ca^{2+} hypothesis for facilitation makes two readily testable predictions, namely;

1) Per cent facilitation with "direct" EPPs should be independent of the intensity and duration of the pulses evoking the EPPs, provided the EPPs are within the range of quantal content where facilitation is independent of external Ca^{2+}.

2) The persisting "active Ca^{2+}" should induce transmitter release at a rate predictable, for any chosen n, from the amplitude of the EPPs.

As shown in Fig. 5 and tabulated in Table 1, facilitation with direct EPPs does not conform to the first prediction. In this experiment the degree of facilitation, seen either as the ratio of quantal content of second EPP to first EPP (m_2/m_1), or as third EPP to second EPP (m_3/m_2), was clearly a function of pulse intensity and duration. Moreover, facilitation was increased by concurrent maintained hyperpolarization, so that very different degrees of facilitation could be obtained with pulses giving (for the first EPP in the train) much the same transmitter release.

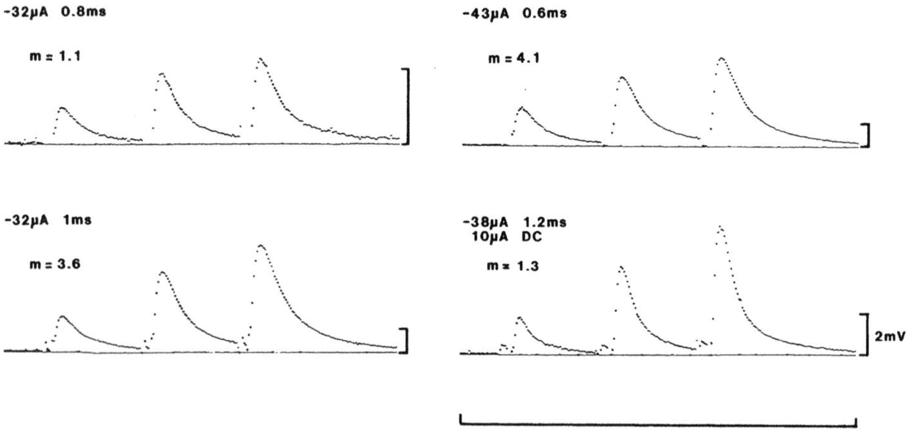

FIGURE 5. Direct EPPs recorded and averaged digitally, scaled to emphasize variation of facilitation with stimulus parameters. All records from one end-plate. Trains of three pulses at 80 Hz were applied every 2 sec. Each trace represents the average of 40 records, and the "stimulus artifact" has in each case been subtracted digitally. Experiment no. 3883-1

Secondly, predicted values of fmepp (L_t) at the time of applying the third pulse, calculated from the observed facilitation of the third EPP (Table 1, k_R/k_L set at 1), with n = 2 or 3, are very much higher than those observed before and after the third pulse. In this experiment the response to each pulse train was examined separately; fmepp was poorly defined because of the paucity of MEPPs but at no time can it have been more than 10/sec. To make the observed fmepp fit the model with n = 3 would require k_R/k_L of about 100, which is much higher than consistent with other data (see THEORY). With n = 4, on the other hand, predicted fmepp falls within the bounds set by the data if k_R/k_L is as low as 15; if we suppose that spontaneous fmepp does not represent C_o and simply adds to the transmitter release produced by "active Ca^{2+}", k_R/k_L could be as low as 3. Of course, if n is chosen sufficiently high, predicted L_t can be made to match that observed with k_R/k_L = 1.

TABLE 1 Facilitation of direct EPPs

Pulse parameters							
current		µA	-32	-43	-28	-32	-38
Δt		msec	0.8	0.6	1.2	1.0	1.2
DC		µA	--	--	--	--	+10
Direct EPP		m_1	1.1	4.1	5.2	3.6	1.3
Facilitation		m_2/m_1	1.71	1.85	2.03	2.17	2.31
		m_3/m_2	1.20	1.28	1.31	1.33	1.48
Predicted fmepp at time of third pulse.	(a)	n = 2	244	1279	2185	1866	1016
		n = 3	62	274	483	450	300
		n = 4	26	86	146	146	117
	(b)	n = 2	207	1191	2069	1759	938
		n = 3	23	153	300	275	170
		n = 4	3	15	32	32	23

Bathing solution contained 1 mM Ca^{2+}, 1 mM Mg^{2+}, 0.5 µM tetrodotoxin and 1 mM 4-aminopyridine. Trains of three focal depolarizing pulses were applied at 80 Hz, every 2 sec. Each value of quantal content (m) is from 40 trials. The spontaneous fmepp (L_o) was 1.6/sec and fmepp after the train always < 10/sec. Predicted fmepp at time of third pulse were calculated (a) assuming L_o represents C_o and (b) assuming C_o negligible and L_o adds to fmepp produced by residual Ca^{2+}. Experiment no. 3883-1.

5. DISCUSSION

The results presented above indicate that the high apparent cooperativity for Ca^{2+} in release of neurotransmitter, at the neuromuscular junction (4, 8, 28) is compounded of two distinct cooperativities. The entry cooperativity would seem to be 2, on the basis of calculations from the observed dependence of transmitter release on external Ca^{2+} (method 1), and from the competition between Ca^{2+} and neomycin (method 2). The latter method is independent of the form of the functional dependence of release on "active Ca^{2+}", and therefore applies even if release grows exponentially with the amount of a bound form of "active Ca^{2+}" as suggested by Cooke et al (4). If the latter is the case, then the interpretation of n as determined using Ba^{2+} ("method 3") becomes somewhat altered; the n of about 2 then becomes simply the slope of log release rate vs log "active Ca^{2+}" at the level of release rate being studied. With the observation that the slope of log quantal content vs log external Ca^{2+} is usually about 4, and at most 5, it is hard to escape the conclusion that the slope of log release vs log "active Ca^{2+}" cannot be more than 3. A slope of 2 to 3 corresponds to the value observed at the squid giant synapse (17). It can be shown that with this limitation the behaviour of the "exponential" model of Cooke et al (4) becomes very similar to that of the "power" model with n = 3, with regard to Ca^{2+} dependence of release and various aspects of facilitation.

As to the role, or lack of role, of residual Ca^{2+} in facilitation processes, the novel finding presented here is that the degree of facilitation is not independent of the intensity and duration of the nerve terminal depolarization producing "direct" EPPs, with EPPs of about the same amplitude as those where facilitation is independent of external Ca^{2+} (20). This is clearly contrary to the residual Ca^{2+} hypothesis, in which variation of Ca^{2+} entry per impulse by varying the impulse should have no more effect than varying the entry by varying external Ca^{2+}.

The failure of fmepp to behave as predicted by the residual Ca^{2+} hypothesis has been noted often (e.g., 13, 19), and deserves some comment, since it could be hypothesized that this observation indicates that a nerve impulse acts largely by enhancing the effectiveness of "active Ca^{2+}" rather than by causing influx of Ca^{2+} into the nerve terminal. In terms of the Parnas-Segel model (equations 1 and 2), this is tantamount to supposing that either λ or K_λ are greatly changed. With n = 3 the factor would have to be at least 100 fold (increase for λ, decrease for K_λ) and this is so large that it would be readily visible as an EPP in the absence of external Ca^{2+}, supported by the internal Ca^{2+} that is responsible for spontaneous release. Such an EPP does not exist in low Ca^{2+}/high Mg^{2+} (e.g. 4), nor have we been able to find any such EPP under conditions where "spontaneous" fmepp is raised in the absence of Ca^{2+}, by La^{3+} or Ba^{2+}, which presumably either mimic Ca^{2+} or cause displacement of Ca^{2+} from binding sites once in the nerve terminal.

The large discrepancy between predicted and observed fmepp between and after EPPs, given that n can be no more than 3, excludes not only residual "active Ca^{2+}" as the basis for facilitation, but also excludes anything that acts in the same way as Ca^{2+}, or by displacing Ca^{2+} from binding sites, since either of these would be manifest in EPPs and fmepp in the same way as residual Ca^{2+}. However, it does not exclude the possibility that facilitation might be due to the entry of an ion for which n is higher than it is for Ca^{2+}. The idea that facilitation represents a persistent sequel to the membrane depolarization itself is strongly supported by the observation by Molgo and Thesleff (29) that a brief electrotonic hyperpolarization of the nerve terminal causes a persistent anti-facilitation with a time course similar to that of facilitation.

With regard to the mechanism of facilitation it may be noted that an alteration in the kinetics of disposal of "active Ca^{2+}" could well cause a parallel change in fmepp and EPPs, as seen experimentally during facilitation, as could an alteration in λ or K_λ. It is notable that in the experiment of Fig. 5 and Table 1, the depolarizations employed were effectively smaller than nerve terminal action potentials; the "direct" EPPs evoked in the presence of tetrodotoxin were much smaller than EPPs produced by nerve impulses in 1 mM Ca^{2+}, 1 mM Mg^{2+}. Nevertheless, the facilitations observed were at least as high as we have been able to find with nerve impulses in low Ca^{2+}/high Mg^{2+}. This observation militates against the proposition that influx of either Na^+ or Mg^{2+} plays an important role in facilitation. If the various phases of facilitation each represents a persistent effect of nerve terminal membrane depolarization *per se*, acting to multiply release rate (14, 27, 30), the generation and/or manifestation of these effects might well be sensitive to external cations (15, 20, 31, 32).

6. SUMMARY

6.1. At mouse motor nerve terminals, a cooperativity of two for Ca^{2+} entry via voltage-gated channels is indicated by the dependence of transmitter release on external Ca^{2+} and from the kinetics of competition of neomycin with Ca^{2+}.

6.2. In the presence of Ba^{2+}, transmitter release following nerve terminal stimulation reflects internal accumulation of Ba^{2+} that enters the nerve terminal via Ca^{2+}-channels; Ba^{2+} disposal has a time constant of several seconds.

6.3. From the intensity-duration relation for depolarizing pulses giving equal after-responses, in the presence of Ba^{2+}, it is possible to estimate the relation between Ca^{2+}-channel opening and presynaptic depolarization. The data indicate that the slope of log release vs. log internal Ca^{2+} is close to 2 and no more than 3.

6.4. With a value of no more than 3 for Ca^{2+} cooperativity in activating transmitter release, the low frequencies of MEPPs observed during facilitation of EPPs are inconsistent with the residual Ca^{2+} model for facilitation.

6.5. With series of focal depolarizing pulses giving "direct" EPPs, facilitation is graded with pulse intensity and duration, i.e., is not independent of Ca^{2+} entry per pulse. This result is also not consistent with the residual Ca^{2+} model for facilitation.

6.6. It is concluded that the relatively long-term effects of nerve action potentials on transmitter release, manifest in the various phases of facilitation, are unlikely to be secondary to persistence within the nerve terminal of the Ca^{2+} that enters via voltage-gated Ca^{2+} channels.

ACKNOWLEDGEMENTS

We are grateful to Dr. R. Miura for his invaluable assistance with the mathematics. This work was supported by grants from the Muscular Dystrophy Association of Canada and from the Medical Research Council of Canada.

REFERENCES

1. Katz B: The Release of Neural Transmitter Substances. Charles C. Thomas, Springfield, 1969

2. Llinas R: Calcium and transmitter release in squid synapse. In: M Cowan and JA Ferrendelli (eds) Society for Neuroscience Symposia. Society for Neuroscience, Bethesda Md. 2:139-160, 1977

3. Reuter H: Divalent cations as charge carriers in excitable membranes. Prog. Biophys. Mol. Biol. 26: 1-43, 1973

4. Cooke JD, Okamoto K, Quastel DMJ: The role of calcium in depolarization-secretion coupling at the motor nerve terminal. J. Physiol. (Lond.) 228: 459-497, 1973

5. Landau EM: The interaction of presynaptic polarization with calcium and magnesium in modifying spontaneous transmitter release from mammalian motor nerve terminals. J. Physiol. (Lond.) 203: 281-299, 1969

6. Liley AW: The effects of presynaptic polarization on the spontaneous activity at the mammalian neuromuscular junction. J. Physiol. (Lond.) 134: 427-443, 1956

7. Jenkinson DH: The nature of antagonism between calcium and magnesium ions at the neuromuscular junction. J. Physiol. (Lond.) 196: 75-86, 1957

8. Dodge FA, Rahamimoff R: Cooperative action of calcium ions in transmitter release at the neuromuscular junction. J. Physiol. (Lond.) 193: 419-432, 1967

9. Del Castillo J, Katz B: Quantal components of the end-plate potential. J. Physiol. (Lond.) 124: 560-573, 1954

10. Del Castillo J, Katz B: Statistical factors involved in neuromuscular facilitation and depression. J. Physiol. (Lond.) 124: 574-585, 1954

11. Dudel J, Kuffler SW: Mechanism of facilitation at the crayfish neuromuscular junction. J. Physiol. (Lond.) 155: 530-542, 1961

12. Hubbard JI: Repetitive stimulation at the mammalian neuromuscular junction, and the mobilization of transmitter. J. Physiol. (Lond.) 169: 641-662, 1963

13. Zengel KE, Magleby KL: Changes in miniature endplate potential frequency during repetitive nerve stimulation in the presence of Ca^{2+}, Ba^{2+}, and Sr^{2+} at the frog neuromuscular junction. J. Gen. Physiol. 77: 503-529, 1981

14. Zengel KE, Magleby KL: Augmentation and facilitation of transmitter release: a quantitative description at the frog neuromuscular junction. J. Gen. Physiol. 80: 583-611, 1982

15. Katz B, Miledi R: The role of calcium in neuromuscular facilitation. J. Physiol. (Lond.) 195: 481-492, 1968

16. Miledi R, Parker I: Calcium transients recorded with arsenazo III in the presynaptic terminal of the squid giant synapse. Proc. Roy. Soc. Lond. B. 212: 197-211, 1981

17. Charlton MP, Smith SJ, Zucker RS: Role of presynaptic calcium ions and channels in synaptic facilitation and depression at the squid giant synapse. J. Physiol. (Lond.) 323: 173-193, 1982

18. Parnas H, Segel LA: A theoretical study of calcium entry in nerve terminals, with application to neurotransmitter release. J. Theor. Biol. 91: 125-169, 1981

19. Miledi R, Thies R: Tetanic and post-tetanic rise in frequency of miniature end-plate potentials in low-calcium solutions. J. Physiol. (Lond.) 212: 245-257, 1971

20. Rahamimoff R: A dual effect of calcium on neurotransmitter facilitation. J. Physiol. (Lond.) 195: 471-480, 1968

21. Landau EM, Smolinsky A, Lass Y: Post-tetanic potentiation and facilitation do not share a common calcium dependent mechanism. Nature New Biol. 244: 155-157, 1973

22. Zengel KE, Magleby KL: Differential effects of Ba^{2+}, Sr^{2+}, and Ca^{2+} on stimulation-induced changes in transmitter release at the frog neuromuscular junction. J. Gen. Physiol. 76: 175-211, 1980

23. Cooke JD, Quastel DMJ: Transmitter release by mammalian motor nerve terminals in response to focal polarization. J. Physiol. (Lond.) 228: 377-405, 1973

24. Elmqvist D, Josefsson, JO: The nature of the neuromuscular block produced by neomycin. Acta Physiol. Scand. 54: 105-110, 1962

25. Silinsky EM: On the role of barium in supporting the asynchronous release of acetylcholine quanta by motor nerve impulses. J. Physiol. (Lond.) 274: 157-171, 1978

26. Nachsen DA, Blaustein MP: Influx of calcium, strontium, and barium in presynaptic nerve terminals. J. Gen. Physiol. 79: 1065-1087, 1982

27. Cooke JD, Quastel DMJ: Cumulative and persistent effects of nerve terminal depolarization on transmitter release. J. Physiol. (Lond.) 228: 407-434, 1973

28. Hubbard JI, Jones SF, Landau EM: On the mechanism by which calcium and magnesium affect the release of transmitter by nerve impulses. J. Physiol. (Lond.) 196: 75-86, 1968

29. Molgo J, Thesleff S: Electrotonic properties of motor nerve terminals. Acta. Physiol. Scand. 114: 271-275, 1982

30. Quastel DMJ: Excitation-secretion coupling at the mammalian neuromuscular junction. In: MVL Bennett (ed) Synaptic transmission and neuronal interaction. Raven Press, New York, 1974, pp 23-43

31. Hurlbut WP, Longenecker HB Jr, Mauro A: Effects of calcium and magnesium on the frequency of miniature end-plate potentials during prolonged tetanization. J. Physiol. (Lond.) 219: 17-38, 1971

32. Rahamimoff R, Lev-Tov A, Meiri H: Primary and secondary regulation of quantal transmitter release; calcium and sodium. J. exp. Biol. 89: 5-18, 1980

DISCUSSION OF THE ARTICLE BY D.M.J. QUASTEL (CANADA)

H. Parnas (Israel): I have a comment on each one of the points of the last slide. The 1st one, the cooperativity of the entry of calcium. From the work of Llinas and Steinberg in the squid synapse, there is no sign of cooperativity in the entry.

D.M.J. Quastel (Canada): I would like to answer it very quickly by saying that they said there was no evidence for cooperativity at all and then had to backtrack on it.

H. Parnas (Israel): O.K., maybe Dr. Steinberg can comment on this if you want. Since you started with our equations, we can go along with them. From our measurements of the duration of facilitation, I believe that the residual calcium is the only explanation for facilitation, it turns out again in agreement with the results in the squid that there is no cooperativity in the entry of calcium with respect to the extracellular calcium. However, even lets say you don't agree to that point. I think the main reason why you have a contradiction between facilitation and release is that you take release as a measure of the intracellular calcium concentration. From this, you want to understand or extrapolate to facilitation. While if this is not the case, then there is no problem with the residual calcium. Lambda that you took as a constant is not a constant.

D.M.J. Quastel (Canada): Excuse me. Lambda was not taken to be a constant.

H. Parnas (Israel): But if lambda is not a constant then many of your calculations are invalid.

D.M.J. Quastel (Canada): No. That is not true. I take lambda to be constant for an epp, for something which repeats and is the same.

H. Parnas (Israel): But its not the same for the mepps.

D.M.J. Quastel (Canada): That's right and what happens then is that if you want to reconcile the mepp frequency with the facilitation data, you have to suppose lambda goes up. However, if lambda goes up as you suggest, that would be signalled by an epp in the absence of extracellular calcium. As I pointed out, you do not get that.

H. Parnas (Israel): No. It does need some calcium.

D.M.J. Quastel (Canada): I want to go back and say: Yes, Lambda does go up and Lambda going up may be facilitation; this is a slow process which develops slowly during a prolonged depolarization. It increments with each depolarization of the terminal, for example, an action potential and it is the slowly decaying process. The fast process is the calcium-dependent process. So in effect we are saying much the same thing but in reverse. I believe you have no proof that facilitation does in fact represent residual calcium. What you are saying is that faciiitation is residual calcium but release is a change of lambda. I do not believe that. I believe that the results of Katz and Miledi explicity deny that. You need calcium at the moment just before you stimulate in order to obtain release.

R. Rahamimoff (Israel): (Chairman)

This was the fifth and final round in this contest.

I. Parnas (Israel): I just want to touch upon the comment of what you said. There is a contradiction between single release and facilitation and the residual calcium theory and to show that there is no dependence between the strength of the depolarizing pulse and the duration of facilitation or level of facilitation. If you measure the duration of facilitation, you find also that when you increase the depolarization pulse, that of the first pulse, there is a range where facilitation will increase and there is a range where facilitation will decrease. If you give a pre-pulse to shut off the release of the first pulse, you still get the same duration of facilitation and the same amount of facilitation.

D.M.J. Quastel (Canada): I think you can make the residual calcium hypothesis work by ignoring the data for release between your pulses. That is the only condition in which it works.

B. Katz (U.K.): I find it very interesting that you get apparent cooperativity of two, for calcium entry; I didn't fully follow but this is still indirect evidence, isn't it?

D.M.J. Quastel (Canada): Basically, it is two independent sets of indirect evidence. If you believe in the "Parnas-Segal equations", which are too flexible, then whatever you take for the internal cooperativity, the data show an external cooperativity of two.

B. Katz (U.K.): I am ignorant of the literature but are there by now various ways in different systems of determining the dependence directly?

D.M.J. Quastel (Canada): With synaptosomes, which have been studied by Blaustein and his colleagues, there is no evidence for cooperativity for entry.

I. Steinberg (Israel): In the measurements of our work, with Llinas, calcium currents were measured directly. Therefore, it is disassociated from the over-all transmitter release. I think it has some more preciseness in the measurements. I'll try to summarize in one sentence what we found. If you measure the current as a function of calcium concentration, you'll find it depends on the membrane potential. The dependence of the calcium current on the calcium concentration depends upon the membrane potential and we could rationalize it very precisely by assuming that the channel saturates.

D.M.J. Quastel (Canada): But there is nothing new about it. Channel saturation is back in Jenkinson (Jenkinson (1957). J. Physiol. 138, 438-444).

I. Steinberg (Israel): But I talk about real data and real fitting to the data, not about the idea. What the analysis gave us is that there is one calcium that flows through the channel. Except when the channel saturates, you have a shift with voltage because it saturates also as function of voltage. So we found no evidence of two calcium ions flowing cooperatively through the channel.

D.M.J. Quastel (Canada): But you're talking about behaviour at relatively high calcium where there is saturation. The cooperativity only shows up in the low range which is the range in which you measure the relation between release and calcium externally (fourth power).

I. Steinberg (Israel): Excuse me, this doesn't fit the squid data.

D.M.J. Quastel (Canada): But you didn't look in the squid. The fourth power is at very low calcium concentrations. Dodge & Rahamimoff as well as our data pertain only to low values of calcium concentration.

I. Steinberg (Israel): In the squid, in low calcium, you'll have no cooperativity in entry.

R.S. Zucker (U.S.): The matter of cooperativity in calcium entry is crucial to all of your conclusions and is indeed the most interesting result. First, it is not only in squid but in a large number of cells, for example, in molluscan neurons and mammalian cultured neurons, that people have looked at calcium channels and they haven't found this cooperativity. It wouldn't fit any kind of reasonable model either. So, given that you don't have a direct measure of calcium entry, it is certainly a puzzling and very interesting conclusion which must be tested further. If you do not have any cooperativity in entry, then you may have internal cooperativity of the release that is as high as 4 or 5. In crayfish neuromuscular junctions with a cooperativity of 5 for release, you get a very good fit between the degree of facilitation in the residual calcium model and the degree of increase in mepp frequency which was your second point. That is not a problem in crayfish.

D.M.J. Quastel (Canada): On the other hand, in squid, from your own data, and in the adrenal gland, the cooperativity for release versus calcium shows an "n" of about 2 and not 4 or 5.

R.S. Zucker (U.S.): There are now some new results. I don't believe 2 is correct. I believe a release cooperativity of 4 or 5 is more consistent with recent measurements, which have not been published yet.

RESTING CALCIUM LEVELS AND EVOKED RELEASE AT THE NEUROMUSCULAR JUNCTION

W. VAN DER KLOOT, I.S. COHEN AND S.B. BARTON

1. INTRODUCTION

In 1967 Dodge and Rahamimoff (1) proposed a model for the relation between $[Ca^{2+}]_{out}$ and transmitter release at the neuromuscular junction that has dominated subsequent work on this problem. We have modified their model by adding one additional parameter: the level of Ca^{2+} in the resting nerve terminal. This single addition substantially alters the behavior of the model and makes it necessary to reinterpret many previous measurements. We have presented the mathematics for the model elsewhere (2,3). Here we will give a more general description of our model, its application to experiments, and some directions for future research.

2. THE MODEL OF DODGE AND RAHAMIMOFF

Dodge and Rahamimoff (1) proposed that quantal release is elicited by the combination of n Ca^{2+} ions with a receptor site (Fig. 1a). In the initial model the Ca^{2+} acted by binding to receptors on the external surface of the nerve terminal. Later it became clear that the Ca^{2+} is entering the terminal, where presumably it binds to intracellular receptors. The assumptions are: (a) Ca^{2+} entry is directly proportional to $[Ca^{2+}]_{out}$; (b) transmitter release is a unique function of $[Ca^{2+}]_{in}$; (c) the binding of one Ca^{2+} does not alter the affinity of the receptor for the next Ca^{2+}; and (d) all transmitter release is Ca^{2+}-dependent. None of these assumptions is likely to be strictly true, but they are reasonable starting points in the absence of firm contrary information.

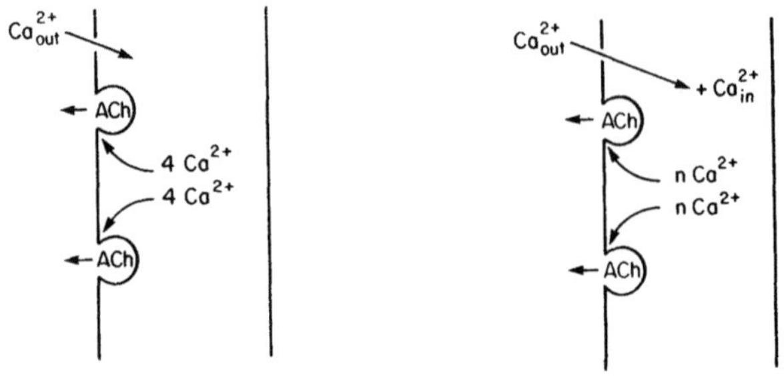

FIGURE 1. Models for the role of extracellular calcium in evoking quantal transmitter release. a) (Left) The model of Dodge and Rahamimoff. Calcium ions enter the nerve terminal following stimulation. Four calcium ions cooperate to trigger a quantal release. b) (Right) The model modified to take into account the calcium already present in the resting nerve terminal. The number of cooperating calcium ions is unknown.

The model is cooperative, because n Ca^{2+} ions work together to elicit release, and the model is saturating, because there is a finite amount of receptor, and therefore a maximum possible rate of quantal release. The release rate,m, depends upon three variables:

1. The $[Ca^{2+}]_{out}$.
2. The maximum possible rate of quantal release, k_{max}, which is a function of the total number of receptors.
3. The degree of cooperativity, n.

The predictions of the model for the effects of changing $[Ca^{2+}]_{out}$ on quantal release are shown as Fig. 2a. Because of the cooperativity, on a log-log plot the slope of the relation between evoked quantal release and $[Ca^{2+}]_{out}$ depends upon the value of n. At low levels of $[Ca^{2+}]_{out}$, where most of the receptor is not bound to Ca^{2+}, the slope of the curve is directly proportional to n. Dodge and Rahamimoff (1) measured quantal release at a series of different $[Ca^{2+}]_{out}$, plotted the data on log-log coordinates and found that the slope approached 4. Consequently, they

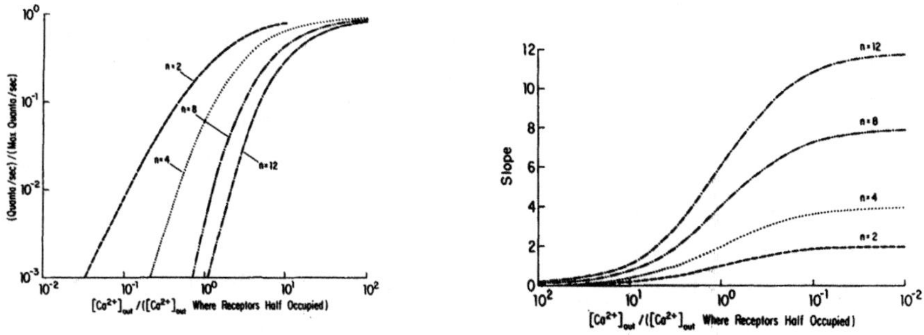

FIGURE 2. Predictions of the Dodge and Rahamimoff model. a)(left) The relations between extracellular calcium concentration and and quantal release for different values of cooperativity, n. b) (right) The relation between extracellular calcium and the slope of the relation between quantal output and the extracellular calcium concentration.

suggested that 4 Ca^{2+}'s cooperate to elicit quantal release. Figure 2b shows that it is essential to measure the slope at low levels of $[Ca^{2+}]_{out}$, because the slope declines as more and more of the receptor sites are occupied. In many preparations this requirement may be difficult to meet.

The success of the Dodge and Rahamimoff model in accounting for the data from the frog led to similar measurements on a variety of preparations and animals (reviewed in 3). The highest reported slope is 5.2 for rat myotubes innervated in tissue culture by sympathetic neurons (4). Slopes of 1 have been reported for norepinephrine release from splenic nerve (5), the squid giant synapse (6,7) and the crayfish neuromuscular junction (8,9,10). Later work at the crayfish neuromuscular junction showed that higher slopes are obtained if the stimuli are spaced far enough apart to prevent any facilitation. Furthermore, it was found that the junctional current elicited by a quantum of transmitter or by the iontophoresis of glutamate is decreased by elevating $[Ca^{2+}]_{out}$ (11,12). When these effects are corrected for, the slope at the crayfish neuromuscular junction is closer to 2 (12,13). This illustrates some of the difficulties in the experimental determination of the slope.

There is uncertainty about the slope at the frog neuromuscular junction at low $[Ca^{2+}]_{out}$. Crawford (14) reported that as $[Ca^{2+}]_{out}$ is lowered the slope abruptly falls from 4 to 1. He suggested that there is a shift in the number of cooperating Ca^{2+} ions. Andreu and Barrett (15) suggest that the slope remains 4, even when $[Ca^{2+}]_{out}$ is reduced to very low levels. These experiments are terribly difficult, because so few quanta are released at low $[Ca^{2+}]_{out}$. The responses to thousands of nerve stimuli must be measured. Furthermore, the evoked release rate is close to the spontaneous release rate. Consequently there is a high degree of statistical uncertainty in these experiments. We believe that neither the continuation of slope 4 down to low $[Ca^{2+}]_{out}$ nor the shift to slope 1 has been conclusively demonstrated.

3. CALCIUM IN THE RESTING NERVE TERMINAL

Obviously there is some finite $[Ca^{2+}]_{in}$ in the resting nerve terminal. Ca^{2+} that enters following stimulation must add onto that already present on the inside and all of this Ca^{2+} must be able to bind to receptor (Fig. 1b). The effect of the resting $[Ca^{2+}]_{in}$ on the model of Dodge and Rahamimoff was first explored by Barton (16). In this model the evoked release is a function of four variables:

1. $[Ca^{2+}]_{out}$.
2. k_{max}, the maximum rate of quantal release.
3. n, the degree of cooperativity.
4. the resting $[Ca^{2+}]_{in}$

One obvious effect of including the resting $[Ca^{2+}]_{in}$ in the model is that, as long as there is cooperativity, the slopes decrease at low $[Ca^{2+}]_{out}$ (Fig. 3a and 3b). This prediction agrees with the observations of Crawford (14), and makes it unnecessary to suppose that there is a shift in the value of n at low $[Ca^{2+}]_{out}$.

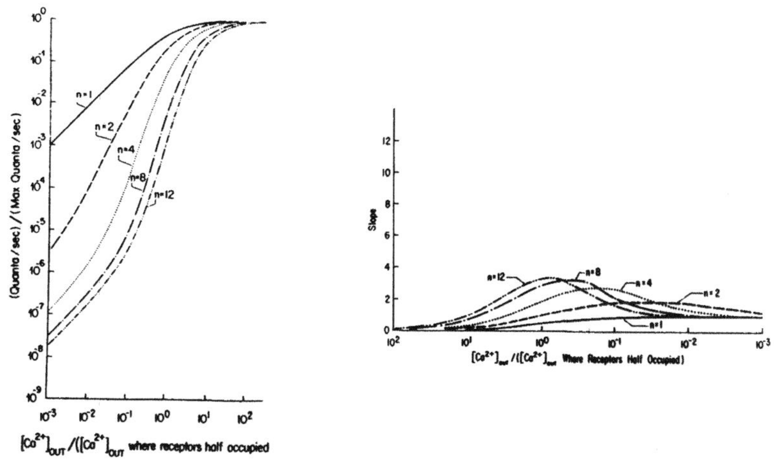

FIGURE 3. Predictions of the model that takes into account the concentration of calcium in the resting nerve terminal. a)(left) The relations between extracellular calcium concentration and quantal release for different values of cooperativity, n. b) (right) The relation between extracellular calcium and the slope of the relation between quantal output and the extracellular calcium concentration. Note that regardless of the value of n, the slope does not exceed 4.

The most striking new property arising from our model is that for higher degrees of cooperativity, the slopes no longer give the value of n (Fig. 3b). It can be shown that if n is greater than 4, the maximum value of the slope will always be 4 or less than 4 (2). Furthermore, as shown in Fig. 3b, the $[Ca^{2+}]_{out}$ at which the maximum slope is reached varies substantially depending on the value of n: with high values of n the slope reaches its maximum at relatively high $[Ca^{2+}]_{out}$, in striking contrast to the Dodge and Rahamimoff model (Fig. 2b).

Therefore, whenever experimental measurements show that the slope is close to 4, the only conclusion possible is that n is 4 or greater than 4. From the slope measurement there is no way of knowing how much greater than 4 the true value of n may be. Regardless of the value of n, at low levels of $[Ca^{2+}]_{out}$ the slope falls to 1.

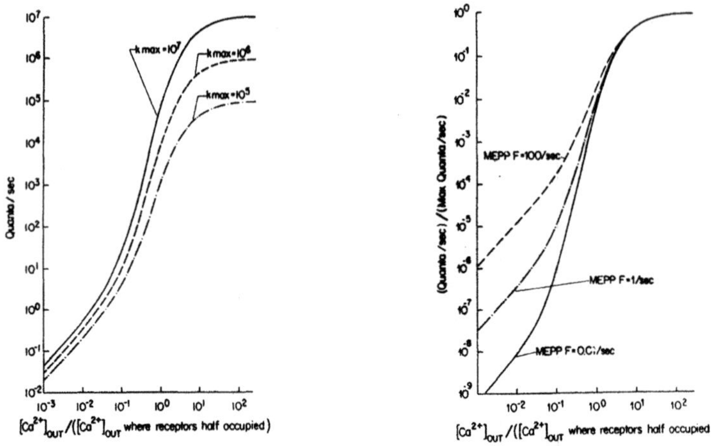

FIGURE 4. Further predictions of the model that takes the calcium concentration in the resting nerve terminal into account. a) (left) The effect of different values for the maximum quantal release rate, k_{max}, on the relation between quantal release and the concentration of calcium in the extracellular solution. The cooperativity is assumed to be 8. Notice that a decrease in the maximum release rate causes a decrease in the slope. b) (right) The effect of different levels of calcium in the resting terminal, which determines the MEPP frequency, on the relation between the quantal release rate and the extracellular calcium concentration, for n=8. Notice that a decrease in MEPP frequency is accompanied by a decrease in the maximum slope.

Consequently, in all instances where the measured slope approaches 4, the only conclusion possible is that n is 4 or greater than 4. The only synapses where the slope appears to be close to 4 is the vertebrate neuromuscular junction. At other synapses the slopes are less than 3 (see 3). This does not mean, however, that at these other synapses the number of Ca^{2+} ions cooperating to cause release is less than at the vertebrate neuromuscular junction. One of the features of the neuromuscular junction is that the number of quanta released following nerve stimulation is high. The peak output from frog motor nerve terminals in normal Ringer is 900 quanta in less than 1 msec (17). Substantially higher outputs can be obtained by raising $[Ca^{2+}]_{out}$ or by prolonging the nerve terminal action potential. The maximum rate of quantal output in the frog is at least 10^6 quanta/sec. At most other synapses the maximum quantal release rate is much lower, presumably because there are far fewer sites where Ca^{2+} can bind to produce release. Fig. 4a illustrates the effect of changes in the maximum rate of quantal release. A decrease in the maximum release rate leads to an appreciable decrease in the slope, without any change in n. Therefore at synapses with substantially lower rates of maximum quantal release than the vertebrate neuromuscular junction, the slope gives an even less accurate measure of n. An estimate of the maximum release rate is necessary before one can evaluate how closely the slope will approximate n. It is quite possible that none of the measurements of the slope has yet given the true value for n at any synapse.

Changes in the number of Ca^{2+} receptor sites may also account for the effects of tetanus and botulinum toxins on the neuromuscular junction. These toxins reduce the maximum release rate and also decrease the slope in the relationship between log(quantal release) and log($[Ca^{2+}]_{out}$) (18,19,20).

4. CONSEQUENCES OF THE MODEL INCORPORATING RESTING CALCIUM

If the Ca^{2+} that enters following nerve terminal stimulation adds to Ca^{2+} already in the terminal to trigger enhanced release, increases in resting $[Ca^{2+}]_{in}$ should increase evoked release. There is abundant evidence at the neuromuscular junction that a rise in resting $[Ca^{2+}]_{in}$ produces an increase in the rate of spontaneous quantal release. One example is the action of calcium ionophores which dramatically increase MEPP frequency when Ca^{2+} is present in the extracellular solution (21,22,23). The ionophores also enhance evoked release. In addition, there are a considerable number of treatments that increase both spontaneous release and the amount of evoked release (see 2, Table 1). Some of these agents, like inhibitors of oxidative phosphorylation, clearly would be expected to elevate $[Ca^{2+}]_{in}$.

However, for many years there appeared to be a striking exception to the correlation between elevated spontaneous and evoked release. Hypertonic extracellular solutions increase miniature end-plate potential (MEPP) frequency substantially (24), but were reported to have no effect on evoked release, except for high tonicities which depress release (25,26,27,28). Eventually, a re-examination showed that moderate elevations in tonicity increase both MEPP frequency and evoked release (29,2). Furthermore, treatment with concanavalin A, in some unknown fashion, substantially depresses the increase in MEPP frequency produced by increases in the tonicity of the extracellular solution (30). The concanavalin A pretreatment also substantially depresses the increase in evoked release produced by elevating the tonicity of the extracellular solution (2). This is additional evidence that the increase in MEPP frequency and in evoked release produced by hypertonic solutions go hand in hand.

Fig. 4b shows the predictions of the model for the effects of increases in MEPP frequency, caused by a rise in $[Ca^{2+}]_{in}$, on the the relation between $[Ca^{2+}]_{out}$ and evoked release. An increase in MEPP frequency, caused by a rise in $[Ca^{2+}]_{in}$, decreases the slope. Hypertonic solutions appear to decrease the slope at the frog neuromuscular junction (31).

5. ATTEMPTS TO DETERMINE n

Since n apparently cannot be determined from the slope of the relation between log(quantal output) and $\log([Ca^{2+}]_{out})$, an alternative method is needed. According to the assumptions and the modified model, there are four variables, three of which can be estimated from measurements:

1. The $[Ca^{2+}]_{in}$ in the resting nerve terminal, which can be estimated from the spontaneous MEPP frequency.
2. The $[Ca^{2+}]_{in}$ in the nerve terminal following the action potential, which can be estimated from the quantal release rate, m.
3. The number of receptors, which can be estimated from the maximum quantal release rate following stimulation in a Ringer containing blockers of K^+ channels (17).
4. The cooperativity, n, for which a method of estimation is needed.

The maximum release rate, k_{max}, was estimated by Katz and Miledi (17) as greater than 10^6 quanta/sec. We can easily measure the MEPP frequency and the evoked release in Ringer's at two different tonicities, produced by adding sucrose. We assume that the elevation of tonicity does not alter the amount of Ca^{2+} entering following the action potential, but only alters the $[Ca^{2+}]_{in}$ in the resting terminal. Then the evoked release at the two tonicities can be modeled by two simultaneous equations (2).

$$m(tonicity_1) \propto [Ca^{2+}]_{out},\ n,\ k_{max},\ \text{MEPP frequency}(tonicity_1)$$

$$m(tonicity_2) \propto [Ca^{2+}]_{out},\ n,\ k_{max},\ \text{MEPP frequency}(tonicity_2)$$

The solution of the simultaneous equations should enable us to estimate n. However, for most of our data the equations have no solution (2). In most of the experiments the rise in evoked output is substantially greater than predicted from the rise in MEPP frequency.

The reason for this discrepancy is uncertain. The most likely explanation is that the assumption that the MEPP frequency is determined solely by the $[Ca^{2+}]_{in}$ in the resting terminal is incorrect. There is evidence that a fraction of the spontaneous release does not depend upon $[Ca^{2+}]_{in}$ (27, 32). If we assume that a fraction of the MEPP frequency is Ca^{2+}-independent, and subtract that value from from the MEPP frequencies at both tonicities, then the two simultaneous equations have solutions. However, since we have no accurate measure of the presumed Ca^{2+}-independent rate of spontaneous release, we still cannot uniquely determine the value of n.

6. WHAT IF n IS GREATER THAN 4

If the measured slope of the relation between log(quantal output) and $\log([Ca^{2+}]_{out})$ is 4, the only conclusion is that n is 4 or greater than 4. The possibility that n is actually greater than 4 has some interesting consequences in synaptic physiology.

6.1 Increases in MEPP frequencies during tetanic stimulation in low Ca^{2+} solutions

In extracellular solutions containing low $[Ca^{2+}]$, tetanic nerve stimulation results in a pronounced rise in the MEPP frequency. Presumably, the rise is owing to an increase in $([Ca^{2+}]_{in})^n$, because a small amount of Ca^{2+} enters with each stimulus or is released from stores within the nerve. Suppose that we try to fit the data by assuming that the $[Ca^{2+}]_{in}$ increases by the same amount following each stimulus and that n=4. The measured MEPP frequency rises considerably more rapidly then predicted by the model (33). There are two obvious possibilities. First, the addition to $[Ca^{2+}]_{in}$ may rise from one stimulus to the next. Second, that n may be greater than 4.

6.2 Effects of varying the tonicity of the extracellular solution

It has been known since Fatt and Katz (24) that the MEPP frequency can be altered by varying the tonicity of the extracellular solution. Increases in tonicity raise the MEPP frequency, decreases in tonicity lower the MEPP frequency. In the frog, the changes in MEPP frequency are sustained for many hours. A simple explanation for the change in frequency would be that the alteration in tonicity moves water in or out of the terminal, thereby altering the $[Ca^{2+}]_{in}$, and thereby changing the rate of spontaneous release in direct proportion to $([Ca^{2+}]_{in})^n$. The changes in MEPP frequency are too large for the predicted changes in $[Ca^{2+}]_{in}$ if n=4. If n=4, doubling $[Ca^{2+}]_{in}$ would produce a 16-fold increase in MEPP frequency. In fact, a change in tonicity that would double the $[Ca^{2+}]_{in}$ produces roughly a 256-fold increase in MEPP frequency (29). To account for the measured change in MEPP frequency would require n=8.

Yet even if n=8 there is difficulty with the idea that changes in tonicity act by altering $[Ca^{2+}]_{in}$ simply by moving water through the nerve terminal membrane. When Ca^{2+} enters the terminal following an action potential its effect is brief, presumably because there are mechanisms that rapidly return $[Ca^{2+}]_{in}$ to the resting level. If a hypertonic extracellular solution moves water out of the terminal and thereby elevates

$[Ca^{2+}]_{in}$, why shouldn't the same mechanisms that operate following the action potential promptly restore $[Ca^{2+}]_{in}$ to the resting level? One possible answer is diagrammed in Figure 5. Suppose that the level of resting $[Ca^{2+}]_{in}$ is regulated by an active transport system that sequesters Ca^{2+} in the lumen of the endoplasmic reticulum. Further suppose that the activity of this system depends on the gradient of Ca^{2+} between the cytosol and the lumen of the endoplasmic reticulum. Moving water out of the terminal would raise the concentration of Ca^{2+} in both the lumen and in the cytosol. This could lead to a sustained elevation in $[Ca^{2+}]_{in}$ and therefore a sustained rise in the MEPP frequency.

Obviously there also must be a mechanism to transport Ca^{2+} that enters the terminal back to the extracellular fluid, otherwise the terminal would eventually fill with Ca^{2+}. Perhaps the active transport system in the cell membrane is not capable of lowering $[Ca^{2+}]_{in}$ to the level found in the resting cell or operates slowly compared to the intracellular sequestering mechanism. In either of these cases, the fine regulation of $[Ca^{2+}]_{in}$ would be done by the endoplasmic reticulum.

Even if changes in tonicity act by shifting $[Ca^{2+}]_{in}$ by altering cell volume, they must also alter the setpoint for $[Ca^{2+}]_{in}$. Therefore the response to changes in tonicity cannot be used to estimate n. Nevertheless, some progress may still be made by studying the changes in MEPP frequency at different tonicities. When preparations in hypertonic solution containing EGTA are exposed to 20 mM $[K^{+}]_{out}$, there is a decrease in MEPP frequency (34). The depolarization is thought to open Ca^{2+} channels in the nerve terminal membrane, so the Ca^{2+} moves down its electrochemical gradient from the cytosol to the extracellular fluid. Using Ca^{2+}-EGTA buffers, it should be possible to estimate the $[Ca^{2+}]_{out}$ at which there is no change in MEPP frequency. By making this estimate at two different tonicities, one could see whether or not there is a change in $[Ca^{2+}]_{in}$ between the two tonicities, and could calculate the value of n necessary to account for the difference in MEPP frequency.

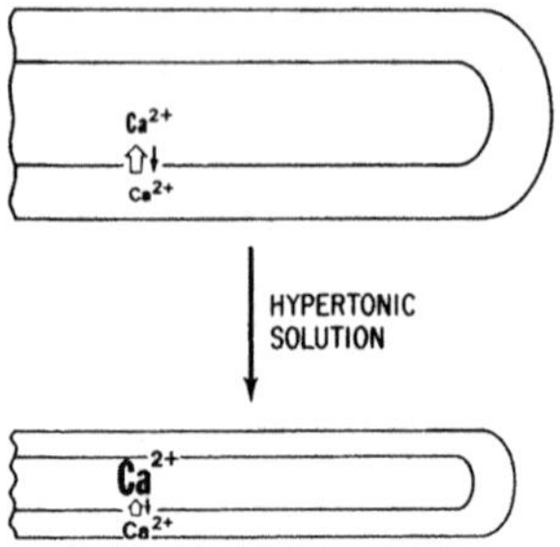

FIGURE 5. A diagram of how changes in nerve terminal volume, caused by a shift in the tonicity of the extracellular solution, might change the setpoint for the concentration of intracellular calcium. The compartment within the terminal might be the lumen of the endoplasmic reticulum. See the text for further explanation.

Stretching frog muscles produces a sustained increase in both MEPP frequency and in evoked release (36,37). Lengthening a cylindrical structure without a change in surface area decreases the volume. Therefore stretch might also act by increasing $[Ca^{2+}]_{in}$. As was the case with changes in tonicity, the stretch must also alter the set point for $[Ca^{2+}]_{in}$, if the change in MEPP frequency is caused by changes in $[Ca^{2+}]_{in}$. Stretch does not affect release from mammalian motor nerve terminals, which are disc-shaped, so stretch would not alter their volume appreciably (37).

7. FACILITATION

The second stimulus of a pair is likely to elicit a larger quantal release-- the response is said to be facilitated (37). The extent of the facilitation can be expressed as a ratio between the amplitudes of the second and the first responses.

$$\text{facilitation} = EPP_2/EPP_1$$

Katz and Miledi (38,39) suggested that facilitation might result from a delay in clearing the Ca^{2+} that entered with the first action potential from the nerve terminal. Therefore, when the second action potential enters the terminal there is a residual elevation of $[Ca^{2+}]_{in}$. The second stimulus brings in the same amount of Ca^{2+} as the first, but it adds onto the elevated level of Ca^{2+} already in the terminal and therefore elicits more release. Obviously, this hypothesis depends on a cooperative action of Ca^{2+} in evoking quantal release.

In support of the residual Ca^{2+} hypothesis, there is a period following each stimulus in which the MEPP frequency is elevated (33, and many subsequent papers). This is called delayed release. The delayed release at any time, t, following the stimulus can be expressed as

delayed release(t) = MEPP frequency(t)/resting MEPP frequency.

Attempts to test experimentally the residual Ca^{2+} hypothesis for facilitation have focused on a quantitative comparison between facilitation and delayed release. Both are assumed to depend upon $([Ca^{2+}]_{in})^n$. The problem is that we do not know the value of n. Usually it is assumed that n=4 (for example, 40,41). However, the experimental test is only as good as the assumed value of n. Since we do not know how to measure n we have looked for ways of testing the residual Ca^{2+} hypothesis that do not depend upon knowing n. Cohen and Van der Kloot (3) show that regardless of the value of n, if delayed release and facilitation both depend upon residual Ca^{2+}, then

delayed release > facilitation.

We have now measured delayed release and facilitation with different intervals between paired stimuli in preparations at room and a low temperatures and in extracellular solutions made with both Ca^{2+} and Sr^{2+} (Sr^{2+} extends the duration of both delayed release and facilitation; see 42 and 43). In every example, delayed release was found to be greater than facilitation (Cohen and Van der Kloot, in preparation). Obviously, these experiments do not prove that the residual Ca^{2+} theory is correct-- they show it plausible. If facilitation had turned out to be greater than delayed release at any time interval in any of the conditions we could have rejected the simple residual Ca^{2+} hypothesis.

Until a method is developed to estimate the true value of n, it will be necessary to devise experimental tests of models of release like that which we have used for residual Ca^{2+}. It is essential that these tests do not rely on assumed values for n.

8. SUMMARY

Dodge and Rahamimoff (1) showed that the non-linear relation between $[Ca^{2+}]_{out}$ and evoked quantal release could be modeled by a cooperative interaction, in which n Ca^{2+} ions bind to a receptor to trigger quantal release. Their model saturates, since there are a finite number of receptors. They showed that the slope of relation between log(evoked quantal release) and $\log([Ca^{2+}]_{out})$ approaches 4. Therefore they proposed that n=4. This model has stimulated much research and the slope has been measured at many synapses; the results range between 1 and 5.

We have added another variable to this model, namely the level of Ca^{2+} in the resting nerve terminal (2). The Ca^{2+} that enters following the action potential adds to that already present to elicit quantal release. This addition was made because more than twenty drugs or treatments have been described that increase both spontaneous and evoked quantal release. These actions could be understood if the treatments raise $[Ca^{2+}]_{in}$ in the resting terminal, to elevate the MEPP frequency, and also raise the amount of Ca^{2+} that adds to that entering following stimulation.

However, the addition of a term for $[Ca^{2+}]_{in}$ in the resting terminal produces a striking change in the predictions of the model-- the slope of the relation between log(quantal output) and $\log([Ca^{2+}]_{out})$ can no longer give an estimate of n. If the slope is 4, then n is 4 or greater than 4. Our model also predicts that if the resting $[Ca^{2+}]_{in}$ is increased there will be a rise in MEPP frequency and a decrease in the slope. This has been observed in preparations in which the MEPP frequency is increased by hypertonic solutions. The slope will also be decreased if there is less receptor-- this may account for the decrease in the slope following treatment with botulinum or tetanus toxins, and for the lower slopes observed at the crustacean neuromuscular junction and at central synapses.

At present, there does not appear to be a satisfactory method for experimentally determining the true value of n.

We discuss the possibilities that increasing the tonicity of the extracellular solution or stretching the muscle act by raising $[Ca^{2+}]_{in}$. This hypothesis requires a change in the setpoint for the regulation of $[Ca^{2+}]_{in}$. It may be possible to determine n by studying the MEPP frequency at different tonicities in preparations depolarized in solutions buffered with Ca^{2+}-EGTA.

We have extended our model to facilitation. Regardless of the value of n, the model predicts that if the residual Ca^{2+} hypothesis is correct, delayed release should always be greater than facilitation. Preliminary results suggest that this is so.

ACKNOWLEDGEMENTS

Assisted by Grant 10320 from the NINCDS.

REFERENCES

1. Dodge FA Jr, Rahamimoff R: Co-operative action of calcium ions in transmitter release at the neuromuscular junction. J Physiol (Lond) 193: 419-432, 1967.
2. Barton SB, Cohen I, Van der Kloot W: The calcium dependence of spontaneous and evoked release at the frog neuromuscular junction. J Physiol (Lond) 337: 735-752, 1983.
3. Cohen I, Van der Kloot W: Calcium and transmitter release. International Rev Neurobiol (in the press), 1984.
4. Nurse CA: Interactions between dissociated rat sympathetic neurons and skeletal muscle cells developing in tissue culture. Devel Biol 88: 71-79, 1981.
5. Kirpekar SM, Misu Y: Release of noradrenaline by splenic nerve stimulation and its dependence on calcium. J Physiol (Lond) 188: 218-234, 1967.
6. Kusano K: Influence of ionic environment on the relationship between pre- and post synaptic potentials. J Neurobiol 1: 435-457, 1970.
7. Llinás R, Steinberg IZ, Walton K: Presynaptic calcium currents and their relation to synaptic transmission: voltage clamp study in squid giant synapse and theoretical model for the calcium gate. Proc Natl Acad Sci USA 73: 2918-2922, 1980.
8. Bracho H, Orkand RK: Effect of calcium on excitatory neuromuscular transmission in the crayfish. J Physiol (Lond) 206: 61-71, 1970.

9. Ortiz CL, Bracho H: Effect of reduced calcium on excitatory transmitter release at the crayfish neuromuscular junction. Comp Biochem Physiol 41A: 805-812, 1972.
10. Staggs DR, Pofcher E, Lheureux R, Ortiz CL, Orkand RK: Excitatory neuro-muscular transmission in crayfish: calcium dependence is unaffected by picrotoxin. J Neurobiol 11: 629-632, 1980.
11. Nickell WT, Boyarsky LL: The effect of calcium at the crayfish neuromuscular junction. Comp Biochem Physiol 664: 259-264, 1980.
12. Dudel J: The effect of reduced calcium on quantal unit current and release at the crayfish neuromuscular junction. Pflugers Arch 391: 35-40, 1981.
13. Parnas I, Parnas H, Dudel J: Neurotransmitter release and its facilitation in crayfish II. Duration of facilitation and removal processes of calcium from the terminal. Pflugers Arch 393: 232-236, 1982.
14. Crawford AC: The dependence of evoked transmitter release on external calcium ions at very low mean quantal outputs. J Physiol (Lond) 240: 255-278, 1974.
15. Andreu R, Barrett EF: Calcium dependence of evoked transmitter release at very low quantal contents at the frog neuromuscular junction. J Physiol (Lond) 308: 79-97, 1980.
16. Barton SB: Facilitation and delayed release of transmitter at the frog neuromuscular junction. D Phil thesis, University of Oxford, 1977.
17. Katz B, Miledi R: Estimates of quantal content during 'chemical potentiation' of transmitter release. Proc R Soc Lond [Biol] 205: 369-378, 1979.
18. Cull-Candy SG, Lundh H, Thesleff S: Effects of botulinum toxin on neuromuscular transmission in the rat. J Physiol (Lond) 269: 177-203, 1976.
19. Thesleff S, Lundh H: Mode of action of botulinum toxin and the effect of drug antagonists. Advances in Cytopharmacol 3: 35-43, 1979.
20. Bevan S, Wendon LMB: A study of the action of tetanus toxin at rat soleus neuromuscular junctions. J Physiol (Lond) 348: 1-17, 1984.
21. Kita H, Van der Kloot W: Calcium ionophore X-537A increases spontaneous and phasic release of acetylcholine at the frog neuromuscular junction. Nature 250: 658-660, 1974.
22. Kita H, Van der Kloot W: Effects of the ionophore, X-537A, on acetylcholine release at the frog neuromuscular junction. J Physiol (Lond) 259: 177-198, 1976.
23. Statham HE, Duncan CJ: The action of ionophores at the frog neuromuscular junction. Life Sciences 17: 1401-1406, 1976.
24. Fatt P, Katz B: Spontaneous subthreshold activity at motor nerve endings. J Physiol (Lond) 117: 109-128, 1952.
25. Furshpan EJ: The effects of osmotic pressure changes on the spontaneous activity at nerve endings. J Physiol (Lond) 134: 689-697, 1956.
26. Thesleff S: Motor end-plate 'desensitization' by repetitive nerve stimuli. J Physiol (Lond) 148: 695-664, 1959.
27. Hubbard JI, Jones SF, Landau EM: An examination of the effects of osmotic pressure changes on the spontaneous activity of motor nerve endings. J Physiol (Lond) 197: 689-697, 1968.

28. Kita H, Van der Kloot W: Time course and magnitude of effects of changes in tonicity on acetylcholine release at frog neuromuscular junctions. J Neurophysiol 40: 212-224, 1977.
29. Kita H, Narita K, Van der Kloot W: The relation between tonicity and impulse-evoked transmitter release in the frog. J Physiol (Lond) 325: 213-222, 1982.
30. Gorio A, Mauro A: Reversibility and mode of action of black widow spider venom at the vertebrate neuromuscular junction. J Gen Physiol 73: 245-263, 1979.
31. Kita H, Van der Kloot W: The quantitative relations between extracellular calcium and acetylcholine release at the frog neuromuscular junction. Brain Res 49: 205-207, 1973.
32. Quastel DMJ, Hackett JT, Cooke JD: Calcium: is it required for transmitter secretion. Nature 267: 170-172, 1971.
33. Narita K, Kita H, Van der Kloot W: Elevated tonicity increases miniature end-plate potential frequency during tetanic stimulation at frog neuromuscular junction in low calcium and manganese saline solutions. Brain Res 289: 79-85, 1983.
34. Shimoni Y, Alnaes E, Rahamimoff R: Is hypertonic neurosecretion from motor nerve endings a calcium-dependent process?. Nature 267: 170-172, 1977.
35. Hutter OF, Trautwein W: Neuromuscular facilitation by stretch of motor nerve endings. J Physiol (Lond) 151: 89-102, 1960.
36. Turkanis SA: Effects of muscle stretch on transmitter release at end-plates of rat diaphragm and frog sartorius muscles. J Physiol (Lond) 230: 391-403, 1973.
37. del Castillo J, Katz B: Statistical factors involved in neuromuscular facilitation and depression. J Physiol (Lond) 124: 574-585, 1954.
38. Katz B, Miledi R: The effect of calcium on transmitter release from motor nerve endings. Proc R Soc Lond [Biol] 161: 469-503, 1965.
39. Katz B, Miledi R: The role of calcium in neuromuscular facilitation. J Physiol (Lond) 195: 481-492, 1968.
40. Younkin SG: An analysis of the role of calcium in facilitation at the frog neuromuscular junction. J Physiol (Lond) 237: 1-14, 1974.
41. Zengel JE, Magleby KL: Changes in miniature endplate potential frequency during repetitive nerve stimulation in the presence of Ca^{2+}, Ba^{2+}, Sr^{2+} at the frog neuromuscular junction. J Gen Physiol 77: 503-529, 1981.
42. Dodge FA Jr, Rahamimoff R: Co-operative action of calcium ions in transmitter release at the neuromuscular junction. J Physiol (Lond) 193: 419-432, 1967.
43. Meiri U, Rahamimoff R: Activation of transmitter release by strontium and calcium ions at the neuromuscular junction. J Physiol (Lond) 215: 709-726, 1971.

DISCUSSION OF THE ARTICLE BY W.G. VAN DER KLOOT (U.S.)

I. Parnas (Israel): You know that in the crayfish it was found that when the slope is measured against extracellular calcium, it gives a value of 1. Some workers on this basis suggested that the residual calcium theory doesn't hold for the crayfish. In fact, when H. Parnas and L. Segel (1980) included the resting calcium to affect the slope we repeated the measurements using maximal facilitation to compute the cooperativity. We found as you did that there is probably a cooperation of 4 also in the crayfish, 4 in the frog, 4 in the mammalian, it seems that the other techniques are more accurate.

W.G. Van der Kloot (U.S.): I appreciate the comment but I don't necessarily agree with the conclusions since as I emphasized, at least using a log-log plot method, 4 doesn't mean 4.

D.M.J. Quastel (Canada): I couldn't agree with you more about the importance of including that resting calcium when dealing with the equations. I would like to point out that when one models what happens with intracellular calcium the 'n' is also critically important. When one comes to facilitation, as one increases 'n' the facilitation of mepp frequency more and more approaches the facilitation of the epp. Therefore, one can for the two values always infer one value for 'n'. Something else happens. As 'n' grows big, one talks about smaller and smaller increments of calcium causing a large change of release rate and something very strange now happens in that facilitation becomes a function of external calcium, when you postulate high 'n's in order to bring your mepp frequency facilitation close to the epp facilitation. That is, what happens is that the facilitation becomes predicted to be very steeply graded with external calcium, which is not observed.

W.G. Van der Kloot (U.S.): I'll think about it. Thank you.

SERENDIPTIC MODULATION OF TRANSMITTER RELEASE: EXTRACELLULAR CALCIUM INHOMOGENEITY

Simona Ginsburg[1,2] and Rami Rahamimoff[1]
[1]Department of Physiology, Hebrew University-Hadassah Medical School
POB 1172, Jerusalem, Israel
[2]Everyman's University, POB 39328, Tel Aviv, Israel

1. SUMMARY

The amplitude and quantal content of synaptic potentials at the neuromuscular synapse of the frog are extremely sensitive to the calcium concentration in the extracellular solution (Jenkinson, 1957; Dodge & Rahamimoff, 1967). When the calcium concentration in the bathing solution is clamped at a constant level by employing a suitable calcium buffer, the average endplate potential amplitude and quantal content are reduced relative to those measured in a Ringer with a similar yet unbuffered free calcium concentration (Ginsburg & Rahamimoff, 1983). This may indicate that the calcium concentration in the extracellular space is inhomogenous, and that normally, its level in the synaptic cleft is probably higher than in the bulk solution. Possible origins for the calcium inhomogeneity may be extracellular matrix components and calcium transporters situated in the synaptic membranes facing the cleft; changes in their distributions and densities may provide the nerve terminal with energetically economical mechanisms for self-regulating synaptic transmission.

2. INTRODUCTION

In the neuromuscular synapse, the arrival of an action potential at the nerve terminal and the subsequent synaptic response involve the movement of sodium, potassium and calcium ions across both the presynaptic and postsynaptic membranes (Fatt & Katz, 1951; Takeuchi & Takeuchi, 1960; Takeuchi, 1963; Katz & Miledi, 1965a,1966). Following transmitter release, molecules of acetylcholine diffuse across the synaptic cleft, bind to postsynaptic receptors and elicit an endplate potential (epp). The transient changes in transmitter concentration within the cleft during synaptic activity have been widely investigated (Magleby & Stevens, 1972; Katz & Miledi, 1973; Anderson & Stevens, 1973; Neher & Sakmann, 1976; Dionne, Steinbach & Stevens, 1978). The possibility of extracellular ion concentration changes in the cleft during transmission was suggested several years ago by Attwell &

Iles (1978). They estimated that the flow of ions through the postsynaptic membrane of an active synapse could cause appreciable extracellular potassium accumulation and sodium depletion near the active zones at the frog neuromuscular synapse and at cat central synapses. This led to the suggestion that such cleft concentration changes may contribute to the time course of the endplate current. On the basis of this proposal, we considered the possibility that during synaptic activity, as calcium enters the nerve terminal, local depletion in extracellular calcium may occur as well. If this were the case, calcium depletion could conceivably contribute to certain synaptic phenomena, e.g., neuromuscular depression and post-tetanic depression in endplate potential amplitude.

To test whether extracellular calcium depletion occurs during synaptic transmission, we searched for a suitable calcium buffer, which would overcome calcium concentration changes, maintaining cleft calcium at a nearly constant level. We found that citrate was a good candidate, since its maximal calcium buffering capacity is between 0.3 to 1.1mM Ca (depending on the value of the dissociation constant). Citrate forms an acidic 1:1 complex with calcium, with reported dissociation constants of 3.16×10^{-4}M (Field et al, 1975), 6.3×10^{-4}M (Sillen & Martell, 1964) or in frog Ringer solution, 1.05×10^{-3}M (Ginsburg & Rahamimoff, 1983). Although these values somewhat differ from one another, they all fall within the range for which transmitter release is strongly dependent on $[Ca]_o$ (Dodge & Rahamimoff, 1967). If calcium depletion occurs during repetitive stimulation of the nerve, it is expected that the average quantal content and amplitude of the endplate potential in a calcium-buffered solution should increase relative to those in an unbuffered Ringer containing the same free calcium: in the unbuffered Ringer, the average quantal content and epp amplitude would reflect the fact that extracellular calcium depletion occurs, while in the presence of a buffered solution, in which no depletion is allowed to take place, these averages would be higher.

3. METHODS

Experiments were performed on the frog sartorius nerve-muscle preparation. We perfused the preparation with unbuffered Ringer containing 1-2mM Mg and 0.25-0.6mM calcium, stimulated the nerve at 0.5 Hz and recorded intracellularly hundreds of epps from a muscle cell. Subsequently, while remaining in the same post-synaptic cell, we changed the perfusing solution to a citrate Ringer containing similar free Ca + Mg concentrations, and continued to record epps. The epps were amplified and sampled by a PDP MINC 11/23 microcomputer. Their individual and average amplitudes and their average quantal contents were analyzed by special programmes. All calcium buffered solutions contained 10mM total sodium citrate introduced by substitution for sodium chloride. Both calcium unbuffered and calcium buffered Ringers contained 116mM Na, 2mM K and 5mM Hepes (pH 7). The introduction of citrate changed the ionic strength of the Ringer by less than 3% and the osmolarity by no more than 6%. The free calcium concentrations in the bathing solutions were measured spectrophotometrically with the dye murexide (Scarpa, 1972). We found that this method for determining free Ca was reproducible to within 5%. Electrochemical measurements of free calcium (Ross, 1967) could not be employed to the same degree of accuracy. It is conceivable that citrate, an organic anion, interacts with the resin present in calcium-selective electrodes; a similar suspicion regarding other organic anions has been confirmed recently (Dani et al, 1983).

4. RESULTS

4.1. Calcium buffered solutions reduce synaptic potentials.

Figure 1 exhibits the results of a typical experiment in which epps were recorded from a single postsynaptic cell perfused first in an unbuffered Ringer and subsequently in a buffered Ringer. In both solutions, the free calcium levels were nearly identical. Clearly, the average epp amplitude in the unbuffered Ringer is higher than the one in calcium-buffered solution, which of course refutes our working hypothesis concerning depletion. This reduction in epp amplitude was observed in over 50 preparations tested and its magnitude varied between 50% and 90%. Thus, by serendipity, we arrived at a new phenomenon which required an altogether different interpretation. We propose that the calcium concentration in the synaptic cleft is different from the concentration in the bulk solution. Several factors may be responsible for elevations in extracellular calcium in the synaptic gap. For example, it is conceivable that mechanisms concerned with calcium extrusion from the nerve terminal, such as the calcium pump and the sodium-calcium exchanger, are localized in the nerve section facing the synaptic gap. These mechanisms, in conjunction with others situated within the muscle membrane, may therefore bring about local transient elevations in cleft calcium. In addition, several structural components inside the cleft may cause permanent inhomogeneities in

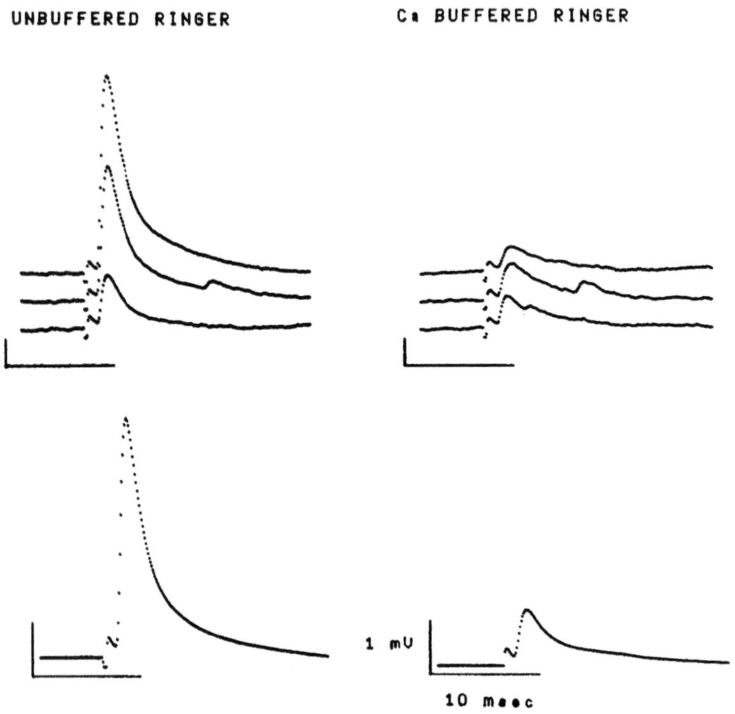

FIGURE 1. The effect of calcium buffering by citrate on epp amplitude. The nerve was stimulated at 0.5 Hz first in control and later in buffered Ringer, and all evoked responses were recorded from a single muscle cell. Top traces represent individual epps and bottom traces are the averages of 250 elicited responses. Left: Individual and average epps recorded in unbuffered Ringer with 0.3mM Ca + 1mM Mg. Right: Epps collected from the same cell as in a, after 20 minutes perfusion with buffered Ringer containing 0.31mM free Ca, 1mM free Mg and 10mM total citrate. In this as in all other experiments, solutions were pH-buffered at 7.0 by 5mM Hepes and the free calcium concentrations were determined by the murexide method.

calcium levels: the most conspicuous candidates of this kind are the glycosaminoglycans which comprise a major part of the extracellular matrix; these highly charged polyanions bind calcium and other divalent cations (Kraemer, 1979) and are specifically stained by ruthenium red (Luft, 1971). The precise types and amounts of glycans in the frog synaptic cleft have not been determined, but their presence has been shown (McMahan et al, 1980) and it is known that they play a role in regulating the arrangement of structural proteins in the synaptic cleft (Yamada et al, 1980; Nitkin et al, 1983;

Sanes, 1983). Other factors that may cause permanent inhomogeneity in extracellular calcium are negative surface charges on the presynaptic and postsynaptic membranes: in certain neurons, up to 10% of the membrane lipid contains negatively charged sialic acid residues (Ledeen, 1978) and in several synaptic membranes 90% of the gangliosides are calcium-bound (Veh & Sander, 1981). Some of the factors which may generate calcium inhomogeneities in the synaptic cleft are illustrated schematically in Fig 2.

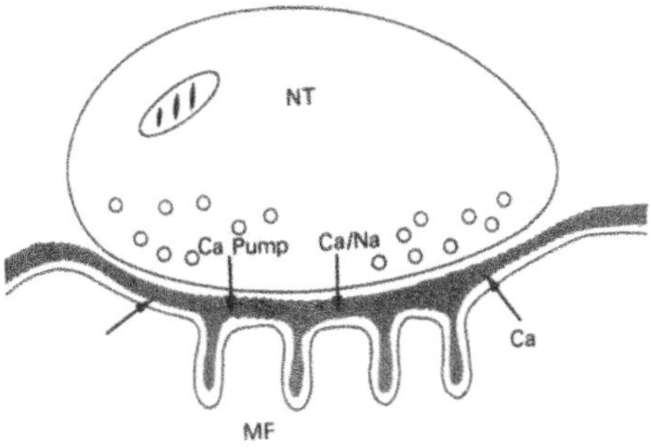

FIGURE 2. Schematic diagram of the synaptic cleft separating the nerve terminal (NT) and the myofiber (MF). Hatched areas indicate the extracellular matrix which extends well into the synaptic folds. Two hypothetical mechanisms may contribute to calcium non-homogeneity in the extracellular space. The negatively charged glycans of the matrix may serve as a calcium reservoir within the cleft. Arrows indicate calcium extrusion mechanisms in both the presynaptic and the postsynaptic membranes. These transporters, if directed towards the cleft, may cause transient elevations in cleft calcium.

Our suggestion then is that one (or more) of these mechanisms and structures produces an inhomogeneity in extracellular calcium such that, normally, the synaptic cleft senses a higher calcium level than that present in the bulk bathing solution. The epp quantal content reflects the calcium concentration within the cleft, or more precisely, the calcium level at the mouth of the voltage-dependent calcium channels of the presynaptic membrane. This level is higher than the concentration of calcium in the bulk Ringer. However, when the preparation is perfused by a calcium-buffered solution, the calcium level at this mouth does not greatly differ from its level in the bulk, since the citrate molecules, acting as a buffer, dissipate at least partially any preexisting calcium gradients in the cleft. The epp amplitude then reflects the calcium concentration of the bulk solution which is also the calcium level near the nerve terminal channels. In other words, although the calcium levels in both buffered and unbuffered <u>bulk</u> solutions are the same, extracellular calcium near the nerve terminal membrane is higher when the preparation is perfused with the unbuffered Ringer. Hence, the amplitude in this case is higher than that recorded in the calcium-buffered Ringer (Fig 1). We estimate that in unbuffered solution, the calcium level at the mouth of the voltage-dependent calcium channels is 1.5 to 2 fold higher than its level in the bulk solution. Our evaluation is based on comparisons between the calcium concentrations in buffered and unbuffered Ringers which are required to produce equivalent epp amplitudes. Such a comparison is illustrated in Fig 3. In this particular experiment, epps of similar average amplitudes were recorded from a single cell when the bathing solution was either an unbuffered Ringer containing 0.35mM calcium or a citrate buffered Ringer containing 0.61mM free calcium.

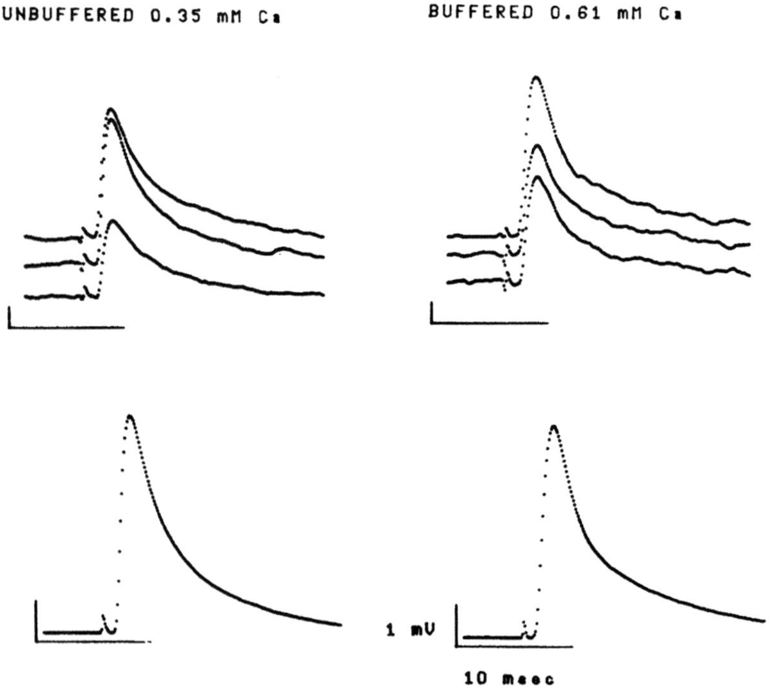

FIGURE 3. Equivalent epp amplitudes produced in unbuffered and buffered Ringers containing different concentrations of free calcium. Left: Epps recorded in unbuffered Ringer with 0.35mM Ca + 1mM Mg. Right: Epps from the same cell following 30 minutes perfusion with buffered Ringer containing 0.61mM Ca + 1mM Mg + 10mM total citrate. Each average epp (bottom traces) represents 200 responses. Note that in order to obtain these comparable amplitudes, the free calcium in the buffered Ringer had to be elevated, in this particular experiment by 74%, relative to its concentration in the unbuffered solution.

4.2. Citrate acts extracellularly: a presynaptic effect.

The effect of citrate on transmission is presynaptic and its mode of action is extracellular. The presynaptic nature of the effect is manifest by the fact that the amplitude of the response to one quantum of transmitter does not change appreciably by citrate. The maximal change in miniature endplate potentials (mepps) found on exposure to citrate was a 10% reduction in average amplitude. This finding, in conjuction with the fact that the average evoked endplate potential amplitude was reduced by 50 to 90%, indicate a presynaptic effect (del Castillo & Katz, 1954). It is highly unlikely that citrate acts pharmacologically by penetrating into the nerve terminal and binding intracellular calcium. Passive diffusion of citrate into the presynaptic cell would have reduced the free internal calcium concentration by no more than 2%. If citrate entered actively by some obscure mechanism and decreased internal calcium appreciably in a specific but unknown manner, this situation would have been reflected as a reduction in miniature endplate potential frequency (Miledi, 1973; Alnaes & Rahamimoff, 1975; Rahamimoff et al, 1978a).

In Figure 4, we present results from an experiment in which mepp frequency was followed from a cell bathing in control and buffered solutions; it is evident that the frequency remained almost constant throughout the experiment. Furthermore, introduction of citrate into the nerve terminal via citrate-loaded liposomes had virtually no effect on either epp amplitude or mepp frequency. Liposomes loaded with other ions transfer their contents into the presynaptic nerve terminal and influence these two parameters (Rahamimoff et al, 1978b). An intracellular effect of citrate is thus ruled out, and as far as we can see, the only other presynaptic action it could produce is extracellular.

We also investigated the possibility that the effect of citrate is mediated through an interference with the nerve terminal spike. For this purpose, we employed extracellular focal recording techniques and utilized the flat cutaneous-pectoris preparation instead of the usual sartorius muscle. Variations in the size of the nerve terminal action potential were followed by measuring the amplitudes of the different phases of the external nerve terminal potential (entp). The latter is the second spatial derivative of the action potential (Katz & Miledi, 1965b) and corresponds to the triphasic membrane current associated with the impulse (see Katz & Schmitt, 1940).

Figure 5 illustrates average entps and endplate currents (epcs) recorded extracellularly from a cutaneous pectoris nerve-muscle preparation bathed in

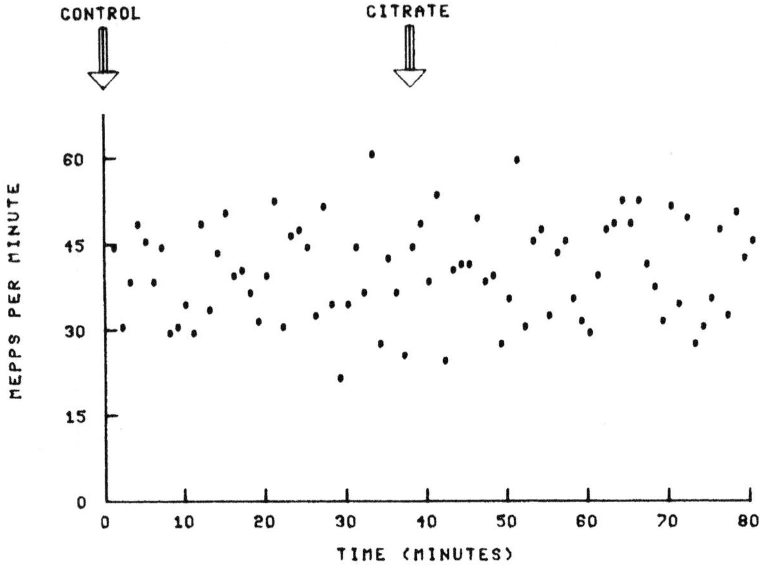

FIGURE 4. Frequency of spontaneous release in unbuffered and buffered Ringers: Mepp frequency is plotted as a function of time; at time 0 (left arrow) the preparation was equilibrated in control Ringer containing 0.3mM Ca + 1mM Mg. 38 minutes later (right arrow), the perfusing solution was changed to a buffered Ringer containing 0.3mM Ca + 1mM Mg + 10mM total citrate. Each point displays the number of mepps recorded during a one-minute interval immediately preceeding the time indicated on the abscisse.

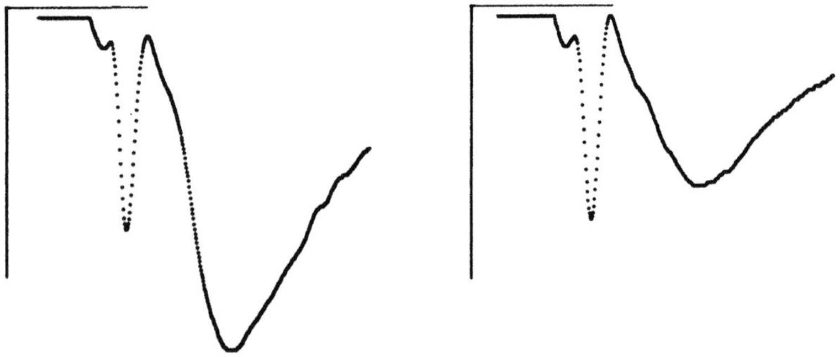

FIGURE 5. Extracellular focal recordings of external nerve terminal potentials (entps) and endplate currents (epcs) from a cutaneous pectoris preparation in control Ringer (left) and following 20 minutes perfusion by citrate Ringer (right). The narrower deflection in each trace represents the entp and the wider one is the endplate current. The usual decrease in quantal content caused by citrate buffering is reflected in these records as a reduction in epc amplitude. This 50% reduction in endplate current was not accompanied by any appreciable change in entp, intimating that citrate does not interfere with the nerve terminal action potential. In this figure, each trace is an average of 50 extracellular records.

unbuffered and citrate buffered Ringers. Clearly, while the average epc amplitude decreased appreciably (by nearly 50%) on changing the bathing solution from unbuffered to buffered Ringer, the amplitude of the entp remained constant, indicating that citrate does not interfere with the action potential.

Citrate forms complexes with both calcium and magnesium ions and this enabled us to devise an amusing experiment which demonstrates in yet another way, that the action of citrate is not pharmacological and that its effect is mediated through extracellular buffering. Normally, i.e., in unbuffered Ringer, addition of magnesium ions leads to a reduction in epp amplitude (del Castillo & Katz, 1954; Jenkinson, 1957) due to competitive binding of magnesium to presynaptic membrane calcium sites (Dodge & Rahamimoff, 1967). This well known effect is demonstrated in Figure 6A, while the result of adding the same amount of magnesium to calcium-buffered Ringer is shown in Figure 6B.

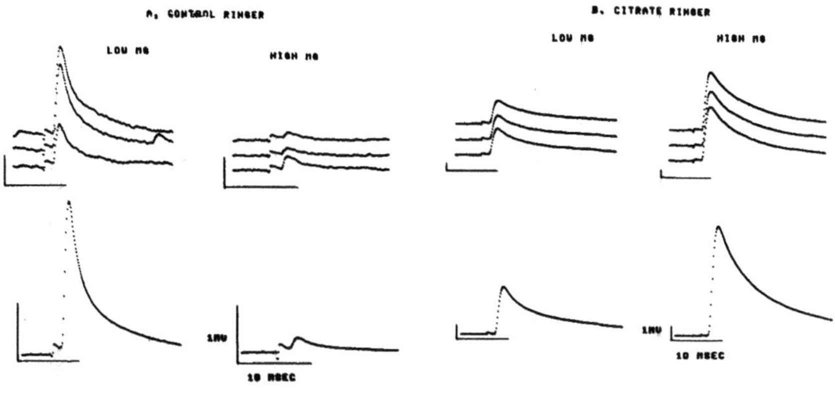

FIGURE 6. Magnesium diminishes and enhances epp amplitudes in unbuffered and buffered Ringers, respectively. A. Individual (top) and average (bottom) epps in unbuffered Ringers containing 0.4mM Ca + 1mM Mg ("low-Mg solution", left) and 0.4mM Ca + 3.57mM Mg ("high-Mg solution", right). The amplitudes are smaller in the high-magnesium Ringer. B. Individual (top) and average (bottom) epps in buffered Ringers containing 10mM total citrate. Left: Epps recorded in citrate solution containing 2.37mM total Ca and 5.93 total Mg. Right: Epps recorded in citrate solution containing 2.37mM total Ca + 8.50mM total Mg. In this case, the amplitudes are larger in the high-magnesium Ringer.
Note that the magnesium concentration difference between the high-Mg Ringer and the low-Mg Ringer was the same (2.57mM) for both buffered and unbuffered solutions. The traces in A were recorded from one single cell while those in B were collected from a second cell. Average epps represent 200 responses.

In the latter case, elevation of magnesium causes an increase in epp amplitude, in contradistinction to the former case. The origin of this behaviour is rather prosaic: when magnesium is introduced into a solution containing set concentrations of free and bound species, a redistribution occurs in accordance to the Le Chatelier Principle; some of the added magnesium binds to free citrate and this induces a certain degree of dissociation of the calcium-citrate complex which was present in the original solution. The overall outcome is an increase in free calcium and hence an augmented epp amplitude. A computation of the calcium distribution in solution (between free and bound species) in the presence of magnesium and citrate is presented graphically in Figure 7. The curves outline how, at a given calcium concentration, its free level increases as magnesium is raised.

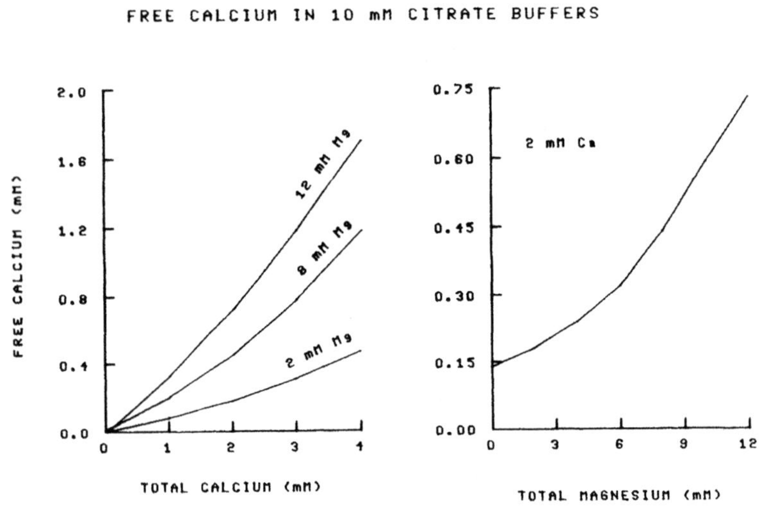

FIGURE 7. Distribution of calcium in solutions of varying magnesium compositions and 10mM total citrate. Left: Relation between free and total calcium in the presence of 2, 8 and 12 mM total magnesium. Right: Free calcium as a function of total magnesium for solutions containing a constant level of total calcium (2mM). The dissociation constants of the citrate complexes used in generating these curves were taken from Sillen & Martell (1964).

In view of the controls and results described in Figures 4-6, it seems safe to conclude that calcium buffering by citrate diminishes the epp quantal content due to a decrease in local extracellular calcium within the synaptic cleft.

The buffering action of citrate is enhanced as the concentration of the anion is raised. The "dose-response" effect thus produced is displayed in Figure 8. At 0.01mM total citrate the reduction in epp amplitude is negligible, but as the concentration is stepped up, the epp decreases progressively down to 25% of its original amplitude recorded in control Ringer. The maximal effect of citrate is reached at 10mM total citrate.

CITRATE DOSE RESPONSE EFFECT

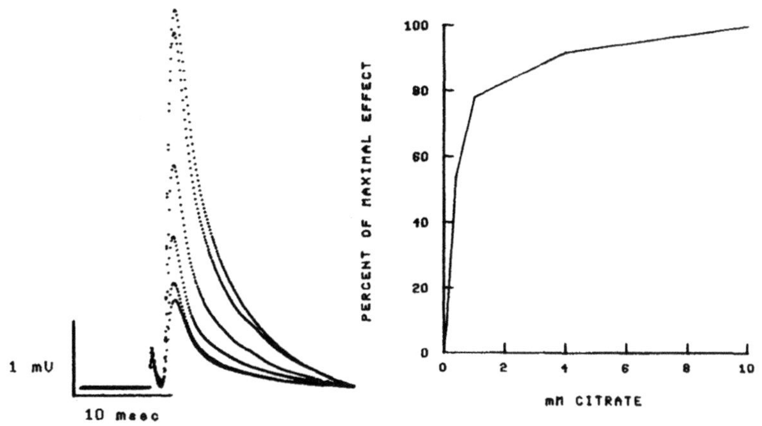

FIGURE 8. "Dose-response" effect of citrate. Left: Average epps recorded from a cell subjected to seven different bathing solutions; the top trace was produced in the unbuffered Ringer and in buffered Ringer containing 0.01mM total citrate; the traces below it (from top to bottom) were recorded in buffered Ringers containing 0.1, 0.4, 1, 4 and 10 mM total citrate. All Ringers contained 0.4mM Ca + 1mM Mg and each average epp represents 200 evoked responses. Right: Percentage of maximal buffering effect (calculated from the epp amplitudes) plotted against citrate concentration.

4.3. Citrate potency in coping with calcium inhomogeneities.

At present, we cannot pinpoint the origin of the calcium inhomogeneity in the synaptic cleft. We can, however, examine the potency of citrate in coping with hypothetical situations in which calcium transients appear. Suppose that the mechanism responsible for elevating calcium is the calcium pump of the presynaptic cell and that the nerve terminal surface carries a 10mM concentration of pump molecules. Assume that the cleft is perfused by an

0.3mM Ca Ringer, that 10% of the pumps are periodically loaded with calcium (on the internal side of the membrane) and that they come into contact with the bathing solution. If the pump molecules, when facing the external solution, have an appropriate dissociation constant for calcium, they will give up some of their calcium load. The extracellular solution near the nerve terminal will then be enriched by calcium, and its concentration will be higher than the original 0.3mM. Reasonable external dissociation constants for the pump are of the order 3-15mM (DiPolo & Beauge, 1980). On the basis of these values, it may readily be shown that on encountering the external 0.3mM Ca Ringer, the pumps would indeed liberate calcium. Figure 9A depicts the outcome of such an encounter, assuming different dissociation constants. When the bathing Ringer contains 0.3mM Ca, the calcium level near the membrane would increase to 0.46, 0.60 and 0.73 mM for pumps loaded to a 10% degree with dissociation constants of 5, 8 and 12 mM, respectively. On the other hand, if the bathing Ringer were a 10mM citrate solution with 0.3mM free calcium, the increases would be greatly dampened, as shown in Figure 9B.

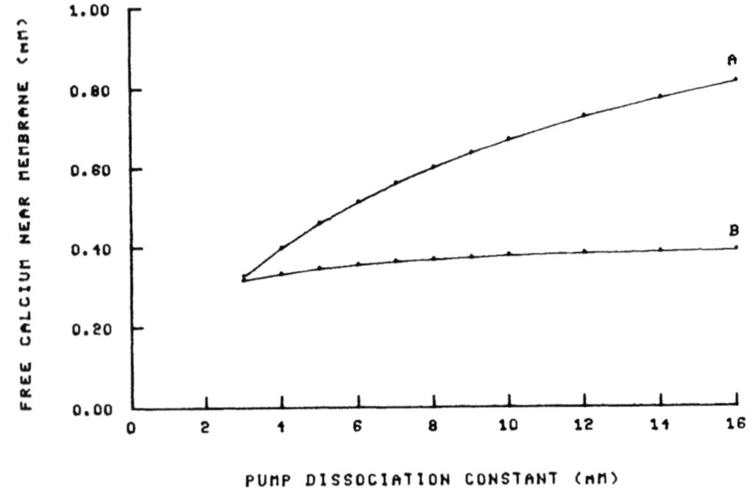

FIGURE 9. Calculated calcium levels near the presynaptic membrane upon contact between bathing solutions and hypothetical calcium-loaded pumps. Curve A: Calcium levels near the membrane as a function of pump dissociation constant when the bulk unbuffered Ringer contains 0.3mM calcium. Curve B: Calcium levels near the membrane when the bathing solution is a 10mM citrate buffered Ringer containing 0.3mM free calcium. Calculations are based on the assumptions that the membrane carries 10mM pumps, 10% of which are loaded with calcium, and that the Ca-citrate dissociation constant is 1mM.

Instead of 0.46, 0.60 and 0.73 mM calcium (in the unbuffered Ringers), the calcium levels near the membrane would be 0.34, 0.37 and 0.38 mM, respectively. In order to reach the level 0.46mM, for instance, as in an unbuffered solution upon contacting the pumps, the citrate Ringer would have to contain originally not 0.3mM but rather 0.43mM free calcium. The hypothetical situations just portrayed fit the experimental data presented in Figures 1 and 3. Under these reconstructed conditions, the epp amplitudes would be reduced by citrate buffering down to 30-90%. The calcium levels in citrate solutions would have to be raised by 50-100% to obtain epp amplitudes which were equivalent to those recorded in unbuffered Ringers.

5. POSSIBLE PHYSIOLOGICAL IMPLICATIONS OF CALCIUM INHOMOGENEITY

The experiments presented here indicate that the calcium content in the synaptic cleft of the neuromuscular junction is higher than the value in the bulk solution of the extracellular space. Since calcium is one of the main regulators of transmitter release, this inhomogeneity opens additional ways for controlling the level of synaptic activity.

We do not know at present how the high gap calcium content is achieved. If we take the extrusion possibility, for example, as a working model, then insertion of calcium pumps and exchangers facing the synaptic cleft can cause an augmentation in the transmission properties of the synapse. Such calcium transporters can have an effect not only if they are inserted in the presynaptic cell, but might be equally effective if inserted in the portion of the postsynaptic element facing the synaptic cleft. One can easily envisage how by a proper insertion of these molecules, the postsynaptic element can also control the amount of transmitter released by the presynaptic nerve terminal and thus be a 'partner' in the determination of the synaptic efficiency.

6. ACKNOWLEDGEMENTS

We thank Mr. Gabi David for the extracellular recordings of entps. We are grateful to Dr. Halina Meiri for her comments on the manuscript.

This work was supported by the Muscular Dystrophy Association, the Fund for Basic Research administered by the Israel Academy of Sciences and Humanities and the U.S.-Israel Binational Foundation.

REFERENCES

1. Alnaes E. and Rahamimoff R. (1975). The role of mitochondria in transmitter liberation from motor nerve terminals. J. Physiol. 248, 285-306.
2. Anderson C.R. and Stevens C.F. (1973). Voltage clamp analysis of acetylcholine produced endplate current fluctuations at frog neuromuscular junctions. J. Physiol. 235, 655-691.
3. Attwell D. and Iles J.F. (1979). Synaptic transmission: ion concentration changes in the synaptic cleft. Proc. R. Soc. Lond. B. 206, 115-131.
4. Dani J.A., Sanches J.A. and Hille B. (1983). Lyotropic anions: Na channel gating and Ca electrode response. J. Gen. Physiol. 81, 255-281.
5. del Castillo J. and Katz B. (1954). Quantal components of the endplate potentials. J. Physiol. 124, 560-573.
6. Dionne V.E., Steinbach J.H. and Stevens C.D. (1978). An analysis of the dose-response relationship of voltage-clamped frog neuromuscular junctions. J. Physiol. 281, 421-444.
7. DiPolo R. and Beauge L. (1980). Mechanisms of calcium transport in the giant axon of the squid and their physiological role. Cell Calcium 1, 147-169.
8. Dodge F.A. and Rahamimoff R. (1967). Co-operative action of calcium ions in transmitter release at the neuromuscular junction. J. Physiol. 193, 419-432.
9. Fatt P. and Katz B. (1951). An analysis of endplate potential recorded with an intracellular electrode. J. Physiol. 115, 320-370.
10. Field T.B., Coburn J., McCourt J.L. and McBryde W.A.E. (1975). Composition and stability of some metal citrate and diglycolate complexes in aqueous solutions. Analytica Chim. Acta 74, 101-106.
11. Gage P.W. (1976). Generation of endplate potentials. Physiol. Rev. 56, 177-247.
12. Ginsburg S. and Rahamimoff R. (1983). Is extracellular calcium buffering involved in regulation of transmitter release at the neuromuscular junction? Nature 306, 62-64.
13. Jenkinson D.H. (1957). The nature of the antagonism between calcium and magnesium ions at the neuromuscular junction. J. Physiol. 138, 434-444.
14. Katz B. and Miledi R. (1965a). The effect of calcium on acetylcholine release from motor nerve terminals. Proc. R. Soc. Lond. B. 161, 496-503.
15. Katz B. and Miledi R. (1965b). Propagation of electric activity in motor nerve terminals. Proc. R. Soc. Lond. B. 161, 453-482.
16. Katz B. and Miledi R. (1966). Input-output relation of a single synapse. Nature 212, 1242-1245.
17. Katz B. and Miledi R. (1973). The binding of acetylcholine to receptors and its removal from the synaptic cleft. J. Physiol. 231, 549-574.
18. Katz B. and Schmitt O.H. (1940). Electric interaction between two adjacent nerve fibres. J. Physiol. 97, 471-488.
19. Kraemer P.M. (1979). Mycopolysacharides: Cell biology and malignancy. In: Surfaces of Normal and Malignant Cells, R.O. Hynes (Ed.), Wiley, N.Y., pp.149-198.
20. Ledeen R.W. (1978). Ganglioside structures and distributions: are they localzed at nerve endings? J. Supramol. Struct. 8, 1-17.
21. Luft J.H. (1971). Ruthenium red and violet II: fine structural localization in animal tissues. Anat. Rec. 171, 369-416.
22. McMahan U.J., Edgington D.K, and Kuffler D.P. (1980). Factor that

influence regeneration of the neuromuscular junction. J. Exp. Biol. 89, 31-42.

23. Magleby K.L. and Stevens C.F. (1972). A quantitative description of endplate currents. J. Physiol. 223, 173-197.
24. Miledi R. (1973). Transmitter release induced by injection of calcium ions into nerve terminals. Proc. R. Soc. Lond. B. 183, 421-425.
25. Neher E. and Sakmann B. (1976). Single channel currents recorded from membrane of denervated frog muscle fibres. Nature 260, 799-801.
26. Nitkin R.M., Wallace B.G., Spira M.E., Godfrey E.W. and McMahan U.J. (1983). Molecular components of the synaptic basal lamina that direct differentiation of regenerating neuromuscular junctions. Cold Spring Harbor Symposia on Quantitative Biology, Vol. XLVIII, part 2, pp. 653-665.
27. Rahamimoff R., Erulkar S.D., Lev-Tov A. and Meiri H. (1978a). Intracellular and extracellular calcium ions in transmitter release at the neuromuscular synapse. Ann. N.Y. Acad. Sci. 307, 583-598.
28. Rahamimoff R., Meiri H., Erulkar S.D. and Barenholz Y. (1978b). Changes in transmitter release induced by ion containing liposomes. Proc. Natl. Acad. Sci. USA 75, 5214-5216.
29. Ross J.W. (1967). Calcium-selective electrode with liquid ion exchanger. Science 156, 1378-1379.
30. Sanes J.R. (1983). Roles of extracellular matrix in neural development. Ann. Rev. Physiol. 45, 581-600.
31. Scarpa A. (1972). Spectrophotometric measurement of calcium by murexide. Meth. Enzym. 24, 343-351.
32. Sillen L.G. and Martell A.E. (1964). Stability constants of metal ion complexes. Chem. Soc. Spec. Publ. 17, London.
33. Takeuchi A. and Takeuchi N. (1960). On the permeability of endplate membrane during the action of transmitter. J. Physiol. 154, 52-67.
34. Takeuchi N. (1963). Effects of calcium on conductance change of the endplate membrane during the action of transmitter. J. Physiol. 167, 141-155.
35. Veh R.W. and Sander M. (1981). Differentiation between gangliosides and sialyllactose sialidases in human tissues. Perspectives in Inherited Metabolism Diseases 4, 71-109.
36. Yamada K.M., Olden K. and Liang-Hsien E.H. (1980). Cell surface protein and cell interactions. In: The Cell Surface: Mediator of Developmental Processes, S. Subtelny and N. Wessel (Eds.), Academic Press, N.Y., pp.43-77.

DISCUSSION OF THE ARTICLE BY S. GINSBURG (ISRAEL)

D.O. Smith (U.S.): I would like to propose an alternative explanation of these data and see if you have reason to exclude it. In our laboratory, we have been studying using ^{45}Ca, calcium binding to the exterior surface of the membrane at the nerve terminal region and we found that if we rinsed the nerve in the terminal region with lanthanum, we can reduce the ^{45}Ca bound to the membrane to about 40% of its original value. The idea is that the lanthanum is displacing the surface bound calcium. Could it be that in your case the citrate is competing with surface binding sites reducing the calcium bound to the membrane, and that, in turn, may reduce the actual amount of calcium entering the cell, assuming that calcium entering may be in a bound state when it goes in. That would explain your results.

S. Ginsburg (Israel): Certainly. In fact, I don't want to commit myself to a specific mechanism. It is definitely possible that these negative charges are located on the plasma membrane.

A. Mallart (France): It is obvious from the recordings you showed that fluctuations in epp amplitude do not occur in buffered solution.

S. Ginsburg (Israel): No. They do occur.

A. Mallart (France):

Very small fluctuations.

S. Ginsburg (Israel): Perhaps in these specific slides but they do occur just as much as in the control.

R. Rahamimoff (Israel): The different estimates of the quantal content, by fluctuations and by direct measurements, agree fairly well so the fluctuations are there.

W.G. Van der Kloot (U.S.): Have people done this sort of experiment with other tissues that are very sensitive to changes in calcium concentration like the frog heart?

S. Ginsburg (Israel): Yes. In fact right now Dr. Shimoni is talking, in another Session, on the effect of citrate buffering on atrial muscle cells. It works in a similar fashion.

A.M. Brown (U.S.): If the citrate is buffering fixed charges in some kind of gel as opposed to competing with fixed charges on the membrane, you might have a shift in the gating parameters of the postsynaptic membrane. Perhaps a test could be that the threshold for the excitation of the muscle might be different in one case while in the other case there will be no difference.

S. Ginsburg (Israel): You mean that if the negative charges are not in the membrane, then the excitability should be affected.

E. Heilbronn (Sweden): Would the intracellular calcium reservoirs, e.g., the mitochondria, play no role at all in the fluctuation of release?

S. Ginsburg (Israel): Not at all. What I was trying to show is that in addition to all the mechanisms which regulate intracellular calcium, which is a very important factor in modulating release, regulation of extracellular calcium is important as well.

CALCIUM INDEPENDENT QUANTAL TRANSMITTER RELEASE AT THE NEUROMUSCULAR JUNCTION

S. THESLEFF and M.T. LUPA
Department of Pharmacology, University of Lund, Lund, Sweden

1. INTRODUCTION

It is generally accepted that transmitter release at the neuromuscular junction is a calcium-requiring process, calcium ions being necessary for depolarization-transmitter secretion coupling to occur (29). Nerve impulse-evoked multi-quantal release of acetylcholine (ACh), giving rise to the end-plate potential (e.p.p.), as well as the spontaneous quantal secretion of ACh, causing the miniature end-plate potentials (m.e.p.ps), depend on extra- and intracellular calcium concentrations (see 20; 28; 40).

The purpose of this presentation is to describe and to characterize a calcium independent quantal secretion of ACh at the neuromuscular junction. Since this secretion is unaffected by calcium it follows that it is not influenced by nerve terminal depolarization and resulting calcium fluxes nor by extra- or intracellular calcium concentrations. This implies that release is mediated by a mechanism different from that accounting for the normal depolarization-evoked transmitter release responsible for synaptic impulse transmission. The calcium independent type of ACh release is particularly prominent at neuromuscular junctions poisoned with botulinal neurotoxins (BoTx) i.e. at junctions where the depolarization, calcium-dependent transmitter release is blocked. However, it also occurs, but at a lower rate, at newly established synapses and at normal synapses. It is proposed to be a more primitive, embryonic type of transmitter release which is suppressed when synaptic impulse transmission is established. The drug 4-aminoquinoline selectively accelerates this type of transmitter release.

2. CHARACTERISTICS AND PROPERTIES OF RELEASE

The unique properties of the calcium independent type of transmitter release were recognized in studies of transmitter release in BoTx poisoned muscles and of the effects of 4-aminoquinoline, see (46). Postsynaptically this type of release gives rise to spontaneous m.e.p.ps with highly variable times-to-peak and amplitudes. Typically, these m.e.p.ps have a prolonged time-to-peak, the mean exceeding more than twice that of calcium dependent m.e.p.ps or nerve impulse evoked quantal e.p.ps. At rat neuromuscular junctions (30°C) m.e.p.ps with times-to-peak between 1 and 10 ms are frequently recorded, while calcium dependent m.e.p.ps do not have times-to-peak longer than 1 ms. Therefore, we designate the former m.e.p.ps "slow" m.e.p.ps (see 24).

The amplitude of "slow" m.e.p.ps is also highly variable with generally a larger, or much larger, amplitude than the relatively uniform-sized normal m.e.p.ps. Amplitudes as large as 10-15 mV have been recorded. Examples of calcium-independent and calcium-dependent m.e.p.ps are shown in Fig. 1A and B, respectively. Record C illustrates quantal e.p.ps from the same fibre as A.

M.e.p.ps resulting from calcium independent release are unaltered in frequency by procedures which affect the extra- or intracellular calcium concentrations of the nerve terminal (Table 1). Thus, they are unaltered in frequency by high or low extracellular potassium or calcium concentrations, by ethanol, ouabain or manganese ions, i.e. by procedures which drastically alter the frequency of m.e.p.ps caused by the calcium dependent transmitter release. Hypertonic solutions, which increase the frequency of calcium dependent m.e.p.ps, decrease the frequency of calcium independent m.e.p.ps. Hypotonic solutions exert an opposite effect on both types of release. Furthermore, the temperature dependence of the two release processes are quite different. Calcium dependent ACh release is enhanced by temperature with a Q_{10} of 2-3, while that of the calcium resistant type increases with a Q_{10} of about 12 (47).

Calcium independent transmitter release is not activated nor influenced by nerve impulses and resulting transmembrane calcium fluxes in the nerve terminal. Consequently, this type of transmitter secretion does not participate in impulse transmission.

The calcium independent type of m.e.p.ps differ not only from the normal type of calcium sensitive m.e.p.ps but also from the so-called sub-m.e.p.ps recorded by several authors at mammalian and frog neuromuscular junctions. The sub-m.e.p.ps are of smaller than usual amplitude and with generally a fast time-to-peak, and their frequency is enhanced in high extracellular calcium as shown by Kriebel & Gross (25).

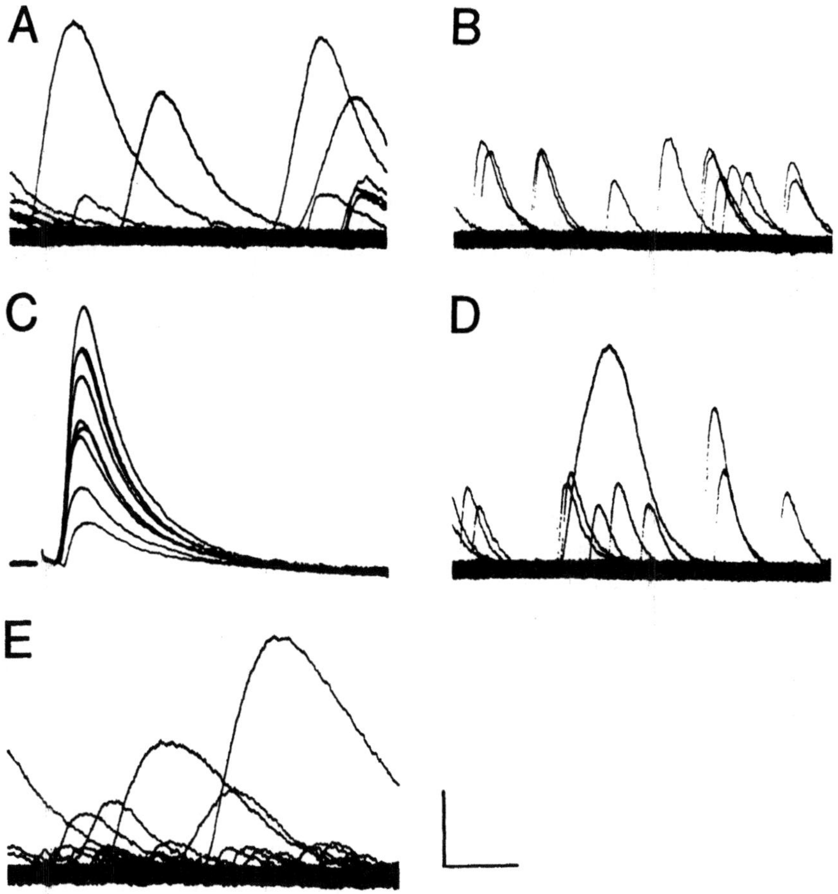

FIGURE 1. Superimposed oscilloscope tracings of calcium-dependent "fast" and calcium-independent "slow" m.e.p.ps recorded from rat extensor digitorum longus muscle. Records B and D show m.e.p.ps recorded from normal untreated end-plates; in B a uniform population of "fast" m.e.p.ps is present, while in D one "slow" m.e.p.p. can be observed among the population of "fast" m.e.p.ps. Records A and C are respectively m.e.p.ps and e.p.ps from the same neuromuscular junction in a botulinum toxin paralyzed muscle (12 days). Note the variably prolonged time-to-peak of many of the m.e.p.ps as compared to the e.p.ps. E is a record of m.e.p.ps in the normal muscle after the application of 200 μM 4-aminoquinoline. The drug increases the frequency of "slow" m.e.p.ps without affecting the uniform population of "fast" m.e.p.ps. Voltage calibrations are 1 mV for all tracings except C for which it is 2 mV. Time calibrations are 2 ms for all tracings except B, D and E, for which it is 4 ms. All records at 30°C.

Table 1. Summary of the effects of various procedures on the frequency of calcium-dependent "fast" and calcium-independent "slow" m.e.p.ps at normal, BoTx poisoned and 4-aminoquinoline-treated end-plates.

Agent or procedure	Normal muscle "fast" m.e.p.ps only	BoTx-poisoned all m.e.p.ps	4-aminoquinoline "slow" m.e.p.ps only
Ca^{2+} 8 mM	Increase 4x*) (8)	No change (8)	No change (33)
K^+ 20 mM	Increase 20x (47)	Increase 2x (47)	No change (33)
K^+ 0 mM	Increase 15x (8)	No change (8)	No change (33)
Ouabain 0.2 mM	Increase 50x (8)	No change (8)	-
Ethanol 0.5 M	Increase 10x (33)	No change (47)	No change (33)
Mn^{2+} 10 mM	Increase 8x (33)	No change (47)	No change (33)
Hypertonicity 2x	Increase 80x (47)	Decrease 0.3x (47)	No change (33)
Hypotonicity 0.5x	Decrease 0.3x (47)	Increase 3x (47)	Increase 3x (47)
Tetrodotoxin 1 μM	No change (47)	No change (47)	No change (33)

*) Approximate change times control.

In fact it is proposed that sub-m.e.p.ps constitute sub-units of calcium dependent m.e.p.ps (26; 48; 51). However, a population of sub-m.e.p.ps may correspond to small amplitude calcium independent m.e.p.ps since it has been reported that some sub-m.e.p.ps are not affected in frequency by changes in extracellular calcium or potassium concentrations (23) or by ethanol or hypertonic solutions (3).

3. RELEASE AT NORMAL NEUROMUSCULAR JUNCTIONS

The first one to observe the calcium independent type of transmitter release was Liley (27), who in 1957 described unusually large m.e.p.ps (giant m.e.p.ps) occurring at a low frequency at rat neuromuscular junctions. These potentials varied in frequency between fibres and were absent in many fibres, but occasionally they constituted up to 20% of all m.e.p.ps recorded at a single junction. Their frequency, as reported by Liley, was unaffected by nerve terminal depolarization and by changes in the extracellular calcium or magnesium ion concentrations. Such potentials also differed from nerve stimulus evoked quantal e.p.ps.

These observations were extended by Jansen and Van Essen (19) who observed that giant potentials consistently were of slower and more irregular shape and that their time course could not be explained by a simple summation of normal m.e.p.ps. Similar observations were made by Heinonen et al. (15) who observed that at room temperature about 4% of all m.e.p.ps released could be classified as giant m.e.p.ps. According to Colméus et al. (7) and Kim et al. (24) giant, slow rising m.e.p.ps constitute on the average 1-3% of all the m.e.p.ps released at rat extensor digitorum longus (EDL) muscle or soleus muscle end-plates at 30°C. Furthermore, these authors emphasize the variability in their occurrence between fibres and the fact that the slow rise time of these m.e.p.ps clearly distinguishes them from the normal type of m.e.p.ps and from quantal e.p.ps (see Fig. 1A and C).

4. RELEASE AT BOTULINUM TOXIN POISONED JUNCTIONS

The application of botulinal neurotoxins to skeletal muscle causes an almost complete block of evoked transmitter release, nerve impulses only occasionally releasing a single quantum of ACh. This effect lasts with type A toxin several weeks (8; 45).

The frequency of m.e.p.ps is concomitantly reduced to about 1% of normal, i.e. to about 1/min or less. The remaining m.e.p.ps have a smaller than normal amplitude classifying them as sub-m.e.p.ps (5; 8; 14; 26; 43). After 3-5 days of poisoning, large sized, slow-rising m.e.p.ps appear and their frequency subsequently increases so that about 10-14 days after poisoning their occurrence reaches 0.3-1 Hz, with large variations between fibres (Fig. 1A). At that time almost all m.e.p.ps recorded at a junction correspond either to a sub-m.e.p.p. or to a "slow" m.e.p.p., the normal type of fast-rising, uniform sized m.e.p.ps being absent (7; 24; 47).

As the effects of the toxin subside and neuromuscular transmission gradually is restored the frequency of the slow-rising m.e.p.ps is reduced while the number of fast-rising, uniformly sized m.e.p.ps is increased (24). It is of interest that the occurrence of slow-rising m.e.p.ps is more marked in muscles in which the dose of BoTx produced only a partial block of evoked transmitter release than in muscles completely paralyzed by the toxin (24).

The slow rising m.e.p.ps have all the previously mentioned characteristics of a calcium independent release of ACh in being uninfluenced in frequency by changes in extra- or intracellular calcium concentrations (Table 1). The difference between "slow" m.e.p.ps and normal "fast" m.e.p.ps becomes particularly evident since at any time after poisoning it is possible to re-introduce at the junction the population of fast-rising uniform sized calcium dependent type of m.e.p.ps by the administration of ATP-blocking agents like dinitrophenol

(42). A similar effect is achieved by the addition to the extracellular medium of a calcium ionophore A 23167 (8). Similarly, it is possible to restore nerve impulse evoked quantal release of ACh by the administration of drugs, such as the aminopyridines, which enhance voltage-dependent calcium currents across the nerve terminal membrane (30). Quantal e.p.ps evoked in such a manner are identical to fast-rising m.e.p.ps and without resemblance to the "slow" m.e.p.ps appearing spontaneously at the junction (7; 41)

5. RELEASE IN OTHER CONDITIONS OF IMPAIRED TRANSMISSION

Slow-rising m.e.p.ps of the calcium independent type are present at early stages of regenerating neuromuscular junctions (1). These m.e.p.ps may constitute up to 60% of all m.e.p.ps and are reduced in number when the frequency of normal fast-rising m.e.p.ps is increased as the synapse matures (7). Such m.e.p.ps are also present in skeletal muscles of chickens curarized during early development (10) and at neuromuscular junctions of dystrophic mice (strain 129/ReJ) as shown by Carbonetto (6).

6. RELEASE INDUCED BY 4-AMINOQUINOLINE AND OTHER DRUGS

4-Aminoquinoline in a concentration of 100-250 µM induces, within minutes of its application to mammalian muscle, the appearance of a population of m.e.p.ps with a larger than normal amplitude and a prolonged time-to-peak (Fig. 1E). This effect occurs without observable changes in the number of normal fast-rising m.e.p.ps. Thus, the drug induces a new population of "slow" m.e.p.ps which constitute 15-45% of all the m.e.p.ps recorded at the junction. The slow m.e.p.ps induced by 4-aminoquinoline have all the characteristics of a calcium independent secretion of ACh. They are unaffected in frequency by nerve terminal depolarization and by increases in intra- or extracellular calcium concentrations as previously described in Table 1. The frequency of slow m.e.p.ps induced by 4-aminoquinoline has a high temperature dependence with a Q_{10} of 12 which is quite different from that of fast-rising m.e.p.ps at the same end-plate which have a Q_{10} of 2-3. The effect of 4-aminoquinoline is long-lasting and is not reversed following washing for several hours with drug-free solution (32; 33).

4-Aminoquinoline exerts a similar effect at BoTx poisoned end-plates where it markedly enhances the frequency of "slow" m.e.p.ps without altering their time course and amplitude distribution. In the early stages of poisoning, when mostly sub-m.e.p.ps are present, it induces large amplitude, "slow" m.e.p.ps similar to those characterizing later stages of poisoning (47).

At muscles with regenerating neuromuscular junctions the drug increases the frequency of "slow" m.e.p.ps about 3 times without affecting the frequency of fast-rising m.e.p.ps (32).

In an attempt to find a morphological correlate to the release of the ACh responsible for the generation of the giant potentials, nerve terminals treated with 4-aminoquinoline were examined for the presence of unusually large synaptic vesicles or sub-sarcolemmal cisternae. No ultrastructural alterations were observed that could explain the release of the amounts of ACh responsible for the observed large potentials (39).

On the basis of this evidence Thesleff and Molgó (46) have concluded that 4-aminoquinoline, by a hitherto unknown mechanism, selectively stimulates the calcium independent mechanism for transmitter release at the mammalian neuromuscular junction.

Several other drugs, ions and procedures are known to induce the appearance of giant m.e.p.ps. Thus it has been reported that giant m.e.p.ps appear at junctions exposed for several hours to isotonic calcium solution (21) or at an extracellular pH of 4.3, or to a hypertonic bathing solution (34; 36; 37). Substitution of Na^+ by Li^+ may also induce giant m.e.p.ps (2) as does prolonged tetanic stimulation (16). During various stages of nerve terminal degeneration giant m.e.p.ps may appear (50). Drugs and toxins such as vinblastine (35; 49), taipoxin and notexin (9) have similar effects. In all these instances the appearance of giant m.e.p.ps requires a long time to develop and is accompanied by morphological changes of synaptic vesicles, i.e. the appearance of large or agglutinated vesicles which might release correspondingly large amounts of ACh. Thus it seems possible that the giant m.e.p.ps recorded in these conditions result from a modification of the calcium dependent release mechanism and not from a stimulation of the calcium independent type of ACh release. However, there have been no reports on the calcium dependency of these types of m.e.p.ps, and additional studies should be directed towards this question before definite conclusions can be made.

7. ORIGIN AND MECHANISM OF RELEASE

The agent responsible for the slow type of m.e.p.p. is presumably ACh since tubocurarine blocks the potentials in a dose-dependent manner (27; 33) and cholinesterase inhibitors increase their amplitude and duration (27).

The release characteristics, i.e. independence of calcium and nerve depolarization, etc. resemble those described for ACh released from Schwann cells in denervated frog muscle (4; 18). One therefore has to consider the possibility that very similar release mechanisms, which do not require calcium, are involved. However, this does not imply that the calcium-independent release, observed at normal and at BoTx poisoned junctions or that induced by 4-aminoquinoline, originates from the Schwann cell and not from the nerve. Presumably the release originates from the nerve as it is abolished by denervation and re-appears only following reinnervation of the end-plate (8; 33). This is the opposite to that seen with m.e.p.ps which originate from Schwann cells. Those potentials only appear after denervation with a delay of a few days (4). A study of the time-course of the disappearance of fast-rising m.e.p.ps and slow-rising m.e.p.ps, i.e. of respectively calcium dependent and calcium independent spontaneous transmitter release, following surgical section of the nerve showed that both kinds of secretion disappeared simultaneously. At normal junctions (rat EDL) in the presence of 4-aminoquinoline the disappearance occurred 17-19 hours after nerve sectioning, and in BoTx poisoned preparations 3 hours earlier (unpublished observation). Furthermore, in so-called Trembler mice which have an inherited Schwann cell abnormality resulting in hypomyelating neuropathy, 4-aminoquinoline induced the same type of m.e.p.ps as in normal animals (46).

It seemed possible that the calcium independent transmitter release could originate from nerve terminal sprouts, particularly since it is known that sprouting is prominent in BoTx poisoning (11) where this type of release dominates. To investigate this possibility BoTx poisoned muscles were directly stimulated electrically in vivo with a frequency pattern known to inhibit sprouting (31). Such stimulation failed to influence the appearance and frequency of "slow" m.e.p.ps (24). X-ray irradiation of the EDL muscle, a procedure which inhibits sprouting and Schwann cell proliferation (13), has no effect on the frequency, amplitude, or time course characteristics of m.e.p.ps at BoTx poisoned end-plates (47). Consequently, there is no evidence indicating a connection between calcium independent transmitter release and nerve terminal growth or sprouting.

The possibility that "slow" m.e.p.ps might result from local differences in cholinesterase activity at the synapse can be excluded since neostigmine (3 µM) only increases their amplitude and duration without influencing their frequency (27 and unpublished observation). Post-synaptic changes are on the whole unlikely as an explanation for such potentials since normal fast-rising m.e.p.ps occur, or can be induced, at the same end-plate without affecting the appearance and time course of "slow" m.e.p.ps (8; 33; 42; 45).

As previously mentioned, no morphological correlates have been observed to explain the release of such large quanta of ACh. Neither in BoTx poisoning nor in the presence of 4-aminoquinoline are abnormally large synaptic vesicles present. One possibility would be that slow, large amplitude m.e.p.ps resulted from the release of a row or a cluster of normal synaptic vesicles. Indirect evidence for this possibility is the observation that "slow" m.e.p.ps sometimes have a notch or a step in their rising phase indicating a composite nature. Molgó and Thesleff (33) have proposed that the release might occur from areas outside of the "active zones", i.e. from areas more distant from the post-synaptic receptor which would account for a more prolonged time-to-peak of the resulting m.e.p.p. (12).

Finally, the possibility should be considered that "slow" m.e.p.ps might represent gated release of ACh from the cytoplasmic pool of free transmitter. Such a mechanism would require that the amount released was voltage-dependent (see 44) and this possibility has so far not been investigated.

8. PHYSIOLOGICAL SIGNIFICANCE

The amount of ACh released by the calcium independent type of transmitter secretion is quite large. When neuromuscular transmission is blocked by BoTx or when the secretion is stimulated by 4-aminoquinoline "slow" m.e.p.ps may occur at a frequency up to 1 Hz. That is close to the frequency of spontaneous calcium dependent fast-rising m.e.p.ps at normal junctions. At the same time the "slow" m.e.p.ps are, on the average, of a much larger amplitude (2-3 times) than "fast" m.e.p.ps. Hence it can be concluded that the amount of ACh released to cause a "slow" m.e.p.p. must be greater than the amount responsible for a "fast" m.e.p.p.

Under resting conditions the spontaneous secretion causing "fast" m.e.p.ps represents about 3% of the total amount of ACh released, with the remaining 97% constituting molecular leakage of ACh (22). Molgó et al. (32) assayed ACh release before and in the presence of 4-aminoquinoline and found no difference. The detection limit of the mass fragmentographic technique used was about 0.3 pmol, or 10% of resting release. Therefore, the additional release induced by 4-aminoquinoline must have been less than that value. One may therefore conclude that the calcium independent ACh secretion constitutes, when maximally activated, 3-10% of the total resting release of transmitter.

The calcium independent secretion of ACh is particularly marked at junctions where synaptic transmission is blocked (BoTx-poisoning) or impaired (regenerating neuromuscular junctions; embryonic development). This suggests that the transmitter release process responsible for "slow" m.e.p.ps is activated when a functional disconnection exists between neuron and muscle. In that respect it is interesting that it has recently been described that growth cones of embryonic neurons in culture spontaneously secrete large ACh quanta (17; 52). On that basis one might propose that calcium independent secretion of ACh constitutes a more primitive, embryonic type of transmitter release (24). Perhaps the muscle, when synaptic transmission operates normally, releases a factor which exerts an inhibitory influence on the mechanism for calcium independent transmitter secretion in the nerve terminal. In favor of such a possibility speaks the observed delay of several days in the onset of this type of secretion following the block of neuromuscular transmission by BoTx.

Nothing is known about the possible physiological role of this type of transmitter secretion; certainly it cannot serve impulse transmission. One can speculate that it in some manner constitutes a chemical signal which assists the axon in finding and maintaining its proper synaptic connection. The possibility should also be considered that agents other than ACh (neuropeptides?) are co-released from the terminal and that such chemicals might have specific neurotrophic functions.

9. SUMMARY

The transmitter release at the normal adult neuromuscular junction, which is responsible for synaptic impulse transmission and a majority of the spontaneous miniature endplate potentials (m.e.p.ps), is a calcium-requiring process. However, there also exists a spontaneous quantal secretion of acetylcholine which is unaffected by increases in the intracellular calcium concentration, and therefore uninfluenced by nerve impulses. This neurosecretion gives rise to m.e.p.ps with a longer mean rise time and generally higher amplitude than the normal calcium-dependent acetylcholine release. Such m.e.p.ps are consequently designated "slow" m.e.p.ps.

The frequency of "slow" m.e.p.ps is low in normal muscle: approximately 0.04 Hz, comprising 1-4% of the total m.e.p.p. population, and with considerable variability between fibres. However, in conditions where the calcium-dependent transmitter release is impaired or blocked this type of release is prominent. Thus, an increased frequency of "slow" m.e.p.ps has been recorded from botulinum toxin poisoned junctions, regenerating neuromuscular junctions, embryonic junctions curarized during development and from neuromuscular junctions of dystrophic mice. The calcium-independent secretion of acetylcholine is selectively activated by the drug 4-aminoquinoline, which does not affect the normal calcium-dependent release process.

It is proposed that the calcium-independent type of transmitter secretion is a primitive, embryonic type of release with a neurotrophic function which is suppressed when synaptic impulse transmission is established.

REFERENCES

1. Bennett MR, McLachlan EM, Taylor RS: The formation of synapses in re-innervated mammalian striated muscle. J Physiol 233:481-500, 1973.
2. Benoit PH, Audibert-Benoit M, Peyrot M: Importance des effets presynaptiques dans le blockage de la jonction neuro-musculaire de grenouille par substitution du lithium ou sodium dans le milieu de survie. Archs ital Biol 111:323-335, 1973.
3. Bevan S: Sub-miniature end-plate potentials at untreated frog neuromuscular junctions. J Physiol 258:145-155, 1976.
4. Bevan S, Grampp W, Miledi R: Properties of spontaneous potentials at denervated motor endplates of the frog. Proc R Soc Lond B 194:195-210, 1976.
5. Boroff DA, del Castillo J, Evory WH, Steinhardt RA: Observations on the action of type A botulinum toxin on frog neuromuscular junctions. J Physiol 240:227-253, 1974.
6. Carbonetto S: Neuromuscular transmission in dystrophic mice. J Neurophysiol 40:836-843, 1977.

7. Colméus C, Gomez S, Molgó J, Thesleff S: Discrepancies between spontaneous and evoked synaptic potentials at normal, regenerating and botulinum toxin poisoned mammalian neuromuscular junctions. Proc R Soc Lond B 215:63-74, 1982.
8. Cull-Candy SG, Lundh H, Thesleff S: Effects of botulinum toxin on neuromuscular transmission in the rat. J Physiol 260:177-203, 1976.
9. Cull-Candy SG, Fohlman J, Gustavsson D, Lüllmann-Rauch R, Thesleff S: The effects of taipoxin and notexin on the function and fine structure of the murine neuromuscular junction. Neuroscience 1:175-180, 1976.
10. Ding R, Jansen JKS, Laing NG, Tönnesen H: The innervation of skeletal muscles in chickens curarized during early development. J Neurocytol 12:887-919, 1983.
11. Duchen LW: Changes in motor innervation and cholinesterase localization induced by botulinum toxin in skeletal muscle of the mouse: Differences between fast and slow muscles. J Neurol Neurosurg Psychiat 33:40-54, 1970.
12. Gage PW: Generation of end-plate potentials. Physiol Revs 56:177-247, 1976.
13. Gomez S, Duchen LW, Homsey S: Effects of X-irradiation on axonal sprouting induced by botulinum toxin. Neuroscience 7:1023-1036, 1982.
14. Harris AJ, Miledi R: The effect of type D botulinum toxin on frog neuromuscular junctions. J Physiol 217:497-515, 1971.
15. Heinonen E, Jansson S-E, Tolppanen E-M: Independent release of supranormal acetylcholine quanta at the rat neuromuscular junction. Neuroscience 7:21-24, 1982.
16. Heuser JE: A possible origin of the "giant" spontaneous potentials that occur after prolonged transmitter release at frog neuromuscular junction. J Physiol 239:106P-108P, 1974.
17. Hume RI, Robe LW, Fischbach GD: Acetylcholine release from growth cones detected with patches of acetylcholine receptor-rich membranes. Nature 305:632-634, 1983.
18. Ito Y, Miledi R: The effect of calcium-ionophores on acetylcholine release from Schwann cells. Proc R Soc Lond B 196:51-58, 1977.
19. Jansen JKS, van Essen D: A population of miniature end-plate potentials not evoked by nerve stimuli. J Physiol 258:103P, 1976.
20. Katz B: The release of neural transmitter substances. Liverpool University Press, 1969.
21. Katz B, Miledi R: Spontaneous and evoked activity of motor nerve endings in calcium ringer. J Physiol 203:689-706, 1969.
22. Katz B, Miledi R: Transmitter leakage from motor nerve endings. Proc R Soc Lond B 196:59-72, 1977.
23. Kelly SS, Robbins N: Bimodal miniature and evoked end-plate potentials in adult mouse neuromuscular junctions. J Physiol 346:353-363, 1984.
24. Kim YI, Lömo T, Lupa MT, Thesleff S: Miniature end-plate potentials in rat skeletal muscle poisoned with botulinum toxin. J Physiol in press, 1984.
25. Kriebel ME, Gross CE: Multimodal distribution of frog miniature end--plate potentials in adult, denervated and tadpole leg muscle. J Gen Physiol 64:85-103, 1974.
26. Kriebel ME, Llados F, Matteson DR: Spontaneous subminiature end-plate potentials in mouse diaphragm muscle: Evidence for synchronous release. J Physiol 262:553-581, 1976.
27. Liley AW: Spontaneous release of transmitter substance in multiquantal units. J Physiol 136:595-605, 1957.

28. Llinas R, Heuser JR: Depolarization-release coupling systems in neurons. Neurosci Res Prog Bull 15:557-687, 1977.
29. Locke FS: Notiz über den Einfluss physiologisches Kochsaltz Lösung auf die Erregbarkeit von Muskel und Nerf. Zbl Physiol 8:166, 1894.
30. Lundh H, Leander S, Thesleff S: Antagonism of the paralysis produced by botulinum toxin in the rat. J Neurol Sci 32:29-43, 1977.
31. Lömo T, Westgaard RH: Further studies on the control of ACh sensitivity by muscle activity in the rat. J Physiol 258:603-626, 1975.
32. Molgó J, Gomez S, Polak RL, Thesleff S: Giant miniature endplate potentials induced by 4-aminoquinoline. Acta physiol scand 115:201-207, 1982.
33. Molgó J, Thesleff S: 4-aminoquinoline-induced "giant" miniature endplate potentials at mammalian neuromuscular junctions. Proc R Soc Lond B 214:229-247, 1982.
34. Pécot-Dechavassine M: Effets conjugués du pH et des cations divalents sur la libération spontanée d'acetylcholine au niveau de la plaque motrice de la grenouille. CR Acad Sci Paris D 271:674-677, 1970.
35. Pécot-Dechavassine M: Action of vinblastine on the spontaneous release of acetylcholine at the frog neuromuscular junction. J Physiol 261:31-48, 1976.
36. Pécot-Dechavassine M, Couteaux R: Recherches sur la signification physiologique et structurale des potentiels miniature d'amplitude anormale observés au niveau de la jonction neuromusculaire de la grenouille dans diverses conditions expérimentales. J Physiol Paris 63: 138A, 1971.
37. Pécot-Dechavassine M, Couteaux R: Potentiels miniatures d'amplitude anormale obtenus dans des conditions expérimentales et changements concomitants des structures présynaptiques. CR Acad Sci Paris D 275: 983-986, 1972.
38. Pécot-Dechavassine M, Couteaux R: Modifications structurales des terminaisons motrices de muscle de grenouille souris à l'action de la vinblastine. C r hebd Séanc Acad Sci Paris 280:1099-1101, 1975.
39. Pécot-Dechavassine M, Molgó J: Attempt to detect a morphological correlate for the "giant" miniature end-plate potentials induced by 4--aminoquinoline. Biol Cell 46:93-96, 1982.
40. Rahamimoff R: The role of calcium in transmitter release at the neuromuscular junction. In: S Thesleff (ed) Motor innervation of muscle. Academic Press, 1976.
41. Sellin LC, Thesleff S: Pre- and postsynaptic actions of botulinum toxin at the rat neuromuscular junction. J Physiol 317:487-495, 1981.
42. Sellin LC, Thesleff S, DasGupta BR: Different effects of types A and B botulinum toxin on transmitter release at the rat neuromuscular junction. Acta physiol scand 119:127-133, 1983.
43. Spitzer N: Miniature end-plate potentials at mammalian neuromuscular junctions poisoned by botulinum toxin. Nature New Biol 237:26-27, 1972.
44. Tauc L: Nonvesicular release of neurotransmitter. Physiol Revs 62: 857-893, 1982.
45. Thesleff S: Supersensitivity of skeletal muscle produced by botulinum toxin. J Physiol 151:598-607, 1960.
46. Thesleff S, Molgó J: Commentary. A new type of transmitter release at the neuromuscular junction. Neuroscience 9:1-8, 1983.
47. Thesleff S, Molgó J, Lundh H: Botulinum toxin and 4-aminoquinoline induce a similar abnormal type of spontaneous transmitter release at the rat neuromuscular junction. Brain Res 264:89-97, 1983.

48. Tremblay JP, Laurie RE, Colonnier M: Is the MEPP due to the release of one vesicle or to the simultaneous release of several vesicles at one active zone? Brain Res Revs 6:299-314, 1983.
49. Turkanis AA: Some effects of vinblastine and colchicine on neuromuscular transmission. Brain Res 54:324-329, 1973.
50. Usherwood PNR, Cochrane DG, Rees D: Changes in structural, physiological and pharmacological properties of insect excitatory nerve-muscle synapses after motor nerve section. Nature 218:589-591, 1968.
51. Wernig A, Stirner H: Quantum amplitude distributions point to functional unity of the synaptic "active zone". Nature 269:820-822, 1977.
52. Young SH, Poo M: Spontaneous release of transmitter from growth cones of embryonic neurones. Nature 305:634-637, 1983.

DISCUSSION OF THE ARTICLE BY S. THESLEFF (SWEDEN)

D.M.J. Quastel (Canada): When we were voltage clamping the mouse, we found variable numbers of these things. They were important to us because you can't average them with ordinary mepps. It seemed to us that there were a higher percentage of them in diaphragms from very young mice. I wondered if you've noticed any correlation with age of the animals.

S. Thesleff (Sweden): We haven't looked at different ages but it has been my experience that the occurrence of these potentials may vary enormously between animals. One rat may have them in very many fibres and another rat may not have them at all.

B. Howard (U.S.): If you treat neuromuscular preparation with both botulinum toxin and black widow spider venom which itself causes a depletion of vesicles, you see that the few vesicles remaining are clustered right at the release sites - the active zones. I wonder if those clustered vesicles might be the source of your giant mepps.

S. Thesleff (Sweden): Many people have studied botulinum poisoned junctions to see if there are clustered vesicles but have failed to detect them. In the case of black widow spider venom, it may be different. Many investigators have demonstrated large, slow rising mepps as a result of drug treatments of various kinds. Vincristine, changes in pH, presynaptically acting snake toxins, etc. In those conditions, some of the large mepps might be similar to what I demonstrated here but so far, I don't know that the calcium-dependence, or independence, of the release has been established and in quite a number of those occasions people have found large vesicles or clusters of vesicles which apparently are absent in my conditions.

B. Howard (U.S.): In one slide, you showed a differential effect of hypertonicity on mepp frequency for 4-amino-quinoline treated junctions as opposed to botulinum toxin poisoned junctions. Did you feel a different mechanism is involved in introduction of those mepps?

S. Thesleff (Sweden): No. It was a small difference that we observed. I am not willing to say whether it would represent a different mechanism.

D.O. Smith (U.S.): In reference to the age dependency: We have looked at 10 and 20 month old rats but we have not seen any change in giant mepps frequency in the older animals. In terms of vesicle arrangement, we see more vesicles packed in the same size terminal.

B. Katz (U.K.): You mentioned the similarity of the original observations of Liley. Liley made a point in saying they are multiquantal and had histograms which indicated that they were multiples but the time course is really against it. Isn't it?

S. Thesleff (Sweden): I think so. The potentials may have a very slow time course. It would be very difficult from the amplitudes to deduce anything regarding quantum content.

I. Steinberg (Israel): Is it sure that the origin of this is vesicular?

S. Thesleff (Sweden): No. We obviously don't know. It could be gated release of cytoplasmic ACh. We have no evidence for it to be vesicular release, nor evidence for it being gated release. For gated release, one would expect a dependency on membrane potential. We have not yet checked this possibility.

B. Ceccarelli (Italy): I am wondering if it couldn't be a completely different phenomenon: the so called compound exocytosis. For compound exocytosis it is meant a fusion with a vesicle that has been already fused with the presynaptic membrane. The compound exocytosis is a process seen in many secretory systems. However, it is rarely seen at frog neuromuscular junction.

S. Thesleff (Sweden): We do not understand the underlying release or fusion mechanism. The point here is that it is not a calcium-dependent mechanism.

R. Rahamimoff (Israel): I am pondering whether this calcium independent release of ACh is not similar to processes that occur in many other cells where Golgi apparatus or other structures fuse with other membranes in order to transfer their contents and components. I wonder, does 4-amino-quinoline express its effect in other types of fusion processes?

S. Thesleff (Sweden): We have not studied it. In this connection, I should like to mention that the amount of ACh released by the calcium independent type of release is just as large as the release accounting for normal mepps at resting condition. So it is quite a large amount of ACh that can be released by this machanism.

ON THE VOLTAGE DEPENDENCE OF NEUROTRANSMITTER RELEASE

H. PARNAS, I. PARNAS AND J. DUDEL*
Dept. Neurobiology, Institute of Life Sciences, Hebrew University of Jerusalem, and * Institute of Physiology, Technical University, Munich.

The process of nerve evoked transmitter release requires Ca^{++} in the extracellular medium (1,2). The specific link for release is the increase in inracellular Ca^{++} concentration (3,4) resulting from terminal depolarization and inflow of calcium. Indeed, this is considered the only role of membrane depolarization; no additional effects are presumed in any of the processes that follow the entry of calcium ions and lead to the fusion of synaptic vesicles with the terminal membrane, exocytosis and release. If it is accepted that Ca^{++} ions are the only trigger for release, it follows that initiation of release results from an increase in intracellular Ca^{++} concentrations near critical sites. Correspondingly, termination of release must result from a drop in Ca^{++} concentration near these critical sites to levels insufficient to support release (5,6,7).

The theory that transmitter release is determined by intracellular Ca^{++} concentration received support from experiments involving two pulse stimulation. At room temperature, the process of evoked release lasts for about 1-2 ms. Following this period, however, spontaneous release frequency (mepp) remains higher than normal declining gradually to resting levels over a period of tens of ms. If during this period a second stimulus pulse is delivered, release in response to the second pulse is increased ("facilitation"). This increase, which may last for several hundreds of ms and even seconds is accounted for by "residual calcium" remaining from Ca^{++} ions which entered following the first impulse (5). The "residual calcium" theory of facilitation which was first suggested for neuromuscular junction, was found to fit experimental results in many preparations (8,9,10,11,12,13,14). Furthermore, a theoretical analysis by H. Parnas and Segel (15) reconfirms the theory and provides a basis for rejecting other theories such as the proposal that extra Ca^{++} enters during the second impulse (16,17).

There is, however, a serious quantitative conflict between the theory that release is determined exclusively by intracellular Ca^{++} concentration and the theory that facilitation results from the presence of residual Ca^{++}. If on the one hand Ca^{++} ions are rapidly removed near critical sites in order to permit the termination of release to occur in 1-2 ms, how can residual Ca^{++} still be available after hundreds of ms in order to enable facilitation (8,18,19,20)? Direct measurements of intracellular Ca^{++} concentration using Ca^{++} sensitive probes (4,21,22) demonstrated that indeed intracellular Ca^{++} remains high for a few hundred ms. This observaton supports the "residual calcium" theory, but conflicts with the claim that the rapid termination of release is due to the rapid removal of Ca^{++}. Katz and Miledi (5) were aware of this conflict and tried to unify the two theories by taking advantage of the sigmodial relation between transmitter release and external Ca^{++} concentration demonstrated by Dodge and Rahamimoff (23). According to Katz and Miledi (5) because of the sigmoidity (cooperativity of four), a small decline in intracellular Ca^{++} will suffice to stop synchronized release while still permitting intracellular Ca^{++} to remain high enough to enable facilitation. If taken for very short intervals and assuming that all Ca^{++} ions from the first impulse remain, maximal facilitation should be 2^n, where n is the cooperativity. If n is taken as four, maximal facilitation should be 16. Using very short intervals and direct polarizaton pulses, Katz and Miledi (5) in fact found facilitation values as high as 70. On the basis of this result they suggested that additional mechanisms may be involved in very short term facilitation, but even without assuming additional mechanisms, if release is determined by intracellular Ca^{++}, then values of facilitation approaching 16 indicates that most of the Ca^{++} from the first impulse still remains at the "release site" at the time of the second impulse. Why then does release stop at all after the first impulse? In order to explain both the time course of release following a single pulse and that of facilitation, Barrett and Stevens (24) claimed that at least two Ca^{++} removal processes are required: A fast and a much slower one. According to this assumption, the fast process accounts for the time course of a single pulse release and the slow process accounts for facilitation. Extending this line of reasoning to a specific removal process, diffusion, Zucker and Stockbridge (6) and recently Stockbridge and Moore (7) were able to explain by computer simulation, the time course of a single pulse release and that of twin pulse facilitation. Unfortunately Zucker (25) later demonstrated that the same assumption makes it impossible to explain the termination of release after a train of impulses. This is because of the accumulation of Ca^{++} near the release sites.

H. Parnas and Segel (26) examined the calcium's role in synaptic release in a more general way. They showed that if the rate of release is determined solely by Ca^{++}, then regardless of the number and nature of the Ca^{++} removal processes two things must be true: a) The duration of the evoked release must increase as the number of pulses increases. b) As extracellular Ca^{++} concentration increases, the duration of release to a single pulse should increase. Neither of these predictions were verified experimentally.

There is obviously inconsistency between the theory that states that release is terminated by rapid removal of Ca^{++} and the one that states that residual Ca^{++} is the basis for facilitation. Either or both theories may be wrong and certainly neither can be used to disprove the other. Rather, independent evidence must be gathered to test each.

One approach is to examine in detail the kinetics of the release of a single quanta. This can be measured directly by constructing delay histograms (e.g. histograms of post-stimulus response time) of single quanta at low temperatures and low quantal content (27). Using this procedure, Datyner and Gage (28) showed that the time course of the delay histograms is not changed by increasing extracellular Ca^{++} concentration or after repetitive stimulation. This finding led Parnas and Segel (26) to conclude that the termination of release cannot be due to exhaustion of intracellular Ca^{++} even locally near the release sites. For this and other reasons these authors suggested (26,29) that in addition to intracellular Ca^{++} release must also depend on another voltage dependent parameter. Experimental support to such a conclusion comes from the work of Llinas et al (30). They demonstrated using voltage clamp in the squid giant synapse that under conditions of constant Ca^{++} current release is higher at higher depolarization.

In the rest of this paper we will lay out our theory which attempts to solve the difficulties mentioned above.

We propose that the action potential reaching the nerve terminal has at least two roles. One is to increase membrane conductance to Ca^{++}. The second is to activate the release machinery for example by producing a conformational change in a membrane protein, rendering it more sensitive to the increased level of intracellular Ca^{++}. Thus, it is the combination of a sensitized release machinery together with the increased level of intracellular Ca^{++} that enables release. Once the action potential is over and the membrane repolarizes, the release machinery inactivates with a time constant of a few ms. This stops synchronized release even though intracellular Ca^{++} concentration may remain high. The residual high Ca^{++} level supports facilitation. Note, that according to our approach, Ca^{++} is a necessary but not sufficient factor for release of neurotransmitter.

In order to support such a theory it is necessary to disassociate Ca^{++} entry from transmitter release at different terminal depolarizations and to directly demonstrate release at various depolarizations while controlling for the effects of Ca^{++} entry. This requires measurements of Ca^{++} entry. Unfortunately, in most nerve terminals Ca^{++} current cannot be measured directly. Indirect methods must be used in order to estimate the amount of Ca^{++} that enters during a depolarizing pulse. One such indirect method is to determine the duration of twin pulse facilitation. Facilitation reflects well the amount of Ca^{++} entering during the first pulse (10,11). However, when release is evoked by variable amplitude depolarization rather than by the constant amplitude of the action potential, the estimate of Ca^{++} entry by this method is confounded by the proposed direct effects of depolarization on the release machinery. This problem was solved by Dudel et al (31). They measured the quantal content for a constant depolarizing test pulse which was preceeded (at various times intervals) by a prepulse of variable amplitude. By this procedure facilitation is measured as the ratio of the quantal content for the test pulse with and without a conditioning prepulse. Figure 1 shows the result of such an experiment (31). Using condition (pre) and test pulse, the quantal content of the prepulse (solid line) increased monotonically with increasing depolarization. At the same time, short term facilitation (F_c=F_{7ms}) and the duration of faciliation (T_F) first increased but then decreased. This means that there exists a range of depolarization in which the quantal content of the test pulse decreases while that of the prepulse continues to increase. Based on the reduction in F_c and T_F we

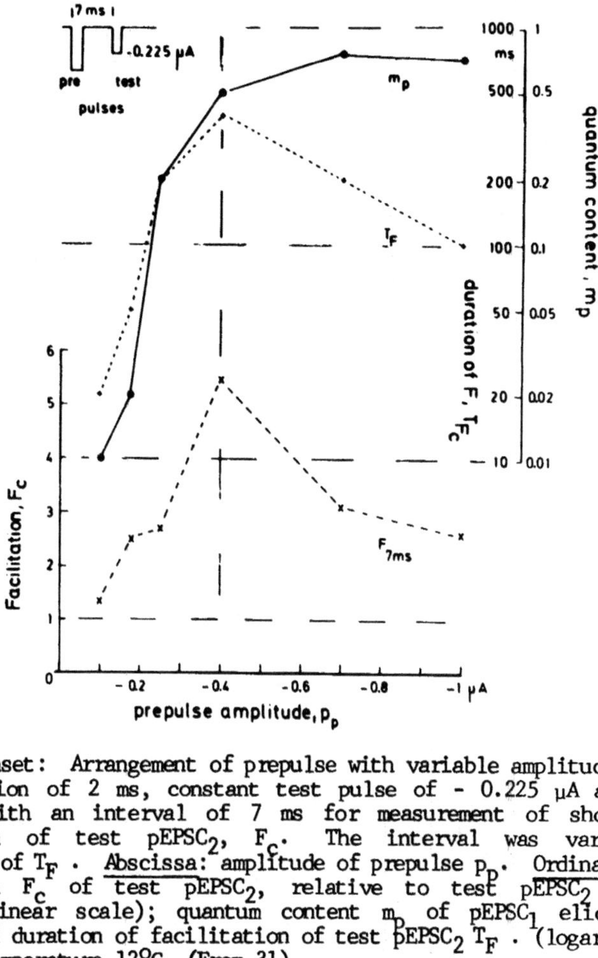

FIGURE 1. Inset: Arrangement of prepulse with variable amplitude p_p and fixed duration of 2 ms, constant test pulse of - 0.225 μA and 2 ms duration, with an interval of 7 ms for measurement of short time facilitation of test $pEPSC_2$, F_c. The interval was varied for measurement of T_F . Abscissa: amplitude of prepulse p_p. Ordinates: facilitation F_c of test $pEPSC_2$, relative to test $pEPSC_2$ without prepulse (linear scale); quantum content m_p of $pEPSC_1$ elicited by prepulse and duration of facilitation of test $pEPSC_2$ T_F . (logarithmic scales). Temperature 12°C. (From 31).

suggest that at the higher prepulse depolarization less Ca^{++} in fact enters into the terminal (because the driving force is smaller) and in spite of this the quantal content of the prepulse continuous to increase due to the activation of the release machinery at high depolarizations. Indeed this result is similar to that of Llinas et al (30) in the giant synapse of the squid, who obtained greater release for the same Ca^{++} current at high depolarization.

If indeed depolarization activates the release machinery, the best way to study the relation between depolarization and this activation is to measure release at different levels of depolarization but with high intracellular Ca^{++} concentration. In this case the unavoidable additional entry of Ca^{++} at the various depolarizations will not alter significantly the Ca^{++} concentration. It is possible to approach this condition by loading the terminal with Ca^{++} by repetitive stimulation. Such an experiment is shown in Figure 2. First the dependence of the quantal content (m) to single pulses at different depolarizations was determined (crosses in Figure 2). The nerve was then stimulated with a train of 6 pulses and the depolarizing test pulse was given again a few

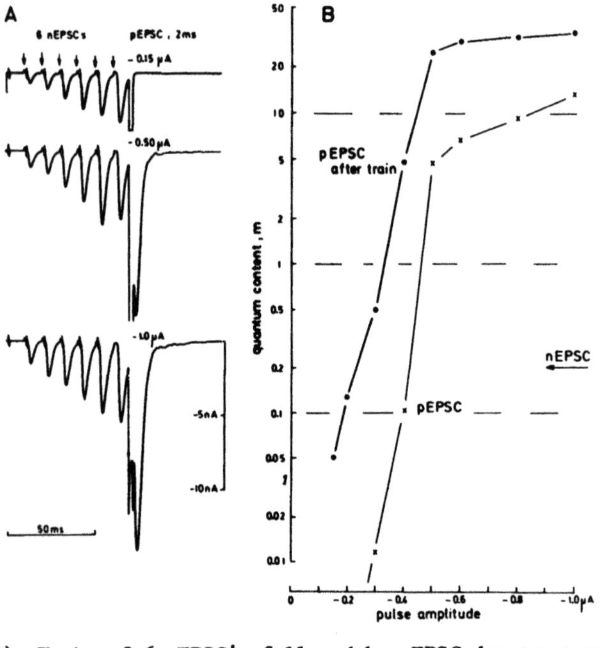

FIGURE 2. A) Train of 6 nEPSC's followed by pEPSC due to a pulse of 2 ms duration and -0.15 µA, -0.50 µA and -1.0 µA amplitude, respectively. The nerve terminal potentials of the train are marked by arrows, the stimulation intervals within the train were 10 ms. The delay betweenlast nerve terminal potential and pulse was 9 ms. The size of the quantum was C_1 = - 431 pA. Each trace average of 32 to 128 single recordings. 12°C. B) Plot of results of same experiment as in a. Abscissa: amplitude of 2 ms pulse. Ordinate: quantum content of pEPSC released by pulse: (x) release by pEPSC alone, (o) released by pEPSC preceded by train, as shown in a. Quantum content of single nEPSC at beginning of experiment marked by arrow. (From 31).

ms later. The dependence of m on depolarization under these conditions was similar to the control, but the whole curve was shifted upwards (circles in Figure 2). The shift reflects the effect of higher intracellular Ca^{++} concentration produced by the pulse train while the unaltered slope reflects the activation of the release machinery by depolarization.

If, as we claim, depolarization has a direct effect on the release machinery, sensitizing it to internal Ca^{++}, then in principle it should be possible to evoke release without the entry of Ca^{++} providing the depolarization is delivered when intracellular Ca^{++} is already high. The most straightforward experiment of this type would be to inject Ca^{++} intracellularly and then to depolarize the terminal in the absence of Ca^{++} in the bathing fluid. Again, such an experiment is impossible in very fine nerve terminals although it may be possible in the squid giant synapse. Therefore, as before we approached this condition by using repetitive stimulation (Figure 3). First the terminal was depolarized with a short (1ms) but relatively strong depolarizing test pulse. Because of the short duration of the pulse, very little Ca^{++} entered and release was negligible, m being 0.006. The nerve was then stimulated 6 times and the evoked synaptic current showed a large facilitation (of 15). Immediately following the synaptic current of the 6th pulse, the same short depolarizing test pulse was given again, but now it elicited a strong response, m being 9.6. This response was in fact larger than the release evoked by the 6th nerve action potential. This can be explained by the strength of the depolarizing pulse which was greater in amplitude than that produced by the action potential. Test pulse facilitation was thus more than 1500.

This high facilitation indicates that Ca^{++} accumulated to high levels during the 6 pulses and yet release stopped after the train to start again only upon depolarization with a test pulse which itself produced only negligible release. We argue that the test pulse depolarization sensitized the release machinery to Ca^{++} that was already in the terminal. This experiment was repeated in the presence of TTX with similar results. Of course in this case the nerve could not be stimulated and therefore the train of 6 pulses intended to load the terminal with Ca^{++} was given by direct depolarizing pulses.

It should be emphasized that a depolarizing pulse is required to evoke synchronized release. Increase in intracellular Ca^{++}

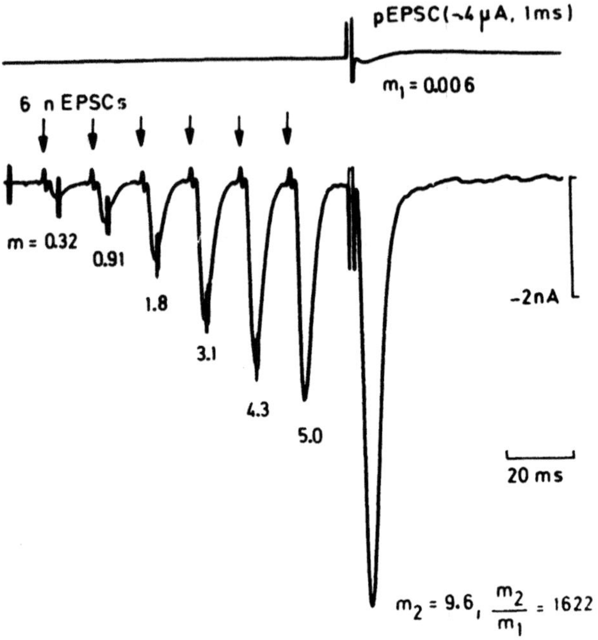

FIGURE 3. Facilitation by train. Upper trace: pEPSC elicited by -0.4 μA, 1 ms pulse. Counted release rate m_1 = 0.006 quanta, the deflection after the pulse is largely artefact. Average of 512 single records at 2.2 s interval. Lower trace: Train of 6 nEPSC's, nerve terminal potentials marked by arrows, interval of nEPSCs 13 ms, quantum content of nEPSCs m marked at maxima. Train followed, 17 ms after last terminal potential, by same pulse as in upper trace, quantum content of pEPSC m_2 = 9.6 Average of 32 single records with 12.2 s interval to avoid increase of facilitation from train to train. Temperature 11°C. Facilitation of pEPSC by train F_c =1622. (From 31).

concentration alone may suffice to increase the frequency of spontaneous release as demonstrated by Miledi (3), but in order to obtain synchronized release, the membrane must also be depolarized. Recently Llinas (32) reported in a short abstract that after injection of Ca^{++} into the squid giant synapse, depolarization evoked release even though no Ca^{++} was present in the bathing fluid which in addition contained EGTA.

As mentioned previously, there is a flaw in the experiments described above, as they are all based on facilitation, specifically on the assumption that facilitation reflects residual Ca^{++}. We can obtain direct insight into the effects of depolarization on the onset and termination of neurotransmitter release by determining the kinetic of a single quanta release under various conditions of membrane potential.

As suggested first by Katz and Miledi (27) and later applied by several authors (24,28), synaptic delay histograms constructed at low quantal content and low temperatures reflect the probability of release as a function of time after the depolarizing pulse. Figure 4 shows such histograms obtained at different levels of depolarizing pulses of 2 ms duration (33). A few points should be emphasized. a) The peak of the histogram at all depolarizations occurs a few ms after the end of the

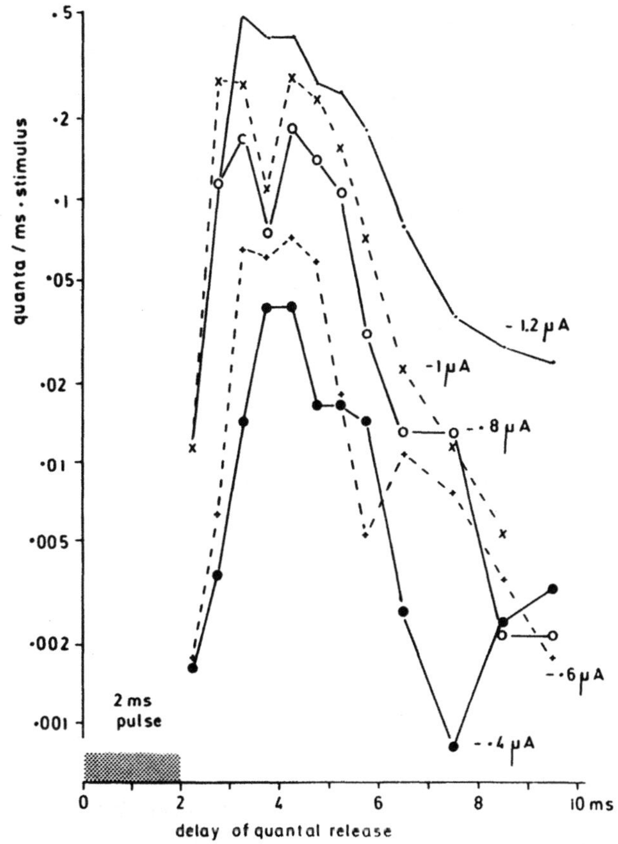

FIGURE 4. Distributions of the delays of quantal release after the start of a 2 ms depolarization pulse, for pulse amplitudes of -0.4 to -1.2 μA, as indicated in each plot. In addition, the average number of quanta, N_q, entered into the respective distribution are indicated. The rate of release in the ordinate (quanta/ms) gives the probability that the delay of one quantum is in the respective time interval of the distribution, i.e. the distributions are scaled to the same area 1. Repetition rate of pulses 0.45/s, temperature 4°C. (From 33).

depolarizing pulse and not during or immediately after it. b) As the depolarization increases more quanta are released at short latencies. c) There is a minimal latency at which no release is seen, and d) the termination of release has about the same kinetics in all depolarizations.

Based on these findings and another result discussed below, Parnas et al (34) suggested a minimal model to describe the basic steps leading to release of a quanta and rejected several possibilities.

The main idea of the model is that one (or more) component is transformed by depolarization to an active form that binds Ca^{++}. In our model we termed the activated component S, for site. For example, S could be release site, a protein that binds Ca^{++}. This complex of S with its bound Ca^{++} then participates in the binding of a vesicle, V, to release its content. There is a need for n such complexes for the fusion of one vesicle. The process of binding the vesicle is the slowest process in the chain of events which lead to release. Upon repolarization, S inactivates with a certain time constant to its inactive form T. At rest most of the sites are in their inactive form. For reasons explained below, we assume two inactive states, T_1 and T. Thus

$$T_1 \underset{k_{-0}}{\overset{k_0}{\rightleftharpoons}} T \underset{k_{-1}}{\overset{k_1}{\rightleftharpoons}} S \qquad (1.1)$$

$$S + Ca^{++} \underset{k_{-2}}{\overset{k_2}{\rightleftharpoons}} (S\ Ca^{++}) \qquad (1.2)$$

$$n(S\ Ca^{++}) + V \xrightarrow{k_3} L \qquad (1.3)$$

where, k_0 and k_1 are rate constants that increase with depolarization, S is thus produced by depolarization. k_{-0} and k_{-1} are rate constants that increase with hyperpolarization, n is the number of complexes that participate in the binding of one vesicle, V. It should be pointed out that with the aid of our model we can not only estimate n, but also distinguish where in the chain of events the Ca^{++} cooperativity takes place. As seen in (1.2) we do not assign cooperativity to the binding of Ca^{++} to S. We propose that n complexes of (S Ca^{++}) bind a vesicle to produce release (1.3). Using this model, we simulated release at low, medium and high amplitude depolarizing pulses (Fig. 5). The essential

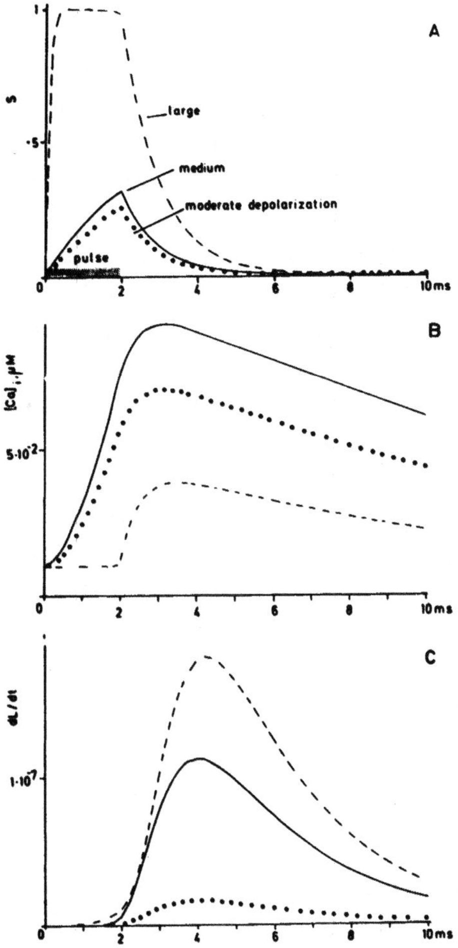

FIGURE 5. Simulation of release reactions according to eq. (1), but including variation in $[Ca]_i$ due to Ca inflow and removal. Effects of moderate, medium and large depolarization pulses of 2 ms duration on S(A), $[Ca]_i$ (B), and the rate of release dL/dt (C). Parameters: $k_1^D=$ 0.15/ms for moderate, $k_1^D = 0.2$/ms for medium, $k_1^D = 10$/ms for large depolarization; $k_1^0 = 0$, $k_{-1}^D = 0.01$/ms, $k_{-1}^0 = 1$/ms. $k_2 = 1$, $k_{-2} = 0.1$; $k_3 = 10$, $k_6^D = 0.5$/ms and $k_m^D = 6\,\mu M$ for medium, $k_6^D = 7$ ms and $k_m^D = 5$mM for large depolarization; $k_6^0 = 0$ and $k_m^0 = 6\ \mu M$, n = 4. (From 33, Equations for Ca inflow are presented in 33).

results (and mentioned above) are fully met. Specifically, the peak of the rate of release (peak histogram) appears few ms after the end of the depolarizing pulse. There is a minimal delay at which no release is expected. This minimal latency becomes somewhat shorter at higher depolarizing pulses and finally termination of release appears at almost the same time for all levels of depolarizations. The details of the model and the experiments as well as the considerations that led us to prefer this particular one upon several other possibilities are beyond the scope of this paper. The reader is referred to our recent articles (33,34,35).

One of the results that the model predicts is that the kinetics of release should be almost the same to a single pulse or after a train of pulses. Indeed this was found to be the case as shown in Figure 6.

If depolarization activates the release machinery (by producing S), is it possible to modulate release by hyperpolarizing pulses? We found that it is. Figure 7 shows that if a hyperpolarizing pulse is given <u>after</u> a depolarizing pulse it inhibits release. Similarly, a hyperpolarizing pulse preceeding the depolarizing pulse renders the latter less effective. Is it possible that the hyperpolarizing pulse blocks release by reducing the entry of Ca^{++}? This is not the case as can be seen in Table 1. Here, facilitation of a second pulse was used as an index for Ca^{++} entry during the first. Two pulses were given to the terminal, each produced release. The 2nd pulse showed facilitation due to residual Ca^{++} from the first. A hyperpolarizing pulse was then given immediately before or after the first pulse reducing its release by as much as 60%. Nonetheless, the second pulse showed the same facilitation as if the original amount of Ca^{++} had entered during the first pulse. Thus modulation of release by the hyperpolarizing pulse was not mediated by a reduction in the entry of Ca^{++} but by blocking the release machinery.

What might the role of hyperpolarizing pulses be? We will first deal with hyperpolarizing pulses that follow the test pulse. During depolarization k_1 is high and S is formed. Post-pulse hyperpolarization increases k_{-1} above is elevated level at rest and S inactivates to become T faster than upon normal repolarization. Postpulse hyperpolarization thus reduces the quantal content of the test pulse. But what is the effect of the hyperpolarizing prepulse? Obviously, S is low at rest, otherwise release would be continuous. Therefore, further lowering of S by increasing k_{-1} cannot be the reason for the block of release by hyperpolarizing prepulse. In order to explain the effect of hyperpolarizing polarizing prepulses we have included a second inactive form of T, T_1, with k_{-0} (like k_{-1}) increasing with hyperpolarization with k_0 increasing with depolarization (eq. 1.1). At rest S is low and

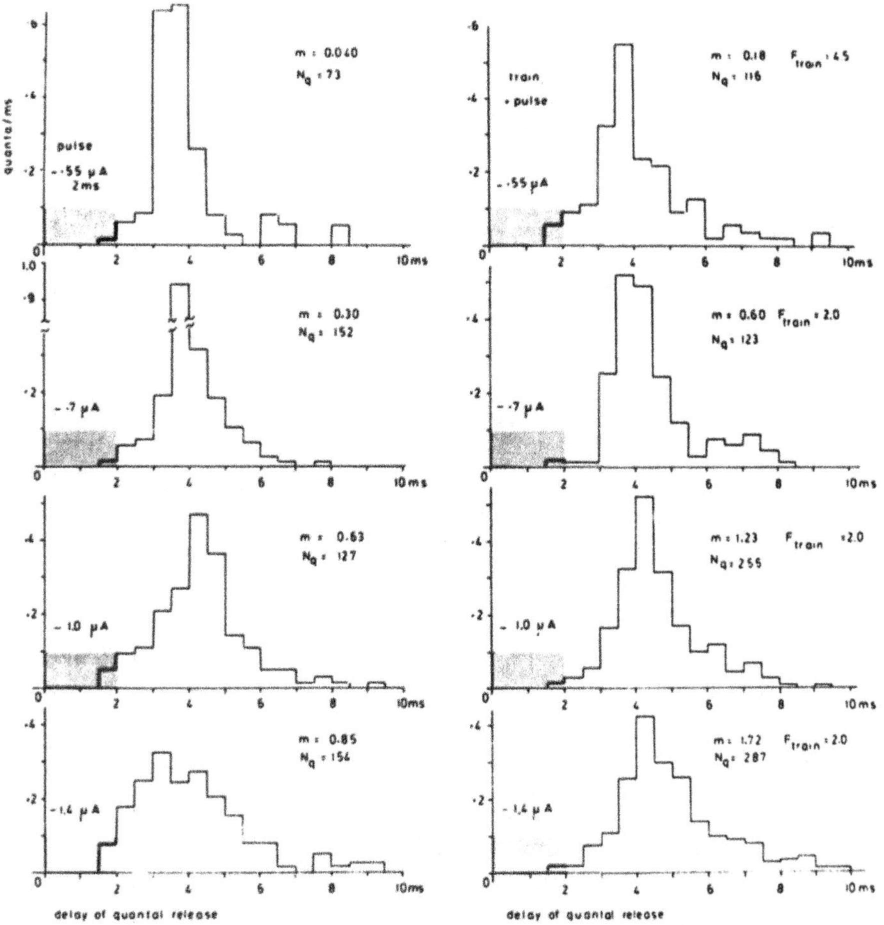

FIGURE 6. Distributions of quantal delays after the beginning of 2 ms depolarization pulses. Ordinate quanta/ms. The pulse amplitude rises from top to bottom from -0.55 μA to -1.4 μA. In the left hand row of plots, only the test pulses were applied, while in the right hand row of plots, the respective test pulses were preceded by 50 ms by conditioning trains of pulses (3 pulses, -1.2 μA, 2 ms, at 100 Hz). F_{train} gives the respective facilitation by the preceeding train. Temperature 4°C. (From 33).

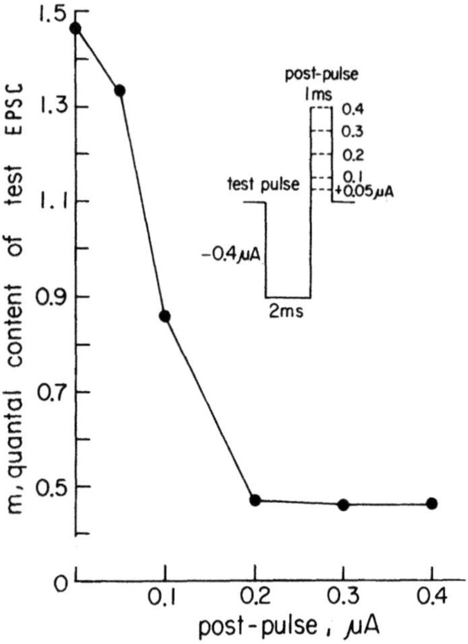

FIGURE 7. Dependence of the average quantum content of a test EPSC, m, on the amplitude of a hyperpolarizing post-pulse. The test pulse which elicits the EPSC has a constant amplitude of -0.4 µA and 2 ms duration, while the amplitude of the post-pulse which folllows the test pulse immediately varies between +0.05 +0.04 µA, as indicated in the inset. The results of at least 256 stimulations were evaluated for each point in the diagram. (From 35).

the release sites are mainly in the form of T and T_1. Prepulse hyperpolarization, due to increase in k_{-0} (as well as k_{-1}) shifts T toward T_1. When the test pulse is now given, it faces a lower fraction of T and therefore less S is produced and consequently release is reduced.

An independent test of the model not based on measurements of quantal content is to check the kinetics of release with and without pre and post hyperpolarizing pulses. This is shown in Figure 8 (see legends for predictions and experimental results). The hyperpolarizing postpulse should not change the beginning of the histogram, but the peak should shift to the left and the histogram becomes shorter (Figure 8 right side). On the other hand, the hyperpolarizing prepulse is expected to reduce the amplitude of the whole histogram without significant changes in the time course (Figure 8 left side). All these predictions were experimentally confirmed (Figure 9).

Facilitation Experiment

	control		+ pre-pulse				+ post-pulse			
	m_1	m_2	m_{1p}	m_{2p}	$\frac{m_{1p}}{m_1}$	$\frac{m_{2p}}{m_2}$	m_{1p}	m_{2p}	$\frac{m_{1p}}{m_1}$	$\frac{m_{2p}}{m_2}$
1	0.27	0.41	0.12	0.48	0.44	1.17				
2	0.30	0.65	0.17	0.56	0.56	0.90				
3	0.43	0.78	0.16	0.72	0.37	0.92				
4	0.33	0.58	0.15	0.59	0.45	1.01				
5	0.37	0.58	0.04	0.47	0.10	0.81				
6	0.49	0.72	0.35	0.75	0.70	1.04				
7	0.68	0.99	0.31	0.99	0.45	1.00				
8	0.66	1.11	0.31	1.01	0.47	0.91	0.17	1.17	0.25	1.05
9	0.59	0.79					0.20	0.84	0.39	1.06
10	0.19	0.55					0.09	0.46	0.47	0.84
11	0.27	0.51					0.13	0.53	0.48	1.04

test pulse: $-0.4 \to 0.6\,\mu A$, 2ms, $n = n^8 \to 2^9$

pre-post pulse: $+0.4 \to 0.6\,\mu A$, $1 \to 2$ms, zero delay, $n = 2^8 \to 2^9$

Table 1: m_1 average quantum content of EPSC elicited by first test pulse, m_2 the respective value for the second test pulse (following with 10 ms delay). m_{1p} and m_{2p} are the analogous values measured when a pre- or post-pulse was given, respectively. The test pulses had amplitudes of +0.4 to 0.6 μA and 1 to 2 ms duration and zero interval with regard to the test pulse. The effects of 2^8 to 2^9 stimulations were evaluated for each value in the table. (From 35).

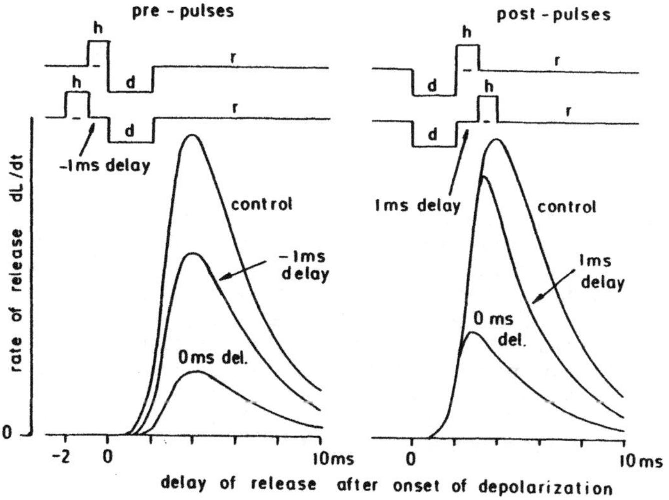

FIGURE 8. Simulated time courses of rate of release, dL/dt, after a test pulse (controls), and with this release modulated by hyperpolarizing pre- or post-pulses, which occurred with 0 or 1 ms delay from the test pulse, respectively. Same representation and conditions of the simulation as in Figure 5. (From 35).

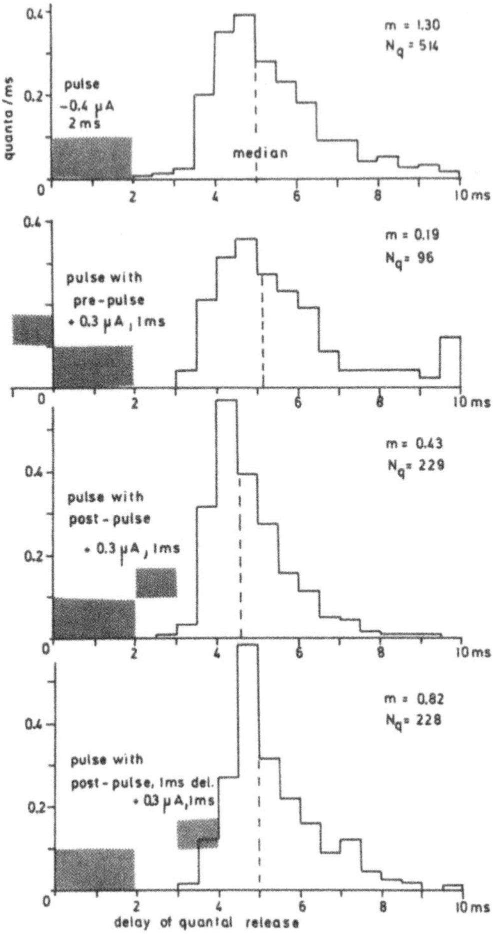

FIGURE 9. Distributions of delays of quantal releases after the onset of the respective depolarizing test pulse. The ordinates give the probability of a quantum to have a delay in the respective interval, i.e. the distributions are normalized to an area 1. The test pulse is the same in all distributions (-0.4 μA, 2 ms), and so are the added modulating pulses (+0.3 μA, 1 ms). The position of the modulating pulses as pre- or post-pulses is indicated in the respective diagram, which also contains the average quantum content, m, of the EPSCs and the number of delays, N_q, evaluated. A dashed vertical line gives the median in each distribution. (From 35).

In summary, we are proposing that depolarization of the presynaptic terminal, in addition to its known role of increasing membrane conductance to Ca^{++}, has the role of activating the release machinery. The onset of release is determined at most levels of depolarizations (excluding maybe the very high levels) by the activation of the S Component (or process) and the termination of release by the inactivation of the S component. Ca^{++} is an essential factor in release, but changes in its intracellular concentration do not determine the kinetic of the synchronized release. This model simultaneously accounts for the mechanisms involved in release after a single pulse, facilitation with twin pulses and after multiple pulses.

Furthermore, we show that release of transmitter can be modulated by slight changes in membrane potential: Hyperpolarization decreases release, slight depolarization increases release. Facilitation of release by S activation differs in mechanisms from that produced by residual Ca^{++} and it falls into the category of "different mechanisms for facilitation" that Katz and Miledi (5) proposed, must exist.

Acknowlegements We would like to thank Marshall Devor, for many helpful comments and reading an earlier version of this manuscript.

References

1. Castillo D, Katz, B: The effect of magnesium on the activity of motor nèrve endings. J. Physiol. 124, 553-559, 1954.
2. Katz, B, Miledi, R: The effect of calcium on acetylcholine release from motor nerve terminals. Proc. Roy. Soc. (London). B. 161, 496-503, 1965.
3. Miledi, R: Transmitter released induced by injection of calcium ions into nerve terminals. Proc. R. Soc. B. 183, 421-425, 1973.
4. Llinas, R, Nicholson, C: Calcium role in depolarization-secretion coupling: An aequorin study in squid giant synapse. Proc. Nat. Acad. Sci. USA. Vol. 72, No. 1 187-190, 1975.
5. Katz, B, and Miledi, R: The role of calcium in neuromuscular facilitation. J. Physiol. 195: 481-492, 1968.
6. Zucker, R.S., and Stockbridge, N: Presynaptic calcium diffusion and the time courses of transmitter release and synaptic facilitation at the squid giant synapse. J. Neuroscience, 3: 1263-1269, 1983.
7. Stockbridge, N, Moore, J.W: Dynamics of intracellular calcium and its possible relationship to phasic transmitter release and facilitation at the grog neuromuscular junction. J. of Neuroscience 4: 803-811, 1984.
8. Rahamimoff, R: A dual effect of calcium ions on neuromuscular facilitation. J. Physiol. 195: 471-480, 1968.
9. Smith, S.J., Charlton, M.P., Zucker, R.S: Arsenazo III reveals long-lasting intracellular calcium transient following the action potential in squid giant presynaptic terminal. Biological Bulletin. Vol. 159, 491-2, 1980.
10. Parnas, H., Dudel, J., Parnas, I: Neurotransmitter release and its facilitation in crayfish. I. Saturation kinetics of release and of entry and removal of calcium. Pflugers Arch. 393: 1-14, 1982.

11. Parnas, I., Parnas, H., Dudel, J: Neurotransmitter release and its facilitation in crayfish. II. Duration of facilitation and removal processes of calcium from the terminal. Pflugers Arch. 393: 232-236, 1982.
12. Parnas, H., Dudel, J., Parnas, I: Neurotransmitter release and its facilitation in crayfish IV. The effect of Mg ions on the duration of facilitation. Pflugers Arch. 395: 1-5, 1982.
13. Parnas, I., Parnas, H., Dudel, J: Neurotransmitter release and its facilitation in crayfish muscle v. Baris for synapse differentiation of the fast and slow type in one axon. Pflugers Arch. 395: 261-270, 1982.
14. Dudel, J., Parnas, I., Parnas, H: Neurotransmitter release and its facilitation in crayfish III. Amplitude of facilitation and inhibition of entry of calcium into the terminal by magnesium. Pflugers Arch. 393: 237-242, 1982.
15. Parnas, H., Segel, L.A.: A theoretical explanation for some effects of calcium on the facilitation of neurotransmitter release.
16. Zucker, R.S: Characteristics of crayfish neuromuscular facilitation and their calcium dependence. J. Physiol. 241: 91-110, 1974.
17. Staggs, D.R. Pofcher, E, L'Heureux, Ramona, Ortiz, C.L. Orkand, R.K: Excitatory neuromuscular transmission in crayfish. Calcium dependence is unaffected by Picrotoxin. J. Neurobiol. Vol. II, No. 6, 629-632, 1980.
18. Mallart, A, Martin, A.R: Two components of facilitation at the neuromuscular junction of the frog. J. Physiol. 191: 191-201, 1967.
19. Balnare, R.J., Gage, P.W: On facilitation of transmitter release at the toad neuromuscular junction. J. Physiol. 239: 657-675, 1974.
20. Balnare, R.J., Gage, P.W: Facilitation of transmitter secretion from toad motor nerve terminals during brief trains of action potentials. J. Physiol. 266: 435-451, 1977.
21. Smith, S.J. Zucker, R.S: Aequorin response facilitation and intracellular calcium accumulation in molluscan neurones. J. Physiol. 300: 167-196, 1980.
22. Miledi, R.F.R.S. Parker I: Calcium transients recorded with arsenazo III in the presynaptic terminal of the squid giant synapse. Proc. R. Soc. Lond. B 212: 197-211, 1981.
23. Dodge, F.A. Jr. Rahamimoff, R: Cooperative action of calcium ions in transmitter release at the neuromuscular junction. J. Physiol. 193: 419-432, 1967.
24. Barrett, E.F., Stevens, C.F: The kinetics of transmitter release at the frog neuromuscular junction. J. Physiol, 227: 691-708, 1972.
25. Zucker, R.S.: Biophys. J. Feb. (abstract), 1984.
26. Parnas, H. Segal, L.A.: Exhaustion of calcium does not terminate evoked neurotransmitter release. J. Theor. Biol. 107: 345-365, 1984.
27. Katz, B., Miledi, R: The measurements of synaptic delay, and the time course of acetylcholine release at the neuromuscular junction. Proc. R. Soc. B 161, 483-495, 1965.
28. Datyner, N.B., Gage, P.W.: Phasic secretion of acetylcholine at a mammalian neuromuscular junction. J. Physiol. 303: 299-314, 1980.
29. Parnas, H., Segel, L.A.: A theoretical study of calcium entry in nerve terminals with application to neurotransmitter release. J. Theor. Biol. 91: 125-169, 1981.
30. Llinas, R. Steinberg, I.Z., Walteon, K: Relationship between presynaptic calcium current and partsynaptic potential insquid giant synapse. Biophys. v. 323-352, 1981.

31. Dudel, J., Parnas, I., Parnas, H: Neurotransmitter release and its facilitation in crayfish muscle. VI release determined by both intracellular calcium concentration and depolarization of the nerve terminal. Pflugers Arch., 1983.
32. Llinas, R., Sugimori, M., Bower, J.M: Visualization of depolarization-evoked presynaptic calcium entry and voltage dependence of transmitter release in giant synapse. Biological Bull. 165: 529, 1983.
33. Parnas, H., Dudel, J., Parnas, I: Neurotransmitter release and its facilitation in crayfish VII. Ca in flow does not determine the time course of phasic release. Pflugers Arch. (submitted).
34. Parnas, I., Parnas, H., and Dudel, J: Neurotransmitter release and its facilitation in crayfish. VIII. Direct activation of phasic release by depolarization. Pflugers Arch. (submitted).
35. Dudel, J., Parna, I., Parnas, H. Neurotransmitter release and its facilitation in crayfish IX. Modulations of release by hyperpolarizing pulses. Pflugers Arch. (submitted).

DISCUSSION OF THE ARTICLE BY H. PARNAS (ISRAEL)

D.O. Smith (U.S.): It appears that a major part of the story depends upon your measurements of the synaptic delay. In this preparation, the synaptic delay is ranging from 4-10 msec. They are very long synaptic delays. Can you rule out the possibility that the calcium entry and the calcium kinetics are fast, in the half msec range, but in acquisition of your data or in your experimental procedure or analysis, you simply miss this kind of resolution. There might be problems when dealing with such long synaptic delays associated with depolarization of the terminal due to a complicated terminal architecture and so forth.

H. Parnas (Israel): I don't see the problem no matter what is the profile of the entry of calcium. If once you measure it in a certain profile of calcium, you then load the terminal with calcium and whatever the length of the synaptic delay is, it should become shorter.

I. Parnas (Israel): First, there is no invasion into the terminal. It is direct depolarization in TTX preparation. Secondly, the effect of the 'binning' does not affect it because we made very short 'bins'. We can do the measurements depending on the 'A to D' converter. You can take 'bins' between 1-20 msec and it comes exactly the same. Secondly, you have to remember that we record with patchclamp. We get very clearcut quanta, the signal to noise ratio is hundreds and the measurements are very precise. You can count the number of quanta, you don't have to average and the measurements of the delay are very precise.

S. Thelseff (Sweden): You talked about a factor which influences release. Do you feel it is the membrane potential of the fibre or a chemical factor?

H. Parnas (Israel): The membrane potential activates it. We suggest that the physical entity is the release sites, a protein or a group of proteins which should become activated in order to bind calcium.

S. Thesleff (Sweden): I am asking that because we have done rather similar experiments at the frog neuromuscular junction and we also used two types of pre-pulses with a varied delay between depolarizing and hyperpolarizing pulses. Under these conditions, we find that hyperpolarization after depolarization is ineffective. So you have observed something that is probably not potential-dependent.

H. Parnas (Israel): Exactly. This was the precise reason why we had to conclude that it is not a direct effect of depolarization but only something which is produced by depolarization and, for example, it is the activation of the release sites. We found exactly what you described, namely that the time course of this activation is 1-2 msec. You put the hyperpolarization away 1-2 msec and the whole effect is over. The same for depolarizing pre-pulse; it has the same time course 1-2 msec and the whole effect is over.

S. Thesleff (Sweden): So you agree. I may just mention that in the frog, at low temperature (+4°C), the time constants are several hundred msec.

D.M.J. Quastel (Canada): The mouse neuromuscular junction behaves as though the time constant of the membrane is at least 1 msec and therefore, your hyperpolarization followed by depolarization is very confusing. With regard to the major scheme, I think we both agree that there must be two effects of depolarization on transmitter release. One is via calcium, the other via some other effect which takes some time to develop. It seems to me that the equations turn out to be very symmetrical. That is, if you attribute facilitation to internal calcium, you obtain a time course for internal calcium and now you have to attribute the fast phase to another effect of depolarization on the membrane. Conversely, if you attribute facilitation to the depolarization, you get a fast time course for the calcium. The only way to distinguish between the two, because they are formally symmetrical in terms of the equations, is to test the calcium sensitivity of the two multiplicative components. We (Cooke and Quastel 1973) reported 11 years ago that in effect the slow phase of response to depolarization expressed as a multiplication, is insensitive to lowering calcium in the medium whereas the fast phase is very sensitive. This led us to conclude that the fast phase is calcium and the slow effect is the other effect of depolarization, presumably through some mechanism other than calcium. Have you tried doing your equations with the opposite assumption - that facilitation is not residual calcium - and seeing what comes out, and secondly, seeing how the two phases are modified by varying calcium?

H. Parnas (Israel): To your first comment, this is done with direct depolarization. The time course of the change of membrane potential is not of the order of 1 msec but even if it would be, it has no effect on our conclusions. Let us say the hyperpolarization reduces the depolarization that the membrane sees. You then expect less calcium to enter and we have shown that this is not the case. Of course, if you give post-pulse hyperpolarization, obviously your comment does not apply. To the rest of your comments, I am not sure if we can go so deeply into this discussion between us, but as you probably know, the main reason that started our thinking that there must be something else that effects release is your work 11 years ago, because you see at very very low calcium concentrations, increase in the baseline of release.

D.M.J. Quastel (Canada): But that takes time to develop.

H. Parnas (Israel): Yes. But also if you measure epps with direct depolarization, this doesn't take time to develop?

D.M.J. Quastel (Canada): No. But the baseline change takes time to develop.

H. Parnas (Israel): To continue this argument, we didn't see any reason and there is no evidence in the literature against the residual calcium quantitatively.

D.M.J. Quastel (Canada): All of Magleby and Zengel's work denies the residual calcium hypothesis (Zengel J.E. and Magleby K.L. (1980). Differential effects of Ba^{2+}, Sr^{2+}, and Ca^{2+} on stimulation-induced changes in transmitter release at the frog neuromuscular junction. J. Gen. Physiol. 76, 175-211; Zengel J.E. and Magleby K.L. (1981). Changes in miniature endplate potential frequency during repetitive nerve stimulation in the presence of Ca^{2+}, Ba^{2+}, and Sr^{2+} at the frog neuromuscular junction. J. Gen. Physiol. 77, 503-529; Zengel J.E. and Magleby K.L. (1982). Augmentation and facilitation of transmitter release: a quantitative description at the frog neuromuscular junction. J. Gen. Physiol. 80, 583-611).

I. Parnas (Israel): About the time constant, Katz and Miledi who did their work on focal depolarization computed that it should be very fast with this technique. Nobody knows what is the time constant in the terminal of one micron, when you record extracellularly what is the rate of depolarization and what is the timecourse of this technique, in any computation. However, if the result would have been changes in the calcium that enters due to hyperpolarization, then why is the delay histogram different for a pre-pulse and a post-pulse? If you say that you lower the quantal content to the same degree, for example, you start with a quantal content of 0.8 or 1.2, and you lower it to 0.25 by a pre-pulse or by post-pulse, if one claims that the effect is due to reduced calcium, why then does one find the difference in the delay histograms?

H. Parnas (Israel): Indeed, the difference in the delay histograms is because the release sites are affected.

R. Rahamimoff: I wonder, you passed current to the nerve terminal; is it known that the same current produces the same degree of depolarization after previous tetanic stimulation and without it? It may be of importance in quantitative estimation.

H. Parnas (Israel): I fully agree. This really can be very important for the interpretation of the experiment in which we gave repetitive stimulation and repeated the same short and high pulse. However, from the last group of experiments in which we had a 10 msec interval between the pulses and gave hyperpolarization and depolarization, we could show that both affect only the first and not at all the facilitation of the second. This means that even if the amount of calcium that enters because of hyper or depolarization changes it did not change so much as to be observed in the second pulse while the reduction in quantal content was very very big or in this case, the increase of quantal content was very big.

A POSSIBLE INVOLVEMENT OF THE Na-Ca EXCHANGER IN REGULATION OF TRANSMITTER RELEASE AT THE FROG NEUROMUSCULAR JUNCTION

Halina Meiri, Julian Zelingher and Rami Rahamimoff
Department of Physiology, Hebrew University-Hadassah Medical School, Jerusalem 91010, Israel

While the important role of calcium ions in transmitter release has been well recognized (del Castillo & Stark, 1952; del Castillo & Katz, 1954; Jenkinson, 1957; Dodge & Rahamimoff, 1967; Katz & Miledi, 1969; Miledi, 1973), the involvement of sodium ions in regulation of the release process - although evident since 1952 (Fatt & Katz, 1952) - is still not completely understood. Sodium ions, although cannot replace calcium in the release process itself, can modulate the number of released transmitter quanta by controlling the metabolism of calcium in the nerve terminal (Kelly, 1965; Birks & Cohen, 1968; Colomo & Rahamimoff, 1968; Lev Tov & Rahamimoff, 1980). Both the influx of calcium ions into the depolarized neuronal axon and their extrusion from the axoplasm, are affected by extracellular sodium.

Calcium influx into the depolarized squid axon is increased when part of the extracellular sodium is replaced by lithium or choline (Baker, Blaustein, Hodgkin & Steinhardt, 1969; Baker & Ridgway, 1971). The competition between the two ions on the influx pathways may be one of the causes for the unusual functioning of the neuromuscular junction in solutions partially deficient in sodium. Under these conditions, it is expected that endplate potentials will be reduced in size, both due to neuronal action potentials of diminished amplitude (Hodgkin & Katz, 1949) and due to a smaller driving force of the postsynaptic current (Fatt & Katz, 1951; Takeuchi & Takeuchi, 1960). Paradoxically, however, stimulation of the motor nerve causes (under appropriate conditions) larger endplate potentials due to augmented release of transmitter (Birks & Cohen, 1965; Kelly, 1965; Colomo & Rahamimoff, 1968). Similarly, the rate of spontaneous release of individual quanta from potassium depolarized nerve endings is increased (Gage & Quastel, 1966; Birks et al., 1968; Statham & Duncan, 1977). However, competition on entry may not be the only process able to explain the paradoxical results in low $[Na]_o$ solutions.

Calcium extrusion which depends on the presence of sodium ions in the extracellular medium, was demonstrated in cardiac cells (Reuter & Seitz,

1968), the squid giant axon (Brinley et al., 1975; Mullins & Brinley, 1975; Baker & McNaughton, 1976; DiPolo & Beauge, 1980) brain synaptosomes and synaptosomal vesicles (Blaustein & Weisman, 1970; Blaustein & Osborn, 1975; Rahamimoff & Spanier, 1979). It was suggested that three or more sodium ions leaking into the cell driven by their electrochemical gradient, are exchanged for one calcium ion extruded from the neuron (Baker, 1972; Mullins, 1977; Requena et al, 1977). This electrogenic model was supported by Blaustein and co-workers (Blaustein et al, 1974; Blaustein, 1976) who noted that the Na dependent Ca efflux is influenced by membrane potential.

Two independent mechanisms of the cell membrane are responsible for the removal of calcium ions from the axoplasm: the ATP utilizing calcium pump and the Na/Ca exchanger (Sulakhe & Louis, 1980). The relative importance of these two parallel modes of calcium extrusion in the regulation of the synaptic activity has not been thoroughly investigated. Most of the calcium extruded from an internally dialized squid axon is apparently uncoupled from any external ion and requires the presence of ATP as long as the intraaxonal calcium concentration is in the low physiological range (0.02-0.06uM) (DiPolo, 1978). Na dependent Ca efflux assumes more significance at higher intracellular calcium concentrations (about 100uM) (DiPolo & Beauge, 1979). Similarly, synaptosomal vesicles are capable of ATP dependent Ca uptake from 0.1uM Ca solutions while the apparent Km for the Na^+ gradient dependent calcium uptake is approximately 60uM (Rahamimoff et al, 1980). It was therefore suspected that the Na-Ca exchanger plays no significant role in maintaining the very low intracellular Ca concentration and thus the low rate of spontaneous transmitter release at rest. The main role of this exchanger seems to be in removing the excessive intracellular calcium following intensive neuronal activity (DiPolo & Beauge, 1983; Lev Tov & Rahamimoff, 1980).

It was the objective of the present work to investigate the release of neurotransmitter while interfering with the Na-Ca exchanger activity, and to establish whether it has a physiological role in the normal function of the neuromuscular junction. Two experimental procedures were employed to slow down the Na-Ca exchanger activity: a) Reduction of sodium concentration in the extracellular solution known to affect the efflux of calcium from the squid axon both in the presence and absence of ATP (see Blaustein, 1977); b) Continuous depolarization of the nerve which affects the electrogenic Ca-Na exchange transport (Mullins, 1977; Mullins & Brinley, 1975).

1. METHODS

The experiments were performed on isolated sartorius and cutaneous pectoris neuromuscular preparations of the Israeli frog Rana ridibunda. Spontaneous miniature endplate potentials (mepps) and stimulation induced endplate potentials (epps) were monitored using conventional electrophysiological techniques. The information was either fed into the PDP 11/23 MINC microcomputer and analyzed "ON-LINE" or taped using H-P taperecorder for subsequent computer assisted analysis. The amplitude and frequency of the mepps and the quantal content of the epps were calculated using Fortran and Macro-Assembler programs devised for this purpose.

2. RESULTS

2.1. Spontaneous release of transmitter in "low-Na" solutions

Mepps were first recorded in preparation continuously perfused by the control Ringer solution containing 116mM NaCl, 2mM KCl, 1mM $MgCl_2$, 5mM Hepes buffer and between 0.2 and 0.4mM $CaCl_2$. At a latter stage of the experiment, the control solution was replaced by a solution containing the same concentrations of KCl, $MgCl_2$, $CaCl_2$ and Hepes in which 40% of the NaCl was replaced by 86.8mM sucrose ("low-Na" solution).

Reduction of the extracellular sodium concentration down to 60% of its normal value induced a gradual increase in the rate of spontaneous release of transmitter. 20-30 minutes after replacement of the control solution by the "low-Na" medium, the frequency of mepps reached a new 2-3 fold higher level (Fig 1A). Removal of 40% of the sodium ions affected not only the release of transmitter from the presynaptic nerve, but also decreased the amplitude of mepps created by individual quanta (Fig 1C). However, these two effects did not develop simultaneously. The mepp amplitude was diminished almost immediately following the onset of perfusion with "low-Na" medium and reached its final size within 3-6 minutes (Fig 1B). This change in mepps amplitude coincided well with the development of postsynaptic hyperpolarization (2-5mV) induced by the "low-Na" solution, but not with the increase in mepp frequency which was unchanged during that initial period, but gradually grew over the next twenty minutes.

Since the sodium concentration in the extracellular fluid reached a new constant level soon after introduction of the "low-Na" solution (as evident from the lower mepp amplitudes), we attributed the delayed increase in mepp frequency to changes in the ionic composition of the nerve terminal. The most

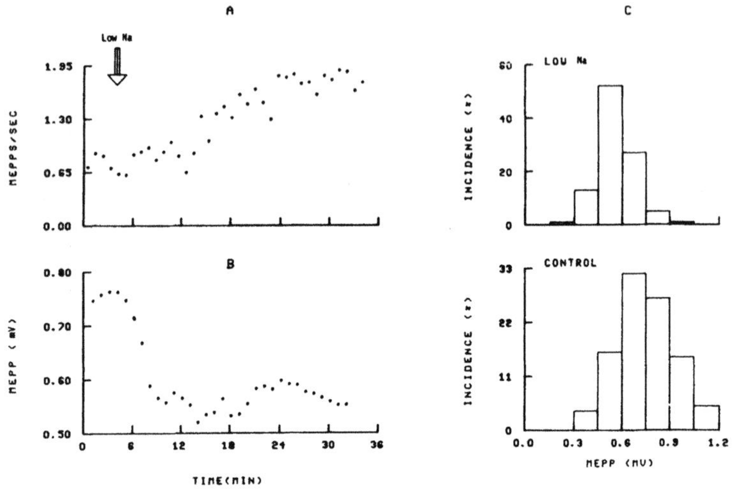

FIGURE 1. Replacement of the control solution (containing 116mM NaCl and 0.3mM $CaCl_2$) by the "low-Na" solution (69.6mM NaCl, 0.3mM $CaCl_2$ and 86.8mM sucrose) increase the frequency of mepps and reduces their amplitude. Mepp frequency is slowly elevated reaching the value 2.5 times the control in 20 minutes (A), while their amplitude drops to 75% of the original size in 6 minutes (B and C).

plausible change is a very slow accumulation of calcium ions inside the nerve terminal induced by the low [Na]out. In order to eliminate possible interference of the experimental treatments with the influx of calcium ions, part of the experiments were performed in the virtual absence of calcium in the Ringer solution. In these experiments, both the control and the "low-Na" medium contained 1mM $MgCl_2$, 1mM EGTA and no added $CaCl_2$. The results shown in Fig 2 suggest that the increase in mepp frequency imposed by reducing the sodium concentration does not require the presence of calcium in the extracellular fluid. Therefore, the modulatory effect of the extracellular sodium on the rate of spontaneous release of transmitter does not reflect enhanced influx of calcium into the nerve terminal but is rather indicative of a partially blocked efflux of this ion.

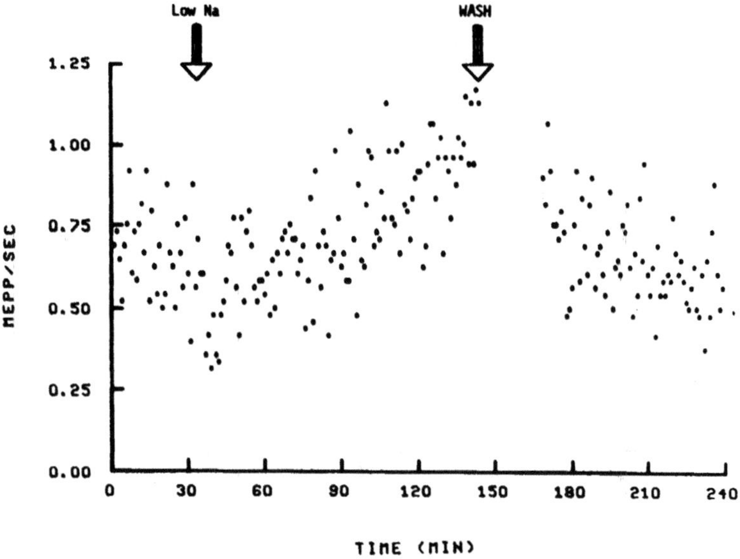

FIGURE 2. Increased mepp frequency evoked by the reduction of the extracellular sodium concentration in the virtual absence of calcium ions. Both the control and the "low-Na" solutions contained 1mM EGTA and 1mM $MgCl_2$.

2.2. Spontaneous release from depolarized nerves in "low-Na" solution

Considering the electrogenic Na-Ca exchanger, which is the candidate of choice for mediating possible interference with calcium extrusion from nerve terminal by reduction of [Na]out, we anticipated an even more pronounced increase in the frequency of mepps in a depolarized nerve terminal. Elevation of the extracellular concentration of KCl from the normal 2mM to 5mM (in the presence of 1mM $CaCl_2$), which depolarized the muscle fiber by 10-15mV, increased the frequency of mepps by 40-50% (Fig 3). Introduction of the "low-Na" solution to such a slightly depolarized preparation evoked a fast increase in mepp frequency which reached a level 5-6 fold higher than the control value within less than 15 minutes. This combined effect of depolarization and partial removal of extracellular sodium on spontaneous release of transmitter is much larger than the algebric sum of their individual contributions. In the experiment shown in Fig 3, depolarization alone induced 50% increase, "low-Na" solution by itself induced 113% increase while the combined procedure caused an increase of 440% in the frequency of mepps.

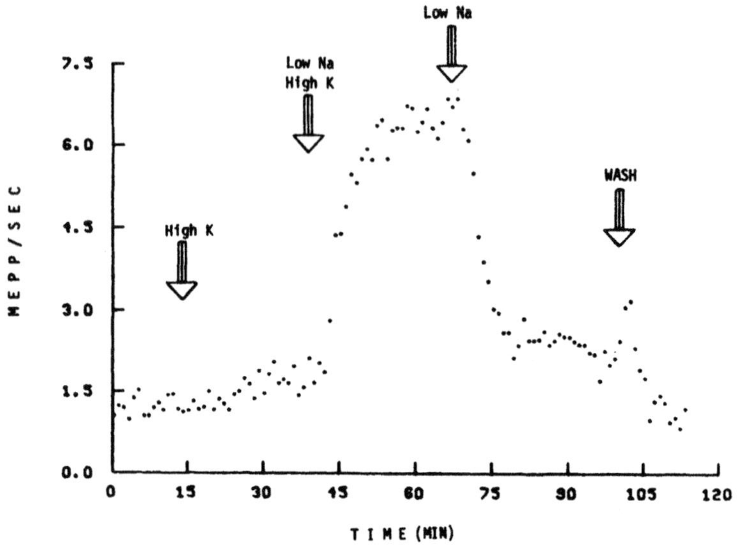

FIGURE 3. The increased rate of spontaneous release in the presence of 60% of the NaCl is highly potentiated by depolarization of the nerve terminal. Elevation of the Ringers KCl concentration from 2mM to 5mM (at 'High K' arrow) augmented only slightly the mepp frequency but made the nerve terminal much more sensitive to the reduction of [Na]out which (at 'low-Na, High K' arrow) increased the spontaneous release by more than 400%. Upon repolarization in the presence of 60% NaCl (at 'low-Na' arrow), the mepp frequency was only 2 times that of the control (1mM $CaCl_2$ present all along the experiment).

Part of the enrichment of the axoplasm with calcium ions and hence the augmented transmitter release under the conditions described above, should be most probably attributed to an increased influx rate of calcium into the depolarized terminal. It has been shown that both sodium and calcium ions can penetrate the nerve cell via voltage dependent late calcium channels and TTX sensitive sodium channels. However, even in the virtual absence of calcium ions in the extracellular medium, when no calcium influx is expected, depolarization of the nerve terminal cooperates with the "low-Na" solution in increasing the frequency of mepps as shown on Fig 4. In the experiment illustrated in this Figure, reduction of the sodium concentration in Ringers containing 1mM $MgCl_2$, 1mM EGTA and no added calcium, caused a 1.8 fold increase in mepp frequency in the presence of a normal concentration of KCl which was further increased, reaching 3.3 times the control value following nerve depolarization. Therefore, we believe that removing of sodium from the extracellular fluid interferes with an electrogenic process which expells calcium ions from the nerve terminal to the extent of affecting the spontaneous release of transmitter at the frog neuromuscular junction.

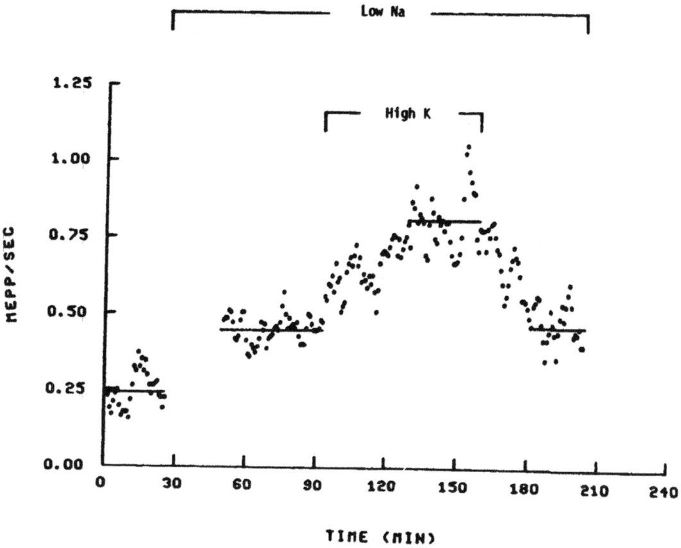

FIGURE 4. The effect of decreased [Na]out and depolarization on spontaneous release in the virtual absence of calcium ions from the extracellular solution. The 1.8 fold increase in mepp frequency observed once the control solution (containing 116mM NaCl) was replaced by the 'low-Na' medium (with only 69.8mM NaCl), was further elevated when 5mM KCl was introduced into the experimental bath (indicated by 'High K'). Since the estimated [Ca]out in the presence of 1mM EGTA and 1mM $MgCl_2$ is about 2.5nM, no interference with the calcium influx but rather impaired efflux of this ion can be responsible for the augmented release.

2.3. Synaptic facilitation low-Na solutions

To test for a possible role of the Na-Ca exchanger in extrusion of calcium following its transient intracellular increase after nerve stimulation, neuromuscular facilitation was examined. Computer programming was used to stimulate the nerve once every 2 seconds with a pair of stimuli 30-200 msec apart in a random schedule of intervals, and for "ON-LINE" analysis of the recorded epps. The degree of synaptic facilitation was calculated as a ratio between the response to the test pulse and the response to the conditioning pulse.

Substitution of 40% of the Ringer's NaCl by sucrose affected the stimulation induced release of transmitter in two ways: the quantal content of the epps was increased and the synaptic facilitation was elevated and prolonged. The extent of facilitation enhancement varied from one experiment to another. Only a minute effect was observed in the experiment shown in Fig 5, in which the ratio between the test epp and the conditioning epp preceding it by 40 msec was 1.24 in control solution and 1.29 in the "low-Na" solution.

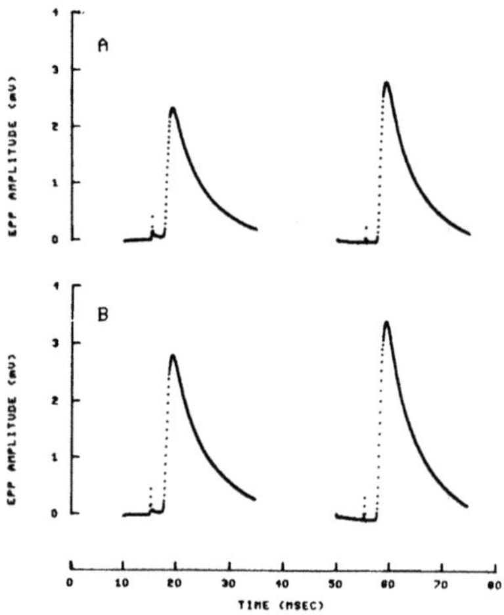

FIGURE 5. Synaptic facilitation is augmented by the removal of sodium ions from the extracellular fluid. The extent of this increase varies from one endplate to another, a relatively low effect is illustrated here. In response to stimulation of the nerve with a pair of stimuli every 2 seconds the second epp was on the average 24% larger than the first in the control solution (A) and was 29% larger in the low Na medium (B). 0.4mM $CaCl_2$ and 1.5mM $MgCl_2$ were present throughout the experiment.

A much more substantial effect of reduced sodium concentration on facilitation is presented in Fig 6A. In this experiment, facilitation decayed from 1.285 to 1.005 when the interval between the test and the conditioning stimulis was increased from 30 to 140 msec in control solution. Following introduction of the "low-Na" solution, facilitation was elevated in all the intervals tested. Very similar augmented and prolonged facilitation could be evoked in the same endplate in control Ringers by doubling the conditional stimulus as shown in Fig 6B. In this part of the experiment, two conditioning pulses 10 msec apart preceded the test pulse. These results interpreted in the light of the residual calcium hypothesis (Katz & Miledi, 1965,1968; Rahamimoff, 1968; Meiri & Rahamimoff, 1972; Rahamimoff & Yaari, 1973; Smith & Zucker, 1980) indicate a slower removal of calcium ions from the activated nerve terminal induced by sodium deficiency in extracellular fluid.

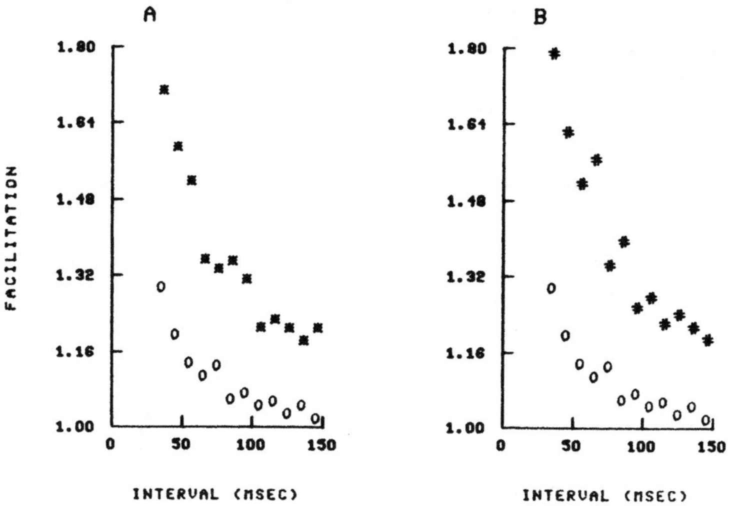

FIGURE 6. Highly augmented and prolonged synaptic facilitation evoked by decreased [Na]out (A) or by a potentiated conditioning stimulus (B).
A. The nerve was stimulated once every 2 seconds with a pair of stimuli 30-140 msec apart. The ratio between the test and the conditioning EPP decayed as the interval between the 2 stimuli was prolonged both in control solution (o) and in the low-Na medium (*), but while in the control solution, most of the 'residual calcium' was removed from the nerve terminal in a period of 140 msec, was not so in the low-Na medium, indicating a slower extrusion process.
B. Simulation of the increased 'residual calcium' was performed in the control solution by doubling the calcium influx into the terminal during the conditioning stimulus period. The nerve was stimulated with 2 conditioning pulses 10 msec apart, followed by a test pulse in a sequence similar to that described in A. The double conditioned facilitation (#) was both higher and more prolonged than the control (o).

Another cellular process obviously affected by reduction of extracellular sodium is the Na-H^+ exchanger of the axolemma but its effects on transmitter release are not easy to envisage. It is known that in crayfish neurons, disruption of the balance of the Na electrochemical gradient shuts down the H^+ extrusion (Moody, 1981), causing acidification of the normally alkaline cytoplasm (Moody 1984). H^+ enrichment of the cytoplasm (by 1.3 pH units) increased [Ca]in in Xenopus embryonic cells (Rink et al, 1980) and in barnacle muscle fibers (Lea & Ashley, 1978). However, a small decrease in [Ca]in was observed in Helix neurons (Alvarez-Leefmans et al, 1984) and Purkinje fibers (Hess & Weingart, 1980).

2.4. A possible contribution of the Na-Ca exchanger to the maintenance of low intracellular concentration of calcium ions - theoretical calculations

Our experimental results suggest that the frog motor nerve terminal utilizes its axolemmal Na-Ca exchanger for net extrusion of calcium both at rest and following nerve stimulation. We therefore calculated the expected consequences of changing the ionic composition of the Ringer solution, assuming that the Na-Ca transporter replaces each calcium ion by 4 sodium ions, as was suggested for the squid giant axon (Mullins, 1977). Given the values of 54mV for the sodium Nernst potential (E_{Na}) (Frankenhauser, 1959) and 116mV for E_{Ca} (DiPolo et al, 1976), the equilibrium potential (E_R) at which the exchanger would introduce into the cell the same amount of calcium it expels from it according to this model, equals -8mV. Since at rest the neuronal membrane potential is -71mV (Huxley & Staempfli, 1951) then sodium ions are expected to be introduced into the nerve cell and calcium ions to be extruded from it, with a driving force of -63mV for the exchanger.

Reduction of the extracellular sodium to 60% of its control value should lower the driving force for the exchanger activity. Although the nerve is hyperpolarized under these conditions by 2-3mV (Huxley & Staempfli, 1951), the driving force of Na-Ca exhange is apparently decreased as a result of a new lower E_{Na}. When [Na]out is reduced to 60%, [Na]in is reduced to only 88% of its control value as extrapolated from results observed by Thomas in snail neurones (Thomas, 1972), therefore, E_{Na} = 44.3mV, E_R = -27.4mV and the driving force is only -46.1mV. A very similar decrease in the driving force can be obtained in normal Ringer solution when the nerve is depolarized to -53mV by 5mM of KCl (instead of the normal 2mM). When both the [Na]out is decreased and [K]out is elevated, the calculated driving force for the exchanger activity is as low as -27.7mV (43.9% of the control value) (Fig 7) and the efflux of calcium ions from the nerve terminal by this route should be impaired.

The activity of the Na-Ca exchanger is strongly dependent upon the concentration of calcium in the extracellular solution, as illustrated on Fig 8. An increase of 200% in the driving force value is predicted upon reduction of the extracellular calcium concentration to 10uM. (It was assumed that [Ca]in under these conditions equals 1nM according to Tsien et al, 1982). It is also evident from this figure that the activity of the Na-Ca exchanger is much less sensitive to alterations in [Na]out in the presence of reduced

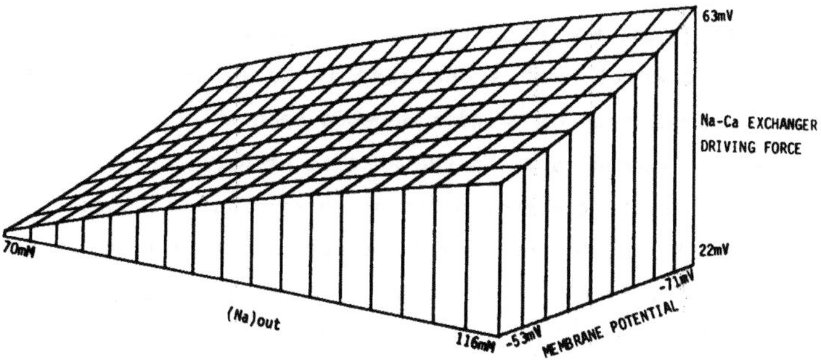

FIGURE 7. The driving force of the Na-Ca exchanger calculated according to the model proposed by Mullins (Mullins, 1977) using frog motor neuron ionic gradients and membrane potential. In control conditions E_{Na} = 54mV, E_{Ca} = 116mV, Resting potential = -71mV. When [Na]out is reduced to 60%, it is assumed that [Na]in is reduced to 88% of its original value (Thomas, 1972) and E_{Na} = 44.3mV. 5mM KCl depolarizes the nerve to -53mV. The calculated driving force is therefore decreased by both the removal of sodium ions and depolarization.

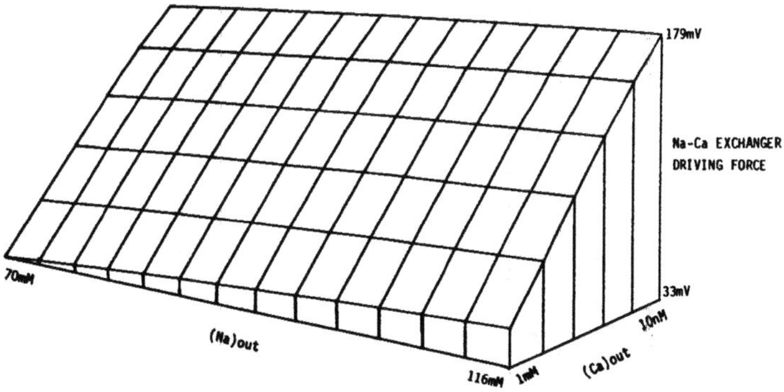

FIGURE 8. The driving force of the Na-Ca exchanger calculated according to the model proposed by Mullins (Mullins, 1977). The driving force is much larger and less sensitive to [Na]out at low extracellular calcium ions concentration, than it is in normal physiological conditions. (Note the logarithmic scale on the calcium concentration axis.)

extracellular calcium than in normal physiological conditions. These predictions of the model are in agreement with our experimental results.

The extent to which such a decrease in Ca efflux would affect the intracellular concentration of calcium depends upon the role of the exchanger in neuronal calcium metabolism. In the absence of data concerning the density of exchanger molecules in the frog motor nerve terminal, we used the available information about the squid axon membrane (being aware that it may not be entirely applicable to frog). Assuming a total calcium efflux of 44 fmole x sec^{-1} x cm^{-2} (following DiPolo, 1979) and supposing that only 20% of it is [Na]out dependent (Baker & McNaughton, 1976) the rate of efflux carried on by the exchanger is 8.8 fmoles x sec^{-1} x cm^{-2}. Considering that the reduction of [Na]in augments the [Na]out dependent Ca efflux in the squid axon (Brinley et al, 1975), this value may be even higher in the frog neuron whose [Na]in is about 0.3 that of squid. If Ca efflux via the Na-Ca exchanger is proportional to the driving force, we expect the efflux in low-Na solution to be only 73.1% as fast as that in the control Ringer. If no other process takes over, calcium ions will be accumulated in the cell with the rate of 2.367 fmole x sec^{-1} x cm^{-2}. A nerve terminal of luM diameter will therefore increase its intracellular calcium ions concentration with the initial rate of approximately 9.5 x 10^{-5} umoles x sec^{-1} x cm^{-3} which will of course slow down as the [Ca]in is elevated (as shown in Fig 9A).

Although our results are basically in accord with the proposed model, it is clear that impairement of calcium extrusion via the Na-Ca exchanger is compensated for by other calcium buffering processes. The frequency of mepps levels off 20-30 minutes after the reduction of [Na]out reaching a rate 2-3 times higher than that of the control, at which time the calculated value of [Ca]in is still climbing up if no compensatory processes are considered (Fig 9A). The axolemmal ATP driven calcium pump is only one of the calcium buffering cellular mechanisms whose activity is augmented by the elevated [Ca]in. Incorporation of this process alone into the theoretical model (using the kinetic parameters of ATP dependent Ca efflux from squid giant axon presented by DiPolo & Beauge in 1980, is sufficient to restrict the [Ca]in level (Fig 9B). For a more realistic estimation of the intracellular concentration of calcium ions, one should introduce all the known calcium buffering processes.

In conclusion, this work shows that procedures known to affect the sodium calcium exchanger, produce the expected changes in transmitter release. Thus, it seems that the Na-Ca exchanger plays a role in maintaining the low probability of quantal release of transmitter at rest and it may participate in the frequency modulation of evoked synaptic activity.

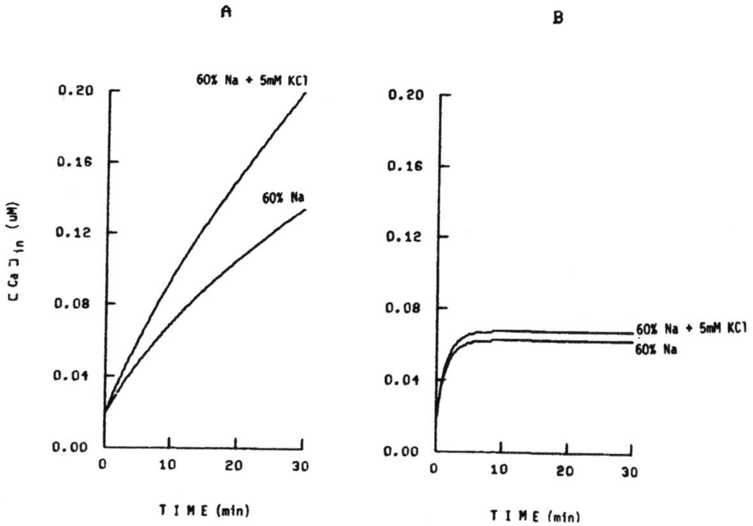

FIGURE 9. The expected increase in [Ca]in calculated on the basis of the decreased driving force for Na-Ca exchange activity folowing partial removal of extracellular sodium and depolarization.
A. Assuming no change in the activity of either intracellular calcium buffering mechanisms or membranal extrusion processes.
B. Accounting for an augmented activity of the axolemmal ATP dependent calcium pump driven by increased [Ca]in, using Michaelis-Menten kinetics with Vmax = 150 fmoles x cm^{-2} x sec^{-1} and $K_{1/2}$ = 0.18uM (DiPolo & Beauge, 1980).

ACKNOWLEDGEMENTS

We thank Dr. Simona Ginsburg for fruitful discussions and Ms. Miriam Silber for unfailing help in the preparation of this manuscript.

This work was supported by the Muscular Dystrophy Association and the U.S.-Israel Binational Science Foundation. Equipment was purchased with the aid of the Bay Foundation.

REFERENCES

1. Alvarez-Leefmans F.J., Rink T.J. and Tsien R.J. (1981). Free calcium ions in neurones of Helix aspersa measured with ion selective electrodes. J. Physiol. 315, 531-548.
2. Baker P.F. (1972). Transport of metabolism of calcium ions in nerve. Prog. Biophys. Molec. Biol. 24, 177-223.
3. Baker P.F. and McNaughton P.S. (1976). Kinetics and energetics of calcium efflux from squid axons. J. Physiol. 276, 127-150.
4. Birks R.I., Burstyn P.G.R. and Firth D.R. (1968). The form of sodium-calcium competition at the frog myoneural junction. J. Gen. Physiol. 52, 887-907.
5. Birks R.I. and Cohen M.W. (1965). Effects of sodium pump inhibitors on myoneural transmission in the frog. Int. Symposium of Muscle. New York Academic Press.
6. Blaustein M.P. (1976). The ins and outs of calcium transport in squid axons: internal and external ion activation of calcium efflux. Fed. Proc. 35, 2574-2578.
7. Blaustein M.P. (1977). Effects of internal and external cations and of ATP on sodium-calcium and calcium-calcium exchange in squid axons. Biophys. J. 20, 79-111.
8. Blaustein M.P. and Osborn C.Y. (1975). The infulence of sodium on calcium fluxes in pinched-off nerve terminals in vitro. J. Physiol. 247, 657-686.
9. Blaustein M.P., Russell J.M. and Dewer P. (1974). Calcium efflux from internally dialized squid axons. J. Supram. Struct. 2, 558-581.
10. Blaustein M.P. and Wiesmann W.P. (1970). Effect of sodium ions on calcium movements in isolated synaptic terminals. Proc. Natl. Acad. Sci. USA 66, 664-671.
11. Brinley F.J., Jr., Spangler S.G. and Mullins L.J. (1975). Calcium and EDTA fluxes in dialyzed squid axons. J. Gen. Physiol. 66, 223-250.
12. Colomo F. and Rahamimoff R. (1968). Interaction between sodium and calcium ions in the process of transmitter release at the neuromuscular junction. J. Physiol. 198, 203-218.
13. del Castillo R. and Katz B. (1954). The effect of magnesium on motor nerve endings. J. Physiol. 124, 553-559.
14. del Castillo R. and Stark L. (1952). The effect of calcium on motor nerve endplate potentials. J. Physiol. 116, 507-515.
15. DiPolo R. (1974). Effect of ATP on the calcium efflux in dialyzed squid giant axons. J. Gen. Physiol. 64, 503-517.
16. DiPolo R. (1978). Ca pump driven by ATP in squid axons. Nature 274, 390-392.
17. DiPolo R. (1979). Calcium influx in internally dialyzed squid giant axons. J. Gen. Physiol. 73, 91-113.
18. DiPolo R. and Beauge L. (1979). Physiological role of ATP driven calcium pump in squid axon. Nature 278, 271-273.
19. DiPolo R. and Beauge L. (1980). Mechanisms of calcium transport in the giant axon of the squid and their physiological role. Cell Calcium 1, 147-169.
20. DiPolo R. and Beauge L. (1983). The calcium pump and sodium-calcium exchange in squid axons. Ann. Rev. Physiol. 45, 313-324.
21. DiPolo R., Requena J., Brinley F.J., Mullins L.J. Scarpa A. and Tiffert T. (1976). Ionized calcium concentrations in squid axons. J. Gen. Physiol. 67, 433-467.

22. DiPolo R., Rojas H. and Beauge L. (1982). Calcium entry at rest and during prolonged depolarization in dialyzed squid axons. Cell Calcium 3, 19-41.
23. Dodge F.A. and Rahamimoff R. (1967). Co-operative action of calcium ions in transmitter release at the neuromuscular junction. J. Physiol. 193, 419-432.
24. Fatt P. and Katz B. (1952). The effect of sodium ions on neuromuscular transmission. J. Physiol. 118, 73-87.
25. Frankenhauser B. (1959). Steady state inactivation of sodium permeability of myelinated nerve fibers of Xenopus Laevidis. J. Physiol. 148, 671-676.
26. Gage P.W. and Quastel D.M.J. (1966). Competition between sodium and calcium ions in transmitter release at mammalian neuromuscular junction. J. Physiol. 185, 95-123.
27. Hess R. and Weingart R. (1980). Intracellular free calcium modified by pH_i in sheep cardiac Purkinje fibres. J. Physiol. 307, 60P-61P.
28. Hodgkin A.L. and Katz B. (1949). The effect of sodium ions on the electrical activity of the giant axon of the squid. J. Physiol. 108, 37-77.
29. Huxley A.F. and Stamfli R. (1951). Direct determination of membrane resting potential and action potential in single myelinated nerve fibers. J. Physiol. 112, 476-495.
30. Jenkinson D.H. (1957). The nature of the antagonism between calcium and magnesium ions at the neuromuscular junction. J. Physiol. 138, 434-444.
31. Katz B. and Miledi R. (1965). The effect of calcium on acetylcholine release from motor nerve terminals. Proc. R. Soc. 161, 495-503.
32. Katz B. and Miledi R. (1968). The role of calcium in neuromuscular facilitation. J. Physiol. 195, 481-492.
33. Katz B. and Miledi R. (1969). Tetrodotoxin resistant electrical activity in presynaptic terminals. J. Physiol. 203, 459-487.
34. Kelly Y.S. (1965). Antagonism between Na^+ and Ca^{2+} at the neuromuscular junction. Nature 205, 296-297.
35. Lea T.Y. and Ashley C.C. (1978). Increase in free calcium in muscle after exposure to CO_2. Nature 275, 236-238.
36. Lev-Tov A. and Rahamimoff R. (1980). A study of tetanic and post-tetanic potentiation of miniature endplate potentials at the neuromuscular junction. J. Physiol. 309, 247-273.
37. Meiri H. and Rahamimoff R. (1972). Fluctuation in endplate potential amplitude and calcium ions. Isr. J. Med. Sci. 8, 4.
38. Miledi R. (1973). Transmitter release induced by injection of calcium ions into nerve terminals. Proc. R. Soc. B. 183, 421-425.
39. Moody W.J. (1981). The ionic mechanism of intracellular pH regulation in crayfish neurones. J. Physiol. 316, 293-308.
40. Moody W.J. (1984). Effects of intracellular H^+ on the electrical properties of excitable cells. Ann. Rev. Neurosci. 7, 257-278.
41. Mullins L.J. (1977). A mechanism for Na/Ca transport. J. Gen. Physiol. 70, 681-695.
42. Mullins L.J. (1979). The generation of electric currents in cardiac fibers by Na/Ca exchange. Am. J. Physiol. 236, C103-C110.
43. Mullins L.J. and Brinley F.J., Jr. (1975). Sensitivity of calcium efflux from squid axons to changes in membrane potentials. J. Gen. Physiol. 65, 135-152.
44. Rahamimoff R. (1968). A dual effect of calcium ions on neuromuscular facilitation. J. Physiol. 195, 471-480.

45. Rahamimoff H., Papazian D., Stanley M., Goldin, Spanier R. and Abramovitz E. (1980). Synaptosome-derived Ca transport systems: Properties and Purification, 28th Int. Adv. Physiol. Soc. Vol. 36, Cellular Analogues of Conditioning and Neural Plasticity. O. Feher, F. Yoo (Eds.)
46. Rahamimoff H. and Spanier R. (1979). Sodium dependent calcium uptake in membrane vesicles derived from rat brain synaptosomes. FEBS Lett. 104, 111-114.
47. Rahamimoff R. and Yaari Y. (1973). Delayed release of transmitter at the frog neuromuscular junction. J. Physiol. 228, 241-257.
48. Requena J., DiPolo R., Brinley F.J., Jr., and Mullins L.J. (1977). The control of ionized calcium in squid axons. J. Gen. Physiol. 70, 329-353.
49. Reuter H. and Seitz N. (1968). Dependence of calcium influx from cardiac muscle on temperature and external ion composition. J. Physiol. 195, 451-470.
50. Rink T.J., Tsien R.Y and Warner A.E. (1980). Free calcium in Xenopus embryos measured with ion selective microelectrodes. Nature 253, 258-260.
51. Smith S.J. and Zucker R.S. (1980). Aequorin response facilitation and intracellular calcium accumulation in molluscan neurones. J. Physiol. 300, 167-196.
52. Sulakhe P.V. and Louis P.Y.S.T. (1980). Passive and active calcium fluxes across plasma membranes. Prog. Biophys. Molec. Biol. 35, 135-195.
53. Takeuchi A. and Takeuchi N. (1960). On the permeability of endplate membrane during the action of transmitter. J. Physiol. 154, 52-67.
54. Thomas R.C. (1972). Intracellular sodium activity and the sodium pump in snail neurons. J. Physiol. 220, 55-71.
55. Tsien R.Y., Pozzan T. and Rink T.J. (1982). Calcium homeostasis in intact lymphocytes: Cytoplasmic free calcium monitored with a new intracellularly trapped fluorescent indicator. J. Cell. Biol. 94, 325-334.

DISCUSSION OF THE ARTICLE BY H. MEIRI (ISRAEL)

I. Steinberg (Israel): The facilitation was on a time scale of hundred milliseconds while the previous experiments were on a time scale of seconds.

H. Meiri (Israel): Yes. There is a difference in the time scale. In most figures, the time axis indicated the time after replacement of the control solution by the low-sodium medium, while in the figures presenting the facilitation it was the interval between the conditioning and the test pulses.

I. Steinberg (Israel): So it means actually that in the facilitation experiment, the resting calcium concentration was lower. The observed effects are not due to changes in calcium removal during the facilitation experiment. Is that right?

H. Meiri (Israel): Yes. The calcium is probably not pumped as efficiently in the low-sodium medium.

I. Parnas (Israel): Also in crayfish, you find similar findings that you prolong the duration of facilitation when you play with sodium, calcium, magnesium. If you block with one the exchange, you cannot affect it more by any other procedure. However, you showed effect of depolarization on the frequency of mepps, on top, after you blocked the pump. When you removed the depolarization, there was an arrow only of low sodium, the frequency was about 10 times or so. I wonder if it is possible that due to the model you presented, some contribution of the effect of depolarization, to increase mepp frequency on top of that produced by the block of the pump. Therefore, the same amount of calcium or the one which goes in becomes more effective. Have you tried to measure the duration of facilitation, not the frequency of mepps, at 5mM potassium and then see whether the duration becomes an indication of the concentration or whether there is another parameter?

H. Meiri (Israel): We haven't measured the duration of facilitation under conditions of both, reduced sodium and increased potassium.

R. Rahamimoff (Israel): If one adds potassium in the absence of extracellular calcium, one sees from the experiments shown here that small changes in potassium cause an increase in transmitter release which corresponds either by direct effect of depolarization or indirect effect of depolarization, for e.g., by acting on the sodium-calcium exchanger.

What makes it of interest is that in 1977, together with Alnaes & Shimoni, we worked on another type of increase of transmitter release, by hyperosmotic neurosecretion similar to what we saw today in Van der Kloot's experiments. If one produces large increases in potassium concentration, one gets after an initial increase - a decrease. Presumably, there is a contribution of opening of calcium channels and since there is a reverse electrochemical gradient for calcium, leakage of calcium to the outside. Therefore, this depends on the degree of change one produces with potassium.

I. Parnas (Israel): By no means it should be clear that calcium is an essential factor for release and its concentration can counteract other affects. For example, if calcium goes down, then even if one increases other factors essential for release, one might still get less release.

J.W. Moore (U.S.): It occured to me that perhaps if this mechanism, sodium-calcium exchange, is working, you might see a change in the time course of the release process. The original figure you've shown for normal sodium and low sodium; are they endplate potentials or endplate currents?

H. Meiri (Israel): These are endplate potentials.

J.W. Moore (U.S.): Probably they may not show the time course change. It seems to me it will be worthwhile to repeat the experiment using a voltage clamp to view endplate currents.

W.G. Van der Kloot (U.S.): Because of changes in the surface potential, as soon as you change the sodium you are also changing the voltage gradient in the membrane through the membrane. I would think it would be nice to do some of the experiments by adding some divalent cation and at the same time take away the calcium so you can more or less keep it balanced.

H. Meiri (Israel): We haven't done it. However, we replaced the sodium by sucrose in some experiments and by glucosamine in others. At least as far as the mepps are concerned, this did not change anything.

W.G. Van der Kloot (U.S.): There are agents like doxirubacin that are supposedly blockers of the sodium-calcium exchanger. They were found ineffective. Did you try them?

H. Meiri (Israel): No, I haven't tried them.

SECTION 3

CALCIUM & EXOCYTOSIS

OPTICAL STUDIES OF EXCITATION AND SECRETION IN VERTEBRATE NERVE TERMINALS

Salzberg, B.M., Obaid, A.L., and Gainer, H. University of Pennsylvania, Philadelphia, Pa., National Institutes of Health, Bethesda, Md., and Marine Biological Laboratory, Woods Hole, Mass. USA

Optical methods that employ potential sensitive molecular probes are shown to monitor rapid changes in membrane potential from populations of nerve terminals in the neurohypophyses of amphibia and mammals. Ca^{++} and Na^{+} contributions to the action potential are demonstrated and, in the mouse, variations in light scattering are detected following membrane potential changes known to trigger the release of peptide hormones. Calcium antagonists such as Cd^{++}, Mn^{++}, Co^{++}, and Ni^{++} block an after-hyperpolarization in the frog that we attribute to the effect of a calcium mediated potassium conductance. Block of sodium and potassium currents with TTX and TEA reveals a pure calcium action potential which is eliminated by 0.5 mM Cd^{++}. In the mouse, the magnitude of the light scattering signal depends upon the frequency of stimulation and the extracellular concentration of Ca^{++}, as well as on the presence of secretagogues such as 4-aminopyridine. A part of this intrinsic optical signal that seems to be correlated with neuropeptide release is blocked by inorganic calcium antagonists, while a faster component related to the action potential itself remains. The weak wavelength dependence of the light scattering signal, contrasted with the strong wavelength dependence of the extrinsic absorption signal provided by linear potentiometric probes (e.g., merocyanines), may permit one to monitor simultaneoulsy, in a stained preparation, the voltage changes in the nerve terminals and the release of secretory products. The inherently fast responses of the two optical measurements may then improve our ability to resolve early events in the coupling of excitation to secretion. Supported in part by USPHS grant NS 16824.

PHYSIOLOGICAL BASES FOR DIFFERING SYNAPTIC EFFICACY IN FROG NEUROMUSCULAR JUNCTIONS

Peter A. Pawson and Alan D. Grinnell

Introduction

Neuromuscular junctions (NMJs) in the frog sartorius muscle differ over a 20-fold range in transmitter release per unit length of nerve terminal (Grinnell & Herrera, 1980). Part of the explanation for these differences in synaptic strength lies in the inherent release properties of different motor units (Grinnell & Trussell, 1983; Trussell & Grinnell, 1984). However, there are still considerable differences between junctions within the same motor unit. Moreover, this observation does not explain the physiological mechanisms responsible. Of further interest, a substantial body of research has shown that the adult neuromuscular junction can exhibit considerable structural and physiological plasticity (Grinnell & Herrera, 1981; Herrera & Grinnell, 1980,1981; Kelly & Robbins, 1983; Rotshenker, 1982; Smith, 1982; Wernig, Pecot-Dechavassine & Stover, 1980). This synaptic plasticity and the apparently dynamic motoneuron regulation that it implies (Nudell & Grinnell, 1982, 1983), suggests a comparison with the previously described plasticity in the central nervous system (CNS) and spinal cord (Cotman, Nieto-Sampedro & Harris, 1981; Mendell, 1984). Since the frog neuromuscular junction has served as an important model for the transmitter release process (Katz, 1969), and because the CNS presents serious obstacles to a cellular analysis of the physiological and structural mechanisms responsible for synaptic plasticity, we have pursued an analysis of this problem at the frog NMJ.

A voltage-dependent Ca^{2+}-influx is essential to transmitter release (Heuser, Reese, Dennis, Jan, Jan, and Evans, 1979; Katz, 1969; Katz & Miledi, 1967; Rahamimoff, Lev-Tov & Meiri, 1980). Both strong and weak NMJs show a similar, approximately 4th power relationship between external Ca^{2+} and release; hence it appears that Ca^{2+} or some Ca^{2+}-activated complex (Ca^*) inside the terminal acts in the same way to evoke release at strong or weak junctions (Grinnell & Herrera, 1980). Moreover, quantal content of evoked endplate potentials (EPPs) and the frequency of miniature EPPs (mEPPs) are well correlated (Grinnell & Herrera, 1980; Kuno, Turkanis and Weakly, 1971). This implies that the differences are not solely associated with the action potential mechanism. Since both EPPs and mEPPs are generally thought to involve the same release mechanism (Katz, 1969); and since mEPP frequency depends on the level of Ca^* inside the terminal (Rahamimoff, Erulkar, Lev-Tov and Meiri, 1978), it can be postulated that the stronger terminals have more release apparatus (active zone) per unit length than weaker terminals and/or that the intraterminal Ca^* is somehow regulated at a higher concentration in stronger junctions. Although it has recently been shown that the amount of active zone per unit terminal length is positively correlated with terminal strength (Grinnell, Herrera & Wolowske, 1983), these differences are relatively small compared with differences in release per unit length. Consequently, it seems likely that there are also differences in the level of intraterminal Ca^* at strong and weak terminals. These could reflect differences in resting and voltage-dependent flux of Ca^{2+} into the terminal and/or differences in buffering of Ca^{2+} within the terminal. To assess the relative importance of these two possible differences, we have used the posttetanic potentiation (PTP) of mEPP frequency (Lev-Tov & Rahamimoff, 1980) in strong vs. weak junctions, since the time constant of the decay of potentiation (τ_p) gives an indication of the kinetics of Ca^{2+} metabolism (Lev-Tov & Rahamimoff, 1980; Zengel & Magleby, 1981). Furthermore, the Ca^{2+} influx can be eliminated during a tetanus by using a Ca^{2+}-free/EGTA Ringer (Erulkar and Rahamimoff, 1978; Lev-Tov & Rahamimoff, 1980). A comparison of the results obtained in Ca^{2+} vs. Ca^{2+}-free/EGTA Ringers indicate that differences in the Ca^{2+} influx contribute importantly to the physiological differences between strong and weak junctions.

Furthermore, there do not appear to be conspicuous differences in buffering kinetics to moderate increases in intraterminal Ca^{2+}.

Methods

The general experimental procedures have been described previously (Grinnell & Herrera, 1980; Pawson & Grinnell, 1983; Pawson & Grinnell, 1984a). Briefly, following the assessment of synaptic strength in a standard low Ca^{2+} Ringer (0.25 mM Ca^{2+}), identified junctions were restudied as described below in one of two Ringer solutions: (1) A Ca^{2+} Ringer (0.1mM Ca^{2+}, 5.0mM Mg^{2+}) whose Ca^{2+}/Mg^{2+} concentraton reduced evoked release sufficiently to prevent contraction and synaptic depression, or (2) A Ca^{2+}-free/EGTA Ringer (1mM EGTA; 2.0 mM Mg^{2+}) (Erulkar & Rahamimoff, 1978; Lev-Tov & Rahamimoff, 1980). In most experiments, two junctions, one strong and one weak, were studied simultaneously. In the Ca^{2+} Ringer, the standard tetanus used was 50 Hz for 40 seconds. Each tetanus was separated by at least an half-hour interval. In some preparations, a "reference" junction was monitored several times during the experiment to ensure that the present stimulus regime did not produce any long-term changes in the release characteristics (see Magleby & Zengel, 1976). We did not observe such effects. All junctions undergo an increase in mEPP frequency during a tetanus. Thereafter, mEPP frequency typically decays back to the resting level in a multiexponential fashion (Lev-Tov & Rahamimoff, 1980; Zengel & Magleby; 1981). In the Ca^{2+}-free/EGTA Ringer experiments, we used a 100 Hz (40 sec) train. There is no detectable Ca^{2+} influx during the tetanus as evidenced by the complete lack of a stimulus-locked transmitter release. It is believed that the associated increase in mEPP frequency is caused by a Na^{+}-induced translocation of Ca^{2+} from intraterminal stores (Lev-Tov & Rahamimoff, 1980; Lowe, Richardson, Taylor and Donatsch, 1976; Meiri, Erulkar, Lerman and Rahamimoff, 1981). mEPP intervals were collected and stored with a microcomputer and subsequently analyzed with a moving bin histogram technique (Rahamimoff & Yaari, 1973).

Range of synaptic strengths

In confirmation of previous findings (Kuno et al, 1971; Grinnell & Herrera, 1980) our data show a positive relationship between synaptic strength and spontaneous mEPP frequency (Fig. 1). The abscissa illustrates that there is an approximately 25-fold range in synaptic strength (release/unit length), which is independent of terminal length. Note that there is a continuum of synaptic strengths. The terms "weak" and "strong" junctions are used relatively.

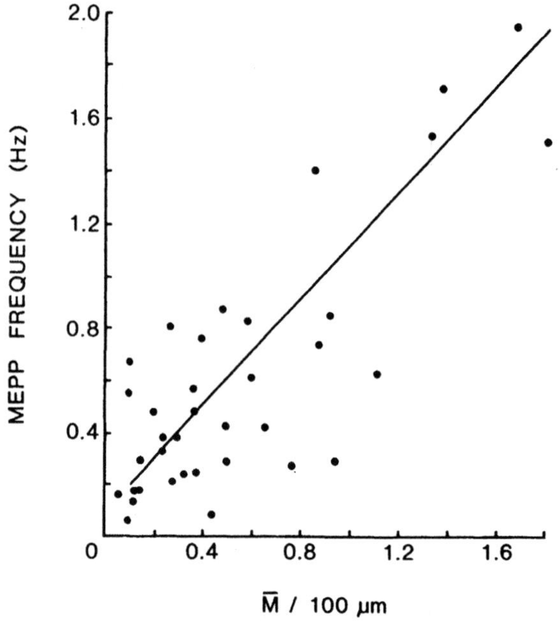

Figure 1. Plot of resting mEPP frequency (Hz) vs. release per unit length (mean quantal content/100 micrometer of terminal) in a low Ca^{2+} Ringer (0.25 mM Ca^{2+}). The straight line represents the least-squares regression of the data.

Tetanic and posttetanic phenomena in a Ca^{2+} Ringer

The relatively long, high frequency trains that we used in the present experiments limited the number of trains that could be delivered to the nerve without causing synaptic depression. Accordingly, the tetanic and PTP data for a given junction were usually collected from a single stimulus train. Therefore it was important to ascertain that the results obtained from single trains were representative of the inherent release properties of a given terminal. This was done in two ways.

First, several junctions were studied repeatedly at 1/2 hr intervals (the standard inter-train interval for all our experiments). We found that with this interval there was no observable change in the tetanic and PTP phenomena at a given junction (Fig. 2). Furthermore, we also monitored a "reference" junction during each experiment. Periodic recording from this junction demonstrated that this experimental approach produced reproducible, unchanging responses throughout the course of our experiments. We have no evidence for long-term changes in transmitter release, as has been described when a greater number of stimuli are delivered to the nerve (Magleby & Zengel, 1976). Finally, since we studied a "strong" and "weak" junction simultaneously, any uncontrolled influences would be expected to be distributed evenly throughout the results.

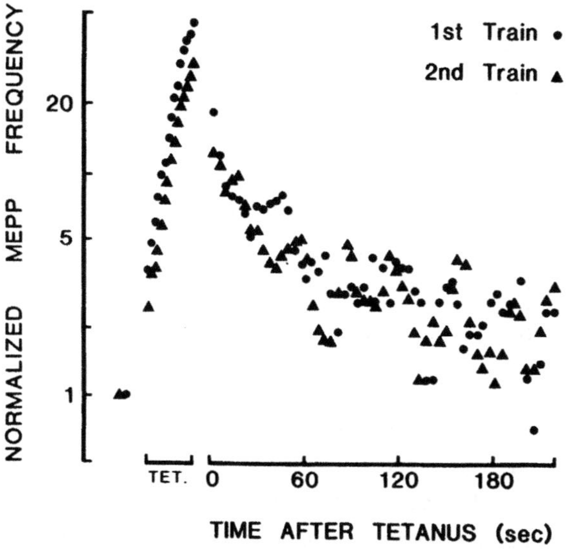

Figure 2. An example of the constancy in transmitter release from a given junction in response to the tetanic train (50 Hz, 40 sec) and intertrain interval (30 min) used in the present experiments. The mEPP frequency was normalized to the average resting mEPP frequency 5 minutes prior to the tetanus. Note the semi-logarithmic plot. Moving bin histogram: 10 sec bin, 1 sec Δbin.

In a Ca^{2+} Ringer, we found that the stronger junctions showed a greater absolute increase in tetanic potentiation and a longer lasting PTP than was seen in weaker junctions (Fig. 3). There was a consistent positive correlation between synaptic strength and peak tetanic mEPP frequency (Fig. 4A).

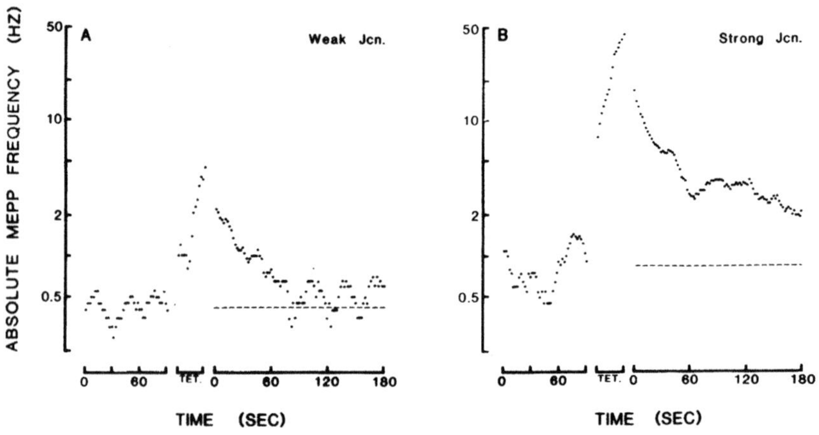

Figure 3. A semi-logarithmic plot of absolute mEPP frequency comparing the responses of (A) a weak (m/100 μm = 0.14) and (B) a strong (m/100 μm = 0.92) junction to the tetanus (50 Hz, 40 sec) used in a Ca^{2+} Ringer. The dashed line represents the average of the resting mEPP frequency for the 5 minutes preceding the tetanus. Strong junctions show a greater tetanic potentiation and a longer lasting PTP. Moving bin histogram: 10 sec bin, 1 sec Δbin.

Our present interest centered on determining the kinetics of Ca^{2+} metabolism, as represented by the posttetanic decay of the mEPP frequency back to the pretetanus resting frequency. The differences between weak and strong junctions in this regard are illustrated more clearly in Fig. 5A,B. In a Ca^{2+} Ringer the time constant of PTP is strongly positively correlated with synaptic strength (Fig. 6A).

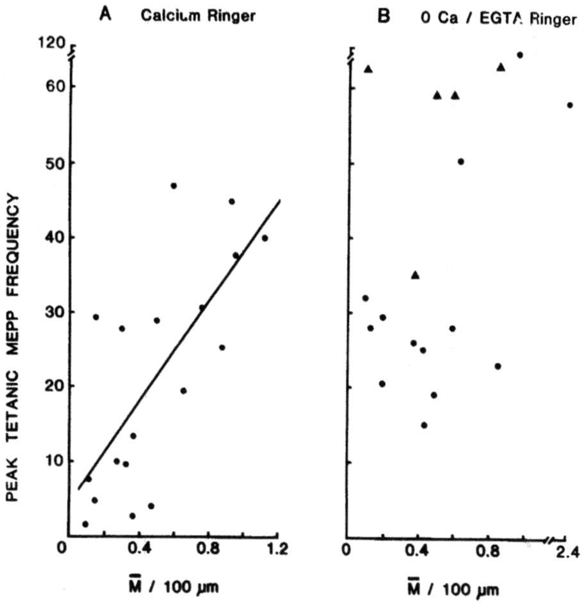

Figure 4. In a Ca^{2+} Ringer (A), peak tetanic mEPP frequency (mEPPs/sec) is positively correlated with synaptic strength. However in the Ca^{2+}-free/EGTA Ringer (B) this does not appear to be the case. Although 2 or 3 of the strongest junctions show an increased tetanic potentiation, this may represent biological noise. Note that following a 100 Hz (80 sec) tetanus (▲), there is no such correlation.

The results were highly reproducible at a given terminal (eg. Fig. 2). Also, when the tetanus duration was increased in 10 second steps from 10 to 50 seconds, a relatively greater increase in both tetanic potentiation and τ_p was produced in strong junctions which were studied concurrently with weaker ones. An increased potential for PTP represents an heretofore unrecognized attribute of strong NMJs. Interestingly, the time constant of augmentation (τ_a) also increases with synaptic strength. Since experimental manipulations that presumably increase the intraterminal Ca^* (eg., increased tetanic duration or frequency) lead to prolongation of both of these values (Lev-Tov & Rahamimoff, 1980), it appears that these two time constants can be used to describe the kinetics of Ca^{2+}-metabolism within the terminal. If our results are due to an accumulation of Ca^* inside the terminal, and if both τ_p and τ_a represent the time course of elimination of this Ca^*, then we would expect to find a good positive correlation between the τ_p and τ_a values at a given junction. The strong correlation ($r=0.73$; $p<0.001$; $n=20$) that we do observe suggests a tight coupling between the underlying processes responsible for the longer τ_p and τ_a values at stronger junctions.

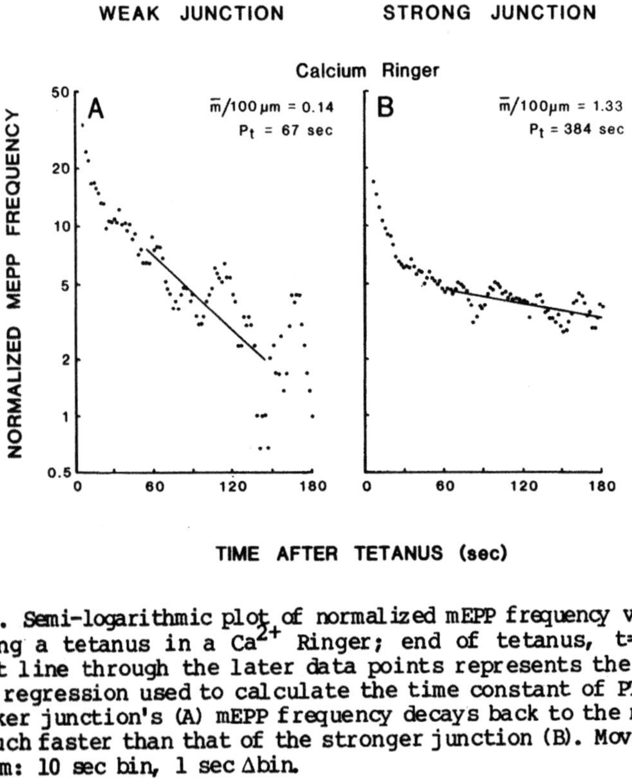

Figure 5. Semi-logarithmic plot of normalized mEPP frequency vs. time following a tetanus in a Ca^{2+} Ringer; end of tetanus, t=0. The straight line through the later data points represents the least-squares regression used to calculate the time constant of PTP (τ_p). The weaker junction's (A) mEPP frequency decays back to the resting level much faster than that of the stronger junction (B). Moving bin histogram: 10 sec bin, 1 sec Δbin.

Tetanic and posttetanic phenomena in a Ca^{2+}-free/EGTA Ringer

The results obtained in a Ca^{2+} Ringer indicate that stronger junctions have a reduced Ca^{2+}-buffering capacity and/or that they have a proportionately greater Ca^{2+}-influx during the tetanus, thereby leading to a saturation of some Ca^{2+}-buffering system(s). In order to determine the importance of Ca^{2+}-buffering systems in our findings, we performed similar experiments in a Ca^{2+}-free/EGTA Ringer where there is no Ca^{2+} influx during the tetanus (Erulkar & Rahamimoff, 1978; Lev-Tov & Rahamimoff, 1980). A greater tetanus frequency (100 Hz) was used in the Ca^{2+}-free/EGTA Ringer in order to produce an adequate increase in mEPP frequency. This accounts for the longer τ_p values than are sometimes seen in a Ca^{2+} Ringer. In contrast to the Ca^{2+} Ringer experiments, there is no consistent correlation between synaptic strength and peak tetanic mEPP frequency (Fig. 4B).

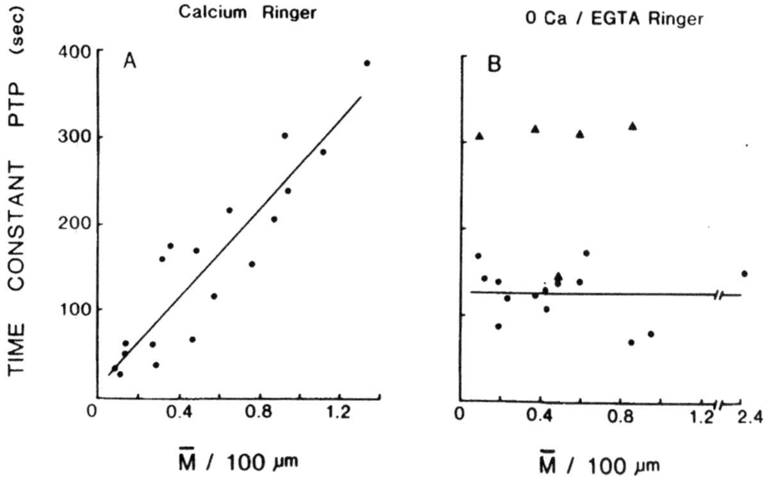

Figure 6. Plot of τ_p for the standard tetanus vs. release per unit length. (A) Results in a Ca^{2+} Ringer, 50 Hz (40 sec) tetanus. Note the strong positive correlation ($r=0.92$; $p<0.001$; $n=18$). (B) In the Ca^{2+}-free/EGTA Ringer, both a 100 Hz (40 sec) tetanus (●) and a 100 Hz (80 sec) tetanus (▲) were used. Under neither condition is there a significant correlation between these 2 variables (40 sec: $r=0.02$; $p>0.50$; $n=14$. 80 sec: $r=0.02$; $p>0.50$; $n=5$).

More importantly, we find that all junctions show a similar time constant for the decay of PTP in the Ca^{2+}-free/EGTA Ringer (Fig. 7). There is no correlation between synaptic strength and τ_p (Fig. 6B). A similar non-correlation Fig. 6B) between release and τ_p was seen following a 100 Hz (80 sec) tetanus, which produces an even greater increase in mEPP frequency. It is noteworthy that the tetani used in the Ca^{2+}-free/EGTA Ringer (100 Hz, 40 & 80 sec) produce mEPP frequencies comparable to those seen in a Ca^{2+} Ringer at junctions of intermediate strength. If mEPP frequency is in this instance a reliable indicator of intraterminal Ca*, we conclude that intraterminal Ca* has undergone a substantial increase in concentration. This conclusion is supported by the long τ_p values produced in the Ca^{2+}-free/EGTA experiments. Therefore, strong and weak junctions do not appear to differ in their buffering kinetics to at least a moderate increase in Ca^{2+} levels from internal stores.

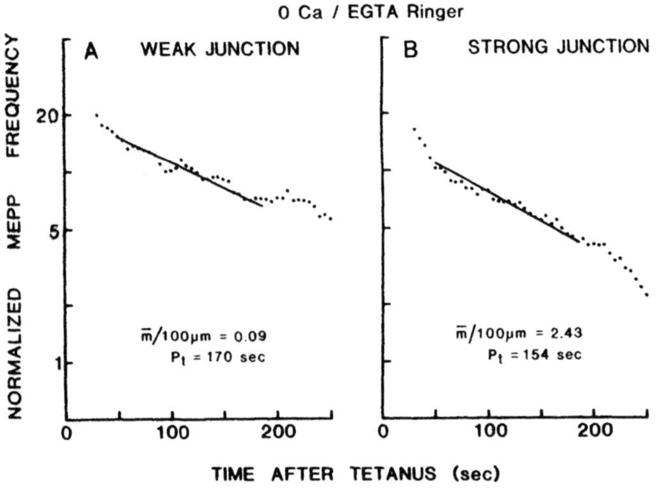

Figure 7. Semi-logarithmic plot of normalized mEPP frequency vs. time following a tetanus in a Ca^{2+}-free/EGTA Ringer (Compare with similar plot in Fig. 5). When there is no Ca^{2+} influx, there are no strength-dependent differences in the time constant of PTP. These are the weakest (A) and the strongest (B) junctions studied under these conditions, representing a 25-fold range in synaptic strength. The 100 Hz (40 sec) tetanus produced an increased mEPP frequency comparable to that seen in a Ca^{2+} Ringer. Moving bin histogram; 25 sec bin, 1 sec Δbin.

Discussion

The differences we see in τ_p might result from a differential Ca^{2+} influx and/or from a differential sensitivity of some intermediate release-related process to Ca^{2+}. For several reasons, we favor the first interpretation. The Ca^{2+}-free/EGTA Ringer experiments indicate that there is no such differential sensitivity to internally released Ca^{2+}. Moreover, if differences in some Ca^{2+}-dependent release process were responsible, one would have to invoke a number of differences between strong and weak junctions. For example, in addition to differences in Ca^{2+} sensititivity of some release process, one would have to propose other counterbalancing differences, such as a differential Na^+-translocation of Ca^{2+}, or differences in effectiveness between entering Ca^{2+} and internally released Ca^{2+}. Furthermore, the obvious importance of Ca^{2+} channel regulation in the release process is well established (Kandel & Schwartz, 1982; Reuter, 1983). Since the manifestation of strength-dependent differences in Ca^{2+}-metabolism (τ_a and τ_p) is dependent upon a Ca^{2+} influx during the tetanus, we believe that a differential Ca^{2+} influx per impulse underlies the differences in synaptic efficacy under normal conditions of evoked transmitter release. Modification of the Ca^{2+} influx during an action potential would provide a direct and efficient method of regulating the release level of a given nerve terminal.

Similar conclusions regarding differences in resting Ca^{2+} influx can be derived from a consideration of the resting mEPP frequency. As described above, we observe a good positive correlation between the spontaneous mEPP frequency and release per unit length (Fig. 1). In going from a low Ca^{2+} to a Ca^{2+}-free/EGTA Ringer, stronger junctions show a substantial decrease in the resting mEPP frequency, whereas weak junctions are relatively unaffected by the withdrawal of external Ca^{2+} (Fig. 8). Apparently, stronger junctions have a greater resting Ca^{2+} leak, consistent with the notion that such junctions may have a greater number of Ca^{2+} channels and/or that each Ca^{2+} channel has in some way a greater conductance.

These findings help explain the variability in the amount of tetanic and posttetanic potentiation that has previously been observed

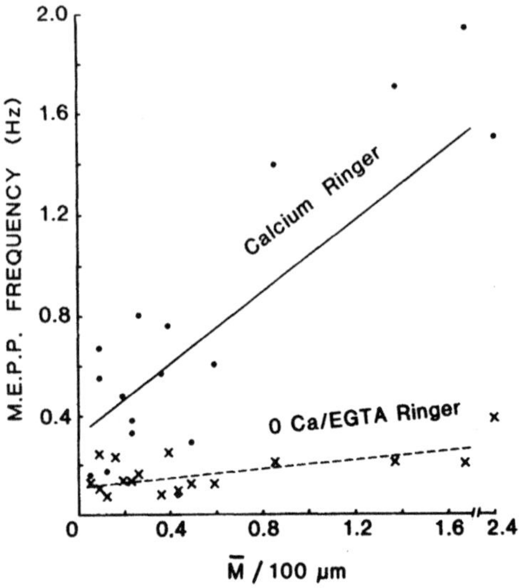

Figure 8. Plot of resting mEPP frequency vs. release per unit length in a Ca^{2+} (● , continuous line) (0.25 mM Ca^{2+}) and in a Ca^{2+}-free/EGTA Ringer (X , dashed line). Each cell was studied in both Ringers (n=17). In each Ringer there is a good correlation in this function, but there is a marked reduction in the slope of the relationship when the external Ca^{2+} is removed; indicating that the resting mEPP frequencies of the stronger junctions are more sensitive to the withdrawal of Ca^{2+}.

at the frog neuromuscular junction (Lev-Tov & Rahamimoff, 1980; Magleby & Zengel, 1976). The results are also of interest in showing that synaptic plasticities such as PTP are closely tied to the inherent release level of the presynaptic terminal in question. Furthermore, the release properties of the frog NMJ have consistently been found to be applicable to the release process at CNS synapses as well. Since frequency coding is an important element of neuronal integration, these results may have important physiological significance not only for the normal functioning of the NMJ but for central synapses as well. Although there are no doubt other possible physiological differences between strong and weak terminals, this demonstration of the importance of differences in Ca^{2+} influx and the differences in intraterminal Ca^{2+} metabolism when Ca^{2+} enters through Ca^{2+} channels, but not when it is released internally, provide us with an important foundation for further study.

The importance of the Ca^{2+} influx in determining parameters of potentiation as well as synaptic strength (Pawson & Grinnell, 1984a; Pawson & Grinnell, 1984c), the relatively minor ultrastructural correlate in a recent serial-section transmission study (Herrera, Grinnell & Wolowske, 1983), and the obvious importance of the active zone in the release process (Heuser, Reese & Landis, 1974), have led us to examine the active zone organization of identified and physiologically characterized terminals (Pawson & Grinnell, 1984b). A simple interpretation of our physiological results might be that strong terminals have a greater calcium channel density than weaker terminals. Since it has been suggested that active zone particles might be the ultrastructural correlate of calcium channels (Pumplin, Reese & Llinas, 1981), we feel that such a study might be very fruitful. We know that active zones are not invariate structures (Ko, 1984; Pumplin, 1983) and we are encouraged by our preliminary studies which suggest that there is considerable variability in the organization of active zone particles from one junction to another. We hope that this work will help to elucidate the structural basis for the observed physiological differencs in synaptic strength.

Acknowledgements

This research was supported by a fellowship from the MDA to P.A.P. and by research grants from the NIH and MDA to A.D.G.

REFERENCES

Cotman, C.W., M. Nieto-Sampedro and E.W. Harris (1981) Synapse replacement in the nervous system of adult vertebrates. Physiol. Rev. 61:684-784.

Erulkar, S.D. and R. Rahamimoff (1978) The role of calcium ions in tetanic and post-tetanic increase of miniature end-plate potential frequency. J.Physiol.278:501-511.

Grinnell, A.D. and A.A. Herrera (1980) Physiological regulation of synaptic effectiveness at frog neuromuscular junctions. J.Physiol. 307:301-317.

Grinnell, A.D. and A.A. Herrera (1981) Specificity and plasticity of neuromuscular connections: long-term regulation of motoneuron function. Prog. Neurobiol. 17:203-282.

Grinnell, A.D., A.A. Herrera and B. Wolowske (1983) Ultrastructural correlates of differences in synaptic effectiveness at frog neuromuscular junctions. Soc. Neurosci. Abstr. 9:1026.

Grinnell, A.D. and L.O. Trussell (1983) Synaptic strength as a function of motor unit size in the normal frog sartorius. J.

Physiol. 338:221-241.
Herrera, A.A. and A.D. Grinnell (1980) Transmitter release from frog motor nerve terminals depends on motor unit size. Nature 287:649-651.
Herrera, A.A. and A.D. Grinnell (1981) Contralateral denervation causes enhanced transmitter release from frog motor nerve terminals. Nature 291:495-497.
Heuser, J.E., T.S. Reese, M.J. Dennis, Y. Jan, L. Jan, and L. Evans (1979) Synaptic vesicle exocytosis captured by quick freezing and correlated with quantal transmitter release. J. Cell Biol. 81:275-300.
Heuser, J.E., T.S. Reese and D.M.D. Landis (1974) Functional changes in frog neuromuscular junctions studied with freeze-fracture. J. Neurocytol. 3:109-131.
Kandel, E.R. and J.H. Schwartz (1982) Molecular biology of learning: Modulation of transmitter release. Science 218:433-443.
Katz, B. (1969) The Release of Neural Transmitter Substances. Liverpool University Press, Liverpool.
Katz, B. and R. Miledi (1967) A study of synaptic transmission in the absence of nerve impulses. J. Physiol. 192:407-436.
Kelly, S.S. and N. Robbins (1983) Progression of age changes in synaptic transmission at mouse neuromuscular junctions. J. Physiol. 343:375-383.
Ko, C.P. (1982) Regeneration of presynaptic active zones at frog neuromuscular junction. Soc. Neurosci. Abstr. 8:27.
Kuno, M., S.A. Turkanis and J.N. Weakly (1971) Correlation between nerve terminal size and transmitter release at the neuromuscular junction of the frog. J. Physiol. 213:545-556.
Lev-Tov, A. and R. Rahamimoff (1980) A study of tetanic and post-tetanic potentiation of miniature end-plate potentials at the frog neuromuscular junction. J. Physiol. 309:247-273.
Lowe, D.A., B.P. Richardson, P. Taylor and P. Donatsch (1976) Increasing intracellular sodium triggers calcium release from bound pools. Nature 260:337-338.
Magleby, K.L. and J.E. Zengel (1976) Long term changes in augmentation, potentiation, and depression of transmitter release as a function of repeated synaptic activity at the frog neuromuscular junction. J. Physiol. 257:471-494.
Meiri, H., S.D. Erulkar, T. Lerman and R. Rahamimoff (1981) The action of the sodium ionophore, monensin, on transmitter release at the frog neuromuscular junction. Brain Res. 204:204-208.
Mendell, L.M. (1984) Modifiability of spinal synapses. Physiol. Rev. 64:260-324.
Nudell, B.M. and A.D. Grinnell (1982) Inverse relationship between transmitter release and terminal length in synapses on frog muscle fibers of uniform input resistance. J. Neurosci. 2:216-224.
Nudell, B.M. and A.D. Grinnell (1983) Regulation of synaptic position, size, and strength in anuran skeletal muscle. J. Neurosci. 3:161-176.
Pawson, P.A. and A.D. Grinnell (1983) Physiological differences between strong and weak neuromuscular junctions. Soc. Neurosci. Abstr. 9:1027.
Pawson, P.A. and A.D. Grinnell (1984a) Posttetanic potentiation in strong and weak neuromuscular junctions: Physiological differences caused by a differential Ca^{2+} influx. Brain Res. (In press).

Pawson, P.A. and A.D. Grinnell (1984b) Freeze-fracture analysis of active zone organization at identified frog neuromuscular junctions. Soc. Neurosci. Abstr. 10 (In press).
Pawson, P.A. and A.D. Grinnell (1984c) Tetanic and posttetanic potentiation at the frog neuromuscular junction: Physiological differences between strong and weak neuromuscular junctions. (Submitted for publication).
Pumplin, D.W. (1983) Normal variations in presynaptic active zones of frog neuromuscular junctions. J. Neurocytol. 12:317-323.
Pumplin, D.W., T.S. Reese and R. Llinas (1981) Are the presynaptic membrane particles the calcium channels? Proc. Natl. Acad. Sci. USA 78:7210-7213.
Rahamimoff, R., S.D. Erulkar, A. Lev-Tov and H. Meiri (1978) Intracellular and extracellular calcium ions in transmitter release at the neuromuscular synapse. Ann. N.Y. Acad. Sci. 307:583-598.
Rahamimoff, R., A. Lev-Tov and H. Meiri (1980) Primary and secondary regulation of quantal transmitter release: calcium and sodium. J. Exp. Biol. 89:5-18.
Rahamimoff, R. and Y. Yaari (1973) Delayed release of transmitter at the frog neuromuscular junction. J. Physiol. 228:241-257.
Reuter, H. (1983) Calcium channel modulation by neurotransmitters, enzymes and drugs. Nature 310:569-574.
Rotshenker, S. (1982) Transneuronal and peripheral mechanisms for the induction of motor neuron sprouting. J. Neurosci. 2:1359-1368.
Smith, D.O. (1982) Physiological and structural changes at the neuromuscular junction during aging. In: The Aging Brain: Cellular and Molecular Mechanisms of Aging in the Nervous System. E. Giacobini et al (ed.), Raven Press, N.Y., pp 123-137.
Trussell, L.O. and A.D. Grinnell (1984) The regulation of synaptic strength within motor units of the frog cutaneous pectoris muscle. J. Neurosci. (In press).
Wernig, A., M. Pecot-Dechavassine and H. Stover (1980) Sprouting and regression of the nerve at the frog neuromuscular junctions in normal conditions and after prolonged paralysis with curare. J. Neurocytol. 9:277-303.
Zengel, J.E. and Magleby, K.L. (1981) Changes in miniature endplate potential frequency during repetitive nerve stimulation in the presence of Ca^{2+}, Ba^{2+}, and Sr^{2+} at the frog neuromuscular junction. J.Gen. Physiol. 77:503-529.

DISCUSSION OF THE ARTICLE BY P. PAWSON (U.S.)

D.M.J. Quastel (Canada): When we look at the frequency of mepps in raised potassium, there is a very large variation at low calcium of the frequency of mepps. However, when one goes up to approximately 8mM calcium, the variation becomes very very small (Gage P.W. and Quastel D.M.J. (1966). J. Physiol. 185, 95-123). It would seem to me that this suggests that the difference is not in the absolute number of calcium channels per nerve terminal per unit length but in how well the calcium is going in for a given concentration outside. Would that fit with what you are saying?

P. Pawson (U.S.): That would be reasonable and consistent with our results. However, in terms of looking at different conductances of a single channel, there seems to be no literature that suggests that one can have significant variance in the unit conductance of the calcium channel.

D.M.J. Quastel (Canada): If you believe the channel is saturable, so to speak, then it would fit with a different affinity of the channel for calcium.

P. Pawson (U.S.): We believe that our results are explainable by a number of situations; either by a different gating of the channel conductance state, by a difference in the channel's voltage dependence, or by a greater number of channels, or perhaps some combination thereof.

R. Rahamimoff (Israel): It has been suggested several times that the double row of particles that one sees in the active zone represent the calcium channels. Do you see a difference in the number of those double rows?

P. Pawson (U.S.): That is exactly what we are doing right now. We have spent the last several months developing a technique to study the freeze-fracture view of active zone organization of identified terminals that we have characterized physiologically. At this point, we have not obtained freeze-fracture views from a significant portion of the active zone of identified terminals. What is of interest and which gives us cause to pursue our present course is that we do observe a fair amount of variability in the active zone organization of randomly studied cutaneous pectoris junctions. In some terminals, one sees the traditional very nice double array of particles, while in other terminals, there exist consistently poorly organized active zones, smaller active zones and up to a 5-fold overall decrease in the total number of active zone particles. Therefore, the potential for establishing an ultrastructural correlate exists, and we are encouraged that our present course will be a fruitful one.

S. Thesleff (Sweden): Have you tried to use an aminopyridine at high concentration to cause maximal transmitter release from junctions and to see if under those conditions there still is a similar difference between weak and strong junctions?

P. Pawson (U.S.): One reason we haven't done that is technical. Under the low Ca^{2+} conditions we are using, some strong junctions go supra-threshold with the tetani we are using. Therefore, it is not possible to record continuously from a junction if a much greater degree of release is evoked.

A. Wernig (West Germany): You have been recording from whole junctions and not from parts of it. In your estimations you are assuming that a junction in itself is homogeneous. Do you have any evidence to support this?

P. Pawson (U.S.): The work of D'Alonzo and Grinnell, which is now in press in J. Physiol., shows that from their analysis of single quantal events and recording from either ends of the junction it appears that the junctions are fairly homogeneous, at least in our hands. Certainly there does not exist the same differences in release levels within a terminal as one observes between terminals.

NEUROTRANSMITTER RELEASE FROM PC12 CELLS

BRUCE D. HOWARD, WILLIAM P. MELEGA, MAYUMI KOIDE,
CATHERINE M. BITLER, MAO-BIN ZHANG AND MELVIN KHAW

PC12 is a clonal line of rat pheochromocytoma cells that upon exposure to nerve growth factor extend neurites and acquire the morphology of neurons (1,2). Both undifferentiated and nerve growth factor treated cells store and secrete dopamine and acetylcholine by a Ca^{2+}-dependent process (1,3-6). Approximately 35% of the cell's acetylcholine and 60% of dopamine is contained within vesicles; the remaining portion is in the cytosol (5). Acetylcholine-containing vesicles are distinct from dopamine containing vesicles as the two types can be physically separated from each other after disruption of the cells (4). These properties makes PC12 useful for pharmacological studies in that one can compare the effects of a particular drug on two different transmitter systems in the same cell. In this paper we discuss the characteristics of the release of these two transmitters from PC12 cells.

Acetylcholine Release. One matter that we have dealt with is the source of the acetylcholine released from stimulated PC12 cells. It is generally agreed that the evoked release of acetylcholine from at least many cholinergic nerve terminals is quantal in nature (7). However, there is controversy as to whether synaptic vesicles or the cytosol is the immediate source of the released acetylcholine (8-13). This matter has been settled for the release of catecholamines from adrenal medullary cells, PC12 cells, and adrenergic neurons. Evidence that the release of catecholamines from these cells originates from vesicles has been obtained from studies with false adrenergic transmitters (14) and from analyses of the substances released along with the catecholamines (15,16); for example, catecholamine release is accompanied by the release of soluble proteins contained within the vesicles but not by release of proteins present in the cytosol. Unfortunately, similar experiments cannot be performed on cholinergic systems because of a lack of a suitable series of false cholinergic transmitters and because cholinergic vesicles do not contain soluble proteins (7).

We have addressed the question of the source of released acetylcholine by examining the effects of 2-(4-phenylpiperidino) cyclohexanol (AH5183) on the release of acetylcholine and dopamine from PC12 cells. AH5183 has been reported to cause neuromuscular blockade by inhibiting the evoked release of acetylcholine from motor nerve terminals (18). The characteristics of this blockade have been interpreted to indicate an AH5183-induced decrease in synaptic vesicle stores of acetylcholine (18). Toll and Howard (19) found that 40 μM AH5183 only slightly affected the snythesis of ACh in intact PC12 cells while markedly inhibiting the loading of acetylcholine into storage vesicles. Anderson _et al_. (20) found that AH5183 inhibited the transport of acetylcholine into isolated synaptic vesicles from _Torpedo californica_ with an IC_{50} of 40 nM.

To examine the effect of AH5183 on the synthesis and vesicular loading of acetylcholine, PC12 cells were incubated for 30 min in the presence of $[^2H_4]$choline and varying concentration os the drug. $[^2H_4]$Cholne was used to label PC12 stores of choline and acetylcholine; choline and acetylcholine levels were measured by combined gas chromatography-mass spectrometry (21). This procedure allows the simultaneous determination of endogenous choline (unlabeled choline), newly accumulated choline ($[^2H_4]$choline), newly synthesized acetylcholine ($[^2H_4[$acetylcholine), and unlabeled acetylcholine

($[^2H_0]$acetylcholine). The $[^2H_0]$acetylcholine largely reflects older acetylcholine stores. These measurements were made on a whole cell extract and on the vesicle fraction of a cell homogenate.

At concentrations of 40 nM to 40 µM, AH5183 did not appear to inhibit the uptake of $[^2H_4]$choline by the cells. Control cells contained 1.36 ± 0.1 nmol of $[^2H_4]$choline per mg of protein while the $[^2H_4]$choline content of AH5183-treated cells ranged from 1.26 to 1.47 nmol per mg of protein.

As shown in Fig. 1, the total cell level of $[^2H_4]$acetylcholine was somewhat reduced by the AH5183 treatment; the $[^2H_4]$acetylcholine content of cells treated with 0.4-40 µM AH5183 was 72-78% of that of control cells, and for cells treated with 40 nM AH5183 it was 88% of control. Fig. 1 also shows that treatment with 40 nM AH5183 resulted in a much greater decrease in the vesicular content of newly synthesized $[^2H_4]$acetylcholine (25% of the control value). Almost no $[^2H_4]$acetylcholine was found in vesicles from cells treated with AH5183 at a concentration of 0.4 µM or greater. AH5183 did not affect the level of $[^2H_0]$acetylcholine in the cells or in the vesicles (Fig. 1). This latter result demonstrates that AH5183 did not cause the vesicles to lyse or to leak previously loaded acetylcholine, conditions that would have caused loss of $[^2H_4]$acetylcholine and $[^2H_0]$acetylcholine equally. Thus, the ability of AH5183 to decrease the vesicular content of $[^2H_4]$acetylcholine specifically means that AH5183 exerted its effect by inhibiting the loading of acetylcholne into vesicles.

The effect of AH5183 on the evoked release of acetylcholine is shown in Fig. 2. For these experiments PC12 cells were incubated for 30 min with $[^2H_4]$choline in the presence or absence of 4 µM AH5183. (Under these conditions little of the newly synthesized $[^2H_4]$acetylcholine would be loaded into vesicles in AH5183-treated cells.) Subsequently, the cells were incubated for 5 min in cell incubation buffer (to monitor spontaneous release)

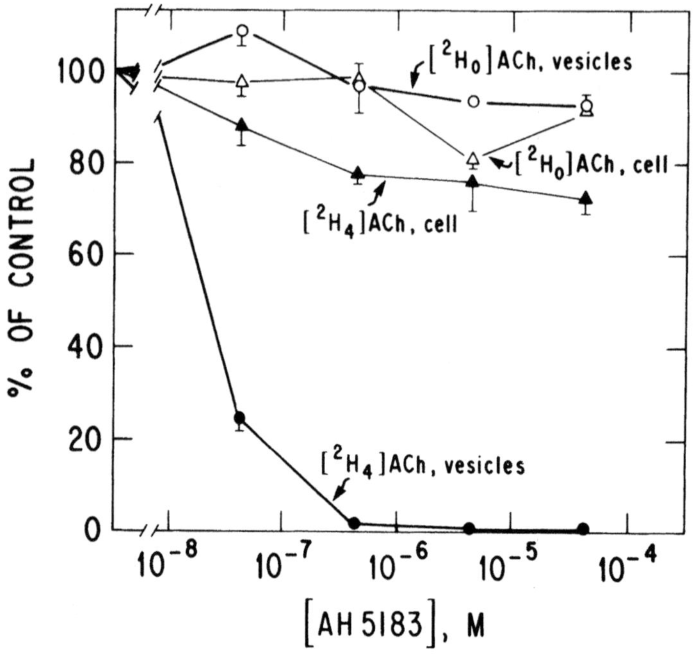

Fig. 1. Effect of AH5183 on the cellular and vesicular content of acetylcholine. PC12 cells were incubated for 30 min with 10 μM [^{3}H]choline. Acetylcholine was extracted from the cells (▲,△) and from a vesicle fraction (●,○) of a cell homogenate, and the levels of [$^2H_4$]acetylcholine (▲,●) and [$^2H_0$]acetylcholine (△,○) were measured. The results are expressed as the percentage of values for controls. Control cell contained, per mg of protein, 1.5 ± 0.05 nmol of [$^2H_0$]acetylcholine and 0.25 ± 0.02 nmol of [$^2H_4$]acetylcholine; control vesicles contained, per mg of protein, 1.1 ± 0.1 nmol of [$^2H_0$]acetylcholine and 0.10 ± 0.01 nmol of [$^2H_4$]acetylcholine. The data points are the means for triplicate incubations and the error bars represent the SD. ACh is acetylcholine.

and then for 5 min in a buffer containing 2 mM Ba^{2+}, which is a secretagogue for several cells (22,23) including PC12 (3). Cells that had not been treated with AH5183 released approximately 17% of their stores of [$^2H_0$]-acetylcholine and 9% of their [$^2H_4$]acetylcholine but only 2% of their [$^2H_4$]-acetylcholine when stimulated with Ba^{2+}. The amount of [$^2H_4$]acetylcholine released from AH5183-treated cells during the incubation with Ba^{2+} was only slightly greater than the spontaneous release. The specific activity

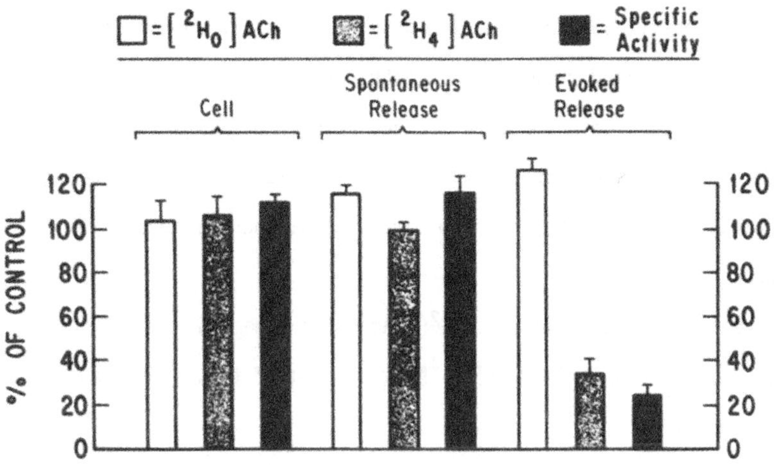

Fig. 2. Acetylcholine release from AH5183-treated cells. Cells were incubated for 30 min with [$^2H_4$]choline in the presence of AH5183, washed, and incubated in cell incubation buffer for 5 min and finally in Ba^{2+} containing buffer for 5 min. The cellular level is the sum of the [$^2H_0$]acetylcholine or [$^2H_4$]acetylcholine released to the buffers during the two 5-min incubations (representing spontaneous release and evoked release, respectively) and the amount remaining in the cells after the incubations. The results, expressed as a percentage of control values for cells that were incubated in the absence of AH5183. The values are the means ± SD for quadruplicate incubations. Per mg protein, the amount of acetylcholine in the cells, released spontaneously, and released by Ba^{2+} were, respectively, 1.104 ± 0.081, 0.008 ± 0.002 and 0.189 ± 0.055 nmol [$^2H_0$]acetylcholine, and 0.352 ± 0.073, 0.004 ± 0.001, and 0.030 ± 0.008 nmol [$^2H_4$]acetylcholine.

([$^2H_4$]acetylcholine/total acetylcholine) of the acetylcholine released by Ba^{2+} from control cells was four fold greater than that of the acetylcholine released from AH5183-treated cells. On the other hand the specific activity of total cellular acetylcholine was almost identical for control and AH5183-treated cells. The fact that AH5183 did not affect the evoked release of [$^2H_0$]acetylcholine shows that AH5183 does not inhibit the acetylcholine release process, itself. Rather, we interpret these data to indicate that, upon exposure to Ba^{2+}, [$^2H_4$]acetylcholine was less readily released from AH5183-treated cells than was [$^2H_0$]acetylcholine because a smaller fraction of the [$^2H_4$]acetylcholine was contained in vesicles. The experiment was repeated and similar results were obtained.

Although 4 μM AH5183 should have almost entirely blocked the vesicular loading of acetylcholine (Fig. 1), note that a small amount of [$^2H_4$]acetylcholine was released from AH5183-treated cells upon exposure to Ba^{2+} (Fig. 2). Some of this was undoubtedly spontaneously released and the remaining portion could have originated from vesicles that were loaded with some [$^2H_4$]acetylcholine during the two 5-min incubations subsequent to removal of AH5183.

The spontaneous release of [$^2H_4$]acetylcholine was not affected by treatment with AH5183 (Fig. 2). The mechanism of this spontaneous release is not known.

Storage and Release of Dopamine. The ability of AH5183 to block vesicular loading of acetylcholine does not apply to dopamine. We examined the effect of AH5183 on dopamine metabolism by incubating PC12 for 30 min with [^3H]-dopamine in the presence and absence of AH5183. AH5183 at 40 μM only slightly inhibited the loading of dopamine into vesicles although it markedly inhibited the transport of dopamine across the plasma membrane. The fraction of accumulated [^3H]dopamine loaded into vesicles in AH5183-treated cells was 78% of that in control cells; however, AH5183-treated cells accumulated only 28% of the [^3H]dopamine accumulated by control cells. AH5183 did not affect the evoked release of recently accumulated [^3H]dopamine. Although the AH5183 treatment resulted in a decrease in the accumulation of [^3H]dopamine by the cells, the percentage of accumulated [^3H]dopamine released from AH5183-treated cells was essentially the same as that of control cells (data not shown). In these studies, release was evoked by incubation in buffers containing either 2 mM Ba^{2+} or 56 mM K^+.

We have given biochemical evidence for vesicular release of acetylcholine from PC12 cells stimulated with the secretagogue Ba^{2+}. Strictly speaking, the data of Fig. 2 show that acetylcholine is released from metabolic pool that acetylcholine does not enter when AH5183 is present. But with respect to acetylcholine entry, storage vesicles constitute the only pool known to be affected by AH5183. In order for the cytosol to be the immediate source of released acetylcholine it would have to contain two separate pools for acetylcholine, and one of these would have to be sensitive to AH5183.

Effect of Phenothiozines. The data presented above indicates that neurotransmitter release from stimulated PC12 cells occurs by exocytosis. The molecular mechanism of this process is not understood. As shown in Table 1, the evoked release of acetylcholine and dopamine from PC12 cells is inhibited by 10 μM trifluoperazine. Chlorpromazine at 10 μM also inhibits the evoked release of these compounds. These results suggest that the release process involves calmodulin since trifluoperazine and chlorpromazine inactivate calmodulin.

TABLE 1. Effect of trifluoperazine on secretion from PC12.

Conditions		Release		
		Undifferentiated PC12		NGF-Treated PC12
K^+ (mM)	TFP (μM)	Dopamine	Acetylcholine	Acetylcholine
6	0	1.0 ± 0.1	1.5 ± 0.1	1.6 ± 0.2
6	10	0.8 ± 0.1	1.2 ± 0.1	1.2 ± 0.2
6	100	1.4 ± 0.1	2.1 ± 0.1	1.2 ± 0.2
55	0	21 ± 2	17 ± 3	19 ± 2
55	10	11 ± 1	8 ± 1	10 ± 1
55	100	5 ± 1	7 ± 1	6 ± 1

Cells were preincubated for 5 min in low K^+ buffer containing the indicated concentration of TFP (trifluoperazine), washed and incubated for an additional 5 min in buffer containing the indicated concentrations of K^+ and TFP. The release is expressed as the percentage of total dopamine or acetylcholine present in the medium at the end of the incubation period. The values are means ± S.D. for triplicate incubations. Cells were induced to extend neurites by a 14 day treatment with (NGF) nerve growth factor prior to measurement of acetylcholine release. For undifferentiated cells, total dopamine and acetylcholine were 1.9 ± 0.3 and 2.2 ± 0.2 nmol, respectively, per mg of protein. For nerve growth factor-treated cells total acetylcholine was 3.3 nmol per mg of protein.

Protein Phosphorylation and Release. The evoked secretion of many compounds from a variety of secretory cells and nerve terminals requires ATP (24-28). In contrast, the secretion of dopamine and acetylcholine from undifferentiated PC12 cells was found not to be dependent on ATP; depletion of PC12 ATP pools to as little as 10% of control values did not result in any inhibition of dopamine or acetylcholine secretion evoked by 56 mM K^+ or by carbachol (29). Reynolds et al., (29) proposed that evoked secretion could occur from ATP-depleted PC12 cells because the cells had circumvented or already accomplished the ATP-dependent reaction(s) involved in secretion from other secretory systems. If this were indeed the case, PC12 would be useful for identifying those reactions.

One postulated role for ATP in secretion is as a substrate for protein kinase; certain proteins are known to be phosphorylated in secretory systems that have been stimulated to secrete (30-34). Perhaps the best characterized of these proteins are M_r 86,000 and M_r 80,000 proteins called proteins Ia and Ib, respectively (32). They are found in nerve terminals of mammalian brain and peripheral nervous systems (35-36).

When PC12 cells were preincubated with $^{32}P_i$ and then stimulated by exposure to 2 mM Ba^{2+}, 56 mm K^+, or 1 mM carbachol, we did not detect any increase in the phosphorylation of proteins corresponding to proteins Ia and Ib. However, PC12 does have proteins similar to proteins Ia and Ib. As shown in the immunoblot of Fig. 3A, antibody raised against proteins Ia and Ib from bovine brain cross reacts strongly only with M_r 80,000 and M_r 86,000 proteins from PC12. PC12 cells contain more of the M_r 86,000 protein than the M_r 80,000 protein. This situation resembles that in rat and bovine brain (32,37).

Although present in PC12 cells the M_r 80,000 protein does not appear to be phosphorylated. We have examined this matter in the following way.

Undifferentiated cells were grown for 21 hr in medium containing $^{32}P_i$. A cell lysate was incubated with antiserum against protein I and Pansorbin (protein A-producing Staph. aureus cells). The immunoprecipitated protein was subjected to $NaDodSO_4$ polyacrylamide gel electrophoresis and the gel was analyzed by autoradiography. As shown in Fig. 3B, a M_r 86,000 phosphorprotein was present in the immunoprecipitate but not when preimmune serum was used. In neither case was a ^{32}P-labeled M_r 80,000 protein detected.

The lack of ^{32}P-labeled M_r 80,000 protein was not due to a failure of antibody and Pansorbin to immunoprecipitate a PC12 M_r 80,000 protein. Using a lysate of unlabeled PC12 cells, this immunoprecipitation procedure was found to selectively precipitate both the M_r 80,000 protein and the M_r 86,000 protein; in fact, the immunoprecipitate, which was analyzed for proteins Ia and Ib by essentially the same technique as described in Fig. 3A, contained more of the M_r 80,000 protein that the M_r 86,000 protein.

PC12 cells contain two proteins that appear to be equivalent, respectively, to proteins Ia and Ib from bovine brain. Results presented here demonstrate that these PC12 proteins comigrate with the bovine proteins Ia and Ib on $NaDidSO_4$ polyacrylamide gel electrophoresis, and the two PC12 proteins specifically cross react with antibody against the bovine proteins. Another similarity is their solubilities under certain conditions of pH shifts. Bovine brain proteins Ia and Ib remain soluble when the pH of a crude protein extract is adjusted to pH 3.0 and then to 6.0 (37). This property is also exhibited by the PC12 M_r 80,000 and M_r 86,000 proteins. In light of these findings it seems justifiable to refer to the two PC12 proteins in question as proteins Ia and Ib.

The phosphorylation of PC12 proteins Ia and Ib differs from that of the corresponding proteins in mammalian brain. Unlike the case with mammalian brain proteins (32), we did not detect an increased phosphorylation of any

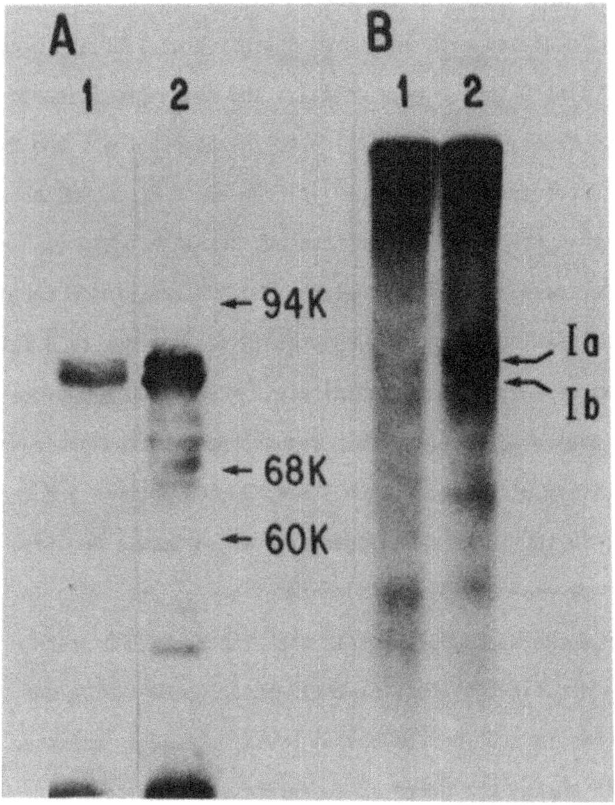

Fig. 3. Autoradiogram of immunoblot and immunoprecipitate of PC12 proteins. (A) PC12 proteins were labeled with antibody to bovine brain protein I and then with ^{125}I labeled protein A. Lane 1 contains proteins from PC12 cells (450 μg of proteins applied to gel). Lane 2 contains purified protein I (Ia plus Ib) standard from bovine brain. The bovine brain protein I standard shows heavily labeled bands at 86,000 and 80,000 daltons with multiple bands at lower molecular weights. These other bands in lane 2 are degradation products of protein I. The positions of molecular weight standards are indicated. (B) PC12 cells were incubated with $^{32}P_i$ for 21 hr, immunoprecipitated with antibody to bovine protein I and subjected to $NaDodSO_4$ polyacrylamide gel electrophoresis. Lane 1 has proteins incubated with preimmune serum and Pansorbin. Lane 2 has proteins immunoprecipitated with antiprotein I antibody and Pansorbin. In each case the protein added to the gel corresponded to 40% of the sample obtained from one flask of cells. The positions of bovine brain proteins Ia and Ib standards are indicated by arrows. Note that a M_r 86,000 phosphoprotein is pronounced in lane 2 but that there was no M_r 80,000 phosphoprotein present.

M_r 80,000 or M_r 86,000 protein when undifferentiated or nerve growth factor PC12 cells were stimulated with 56 mM K^+ or 1 mM carbachol. In itself this finding is not particulary remarkable because the content of proteins Ia and tb in PC12 may be too low to detect an increased phosphorylation under the conditions used. However, what is unusual is that when the cells were PC12 cells were incubated in $^{32}P_i$ for 21 hr and proteins Ia and Ib purified by immunoprecipitation, only protein Ia exhibited phosphorylation. This finding occurred in spite of the fact that there is more protein Ib than Ia in PC12 (Fig. 3A) and in the immunoprecipitate.

In mammalian brain nerve terminals, both proteins Ia and Ib undergo some phosphorylation even in the absence of stimulation of the terminals (32). The 21 hr period used for the labeling of PC12 in this experiment was greater than half of the cell doubling time so there should have been substantial synthesis and phosphorylation of protein Ib during this period.

The unusual phosphorylation pattern of protein I in PC12 makes PC12 an attractive system for determining whether protein I functions in transmitter release. It is tempting to speculate that the lack of detectable phosphorylation of protein Ib is related to the fact that the evoked release of dopamine and acetylcholine from undifferentiated PC12 cells is independent of ATP stores (29). An interesting possiblity is that some modification of protein Ib in PC12 not only prevents phosphorylation of the protein, but allows protein Ib to perform whatever function it has in transmitter release even in the absence of phosphorylation. A speculation by Reynolds _et al._ (29) may be relevant in this regard. They found that unlike the case with undifferentiated PC12 cells, acetylcholine release from nerve growth facter treated PC12 was partially inhibited when cellular ATP pools were depleted; and they suggested that the protein modification, which they postulated to account for the ATP independence of release from undifferentiated PC12 cells, may be present to a lesser extent at release sites in differentiated cells.

Phorbol Esters and Release. In recent years it has been demonstrated that certain phorbol esters stimulate secretion from a variety of secretory cells (34,38). This property is interesting because phorbol esters are potent activators of a Ca^{2+}-activated, phospholipid-dependent protein kinase, which is called protein kinase C (34). The Ca^{2+} activation of protein kinase C is not mediated by calmodulin. As shown in Fig. 4, the phorbol esters 12-O-tetradecanoyl phorbol-13-acetate (TPA) and phorbol-12,13-dibenzoate (PDB) stimulate the release of dopamine from PC12 cells. PDB is more effective than TPA in this regard, and each apparently requires some increase in the level of intracellular free Ca^{2+} to stimulate dopamine release. Neither phorbol ester stimulated dopamine release unless the cells were treated with the Ca^{2+} ionophore at a concentration just below that which evokes dopamine release by itself.

The data of Fig. 5 indicate that PDB stimulates transmitter release by reducing the level of free Ca^{2+} required to trigger release. These studies were done with digitonin-permeabilized cells (39,40), in which the intracellular level of free Ca^{2+} reflects that of the incubation buffer. The results of Fig. 5 are similar to those obtained for permeabilized chromaffin cells (38), and they indicate that stimulation of dopamine release by PDB is due to activation of protein kinase C; phorbol esters increase the affinity of the enzyme for Ca^{2+} (41).

The finding that transmitter release from PC12 can be evoked when ATP stores are drastically depleted, and yet can be stimulated by phorbol esters suggests that activation of protein kinase C may trigger release but normal transmitter release from stimulated PC12 does not require activation of this

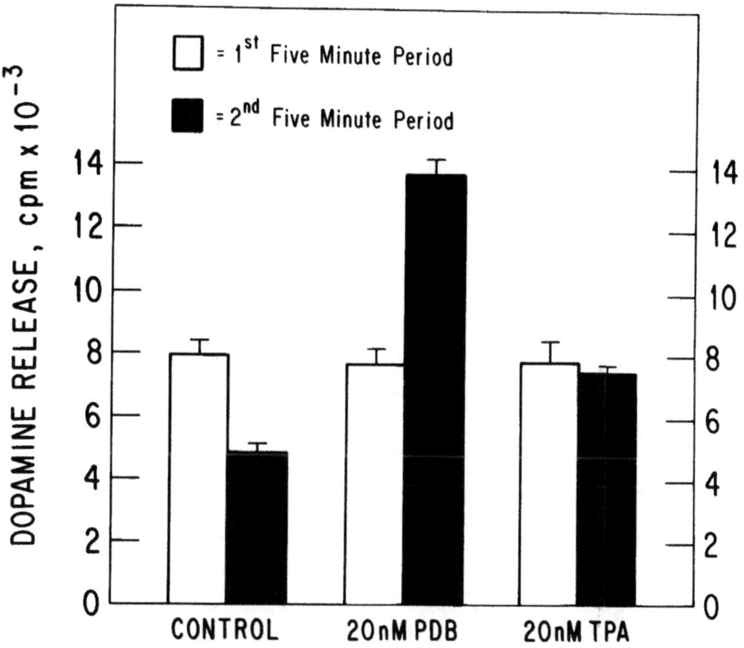

Fig. 4. Phorbol ester-induced release of [^{3}H]dopamine. Cells were incubated with [^{3}H]dopamine for 30 min, washed and incubated for 5 min in Ca^{2+}-free buffer containing 80 ng of A23187 per ml and then for 5 min in Ca^{2+}- and A23187-containing buffer that had 20 nM PDB or TPA as indicated. The results are expressed as radioactivity released per Petri dish of cells during the two 5-min incubation periods. The cells released 6-20% of the accumulated dopamine during the two incubation periods.

enzyme. Otherwise, evoked release should decrease in stimulated cells depleted of ATP.

Genetic Variants of PC12. One advantage of the PC12 system is that it is possible to use genetic approaches to study transmitter release. For this purpose we have been attempting to isolate PC12 variants defective in transmitter storage and release. Most of our studies to date have been performed with a PC12 variant that was selected by virtue of its resistance to the compound N-methyl-4-phenyl-1,2,3,6-tetrahydropyridine (NMPTP). This NMPTP-

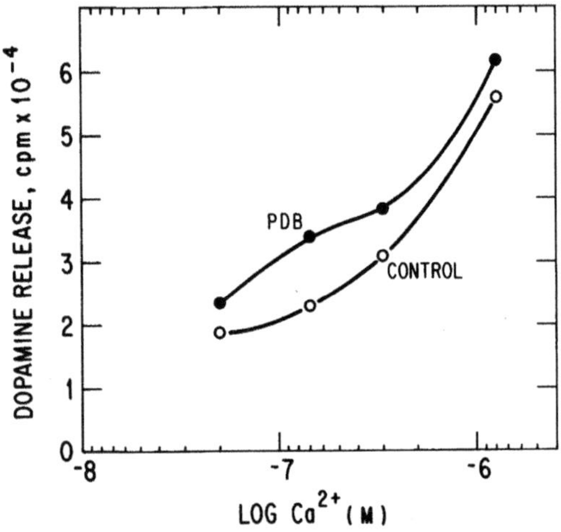

Fig. 5. Effect of PDB on [^{3}H]dopamine release from digitonin treated cells. Cells were incubated for 30 min with [^{3}H]dopamine and then incubated as described (39,40) for 5 min with a Ca^{2+}-free buffer containing 3 µM digitonin, and then for 5 min in buffer containing digitonin, varying concentrations of Ca^{2+} and no (O) or 60 nM (●) PDB. The values are the radioactivity released during the second 5-min incubation period, and the data points are the means for duplicate incubations. A maximum of 40% of the accumulated dopamine was released during this time.

resistant variant contains no acetylcholine or dopamine and fails to extend neurites upon treatment with nerve growth factor (42). As shown in Fig. 6, NMPTP-resistant cells apparently fail to produce several proteins found in wild type PC12 cells and in a guanethidine-resistant variant that stores and releases neurotransmitter like wild type cells. In this study homogenates of wild type, NMPTP-resistant and guanethidine-resistant cells were incubated with [^{3}H]xylamine, a nitrogen mustard that binds covalently to protein.

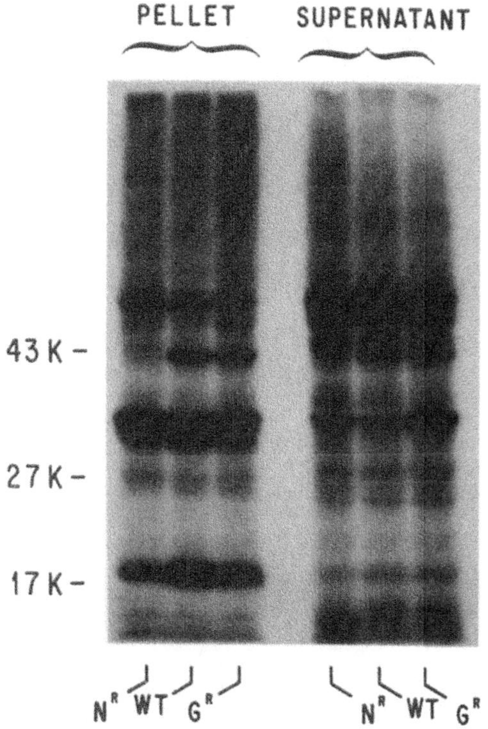

Fig. 6. $[^3H]$Xylamine-labeled proteins of wild type and variant PC12 cells. Homogenates of wild type (WT), NMPTP-resistant (N^R) and guanethidine-resistent (G^R) cells were incubated with $[^3H]$xylamine and centrifuged at 12,000x g for 20 min. The pellet and supernatant fractions were subjected to $NaDodSO_4$ polyacrylamide gel electrophoresis, and an autoradiograph of the gel was obtained. Molecular weight standards were used to determine the size of the labeled proteins.

Xylamine has been used to study catecholamine transport into nerve terminals (43). Autoradiographs of xylamine-labeled proteins that had been subjected to polyacrylamide gel electrophoresis revealed M_r 43,000, M_r 27,000 and M_r 17,000 proteins labeled from wild type and guanethidine-resistant cells but not from NMPTP-resistant cells. Thus, the NMPTP-resistant variant offers the opportunity to identify several hertofore unrecognized proteins involved in neurotransmitter storage and secretion.

Some relevant proteins missing in NMPTP-resistant cells may be present in wild type cells in insufficient quantity to identify and characterize readily. Therefore, we have begun to examine whether the variant is missing some mRNA molecules present in wild type cells, presumably because of reduced expression of certain genes. We have found that a cDNA clone (p1A75) that specifically hybridizes to mRNA from rat brain but not from other tissues (44) hybridizes differentially to mRNA from wild type and NMPTP-resistant cells (Fig. 7). Compared to wild type PC12, the NMPTP-resistant variant produces much less mRNA that hybridizes to the p1A75 cDNA clone. Sutcliffe et al. have found that p1A75 codes for a M_r 28,000 protein. The function of this protein is not known.

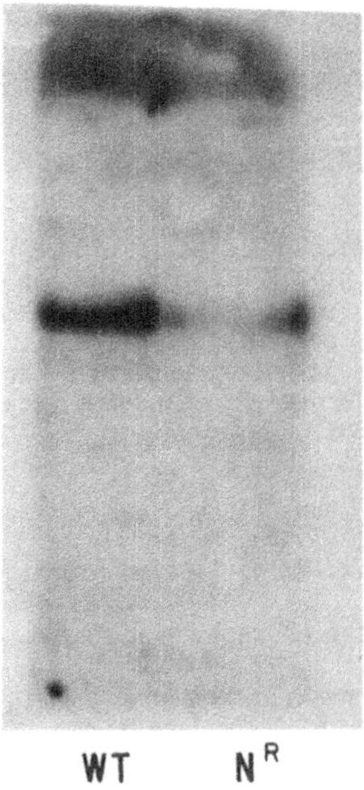

Fig. 7. Northern blot hybridization of p1A75 cDNA to mRNA from wild type (WT) and NMPTP resistant (N^R) PC12 cells.

References

1. Greene LA, Tischler AS: Establishment of a norodrenergic clonal line of intadrenal pheochromocytoma cells which repond to nerve growth factor. Proc. Natl. Acad. Sci. U.S.A. 73:2424-2428, 1976.
2. Tischler AS, Greene LA: Morphological and cytochemical properties of a clonal line of rat adrenal pheochromocytoma cells which respond to nerve growth factor. Lab. Invest. 39:77-89, 1978.
3. Greene LA, Rein G: Release, storage and uptake of catecholamines by a clonal cell line of nerve growth factor (NGF) responsive pheochromocytoma cells. Brain Res. 129:247-263, 1977.
4. Schubert D, Klier FG: Storage and release of acetylcholine by a clonal cell line. Proc. Natl. Acad. Sci. U.S.A. 74:5184-5188, 1977.
5. Rebois RV, Reynolds EE, Toll L, Howard BD: Storage of dopamine and acetylcholine in granules of PC12, a clonal pheochromocytoma cell line. Biochemistry 19:1240-1248, 1980.
6. Melega WP, Howard BD: Choline and acetylcholine metabolism in PC12 secretory cells. Biochemistry 20:4477-4483, 1981.
7. Martin AR: Junctional transmission. Presynaptic mechanisms. In: JM Brookhart and VB Mountcastle (eds). (American Physiological Society, Bethesda, MD) Sect. 1, pp. 329-356, 1977.
8. Marchbanks RM: The subcellular origin of the acetylcholine released at synapses. Int. J. Biochem. 6:303-312, 1975.
9. Marchbanks RM: The vesicular hypothesis questioned. Trends Neurosci. 1:83-85, 1978.
10. Durant Y, Israel M: When the vesicular hypothesis is no longer the vesicular hypothesis. Trends Neurosci. 2:130-132, 1979.
11. Zimmermann H: On the vesicle hypothesis. Trends Neurosci. 2:282-285, 1979.
12. Collier B: Synaptic vesicles and acetylcholine quanta. Trends Neurosci. 2:285-287, 1978.
13. Ceccarelli B, Hurlbut WP: Vesicle hypothesis of the release of quanta of acetylcholine. Physiol. Rev. 60:396-441, 1980.
14. Kopin IJ: Biochemical aspects of release of norepinephrine and other amines from sympathetic nerve endings. Pharmacol. Rev. 18:513-523, 1966.
15. Geffen LB, Livett BG: Synaptic vesicles in sympathetic neurons. Physiol. Rev. 51:98-157, 1971.
16. Kirshner N, Viveros OH: The secretory cycle in the adrenal medulla. Pharmacol. Rev. 24:385-398, 1972.
17. Morris SJ: The structure and stoichiometry of electric ray synaptic vesicles. Neuroscience 5:1509-1516, 1980.
18. Marshall IG: Studies on the blocking action of 2-(4-phenyl piperidino) cyclohexanol (AH5183). Br. J. Pharmacol. 38:503-516, 1970.
19. Toll L, Howard BD: Evidence that an ATPase and a protonmotive force function in the transport of acetylcholine into storage granules. J. Biol. Chem. 255:1787-1789, 1980.
20. Anderson DC, King SC, Parsons SM: Pharmacological characterization of the acetylcholine transport system in purified Torpedo electric organ synaptic vesicles. Molec. Pharmacol. 24:48-54, 1983.

21. Freeman JJ, Choi RL, Jenden DJ: Plasma choline: its turnover and exchange with brain choline. J. Neurochem. 24:729-734,1975.
22. Douglas WW, Rubin RP: Stimulant action of barium on the adrenal medulla. Nature (London) 203:305-307, 1964.
23. Douglas WW, Rubin RP: The effects of alkaline earths and other divalent cations on adrenal medullary secretion. J. Physiol. (London) 175:231-241, 1964.
24. Kirshner N, Smith WJ: Metabolic rquirements for secretion from the adrenal medulla. Science 154:422-423, 1966.
25. Rubin RP: The role of energy metabolism in calcium-evoked secretion from the adrenal medulla. J. Physiol. 206:181-192, 1970.
26. Nelson-Krause DC, Howard BD: Energy utilization in the induced release of gamma aminobutyric acid from synaptosomes. Brain Res. 147:91-105, 1978.
27. Baker PF, Knight DE: The 'leaky' adrenal medullary cell. Trends Neurosci. 2:288-290, 1979.
28. Johansen T: Adenosine triphosphate levels during anaphylactic histamine release in rat cells in vitro. Effect of glycolytic and respiratory inhibitors. Eur. J. Pharmacol. 58:107-115, 1979.
29. Reynolds EE, Melega WP, Howard BD: Adenosine 5'-triphosphate independent secretion from PC12 pheochromocytoma cells. Biochemistry 21:4795-4799, 1982.
30. Krueger BK, Forn T, Greengard P: Depolarization-induced phosphorylation of specific proteins, mediated by calcium ion flux, in rat brain synaptosomes. J. Biol. Chem. 252:2764-2773, 1977.
31. DeLorenzo RJ, Freedman SD: Calcium-dependent phosphorylation of synaptic vesicle proteins and its possible role in mediating neurotransmitter release and vesicle function. Biochem. Biophys. Res. Commun. 77:1036-1043, 1977.
32. Sieghart W, Theoharides TC, Alper SL, Douglas WW, Greengard P: Calcium-dependent protein phosphorylation during seretion by exocytosis in the mast cell. Nature 275:329-331, 1978.
33. Amy CM, Kirshner M: Phosphorylation of adrenal medulla cell proteins in conjunction with stimulation of catecholamine secretion. J. Neurochem. 36:847-854, 1981.
34. Nishizuka Y: Protein kinases in signal transduction. Trends Biochem. Sci. 9:163-166, 1984.
35. DeCamilli P, Ueda T, Bloom FE, Battenberg E, Greengard P: Widespread distribution of protein I in the central and peripheral nervous systems. Proc. Natl. Acad. Sci. U.S.A. 76:5977-5981, 1979.
36. Bloom, FE, Ueda T, Battenberg E, Greengard P: Immunocytochemical localization, in synapses, of protein I, an endogenous substrate for protein kinases in mammalian brain. Proc. Natl. Acad. Sci. U.S.A. 76:5982-5986, 1979.
37 Ueda T. Greengard P: Adenosine 3':5'-monophosphate-regulated phosphoprotein system of neuronal membranes. J. Biol. Chem. 252: 5155-5163, 1977.
38. Knight DE, Baker PF:The phorbal ester TPA increases the affinity of exocytosis for calcium in 'leaky' adrenal medullary cells. FEBS Lett. 160:98-100, 1983.

39. Dunn LA, Holz RW: Catecholamine secretion from digitonin-treated adrenal medullary chromaffin cells. J. Biol. Chem. 258:4989-4993, 1983.
40. Wilson SP, Kirshner N: Calcium-evoked secretion from digitonin-permeabilized adrenal medullary chromaffin cells. J. Biol. Chem. 258:4994-5000, 1983.
41. Castagna M, Takai Y, Kaibuchi K, Sano K, Kikkowa U, Nishizuka Y: Direct activation of calcium-activated, phospholipid-dependent protein kinase by tumor-promoting phorbal esters. J. Biol. Chem. 257:7847-7851, 1982.
42. Denton T, Howard BD: Inhibition of dopamine uptake by N-methyl-4-phenyl-1,2,3,6-tetrahydropyridine, a cause of parkinsonism. Biochem. Biophys. Res. Commun. 119:1186-1190, 1984.
43. Kammerer RC, Amiri B, Cho AK: Inhibition of uptake of catecholamines by benzylamine derivatives. J. Med. Chem. 22:352-355, 1979.
44. Milner RJ, Sutcliffe JG: Gene expression in rat brain. Nucleic Acids Res. 11:5497-5520, 1983.
45. Sutcliffe JG, Milner RJ, Shinnick TM, Bloom FE: Identifying the protein products of brain-specific gene with antibodies to chemically synthesized peptides. Cell 33:671-683, 1983.

DISCUSSION OF THE ARTICLE BY B. HOWARD (U.S.)

D.O. Smith (U.S.): You mentioned in your dopamine release, it did not require ATP and as evidence for this, you depressed the ATP levels to as little as 10% of the controls, yet you still saw dopamine release. Most ways I can think of depressing ATP levels also markedly affect calcium metabolism. Can you comment as to whether or not you regulated or controlled for this effect upon calcium?

B. Howard (U.S.): We haven't done that directly. One slide shows that there is an increase in spontaneous release under depletion of ATP stores which we attribute to possible inhibition of calcium pumps. That is the reason for the increase.

S. Thesleff (Sweden): Where did you get the drug which blocks the uptake of ACh into the vesicles?

B. Howard (U.S.): The drug was made available to us by Dr. Brittain. A.H. stands for Allen and Hanburys, Ltd., which is a company in England. They have sent it to several people so I think there is no problem in attaining it.

SUB-MEPPS, SKEW-MEPPS AND THE SUBUNIT HYPOTHESIS OF QUANTAL TRANSMITTER RELEASE AT THE NEUROMUSCULAR JUNCTION

M.E. KRIEBEL AND C. ERXLEBEN

sub-MEPPs, skew-MEPPs and bell-MEPPs. In 1973 we reported that there is a class of small MEPPs (sub-MEPPs) at the normal adult frog neuromuscular junction and that the normally low percentage of these small MEPPs is temporarily increased after a heat challenge or nerve stimulation (Gross and Kriebel, 1973; Kriebel and Gross, 1974; Bevan, 1976) (Fig. 1) or with focal depolarization (Cooke and Quastel, 1973). Sub-MEPPs are absent with curare (Cooke and Quastel, 1973) and potentiated by eserine similarly to the larger MEPPs that were initially described by Fatt and Katz (1952). The larger MEPPs are termed bell-MEPPs because the amplitude distribution is bell-shaped (Fig. 2). Since both sub-MEPPs and the larger MEPPs are found in the normal preparation, the term normal-MEPP or classical-MEPP is inaccurate. The sub-MEPPs form a distinct class separate from the bell-MEPPs as indicated by the discontinuity in the profile of the MEPP amplitude histogram. In the frog neuromuscular junction, the small MEPP amplitudes usually form a bell distribution and are termed sub-MEPPs or s-MEPPs. However, in the mouse neuromuscular junction, the class of small MEPPs shows an overall skew distribution with a mode near the noise level (hence termed skew-MEPPs; term from Harris and Miledi, 1971). The skew class of MEPPs may show integral peaks providing that enough MEPPs are present in each histobar and if adequate resolving power of the histogram is used (i.e., bins between peaks; Fig. 2; Matteson et al., 1979). When the skew-class shows multiple peaks, the first peak is the sub-MEPP (s-MEPP) and its modal value is the same as that of the overall skew-class. The skew-class thus includes the

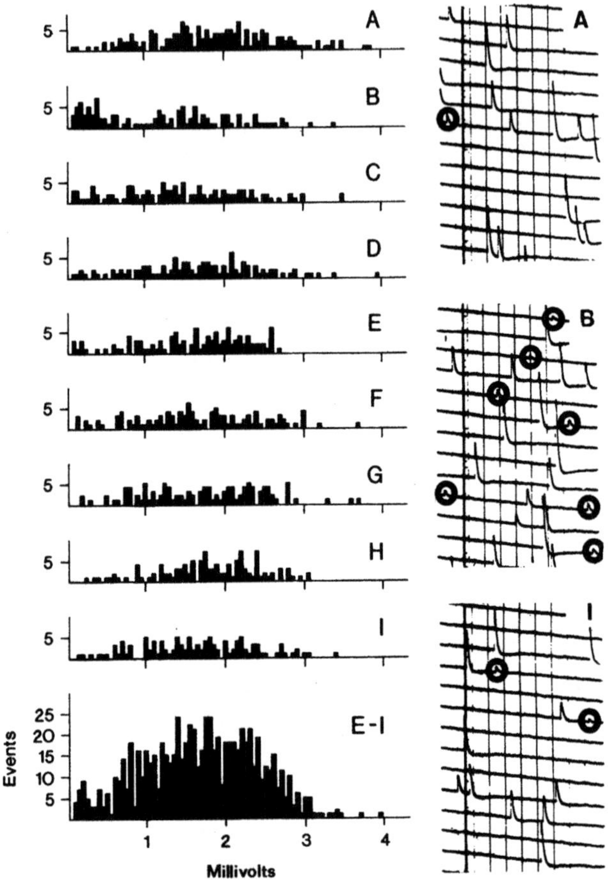

Fig. 1. MEPP amplitude histograms showing effect of a 1-min temperature challenge (38°) which generated about 4,000 MEPPs on subsequent MEPP amplitudes (challenge between plots A and B). (A) Control 18°, about 1 MEPP/s, resting potential fell from -90 to -85 mV. (B-I) Sequential plots after the challenge. About 18° and about 1 MEPP/s. Resting potential -83 mV in B, -85 for other plots. Each plot contains 113 MEPPs which were generated in sequential periods (each period about 2 min). The last plot, E-I, shows sub-MEPPs which correspond to those in B just after the challenge. Sub-MEPPs are circled in the samples of records used to plot histograms. (1 s per sweep, 1.4 mV between traces, 18°C). (Top) Control: sample of MEPPs used to plot histogram A. (Middle) About 1 min after the heat challenge. Sample of MEPPs used to plot histogram B. (Bottom) About 16 min after the challenge. Sample of MEPPs used to plot histogram I. (From Kriebel and Gross, 1974).

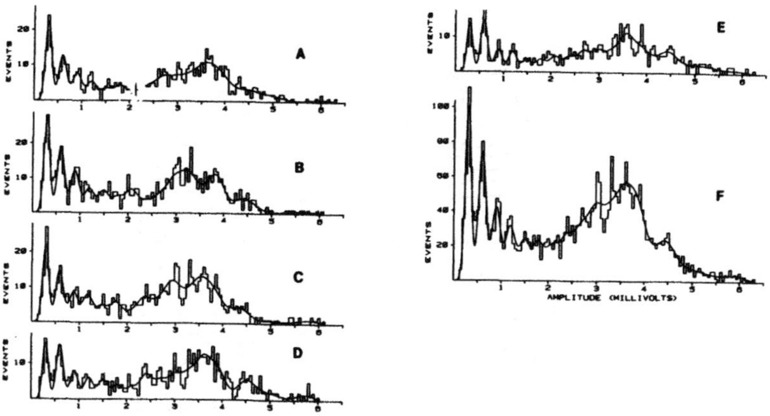

Fig. 2. Control histograms demonstrating stationarity of sub-MEPP mean and peak intervals. (A-E) Successive 12-min sections. (F) Summed histogram. (A) 612 MEPPs 51 MEPPs/min, (B) 734 MEPPs, 61 MEPPs/min, (C) 725 MEPPs, 60 MEPPs/min, (D) 634 MEPPs, 53 MEPPs/min, (E) 591 MEPPs, 49 MEPPs/min and (F) 3297 MEPPs, 55 MEPPs/min. Resting membrane potential went from -80 to -78 during 1 h of recording. Mean MEPP amplitude (A) 2.46 mV, (B) 2.28 mV, (C) 2.42 mV, (D) 2.65 mV, (E) 2.75 mV and (F) 2.50 mV. Note that the estimated sub-MEPP mean, and therefore, the estimated peak interval, is the same in (F) as in each subsection. 18-day-old mouse. The calculated noise and measurement error variance for this cell was 0.044 mV, variance of subunit 0.049 mV, and the subunit mean is 0.30 mV. (From Matteson, et al., 1979).

sub-MEPPS (s-MEPPs). Doublet MEPPs (two bell-MEPPs, Kriebel and Stolper, 1975), are not discussed here.

A word of caution is appropriate in regards to investigations concerning skew-MEPPs. In order to define the mode, the noise-to-signal (bell-MEPP) ratio must be under 2%. For example, a record of 2 mV bell-MEPPs should have a peak-to-peak noise level of 80 µV or better. It should be pointed out that small MEPPs in histograms published by others skew into the noise. Nevertheless, we propose that the skew-class of MEPPs that we have described is the same as reported by Dennis and Miledi (1974a,b) at the reinnervating junction, Harris and Miledi (1971) with botulinum toxin (BTX) poisoning, Bevan (1976) at the normal frog junction, and purported by Magleby and Miller (1981) at the mouse and frog junctions, although the latter authors did not show an

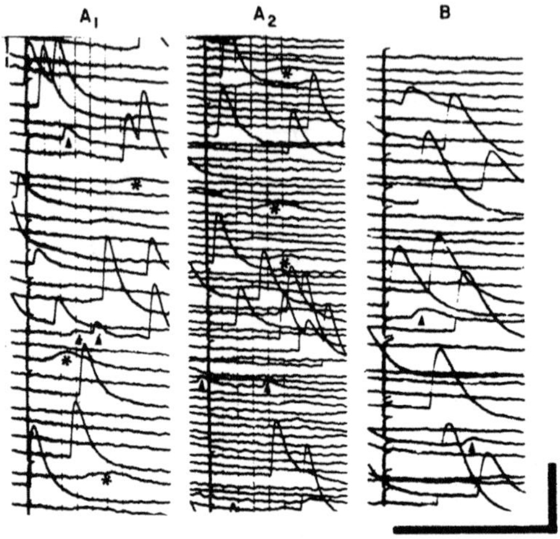

Fig. 3. Sample records of frog MEPPs: (A) Doubly innervated muscle cell. Camera film speed is slower in A than in A . MEPPs generated by a remote junction have longer time characteristics and are indicated with an asterisk. Sub-MEPPs with shorter time characteristics are indicated with arrowheads. Mean of sub-MEPPs is one-fifth that of the bell-MEPP (3.0 mV). Sub-MEPPs composed 10% of amplitude histogram. Frequency was increased from less than 1/s to 5/s with 4 X calcium. Calibration: 200 ms and 4 mV; -92 mV resting potential. 12μM cell. (B) Rise times of sub-MEPPs (indicated with arrowheads) are similar to those of the bell-MEPPs. Mean of sub-MEPPs is 1/7 that of the bell-MEPPs (3.8 mV). Sub-MEPPs composed about 2% of the amplitude histogram. Calibration of AC coupled trace is 100 ms and 4 mV (-90 mV), 10-μM cell. (From Kriebel and Gross, 1974).

example of sub-MEPPs and stated that they could not distinguish sub-MEPPs from noise (page 272, para. 2). We also propose that some smaller skew-MEPPs reported by Sellin and Thesleff (1981) and Thesleff, et al. (1983) in the botulinum toxin injected rat muscle are the same skew-MEPPs that we have described in normal mouse fibers and that persist with BTX poisoning (Kriebel, et al., 1976, see Thesleff, this volume). Larger skew-MEPPs may have slower time constants which we attribute to summated sub-MEPPs (Kriebel, et al., 1976) whereas Colméus, et al.. (1982) propose that the slow, small quanta were released at a greater distance from receptors. Dennis and Miledi (1974b) also proposed that the skew-MEPPs observed during regeneration are the same as those remaining after BTX poisoning. We have found skew-MEPPs in normal mouse, snake, frog and human junctions as well as in the skate electrocyte.

In all preparations, the sub-MEPPs are usually 1/7th to 1/10th the size of the bell-MEPP class, and they have the same voltage time characteristics (Fig. 3) (Cooke and Quastel, 1973; Kriebel and Gross, 1974; Kriebel, et al., 1976; Muniak, et al., 1982; Bevan, 1976). Recently, we have demonstrated that the time characteristics of the skew-MEP currents at the mouse diaphragm junction are also similar to that of the bell-MEP currents (Erxleben and Kriebel, 1983; Erxleben, Carlson and Kriebel, 1983; Fig. 4).

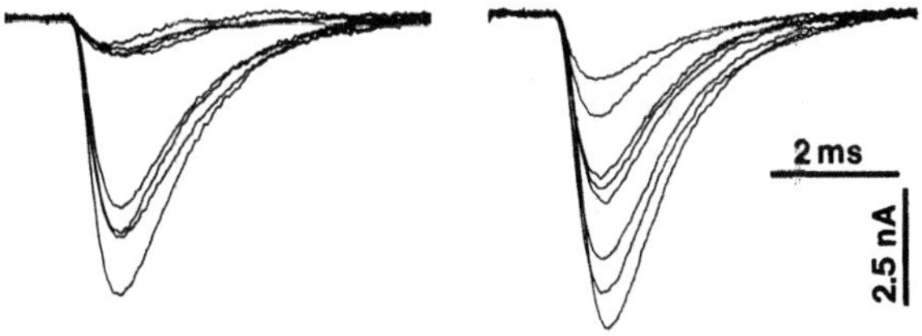

Fig. 4. Superimposed MEPCs. The holding potential was -140 mV; temperature was 30°C. The left panel shows 4 sub-MEPCs. Note that the time characteristics are the same as the larger MEPCs.

<u>Schwann-cell MEPPs</u>. Initially, we thought that the sub-MEPPs could be the Schwann cell potentials found by Birks et al. (1960) and Thesleff, (1960) after denervation. It could be argued that treatments which increase quantal release from the nerve such as heat, calcium and nerve stimulation either directly or indirectly influence the Schwann cell to increase the frequency of release which is normally so low as to have gone undetected in the normal neuromuscular junction (cf. Harris and Miledi, 1971). We found that the small MEPPs remaining after botulinum toxin poisoning are exactly the same as the normal skew class (next section) and we believe that they are the same as described by Harris and Miledi (1971) after BTX poisoning. They also concluded that their small MEPPs were probably not from Schwann cells but from the axon since the frequency was increased by elevated $[K]_o$ or tetanic nerve stimulation whereas hyposmotic

saline or La^{+++} did not increase the frequency as found in denervated preparations. (See also Kriebel, Hanna and Pappas, 1980). We found that bell-MEPPs, just as reported by Birks et al. (1960), ceased along with the sub-MEPPs after denervation, and a quiet period existed for several days as the Schwann cell-muscle junction formed (Kriebel and Gross, 1974). Moreover, we have recorded "quiet periods" up to 30 min from newly formed, well-defined Schwann cell-muscle junctions (Kriebel, et al., 1978).

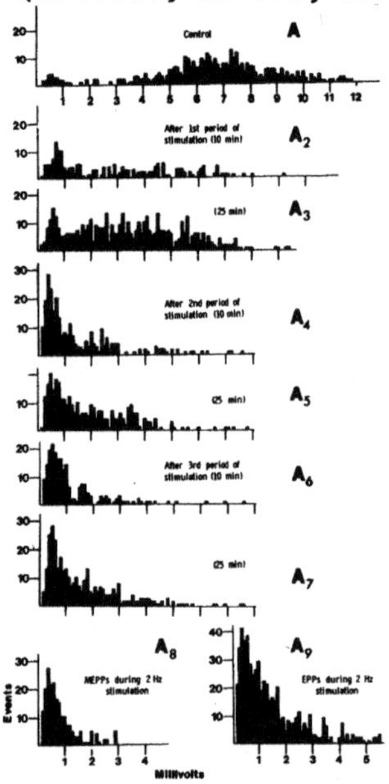

Fig. 5. Effect of short periods of tetanic nerve stimulation on subsequent MEPP and EPP amplitude histograms. (A) Control MEPP histogram -68 mV resting potential. MEPP frequency 15/min. Muscle cell diameter 6μ. (2) MEPP histogram of first 10 min after first period of nerve stimulation (40 Hz, 30 sec). (3) MEPP histogram of subsequent 10-35 min recovery period. (4) MEPP histogram during the first 10 min after the second tetanus. (5) MEPP histogram of subsequent 10-35 min recovery period. (6) MEPP histogram during the first 10 min after the third tetanus. (7) MEPP histogram of subsequent 10-35 min recovery period. (8) MEPP histogram during 2 Hz stimulation following the fourth period of 10 Hz stimulation. (9) EPPs recorded during 2 Hz stimulation (Over 90% failures) (-65 mV at the end of the experiment). (From Kriebel, 1978).

<u>Quanta of the skew class are usually not evoked</u> at the frog junction (Bevan, 1976; Kriebel, 1978) except after intense

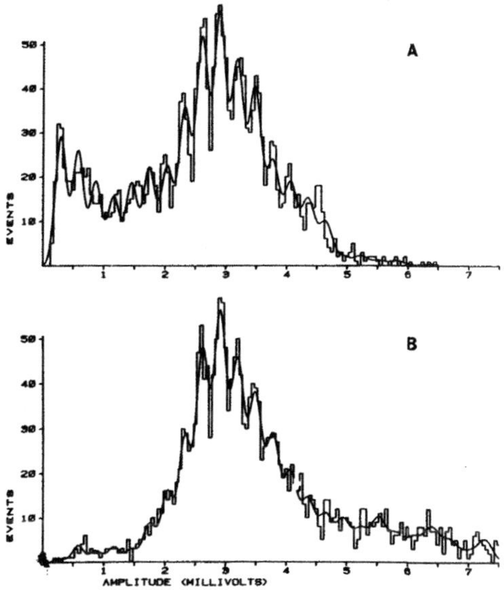

Fig. 6. MEPP and EPP amplitude distributions from a 14 day old mouse diaphragm junction at 25°C. No neostigmine. The phrenic nerve was stimulated at 2 Hz and all MEPPs and EPPs were measured during the 70 min of continuous recording. The continuous lines show the predicted distributions based on the subunit hypothesis. The noise and measurement standard deviation was 0.09 mV. Membrane potential was -64 mV, Co^{++} 4 mM. A, MEPPs. $2.4x10^3$ MEPPs (34/min) 20% are skew-MEPPs and 5% are sub-MEPPs. The estimated value of subunit size is 0.29 mV (±0.016). B, EPPs; Since failures were 75%, most EPPs are unitary. The few larger EPPs (6 mV) represent two unitary potentials. The EPP histogram shows multiple peaks which are in register with those in the MEPP amplitude distribution. $2.5x10^3$ EPPs subunit=0.29 mV (±0.018). (From Kriebel, et al., 1982).

nerve stimulation (Fig. 5); and, in some junctions in a high Mg^{++}, low Ca^{++} saline (Kriebel, 1978). We have found a similar wide range of observations at the mouse diaphragm. In those junctions that demonstrate a marked discontinuity between the bell-MEPP and overall skew-MEPP class, only quanta of the bell-MEPP size are evoked (Fig. 6) (Kriebel et al., 1982). In those junctions that generate more skew-MEPPs than bell-MEPPs, which produce a histogram with an overall MEPP amplitude profile that is both relatively small and skewed, unitary evoked quanta the size of skew-MEPPs have been observed (Figs. 7 and 8) (Kriebel et al., 1982).

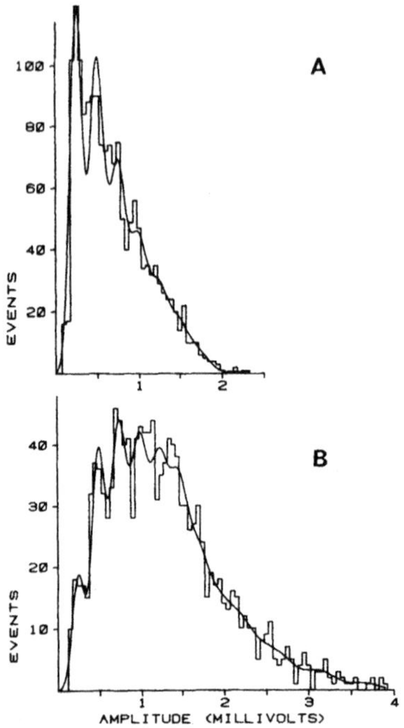

Fig. 7.

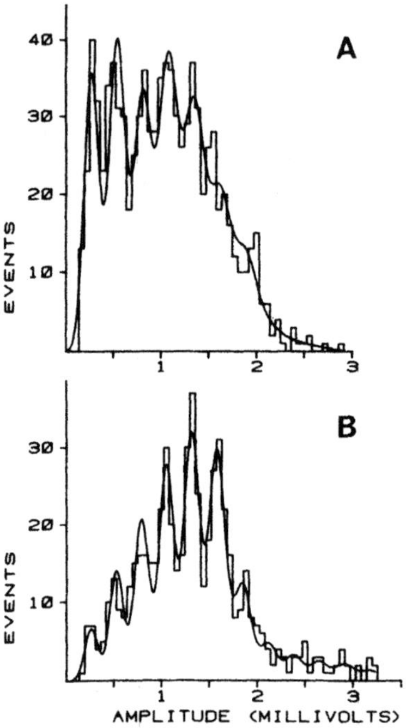

Fig. 8.

Fig. 7. MEPP and EPP amplitude distributions from a 15-day-old mouse diaphragm 35°C. Mainly skew-class. Neostigmine. The phrenic nerve was stimulated at 2 Hz. The continuous lines are predicted from the subunit hypothesis. The standard deviation of the noise and measurement is 0.067 mV. -62 mV Co^{++} 4 mM. A) MEPPs. The peak intervals are 7.0 histobars. $1.8x10^3$ MEPPs (72 MEPPs/min). Subunit is 0.29 mV (+0.053). B) EPPs. Failure was 80%. Peaks match intervals in the MEPP distribution. Subunit is 0.29 mV (±0.04). $1.2x10^3$ EPPs.

Fig. 8. MEPP and EPP amplitude distributions. 13-day-old mouse. 35°C, neostigmine. The standard deviation of the noise and measurement error is 0.067 mV. The resting potential -67 to -65 mV for the 35 min of recording time, Co^{++} 4 mM. A) MEPPs. Ratio of skew- to bell-MEPPs about one. The peak intervals are 4.75 histobars. $9.0x10^3$ MEPPs (24 MEPPs/min). Subunit is 0.27 mV (±0.037 mV). B) EPPs. 80% failure. Subunit is 0.26 mV (±0.021). (From Kriebel, et al., 1982).

Figures 7 and 8 show that profiles of MEPP amplitude distributions are not always similar to the bell-MEPP profile. These differences depend on the percentage of skew-MEPPs to bell-MEPPs and the amplitude ratio. We found that the amplitude profile of the unitary evoked potential was always more-or-less bell-shaped and represented the distribution of the bell-MEPP class of quanta. However, Kelly and Robbins (1984) report yet another class of MEPPs 1/3 the size of the "classical" MEPP and these were evoked (mouse, CBF-1, Charles River). Moreover, Cull-Candy, Lundh and Thesleff (1976) found that small quanta remaining after BTX poisoning were evoked (injected rat) but after 4 hr of stimulation only the larger unitary EPPs were evoked.

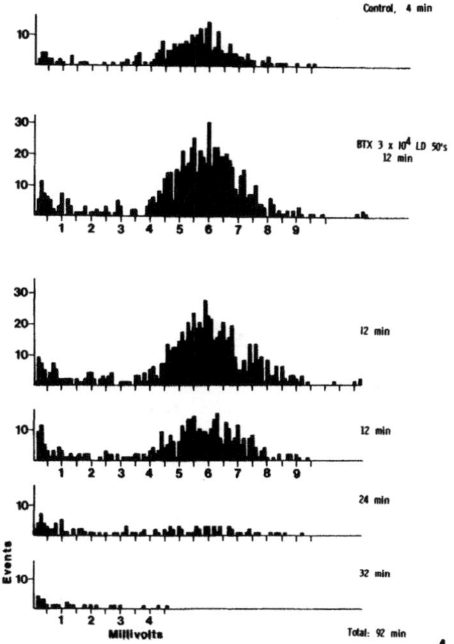

Fig. 9. Effect of botulinum toxin type A, 3×10^4 intraperitoneal median lethal doses (I.P. LD_{50})/ml., on MEPP amplitude distributions. These histograms show all recorded MEPPs during 96 min (4 min of control). 33°C; 23-day-old mouse; neostigmine. Resting potential at start, -64 mV; at end -60 mV. Control histogram 266 MEPPs/4 min. Note that the means of bell-MEPPs and sub-MEPPs remain the same although the frequency was dropping through the sequence of histograms. (Kriebel, et al., 1976).

Pharmacological differences between skew- and bell-MEPPs. In the mouse, botulinum toxin (*in vitro*) first blocks the spontaneous release of the bell-MEPP leaving the skew class (Harris and Miledi, 1971, frog) which is finally blocked after several hours (Fig. 9) (Kriebel, et al., 1976, on mouse). We have recorded all MEPPs from a single junction during progressive BTX poisoning and found neither change in

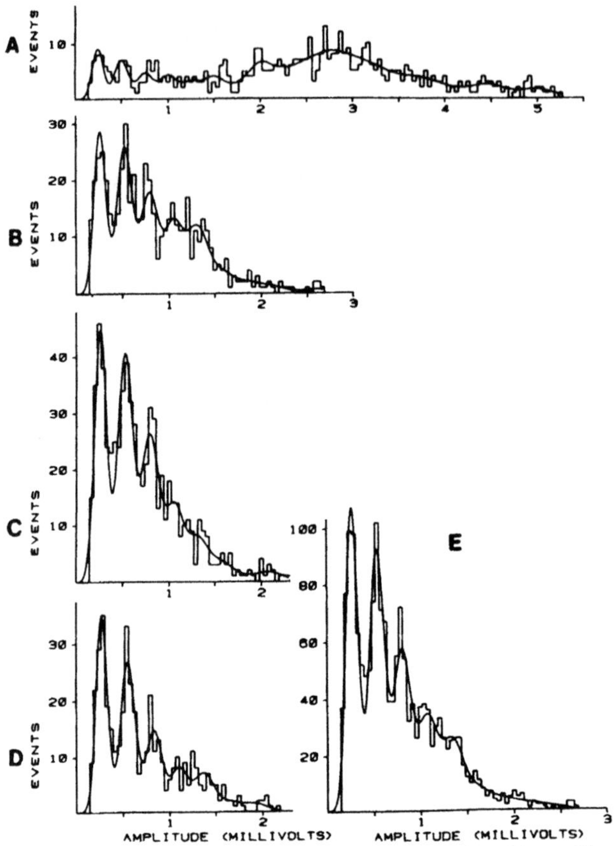

Fig. 10. Effect of β-BuTX on MEPP amplitude distribution. A: control histogram, 20 min of recording. After addition of 20 μg/ml β-BuTX there was 5 min of depression followed by a 4-fold increase in frequency for 40 min. B,C and D: successive 12 min periods 1 h after β-BuTX. E: B,C and D combined. 26°C, -70 mV to -66 mV. 21-day-old mouse. The curves are predicted from a subunit model based on the following assumptions: (1) sub-MEPPs result from the release of a subunit of transmitter. The subunit amplitude is assumed to be normally distributed (0.26 mV ± 0.046) and does not include measurement and noise error; (2) larger amplitude MEPPs result from the synchronous release of two or more subunits which sum linearly. (From Llados, et al., 1980).

the bell-MEPP mean nor skew-MEPP mode. The BTX blocking effect occurred during the generation of only a few hundred MEPPs. During initial stages of block at a time when only skew-MEPPs were generated, nerve stimulation initiated the reappearance of the bell-MEPP class although the quanta were not time-locked to the stimulus (also in frog, Harris and Miledi, 1971). Harris and Miledi (1971) concluded, as we have, that the small MEPPs occurring at low frequencies may have escaped notice in previous studies.

Duchen and Tonge (1973) found skew-MEPPs after tetanus toxin poisoning and the bell-MEPP frequency was increased with nerve stimulation. They also found the skew-MEPPs before nerve sprouting. They proposed that the effect was different from BTX because an elevated K^+ saline had little effect on frequency but it should be pointed out that Molgó and Thesleff (1982) found that an elevated K^+ saline did not increase the skew-MEPP frequency as it did the bell-MEPP frequency.

β-Bungarotoxin also blocks the generation of bell-MEPPs but with little effect on the skew-class (Fig. 10) (Llados, et al., 1980). This blockage of bell-MEPPs occurs after 10-30 minutes of high frequency (20-100 times normal) MEPPs.

Lanthanum ions greatly increase the bell-MEPP frequencies (Heuser and Miledi, 1971; 300 to 700 times). Initially there is little change in sub-MEPP frequencies and nerve release is blocked within a few minutes; and after an hour or two with the generation of 2×10^5 MEPPs, mainly sub-MEPPs remain (Fig. 11; Kriebel and Florey, 1982; Bevan, 1976). The La^{+++} effect is similar to that of black widow spider venom (Kriebel and Stolper, 1975). Augustine and Levitan (1983) report that erythrosin B increases both skew- and 'giant' MEPPs and conclude that the effect is similar to La^{+++}, spider venom and ionophores.

Skew- and bell-MEPPs are released by a different presynaptic mechanism because skew- and bell-MEPP frequencies have very different temperature characteristics. Below 24°C, skew-MEPPs had a higher Q_{10} than bell-MEPPs whereas 24°C to 34°C greatly increased bell-MEPP frequencies with little effect on skew-MEPPs (Carlson, Kriebel and Muniak, 1982). This is exactly the effect of temperature on skew-MEPPs (after BTX) in the temperature range of 10-20°C reported by Thesleff, Molgó and Lundh (1983).

Finally, the Ca^{++} ionophore (X-537A) shows a spectacular effect in that the frequency of the skew class

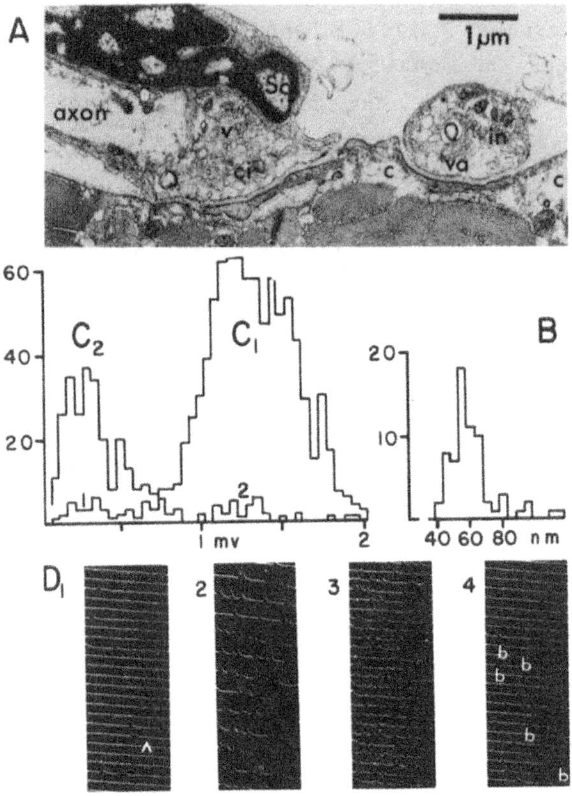

Fig. 11. 1 3/4 hour La^{3+} treatment (24°C) on frog junction. (A) Nerve terminal shows few vesicles (v); vacuoles (va); infoldings (in); vacuoles in muscle fiber (c); cisternae (ci). (B) Vesicle diameters (60.5±14.8 nM). Control (61±8 nM). (C1) Control MEPP histogram, sub-MEPPs (0.25 mV) are 1/5 bell-MEPPs (1.25 mV). (C2) MEPP histogram at end of La^{+++} treatment. Note that most MEPPs are of the sub-MEPP class although enough bell-MEPPs remain to establish their mean of 1.25 mV (same as C1). (D) MEPPs. Traces are 0.5 s long, gain is same in all strips. (D1) Control: sub-MEPP indicated with arrowhead. (D2) MEPPs 5 min after start of La^{+++}. 25 MEPPs/s. (D3) 1 h of La^{+++} treatment. Many are sub-MEPPs. 25 MEPPs/s. $1x10^4$ MEPPs have been generated by 1 h. (D4) After 1 3/4 h of La^{+++} treatment. Most MEPPs are sub-MEPPs. The few remaining larger MEPPs (b) are bell-MEPPs. The sub-MEPPs do not skew into the noise. About $1.4x10^4$ MEPPs were generated during treatment. (From Kriebel and Florey, 1982).

increases until histogram profiles are composed of mainly sub-MEPPs (Fig. 12; Kriebel, Llados and Carlson, 1980). Kita and Van der Kloot (1976) attributed the reduction in

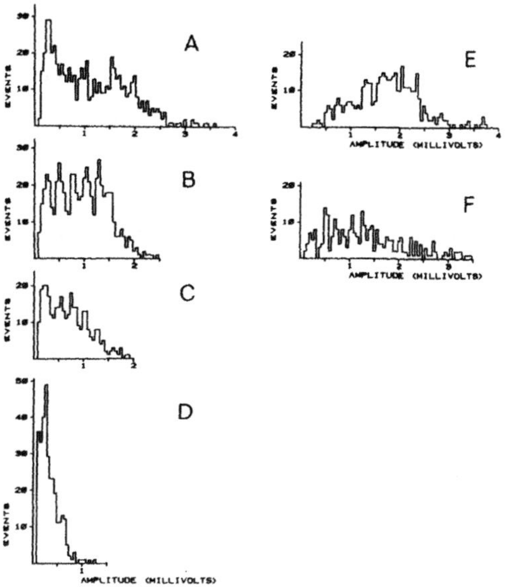

Fig. 12. Effect of x-537A ionophore on MEPP and EPP distributions. Eighteen day old mouse. 26°C. -62 mV. 70% failures, Co^{++} 4 mM. Prostigmine (10^{-6} g/ml). A: Control MEPPs, 30 min, 0.33/s. B-D: MEPPs recorded after the addition of the ionophore. All MEPP amplitudes are shown in the 3 contiguous sections. B: 10 min, frequency 1.0/s. C: 4 min, frequency 1.4/s. D: 4 min, frequency 1.3/s. E: Unitary evoked potentials recorded before ionophore was added. Same time as MEPPs shown in A. Note that few potentials the size of skew-MEPPs were evoked. F: Unitary evoked potentials recorded after ionophore was added. Same time as MEPPs shown in B. (From Kriebel, et al., 1980).

MEPP amplitudes (<50%) to a depression of ACh sensitivity and found that X-537A depolarized frog fibers by 15mV (we did not observed this at the diaphragm). We found the changes in the MEPP amplitude profile are quite variable, and this variation in the ionophore effect may explain the extremely wide range of changes in the MEPP amplitude profiles that are found with 2 to 10 times normal calcium. Increased calcium (10 mM) decreases the channel conductance by 18% (Magleby and Weinstock, 1980) and McLarnan and Quastel (1983) found a 25% decrease with 20 mM Ca^{++} but in many experiments we have observed a MEPP amplitude decrease of 2 to 10 times such that mainly skew-MEPPs or sub-MEPPs remained. These changes in MEPP amplitude profiles can be explained with a presynaptic change in the skew- to bell-MEPP ratio (Showalter, unpublished; Kriebel and Gross, 1974).

Synaptogenesis and denervation. There are dramatic changes in the ratio of sub-MEPPs:bell-MEPPs during morphogenesis. Matthews-Bellinger and Salpeter (1983) found that the junctional ACh receptors are correlated with the specialized membrane and the receptor density of neonates is comparable to the adult. Thus, changes in receptor density probably do not account for changes in MEPP amplitude histograms. In the one day old neonate mouse diaphragm, there are mainly skew-MEPPs and sometimes the skew-class is composed of only sub-MEPPs (Muniak, et al., 1982; see Tonge, 1974). Yet the evoked unitary EPPs are larger and appear to be of the bell-MEPP class (Dennis and Miledi, 1974a,b). Similarly, the first MEPPs to appear at the regenerating junction are sub-MEPPs (Muniak, et al., 1982) and in tissue culture (Kidokoro, 1984). During synaptogenesis, the percentage of bell-MEPPs increases until the adult MEPP amplitude profile is reached. The percentage of sub-MEPPs increases with distance from the growing phrenic nerve until most MEPPs are of the sub-MEPP class (Fig. 13; Muniak, et al., 1982). Dennis and Miledi (1974b) proposed that the skew class at the regenerating frog junction is the same class remaining after BTX and with extracellular focal recording concluded that both the bell- and skew-MEPPs are from the same point on the nontransmitting terminal. Cooke and Quastel (1973) and Carlson (unpub.) also found both classes of MEPPs at one focal recording site.

Finally, at the frog junction, the two classes of MEPPs are affected differently during degeneration in that sub-MEPPs increase in frequency and after a high frequency burst of MEPPs, sub-MEPPs remain for an hour or so (Kriebel and Gross, 1974). This was usually not observed at the mouse diaphragm during phrenic nerve degeneration (Muniak, et al., 1982). We have examined identified junctions that had just

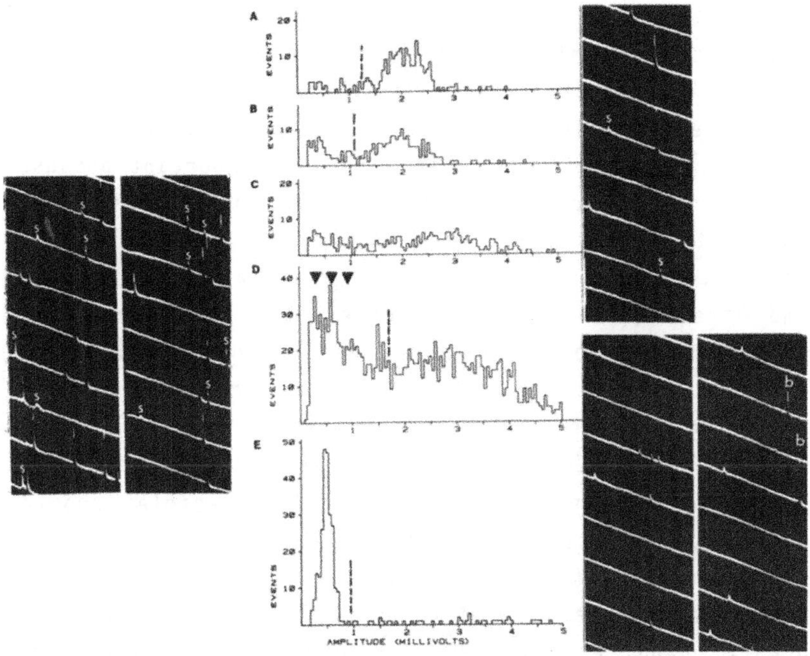

Fig. 13. MEPP amplitudes from re-innervated junctions 5 days after the phrenic nerve was crushed. This series of histograms demonstrates the increase in bell-MEPPs during re-innervation. Fiber A was closest to the point of nerve entry and fiber E was farthest away (4 mM Co^{++}, 22°C). Dashed lines divide skew- and bell-MEPPs. A: Ratio of bell-MEPPs to skew-MEPPs is in the control range (records, right top). Nerve stimulation evoked EPPs of the bell-MEPP size. -62 mV, 60 MEPPs/min. B: Junction shows more skew-MEPPs than control. Nerve stimulation evoked EPPs. -60 mV, 7 min, 45 MEPPs/min. C: uniform MEPP amplitude distribution of skew- and bell-MEPPs. Nerve stimulation evoked EPPs. -80 mV, 30 MEPPs/min, 11 min. D: overall MEPP distribution is skewed since there is an equal number of skew- and bell-MEPPs. The exact demarcation between the 2 classes is not clear. Skew-MEPPs show 3 integral peaks (arrowheads). -78 mV, 30 min, 60 MEPPs/min. records left; s, skew-MEPPs. E: The skew-MEPP amplitudes form a bell-shaped distribution (sub-MEPPs) and the bell-MEPP amplitudes form a flat distribution. Nerve stimulation did not evoke EPPs. -60 mV, 14 min, 20 MEPPs/min. records right, bottom; b, bell-MEPPs. (Muniak, et al., 1982).

ceased to generate sub-MEPPs and found that the nerve endings exhibited "clumped" vesicles and disrupted mitochondria (Kriebel, et al., 1980).

<u>The Sub-unit Hypothesis</u> states that the bell-MEPP is

composed of 7-10 subunits and that the sub-MEPP represents the release of a single subunit. Harris and Miledi (1971) and Dennis and Miledi (1974b) proposed this hypothesis to account for the two classes of MEPPs, skew and bell. Dennis and Miledi (1974b) stated that the finding of skew-MEPPs which are smaller than unitary EPPs is important because it was the strict correspondence in mean amplitudes of MEPPs and unitary EPPs that led del Castillo and Katz (1956) to formulate the quantum hypothesis (see Katz, 1962). Dennis and Miledi (1974b) also suggested that the skew-MEPPs may arise from the top of the axon, partially filled vesicles or membrane gating mechanisms with two durations. The discontinuity between skew- and bell-MEPPs and the consistant size of sub-MEP currents argues against the latter suggestions. Moreover, partially filled vesicles should be evoked. The observation that sub-MEPP (and sub-MEPC) time characteristics are the same as the bell-MEPPs implies that both classes are generated by the same postsynaptic membrane. Land, et al. (1980) demonstrated a linear relation between receptor density and the MEPC rise time.

We originally favored the subunit hypothesis in order to explain the integral peaks that are multiples of the sub-MEPP on MEPP amplitude histograms (Gross and Kriebel, 1973; Kriebel and Gross, 1974; Kriebel, et al., 1976; Kriebel, 1978) (Figs. 2,6,7,8,10). We only observed these peaks from relatively small diameter muscle fibers (small frogs) which generated relatively large MEPPs (2-4 mV) (see Katz and Thesleff, 1957) such that sub-MEPPs yielded a distinct mode separated from background noise. Moreover, we reported that the peaks occurred in the same place in sequential plots (Fig. 2) (Matteson et al., 1979; Kriebel et al., 1976, 1982). Not all of our MEPP amplitude distributions show integral peaks because the sample size is small, the noise level is too high or there was a change in the muscle fiber resting potential. Many of the histograms were published to demonstrate changes in the ratio of skew- to bell-MEPPs with different treatments and were <u>not</u> published to demonstrate integral peaks.

The distributions of skew-MEPPs do not always show integral peaks as shown in Figs. 2 and 10. However, when present, the peaks are sharp and show the same interval as the bell-MEPPs. Therefore, both the skew- and bell-MEPPs appear to be composed of the same subunits. We found variability in the time-to-peak of the large skew-MEPPs (and skew-MEPCs) even though the time characteristics of the sub-MEPPs were the same as the bell-MEPPs (Kriebel, et al., 1976; Muniak, et al, 1982). Larger skew-MEPPs often showed breaks on their rising phases indicating two or more events and these increased after a heat challenge (Carlson, et al., 1982). These slow skew-MEPPs were not doublet-MEPPs which are 2 bell-MEPPs, (see Kriebel and Stolper, 1975), since the amplitude distributions of the "breaks" correspond to the skew-class. Slow skew-MEPPs with "breaks" on their rising phase are explained by the subunit hypothesis.

The subunit composition of unitary evoked potentials was demonstrated when the evoked response (diaphragm) was reduced with cobalt ions until most nerve stimuli (70%) resulted in failure of release indicating that most EPPs were unitary potentials. The amplitude distributions of unitary EPPs showed integral peaks which were the same as those in the MEPP distributions (Figs. 6, 7 and 8) (Kriebel et al., 1982). The MEPP amplitude histogram profile and the unitary EPP profile as well as mean size were sometimes quite different (Figs. 7 and 8). This discrepancy between MEPP and unitary EPP amplitude profiles is common at the regenerating synapse (Dennis and Miledi, 1974b; Muniak, et al., 1982; Colméus, et al., 1982); and, bell-MEPPs are increased with nerve stimulation during BTX poisoning (Harris and Miledi, 1971).

We have digitized and measured the MEPC amplitudes with a computer and have found integral peaks on the histograms that are the same under similar conditions regardless of the muscle fiber diameter (Erxleben and Kriebel, 1983). The number of peaks, peak intervals and peak positions persisted

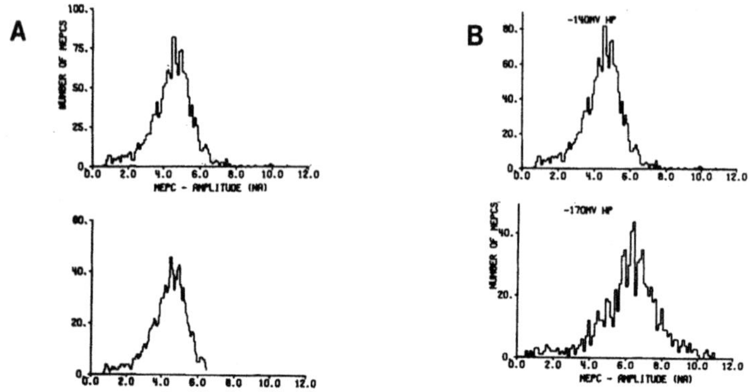

Fig. 14. A: Stationarity of peaks with different bin sizes. Holding potential -140 mV, 30°C. No eserine. Top: 0.058 nA/bin. Bottom: 0.09 nA/bin. Note three central peaks are in same position for both bin sizes. 0.41 nA between peaks. The modal peak is 11 times the peak interval. B: Effect of different holding potentials on peak intervals. Top: -140 mV holding potential. The 3 central peaks are integrals of 0.41 nA and the modal peak is 11 times the peak interval (same junction as A). Bottom: -170 mV holding potential. The central peaks are integrals of 0.575 pA. The modal peak is 11 times the peak interval (as at -140 mV). Curves were smoothed with a 3 point moving averge routine.

with a change in bin size (Fig. 14A).

Experimental evidence for subunits has also been obtained with conditions that altered the post-synaptic response. Wernig and Motelica-Heino (1978) reported that eserine increased the peak intervals at the frog neuromuscular junction and we have obtained the same results at the mouse diaphragm (Erxleben, et al., 1983). We use a two electrode Dagan voltage clamp, and we have found that a 25 mM calcium saline greatly improves the membrane seal around the electrodes. McLarnon and Quastel (1983) found that 20 mM Ca^{++} decreased the MEPC amplitude 25% and had no effect on the time course. Our standard clamp condition (-140 mV, 30°C, 25 mM Ca^{++}) yielded a peak interval current of 0.44 nAmp. This value was significantly increased with eserine, or a -170 mV holding potential (Fig. 14B), or with a higher bath temperature. The peak intervals were reduced with a cooler bath temperature.

<u>Requirements to observe sub-MEPPs and integral peaks.</u> Wernig and Motelica-Heino (1978), Wernig and Stirner (1977) and Vautrin and Mambrini (1981) have published histograms

showing integral peaks and these are comparable with our work on the frog such that the mode was at the 7th or 8th peak. However, Bevan (1976) and Magleby and Miller (1981) found neither integral peaks nor stationary peaks. We propose that the following criteria must be rigidly adhered to in order to obtain integral peaks. In regards to MEPP amplitude distributions the muscle fiber resting potential must not change over 5%. The second requirement is that the noise (RMS) must be no greater than the bin size. This was achieved with peak-to-peak noise values of 40-80 μV and sub-MEPP modes of 250-500 μV. Thirdly, there must be enough MEPPs so that the sample sizes in the central bins overcome random variations resulting from small sample size. We have found with computer modeling that 40-60 MEPPs in the central bins are adequate to reveal integral peaks (unpub.). A concurrent condition is that there should be over 4 bins between peaks. We have found it to be difficult to collect more than $2\text{-}5 \times 10^3$ events with stable conditions so the 500 classes that we use are too many for our data sizes. When grouped by 5's, peaks are visable but ragged (Fig. 15A) although a curve smoothing routine (3 point moving average) clarifies the central integral peaks (unpub). Grouping by 7's shows smooth peaks in the raw data and by 9's only two central peaks remain. Note that the positions of the peaks did not change with different bin groupings. Finally, a grouping by 13s yielded a Gaussian curve. It is necessary to have $2\text{-}4 \times 10^3$ MEPCs to yield 40-60 MEPCs in the central histobars and yet have at least 40 histobars to the central bell-MEPC mode. We have modeled integral peak distributions as shown in Fig. 6 and have found it necessary to have $6\text{-}8 \times 10^3$ MEPPs in order to reproduce Figure 6 every time. However, we did obtain histograms showing all integral peaks about 1/3 of the time with $2\text{-}4 \times 10^3$ MEPPs and this sample size was always enough to yield a histogram showing the position of the central 3 peaks. A sample size of $1\text{-}2 \times 10^3$ MEPPs yields histograms with central peaks 1/3 of the time and only suggests central peaks 1/3 of the time (Kriebel and

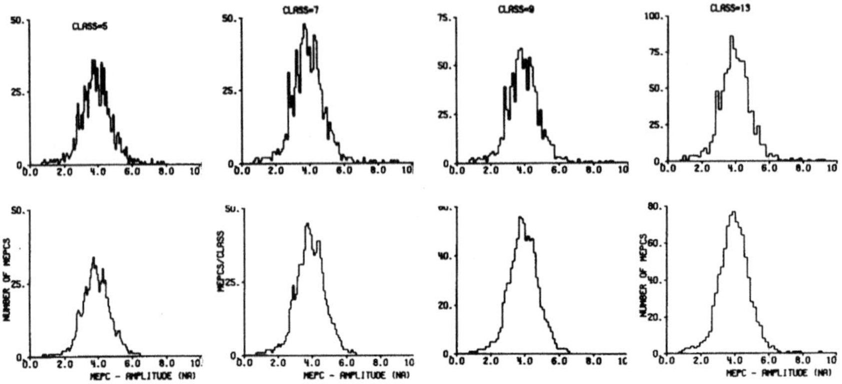

Fig. 15. Effect of bin size and a curve smoothing routine on integral peaks. Holding potential -140 mV. 30°C. No eserine. The MEPC amplitudes were measured to 500 classes. With classes grouped by 5's the sample size is relatively small for the central histobars and the profile is ragged. A curve smoothing routine (3 point moving average) shows peaks separated by 7.8 histobars. Modal peak is 8.0 times peak interval. 56 histobars to central peak. The classes grouped by 7's yield peaks separated by 4.75 histobars. 38 histobars to central peak. The curve smoothing routine indicates the same interval spacing although one of the peaks is represented by a shoulder. Classes grouped by 9's. 32 histobars to central peak, peak intervals are 4.0 histobars. The curve smoothing routine leaves only the 2 central peaks. Classes grouped by 13's gives smooth envelope. 22 histobars to central peak. Integral peaks not clear but would be separated by only 2.87 histobars. Curve smoothing yields a smooth bell-curve.

Llados, unpub.). We have also modeled a Gaussian distribution and have found that profiles of amplitude distributions of under 10^3 events sometimes (10%) showed integral peaks. However, subsequent distributions never yielded the same peaks. Moreover, $2\text{-}4 \times 10^3$ events were almost always (90%) enough to show a more-or-less smooth distribution.

Model of subunit hypothesis. We proposed that both the skew-MEPP and bell-MEPP classes are composed of the same subunits (sub-MEPPs) because the peak intervals were the same (Matteson, et al., 1981). Magleby and Miller (1981) did not find integral peaks but did find multiple peaks with rather small sample sizes that were not stationary which they ascribed to variations in small sample sizes. The

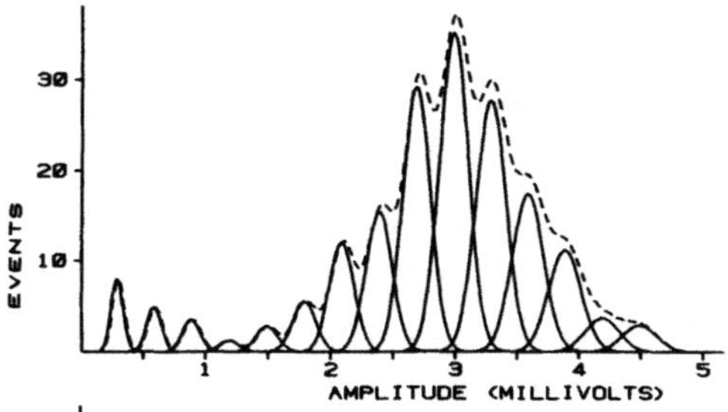

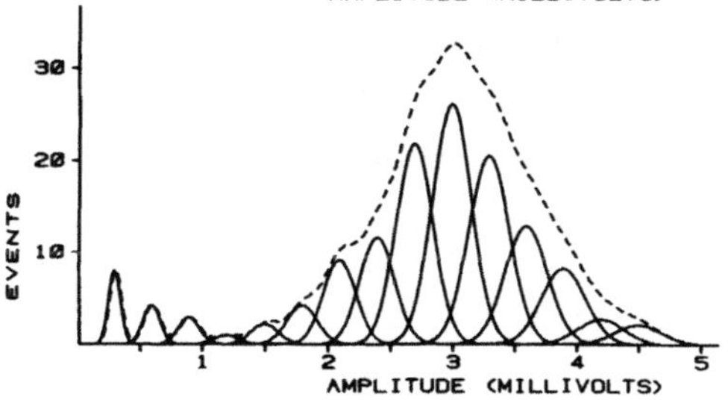

Fig. 16. Top: These curves represent theoretical MEPP amplitude distributions based on the multimodal model. The ordinate is scaled in number of events to be compatible with the scale used for amplitude histograms. The values used to plot these curves were computed by multiplying the appropriate probability density function by the total number of observed events. Each solid curve represents a subpopulation of MEPP amplitudes resulting from the synchronous release of a constant number of subunits. The dashed curve is a mixture of all subpopulations; i.e., the overall theoretical MEPP amplitude distribution predicted by the multimodal model. Mean and standard deviation values were: 0.3 mV; 0.035 mV. Bottom: These curves are also generated from the multimodal model using the same total sub-MEPP variance as above. However, measurement error variance was assumed to be zero. Mean and S.D. values used were 0.3 mV; 0.049 mV. (From Matteson, et al., 1981; data from Kriebel, et al., 1976).

subunit hypothesis predicts that the variance of each class of summed subunits should increase and be calculated by multiplying the variance of the subunit times the class (peak) number plus the recording system noise and measurement error. If the variance resulting from noise and measurement error is not subtracted from the measured subunit variance, (i.e., the variance of the plotted s-MEPPs) we could expect no more than 5 or 7 integral peaks and would not observe integral peaks in the central region of the amplitude distribution that are 10 times the sub-MEPP (Fig. 6). When the variance of the plotted sub-MEPP distribution is used to calculate the variance of summed subunits, no central peaks are observed (Fig. 16 bottom) as suggested by Katz (1977). However, when the variances due to measurement and noise (RMS) are subtracted, multiple peaks are predicted (Fig. 16 top). Boyd and Martin (1956) noted that their histogram of evoked potentials had sharper peaks than the model based on the observed variance of the unitary evoked potential, and they stated that the theoretical fit would be better if the noise contribution were subtracted.

The accuracy of the subunit model was evaluated by a Chi-square goodness of fit test (Matteson et al., 1981). The generalized likelihood ratio test was used to test the subunit model against a reduced bimodal model. The fit produced by the subunit model was accurate for 90% of the 68 histograms tested and was significantly better than the fit produced by the bimodal model (skew-MEPPs and MEPP) for 93% of the 68 histograms tested (Fig. 17) (Matteson, et al., 1981).

Effect of random variations in sample size on histogram profiles. Initially we did not keep count of the percentage of junctions that yielded MEPP amplitude profiles showing integral peaks because there are so many reasons why integral peaks would be absent. For example, a slight drift in muscle fiber resting potential, too much noise or too few MEPPs can obscure peaks. Magleby and Miller (1981) have found only 2 histograms out of 46 that show integral peaks (skew class). We performed 2 sets of experiments designed to determine the percentage of histograms that show integral peaks in the central region of MEPP amplitude histograms from junctions that met the rigorous criteria described above. In one set of experiments we asked: what is the percentage of sub-MEPPs in large adult frogs and do histograms show the same integral peaks as we have reported in small frogs? The muscle fibers were relatively large such that the bell-MEPPs were only 0.4 mV. The peak-to-peak

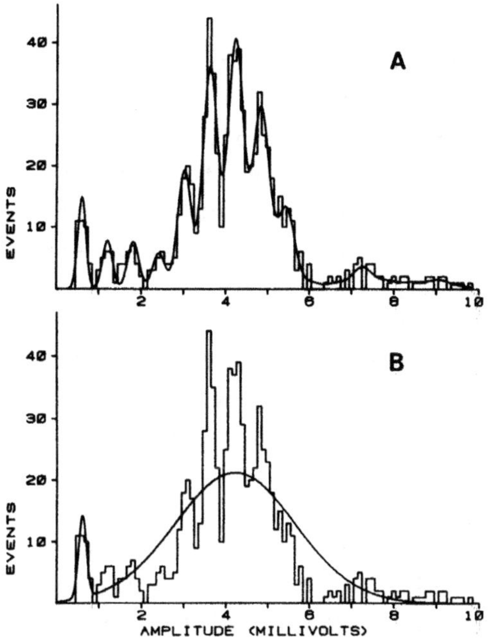

Fig. 17. (a) The fit produced by the multimodal model. The number of subpopulations in this case was 16. For each histogram this number was simply determined by the amplitude range of the histogram, under the assumption that the weighting factors of any larger amplitude subpopulations are zero. The maximum likelihood estimates were determined. The subunit was 0.60 mV, variance 0.065, noise and measurement 0.071 mV. (b) The fit produced by the bimodal model. Maximum likelihood estimates of the parameters are as follows: $W=3.96 \times 10^{-2}$, first peak, 0.61 mV, var 0.094, second peak, 4.2 mV, var 1.4 mV. Using the likelihood ratio test criterion the fit in (a) is significantly better ($P<0.001$). (Matteson, et al., 1981; data from Kriebel, et al., 1976, Fig. 6L).

noise was less than 30 µV, at least 2×10^3 MEPPs were recorded, no more than a 2 mV change in muscle fiber resting potential (80-95 mV) was tolerated and finally, the mean MEPP had to be stationary throughout the recording period. We analyzed 8 fibers that we initially thought met these criteria. The data were recorded on magnetic tape and subsequently filmed at two different gains, and MEPP amplitudes from both films were measured. MEPP amplitude histograms from 4 of the 8 juntions showed the same central,integral peaks from both films. Two of the 8 junctions showed multiple peaks which were rated as only suggestive of integral peaks; and, 2 of the 8 junctions showed histograms with irregular peaks. One of the latter did not have a sufficiently low noise-to-signal ratio throughout the recording period, and the second fiber had a 5% change in the mean MEPP amplitude with no apparent change in muscle fiber resting potential. The sub-MEPPs were present in all adult junctions but composed only 1% (Kriebel and Motelica-Heino, unpublished).

In the second class of experiments, we used the diaphragms from six 20-25 day old mice. We recorded control MEPPs and then added eserine and recorded the potentiated MEPPs. We also asked: Are skew-MEPPs potentiated like bell-MEPPs with eserine Two readers read the film strips at different film enlargements. Three of the 6 junctions showed the same number of central integral peaks on the control and eserine histograms with both film enlargements (Erxleben, et al., 1983). Histograms of two junctions showed peaks that were suggestive of subunits, and one junction generated a smooth histogram (even though the signal-to-noise ratio was extremely favorable and there was no change in the muscle fiber resting potential). This set of 6 experiments also established the fact that eserine increases the skew-MEPP mode exactly like the bell-MEPP mode. The MEPP sample sizes of the 6 junctions ranged from 2 to $6x10^3$ MEPPs and from our modeling analysis we could not expect to find a higher percentage of histograms that showed integral, clearly defined peaks. In conclusion, when data meet rigorous criteria more often than not the histograms will show well-defined integral peaks in the central region of the amplitude distribution.

<u>Morphological basis of the sub-MEPP and MEPP subunit</u>. The observation that the peak intervals of both the skew- and bell-MEPP distribution are the same indicates one presynaptic subunit class. Dennis and Miledi (1974b) report that both the skew-MEPPs and bell-MEPPs at the regenerating synapse are recorded extracellularly at the same point.

And, Cooke and Quastel (1973) found s-MEPPs at the same extracellular site as the bell-MEPPs at the mammalian junction. In addition, Thesleff, et al. (1983) found that the slow and fast rising MEPCs were generated at the same focal electrode. However, Colméus, et al. (1982) propose that the slow quanta are released from sites at a greater distance from the postsynaptic receptors (see Thesleff, this volume).

Pharmacological differences in skew- and bell-MEPPs and the fact that the skew class is not normally released with nerve stimulation supports a hypothesis based on two different release mechanisms capable of releasing the same subunits. The morphological or biochemical mechanism for the subunit must account for the small variance of the subunit which ranged from 6-18% (Fig. 2, MEPP subunit-16%; Fig. 6, MEPP subunit and EPP subunit-6%, no eserine; Fig. 7, MEPP subunit-16%, EPP subunit-18%; Fig. 9, MEPP subunit-14%, EPP subunit-8%; Fig. 10, MEPP subunit-18%). These values compare favorably to those of Wernig and Motelica-Heino (1978). Assuming that the synaptic vesicles represent the transmitter storage sites, there could be 10-15 classes of vesicles each generating one peak on the histogram. In the adult nerve terminal we could not expect to distinguish 15 vesicle classes based on volume and equal concentrations of ACh because the differences in vesicular diameters between the classes would be too small. However, in the one day old neonate mouse diaphragm, there are mainly sub-MEPPs (Muniak et al., 1982) which are 1/10 the bell-MEPP. Assuming that all vesicles contained the same concentration of ACh and that the basis for sub-MEPPs and bell-MEPPs was vesicular release we could have expected to find two classes of vesicles based on volumes (25 nM and 50 nM dia) but we found neonate vesicles to be like adult vesicles. In regards to different classes of vesicles, lanthanum experiments are interesting because only sub-MEPPs remain after intensive quantal release (Kriebel and Florey, 1983); yet, the vesicle diameter did not change even though few remained.

The observations that the time characteristics of sub-MEPPs (and sub-MEPCs) and bell-MEPPs (MEPCs) are the same, that it is easy to experimentally change skew-MEPP to bell-MEPP ratios and that the effect of temperature on skew- and bell-MEPP frequencies is different (Carlson et al., 1982) demonstrate that a post-synaptic basis for different classes is unlikely. However, Sellin and Thesleff (1981) conclude that the increase in the decay phase of MEPCs during BTX poisoning indicates a new population of receptors. Also, Colméus, et al. (1982) explain the slow time-to-peak MEPPs as quanta resulting from release sites far from the receptors.

Since there is only one class of synaptic vesicles based on diameter, either subunits are contained within vesicles or the sub-MEPP results from the release of one vesicle and synchronized vesicular release generates the bell-MEPPs (Kriebel and Gross, 1974). There are several problems with the latter hypothesis. Firstly, the variance of vesicle volumes (50-80%, Kriebel and Hanna, 1983) is much greater than the variance of the subunit (6-18%, Matteson et al., 1981) and, for that matter, greater than the variance of the overall bell-MEPP distribution (30%). It is possible that vesicular diameter variance may reflect the effects of fixation and embedding. Katz and Miledi (1979) demonstrated that over 10^4 quanta are released with one nerve impulse during diaminopyridine treatment. If each classical quantum represented the synchronized release of 7 vesicles, there simply are not that many vesicles near the presynaptic membrane. In addition, Heuser, et al. (1979) have shown with freeze fracture studies of 4-aminopyridine treated junctions that one vesicle pit (plus clusters of particles representing a heretofore pit) occurs with the generation of one electrophysiological quantum. Heuser, et al. (1979) also noted that vesicle pits do not appear to be grouped which would be expected with a synchronous release of vesicles. Yet, Heuser, Reese and Landis (1974) expected 1 pit per ridge but found 4. It is possible that each "release site" functions as a unit, as proposed by Wernig and Stirner (1979), from focally recording experiments, and that only 7 of the 20-30 vesicles at the ridge are released together.

In order to preserve the vesicle hypothesis we speculate that there are two classes of vesicles composed of the same subunits and two release mechanisms. One generates the skew-MEPP class and a second generates bell-MEPPs and is activated with nerve stimulation.

Supported by N.S.F. BNS-8219694

REFERENCES

Augustine, G.J. and H. Levitan. Neurotransmitter release and nerve terminal morphology at the frog neuromuscular junction by the dye Erythrosin B. J. Physiol. 334:47-63, 1983.

Bevan, S. Sub-miniature end-plate potentials at untreated frog neuromuscular junctions. J. Physiol. 258: 145-155, 1976.

Birks, R., B. Katz and R. Miledi. Physiological and structural changes at the amphibian myoneural junction in the course of nerve degeneration. J. Physiol. 150: 145-168, 1960.

Boyd, I.A. and A.R. Martin. The end-plate potential in mammalian muscle. J. Physiol. 132: 74-91, 1956.

Carlson, C.G., M.E. Kriebel and C.G. Muniak. The effect of temperature on MEPP amplitude distributions in the mouse diaphragm. Neuroscience 7: 2537-2549, 1982.

Colméus, C., S. Gomez, J. Molgó and S. Thesleff. Discrepancies between spontaneous and evoked synaptic potentials at normal, regenerating and botulinum toxin poisoned mammalian neuromuscular junctions. Proc. R. Soc. Lond. B 215; 63-74, 1982.

Cooke, J.D. and D.M.J. Quastel. Transmitter release by mammalian motor nerve terminals in response to focal polarization. J. Physiol. 228: 377-405, 1973.

Cull-Candy, S.G., Lundh, H., Thesleff, S. Effects of botulinum toxin on neuromuscular transmission in the rat. J. Physiol. 260: 177-203, 1976.

del Castillo, J. and B. Katz. La base 'quantale' de la transmission neuro-musculaire. In: 'Microphysiologie Comparée des éléments excitables'. Coll. Internat. C.N.R.S. Paris No. 67: 245-258, 1957.

Dennis, M.J. and R. Miledi. Non-transmitting neuromuscular junctions during an early stage of end-plate reinnervation. J. Physiol. 239: 553-570, 1974a.

Dennis, M.J. and R. Miledi. Characteristics of transmitter release at regenerating frog neuromuscular junctions. J. Physiol. 239: 571-594, 1974b.

Duchen, L.W. and D.E. Tonge. The effects of tetanus toxin on neuromuscular transmission and on the morphology of motor end-plate in slow and fast skeletal muscle of the mouse. J. Physiol. 228: 157-172, 1973.

Erxleben, C., G. Carlson and M.E. Kriebel. Studies of miniature endplate currents show that the quantum of release is composed of subunits. Neuroscience Abstracts, 1983.

Erxleben, C. and M.E. Kriebel. Characteristics of miniature and sub-miniature endplate currents at the mouse diaphragm endplate. Naunyn-Schmiedeberg's Arch. of Pharmacology. Suppl. to Vol. 322, R. 62, 1983.

Fatt, P. and B. Katz. Spontaneous subthreshold activity at motor nerve endings. J. Physiol. 117: 109-128, 1952.

Gross, C.E. and M.E. Kriebel. Multimodal distribution of MEPP amplitudes: the changing distribution with denervation, nerve stimulation and high frequencies of spontaneous release. J. Gen. Physiol. 62: 658-659, 1973.

Hanna, R. and M.E. Kriebel. Relationship between synaptic vesicles and miniature endplate potentials in adult and neonatal mouse diaphragm. Neuroscience Abstracts, 1983.

Harris, A.J. and R. Miledi. The effect of type D botulinum toxin on frog neuromuscular junction. J. Physiol. 217: 497-515, 1971.

Heuser, J. and R. Miledi. Effect of lanthanum ions on function and structure of frog neuromuscular junctions. Proc. R. Soc. Lond. B 179: 247-260, 1971.

Heuser, J.E., T.S. Reese, M.J. Dennis, Y. Jan, L. Jan and L. Evans. Synaptic vesicle exocytosis captured by quick freezing and correlated with quastal transmitter release. J. Cell Biol. 81: 275-300, 1979.

Heuser, J.E., T.S. Reese and D.M.D. Landis. Functional changes in frog neuromuscular junctions studied with freeze-fracture. J. Neurocytology 3: 109-131, 1974.

Katz, B. The transmission of impulses from nerve to muscle, and the subcellular unit of synaptic action. Proc. R. Soc. B 155: 455-477, 1962.

Katz, B. "Prologue", In: Synapses. eds. G.A. Cottrell and P.N.R. Usherwood. Academic Press. 1977. Chapter 1, pp. 1-5.

Katz, B. and R. Miledi. Estimates of quantal content during 'chemical potentiation' of transmitter relese. Proc. R. Soc. Lond. B 205; 369-378, 1979.

Katz, B. and S. Thesleff. On the factors which determine the amplitude of the 'miniature end-plate potential'. J. Physiol. 137: 267-278, 1957.

Kelly, S.S. and N. Robbins. Bimodal miniature and evoked end-plate potentials in adult mouse neuromuscular junctions. J. Physiol. 346; 353-363, 1984.

Kidokoro, Y. Two types of miniature endplate potentials in Xenopus nerve-muscle cultures. Neuroscience Research 1: 157-170, 1984.

Kita, H. and W. Van der Kloot. Effects of the ionophore X-537A on acetylcholine release at the frog neuromuscular junction. J. Physiol. 259: 177-198, 1976.

Kriebel, M.E. Small mode miniature endplate potentials are increased and evoked in fatigued preparations and in high Mg^{2+} saline. Brain Research 148: 381-388, 1978.

Kriebel, M.E. and E. Florey. Effect of lanthanum ions on the amplitude distributions of miniature endplate potentials and on synaptic vesicles in frog neuromuscular junctions. Neuroscience 9: 535-547, 1983.

Kriebel, M.E. and C.E. Gross. Multimodal distribution of frog miniature endplate potentials in adult, denervated, and tadpole leg muscle. J. Gen. Physiol. 64: 85-103, 1974.

Kriebel, M.E., R.B. Hanna and G.D. Pappas. Spontaneous potentials and fine structure of identified frog denervated neuromuscular junctions. Neuroscience 5; 97-108, 1980.

Kriebel, M.E., F. Llados and C.G. Carlson. Effect of the Ca^{++} ionophore X-537A and a heat challenge on the distribution of mouse MEPP amplitude histograms. J. Physiol. Paris 76; 435-441, 1980.

Kriebel, M.E., F. Llados and D.R. Matteson. Spontaneous subminiature end-plate potentials in mouse diaphragm muscle: evidence for synchronous release. J. Physiol. 262: 553-581, 1976.

Kriebel, M.E., F. Llados and D.R. Matteson. Histograms of the unitary evoked potential of the mouse diaphragm show multiple peaks. J. Physiol. 322: 211-222, 1982.

Kriebel, M.E. and D.R. Stolper. Non-Poisson distribution in time of small- and large-mode miniature end-plate potentials. Am. J. Physiol. 229: 1321-1329, 1975.

Land, B.R., E.E. Salpeter, M.M. Salpeter. Acetylcholine receptor site density affects the rising phase of miniature endplate currents. Proc. Natl. Acad. Sci. U.S.A. 77(6): 3736-3740, 1980.

Llados, F., M.E. Kriebel and D.R. Matteson. β-Bungarotoxin preferentially blocks one class of miniature endplate potentials. Brain Research 192; 598-602, 1980.

Magleby, K.L. and D.C. Miller. Is the quantum of transmitter release composed of subunits A critical analysis in the mouse and frog. J. Physiol. 311: 267-287, 1981.

Magleby, K.L. and M.M. Weinstock. Nickel and calcium ions modify the characteristics of the acetylcholine receptor-channel complex at the frog neuromuscular junction. J. Physiol. 299: 203-218, 1980.

Matteson, D.R., M.E. Kriebel and F. Llados. A statistical model supports the subunit hypothesis of quantal release. Neuroscience Letters 15: 147-152, 1979.

Matteson, D.R., F. Llados and M.E. Kriebel. A statistical model indicates that miniature endplate potentials and unitary evoked endplate potentials are composed of subunits. J. Theort. Biol. 90: 337-363, 1981.

Matthews-Bellinger, J.A. and M.M. Salpeter. Fine structural distribution of acetylcholine receptors at developing mouse neuromuscular junctions. J. Neuroscience 3: 644-657, 1983.
McLarnon, J.G. and D.M.J. Quastel. Postsynaptic effects of magnesium and calcium at the mouse neuromuscular junction. J. Neuroscience 3: 1626-1633, 1983.
Molgó, J. and S. Thesleff. 4-Aminoquinoline-induced 'giant' miniature endplate potentials at mammalian neuromuscular junctions. Proc. R. Soc. Lond. B 214: 229-247, 1982.
Muniak, C.G., M.E. Kriebel and C.G. Carlson. Changes in MEPP and EPP amplitude distributions in the mouse diaphragm during synapse formation and degeneration. Devel. Brain Research 5: 123-138, 1982.
Sellin, L.C. and S. Thesleff. Pre- and post-synaptic actions of botulinum toxin at the rat neuromuscular junction. J. Physiol. Lond. 317: 487-495, 1981.
Thesleff, S. Supersensitivity of skeletal muscle produced by botulinum toxin. J. Physiol. 151: 598-607, 1960.
Thesleff, S., J. Molgó and H. Lundh. Botulinum toxin and 4-aminoquinoline induce a similar abnormal type of spontaneous quantal transmitter release at the rat neuromuscular junction. Brain Research 264: 89-97, 1983.
Tonge, D.A. Physiological characteristics of re-innervation of skeletal muscle in the mouse. J. Physiol. 241: 141-153, 1974.
Vautrin, J. et J. Mambrini. Caractéristiques du potentiel unitaire de plaque motrice de la grenouille. J. Physiol., Paris 77: 999-1010, 1981.
Wernig, A. and I. Motelica-Heino. On the presynaptic nature of the quantal subunit. Neuroscience Letters 8: 231-234, 1978.
Wernig, A. and H. Stirner. Quantum amplitude distributions point to functional unity of the synaptic 'active zone'. Nature, Lond. 269: 820-822, 1977.

DISCUSSION OF THE ARTICLE BY M.E. KRIEBEL (U.S.)

B. Katz (U.K.): I gather you say that the subunits have a coefficient of variation of 16%.

M.E. Kriebel (U.S.): Yes, that is corrected data. After subtracting the noise, we are left with a coefficient of variation of 16% (Matteson et al (1981). J. Theoret. Biol. 90, 337-363).

A. Wernig (West Germany): Did you ever see in your current measurements something like a break of the upstroke?

M.E. Kriebel (U.S.): Yes, there are breaks on the rising phase of the regular mepp. If you plot the amplitude of the break, one finds a skew mepp distribution and that in itself is an argument in favor for a subunit structure to the mepps. Otherwise, if you simply have two bell mepps, the distribution to the first break should mirror the overall distribution of the mepp (Carlson C.G., Kriebel M.E. and Munick C.G. (1982). The effect of temperature on mepp amplitude distributions in the mouse diaphragm. Neuroscience 1, 2537-2549).

S. Thesleff (Sweden): The sub-mepps seemed to me smaller than the mepps I was speaking about. Do you believe they are of the same kind?

M.E. Kriebel (U.S.): Yes. After adding butulinium toxin to the bath, we see just the skew class remaining. I think some of yours look exactly like that. If the signal to noise ratio isn't good enough, we see the distribution skewing to the noise level like Miledi showed (Kriebel et al. (1976). J. Physiol. 262, 553-381).

FREQUENCY OF MINIATURE END PLATE POTENTIALS DETERMINED BY AN EXTENSION OF CAMPBELL'S THEOREM

Segal, J.R., Ceccarelli, B., Fesce, R., and Hurlbut, W.P. Veteran's Administration Medical Center, New York, USA; CNR Center of Pharmacology, Milan, Italy; Rockefeller Univ., New York, USA

A method based upon an extention of Campbell's theorem is used to measure the time course, amplitude and frequency of occurence of miniature end plate potentials (mepp) at rapidly secreting neuromuscular junctions in frog cutaneous pectoris muscles. Measurements of the SKEW (third central moment), variance and power spectrum of the fluctuations in membrane potential are used to determine the mepp parameters. The resulting estimates of mepp amplitude and frequency are insensitive to slow drifts in membrane potential that preclude the conventional application of Campbell's theorem, which uses the MEAN and variance. The new method fails at very high mepp frequencies because the distribution of the values of the membrane potential approaches a Gaussian, and the measurements of the SKEW lose statistical significance. The frequency at which this failure occurs is raised to about $10^4 s^{-1}$ if the data are high-pass filtered. The method has been tested on computer-simulated data, applied to junctions exposed to La^{3+}, and the effects of Ca^{2+} on the La^{3+}-induced secretion have been explored. Some muscles were fixed after treatment with La^{3+}, and the changes in nerve terminal intrastructure were assessed by morphometric analysis of electron micrographs. Horseradish peroxidase (HRP) was used to obtain direct information about vesicle recycling. Most experiments were done at room temperature (22-25°C) using solutions that contained 0.1mM La^{3+}, a concentration that caused enormous increases in mepp frequency. When the rate of secretion was at its peak, these frequencies sometimes exceeded $5 \times 10^{+3} s^{-1}$ and mepp amplitudes were reduced by over 60%. Several million quanta were secreted from single nerve terminals during an hour, ie several times the number of synaptic vesicles in control terminals. The number of vesicles in the experimental terminals fell by 75% or more, the axolemma expanded to form extensive folds that penetrated deep into the axoplasm, and some of the remaining vesicles were labelled with HRP if this extracellular tracer had been present in the bathing solution. Ca^{2+}-free solutions increased the rates at which these changes developed, but had relatively little effect on the final state of the terminals.

DISCUSSION OF THE ARTICLE BY W.P. HURLBUT (U.S.)

M.E. Kriebel (U.S.): I wanted to point out that we did similar experiments. We used very small fibres, and counted the mepps directly. At rest, the mepp frequency was 10/min. With La^{3+}, the frequency went up by 500-700 times but it was still possible to count the mepps directly. If we consider junctional size, we would have the same number of mepps released with La^{3+} treatment as you found (Kriebel M.E. and Florey E. (1983). Effect of lanthanum ions on the amplitude distributions of miniature endplate potentials and on synaptic vesicles in frog neuromuscular junctions. Neurosci. 9, 535-547).

MARKERS FOR MEMBRANE COMPONENTS TO STUDY "RECYCLING" AT NEUROMUSCULAR JUNCTIONS

Ceccarelli, B.

We recentlly applied an improved quick-freezing technique at frog nerve muscle preparations and were able to show that fusion of synaptic vesicles with the axolemma and quantal release of ACh are coincident. This result fulfills an important requirement of the vesicle hypothesis of transmitter release. The continued and intense secretion of quanta over long period of time requires some means of maintaining the population of synaptic vesicles. Over the short term this is done by the endocytosis of membrane from the axolemma. However, almost nothing is known about this important membrane conservation process. In particular, it is not known how rapidly endocytosis follows upon exocytosis, whether the sites of endocytosis are distant from the sites of exocytosis or what factors control the rate of endocytosis. Much of our ignorance concerning endocytosis stems from the difficulty in tracing the movements of vesicle membrane. Specific markers for vesicle membrane that could be localized at the E.M. level allow many of the controversial questions mentioned above to be settled. We have therefore developed immunohistochemistry techniques in order to visualize at frog neuromuscular junctions specific markers for the synaptic vesicle and nerve terminal membranes. Synapsin I has been shown by immunohistochemistry to be present in many mammalian terminals and in brain synaptosomes it is associated with the outer half of the synaptic vesicle membrane. Our work shows that Synapsin I is also present in frog nerve terminals. We have applied similar immunofluorescence techniques to study the binding of -latrotoxin, the active component of black widow spider venom, to nerve muscle preparations. The binding is highly localized and seems to be present only at the nerve endings. If these two proteins prove to be specific markers for the synaptic vesicle membrane and axolemma, respectively, it would be of immense value to study membrane turnover at neuromuscular junction.

DISCUSSION OF THE ARTICLE BY B. CECCARELLI (ITALY)

E. Zlotkin (Israel): I would like to clarify the connection between the binding assay which you have demonstrated and the fluorescent marking of the presynaptic sites with alpha latrotoxin. In what preparation was the binding assay performed?

B. Ceccarelli (Italy): It was done in the cutaneous pectoris. The same preparation that we used for immunohistochemistry.

E. Zlotkin (Israel): That means that the preparation used contained presynaptic, postsynaptic, and neuronal elements.

B. Ceccarelli (Italy): Yes.

E. Zlotkin (Israel): However, you attributed the specific binding to the presynaptic sites.

B. Ceccarelli (Italy): The reason we feel it is presynaptic is because immunofluorescence was localized exclusively at the endplate region, and both specific binding and immunofluorescence disappeared after denervation. In addition, when the toxin was applied at room temperature (a condition in which massive fusion of synaptic vesicles takes place) the terminal branches were swollen, as expected, without concomitant changes of the other structures.

E. Zlotkin (Israel): In the second experiment, with the fluorescent marking, did you displace this marking with cold toxin?

B. Ceccarelli (Italy): Your question makes me wonder whether I was able to explain those experiments adequately; what we did was to bind the toxin to the muscle, then wash, bind a monospecific rabbit antitoxin antibody and then a fluorescent antirabbit Ig antibody. When the toxin was left out, or a preimmune serum used, no fluorescence appeared.

E. Zlotkin (Israel): What attracted my attention was the fact that there was a very low capacity in your binding assay. On the other hand, in your immunocytochemistry study, I got the impression that there was very dense saturation.

B. Ceccarelli (Italy): You have to realize that presynaptic terminals account for only an exceedingly small proportion of the muscle used. If binding is restricted to the surface of terminals only, its density (number of sites/ m^2) can be calculated to be indeed considerable, of the order of 2,000/ m^2. In the other preparations in which specific binding of -LTz has been detected, the density of the sites was much lower (35 and 12 sites/ m^2 in synaptosomes and PC12 cells respectively (Meldolesi J. (1982) J. Neurochem. 38, 1559-1569; Meldolesi J. et al. (1983) Neurosci. 10, 997-1009).

R. DeLorenzo (U.S.): These are very difficult studies to do and I admire the work you are doing. You wetted my appetite to find out whether you see any evidence of recycling. You implied from the fluorescent studies that when you deplete the vesicles, some of the labeling may be going into the membrane on the outside. Do you see that with HRP or is that something you haven't done yet?

B. Ceccarelli (Italy): No. In our studies at the electrone microscopy level we have not so far used HRP conjugated antibodies.

R. DeLorenzo (U.S.): So that is something you will be doing that will be very exciting.

B. Ceccarelli (Italy): In general we prefer to use as a second antibody the ferritine conjugated rather than the HRP conjugated. What we showed with P36 is an extreme condition. We treated the preparation with black widow spider venom; a toxin that causes a profound depletion of synaptic vesicles that is accompanied by a profound swelling of the nerve endings. At the light level, we can detect only this terminal swelling because of the increase in the transverse diameter of the nerve terminal branches.

R. DeLorenzo (U.S.): That is the conditions Heuser and Reese used to get their recycling. Your tool would be a very interesting way to look at that story in terms of seeing whether the vesicle membrane actually does go into the axolemma.

B. Ceccarelli (Italy): We hope so. However, we have to apply similar immunohistochemical techniques at the E.M. level, possibly under more physiological conditions of stimulation. We feel in fact that by using black widow spider venom, we can achieve a quick depletion of synaptic vesicles because venom indeed blocks the recycling mechanism.

THE ROLE OF CALCIUM IN TRIGGERING THE RELEASE OF TRANSMITTER AT THE VERTEBRATE NEUROMUSCULAR JUNCTION. AN ANALYSIS OF THE EFFECTS OF DRUGS

C.J. DUNCAN

Katz (1) in his Sherrington lecture summarized our understanding of the microphysiology of the presynaptic terminals of the vertebrate neuromuscular junction (nmj). The arrival of a nervous impulse causes a transient increase in the calcium permeability of the plasma membrane and the associated entry of calcium ions triggers the synchronized release of quanta of transmitter. The magnitude of the evoked response, recorded as the endplate potential (EPP) is therefore primarily dependent on the concentration of extracellular calcium (2). There is also a spontaneous release of quantal packets of acetylcholine (ACh), recorded as the miniature endplate potentials (MEPPs), and the MEPP frequency is determined, at least in part, on the other hand, by the concentration of intracellular calcium at the terminals (3, 4). However, we still lack a firm conclusion as to how the intracellular calcium ions trigger both the spontaneous and the evoked release of transmitter. Since the earlier biophysical studies of Katz and Miledi a large number of pharmacological agents have been tested on the vertebrate nmj, and the purpose of this very brief review is to attempt to determine their sites of action in the hope of discovering some of the biochemical events of exocytosis.

1. TYPES OF TRANSMITTER RELEASE

There now appears to be at least three types of transmitter release at the nmj (5):

Type I is the calcium-dependent, quantal release described above. One quantum (a MEPP) almost certainly represents the exocytosis of one vesicle.

Type II is non-quantal and consists of the molecular leakage of substantial quantities of ACh (6); the physiological role of non-vesicular release is uncertain.

Type III is quantal and spontaneous, producing MEPPs, but is independent of calcium entry at the terminals. Type III release is revealed following botulinum poisoning or nerve regeneration and has particularly interesting properties (5).

This review concentrates solely on Type I release, where exocytosis involves the fusion of the vesicle membrane with the inner face of the terminal plasmalemma and its subsequent fission.

2. DRUGS AFFECTING TYPE I TRANSMITTER RELEASE AT THE VERTEBRATE NMJ

Since both evoked and spontaneous release can be modified by local changes in calcium concentration close to the vesicles and plasmalemma, it is evident that pharmacological treatments may be classified into those that, directly or indirectly, affect calcium fluxes and those that affect the calcium-vesicle-plasmalemma interaction, i.e. the molecular events of calcium-triggered fusion-fission. It is this interaction that is of particular interest at present. However, firstly we need to consider those treatments that act via a modification of intracellular Ca concentration. Some drugs can modify intracellular calcium in different ways and sometimes in different directions; some may affect both calcium levels and exocytosis.

3. CONTROL OF INTRACELLULAR CALCIUM AT THE PRESYNAPTIC TERMINALS

The intracellular calcium concentration represents the steady-state position of a number of interacting mechanisms; it is a system that is regularly challenged by the calcium influx associated with the depolarization of the action potential. We have only limited knowledge of the detailed properties of these control mechanisms at the nmj, but substantial information is available for the terminals of the CNS (particularly the synaptosome preparation, 7, 8). Combining this information with the effects of different pharmacological agents at the nmj (sections 5-10 below), it is probable that the different elements of the calcium control mechanism at the nmj are as follows:

a) Voltage-dependent calcium channels in the plasmalemma which serve as the major influx channel and which in brain cells are metalloproteins (9) and have a molecular weight of 185K (10). Additionally, there is evidence of a slow, constant Ca influx in synaptosomes (11). Passive Ca efflux via either of these channels against a substantial electrochemical gradient will be negligible.

b) The major pathway for Ca efflux has been considered to be a Na^{+}/Ca^{2+} exchange mechanism, in which Ca efflux is decreased by removal of extracellular sodium, intracellular calcium concentration may be predicted by the sodium concentration gradient and the energy of the Na electrochemical gradient may be used to power Ca extrusion (7). However, no Ca efflux driven by a Na electrochemical potential could be detected in a synaptosome preparation (11) and there is increasing evidence that Ca-extrusion at synaptic terminals is largely linked to ATP hydrolysis and catalysed by a (Ca+Mg)-ATPase (12, 13). This enzyme is inhibited by La (14) and may be regulated by phosphorylation and stimulated by calmodulin (15).

c) The mitochondria represent a potentially important intracellular mechanism for sequestering Ca. Their affinity is relatively low (half maximal uptake at 1-10μM Ca and possibly higher at physiological Mg concentrations) and hence uptake will be small under normal cytosolic conditions (intracellular Ca approximately 0.1μM). Campbell (16) suggests that under conditions of low cytosolic Ca there is net efflux of mitochondrial Ca, and net uptake above a balance point (perhaps 0.1-0.3μM). Thus, the mitochondria may well not be involved in normal regulation of intracellular Ca. On the other hand, the mitochondria have a relatively large Ca capacity.

d) Other intracellular Ca-storage sites undoubtedly exist in the presynaptic terminals and these probably include the endoplasmic reticulum (16, 17) and the inner face of the plasma membrane (18, 19). These non-mitochondrial stores are insensitive to mitochondrial poisons and have a high affinity (uptake half-saturated at approximately 0.35μM) and a rather low capacity for Ca (17, 20). Although Ca regulation by the endoplasmic reticulum was reported to be of minor significance in a synaptosomal preparation (11), these non-mitochondrial Ca-sequestering sites are increasingly being recognized as being of major importance during normal cell functioning (21) and the Ca stores associated with the inner face of the plasma membrane are probably the site of action of Botulinum type A toxin (22).

e) Isolated cholinergic synaptic vesicles from mammalian brain and from Torpedo display a (Mg-ATP)-dependent uptake of Ca (23, 24) that is stimulated by calmodulin (25); it is probable that the vesicles at the nmj can also take up Ca which is then released with the ACh at exocytosis.

All these intracellular storage sites can act only to maintain cytosolic Ca at a low level on a temporary basis and ultimately the steady-state position is determined by the rates of influx and efflux of Ca across the plasmalemma.

4. DRUGS THAT ACT TO MODIFY INTRACELLULAR Ca AT THE TERMINALS

Inspection of published results reveals that the majority of pharmacological treatments that modify or block transmitter release act via an alteration in cytosolic Ca. Their action can be tentatively assigned to the sites described below; (+) indicates a rise in intracellular Ca and an augmentation of release, whilst (-) indicates the reverse effect. Shortage of space prevents a full analysis and also detailed references to the many results available, which will be given in detail elsewhere (Duncan, in preparation).

5. CHANGES IN Ca ENTRY VIA THE VOLTAGE-DEPENDENT Ca-PERMEABILITY CHANNELS

These treatments are dependent on extracellular Ca and usually affect evoked release, although spontaneous release is frequently modified also:

- Depolarization of the presynaptic terminals (+)
- Elevation of extracellular K concentration (+)
- Decrease in extracellular pH (depolarization via change in chloride permeability) (+)
- Post tetanic potentiation (+)
- Elevation of extracellular Ca concentration (+)
- Elevation of extracellular Mg concentration (-)
- Increase in Na permeability (and hence depolarization):
  - Ionophore X537A (+)
  - Hexachloro-cyclohexane (lindane) (+)
  - *Anemonia sulcata* toxin (ATX II) (+)
  - Monensin (+)
  - α-dihydrograyanotoxin II (+)
- Inhibitors of K conductance:
  - 4-Aminopyridine (and analogs) (+)
  - 3,4-Diaminopyridine (and analogs) (+)
  - Uranyl ions (+)

Tetraethylammonium (+)
Very high extracellular Ca (via A.P. block) (-)
Aldrin-transdiol (+)
Ruthenium red (-)
? Botulinum toxin, types A and D, early phase (-)
Bekanamycin (-)
Enkephalin (-)
Mn, Co, Ni (-)
Hg; methyl mercury; mersalyl (-)
Cd (less than 0.1mM) (-)
Pb (-)
La, Er, Pr, Y (-)
Amino-glycosidic-ammocyclitolic antibiotics (-)
Reduction of extracellular Na (+) (? Na-Ca competition at entry)
β-Leptinotarsin-h, initial phase (+)
Verapamil (at 10^{-5}M) (-)

6. CHOLINERGIC FEEDBACK

The evidence suggests that non-specific cholinesterases are present on the amphibian presynaptic terminals and that these can act as muscarinic cholinergic receptors; cholinergic agents interact with these receptors and reduce Ca-permeability directly. ACh therefore has a presynaptic inhibitory effect and in this preparation the following agents all reduce spontaneous release (26):

Cholinesterase inhibitors (e.g. neostigmine, edrophonium, eserine) (-)
Tetraisopropylpyrophosphoramide (-)
Carbachol, muscarine, metacholine (-)

However, at the mammalian nmj these receptors are blocked by D-tubocurarine and are excitatory (27).

7. MODIFICATION OF Ca^{2+} ENTRY AT THE PLASMA MEMBRANE (not necessarily via voltage-dependent channels). Again the following treatments are usually dependent on extracellular Ca:

Black widow spider venom, α latrotoxin (?minor part of action, perhaps via voltage-dependent channels) (+)
Ionophore A23187 (+)
Salmonella typhimurium (SR II) (initially +)
Endotoxin (initially +)

Lipopolysaccharides from bacterial cell walls (initially +)
B-Bungarotoxin (initially +) - probably a non-specific increase in ion permeability
Nystatin, filipin (polyene antibiotics) (+)
PLA_2 (from Pseudocrastes fieldi venom), initial action (+)
Primycin (+)
Complement (+)
Glycera venom (+)

8. PROMOTION OF NET Ca-EFFLUX FROM THE MITOCHONDRIA

The action of these drugs is usually independent of extracellular Ca; MEPP frequency is a sensitive indicator of this site of action. Because of the rather high Ca-capacity of the mitochondria, the response to these agents is often a steady rise, culminating in a very high MEPP rate (although there is no synchronization of release, as in the EPP) followed by a (frequently irreversible) decline as the Ca-stores are emptied.

Classical uncoupling agents and mitochondrial poisons (+) (E.g. dicoumarol, antimycin, rotenone, cyanide, 2,4 DNP, carbonylcyanide-p-trifluoromethoxy-phenylhydrazone, azide)
? Barbiturates (+)
Tetraphenylboron (+)
Acetaldehyde (+)
Erythrosin B (+)
Guanidine (+)
? Ethanol (+)
Hg; methyl mercury; mersalyl; initial effects (+)
Mn, Co, Ni (+)
Pb (+)
Primycin (+)
? Cd (at 0.1-1mM) (+)
N-ethyl maleimide (+)
p-Chloromercuribenzoate (+)
Elevated partial pressure of CO_2

Additionally, rises in intracellular Na or Li cause rises in MEPP frequency, probably by accelerating Ca-efflux from the mitochondria. Treatments acting in this way include:

Ouabain (+)
Digoxin (+)

Incorporation of Na via liposomes (+)
Li saline (+)

The following treatments probably act by specifically inhibiting the Ca-uptake by the mitochondria:

Ruthenium red (+)
La, Er, Pr, Y (+)
? Heavy metal ions

Furthermore, in the mammalian nmj, cellular oxygen supply can modify mitochondrial function, e.g.:

Hypoxia (+)
Fluorocarbons (-)

9. RELEASE OF Ca FROM OTHER INTRACELLULAR STORES

Because of the small Ca-capacity of these stores, pharmacological challenge usually results in a more modest alteration in MEPP frequency and the response is again independent of extracellular Ca. Some of the treatments listed under mitochondrial effects may well act at other intracellular Ca-storage sites also.

Ultraviolet light (+)
? Adrenocorticotrophic hormone (+)
Caffeine (+)
Theophylline (+)
Dantrolene sodium (-)
Mersalyl, Hg (+)
? Ionophore A23187
? Botulinum toxin, type A (-) (? improves buffering of intracellular Ca)

10. MODIFICATIONS OF THE Ca-PUMPS AT THE PLASMALEMMA

(Mg+Ca) ATPase:

Ruthenium red (+)
? Phosphorylation via cAMP/adrenaline
La
? Calmodulin blockers

Na-Ca Exchange:

La
? Ethanol

It is clear from the foregoing that many agents have more than one site of action and, indeed, some may have apparently opposing effects, blocking evoked release and accelerating spontaneous release (e.g. ruthenium red, Pb).

11. EFFECT OF TEMPERATURE

Temperature will probably affect many of the elements of intracellular control, in particular the activity of the Ca-pump, so explaining, for example, the temperature-sensitivity of the transient increase in MEPP frequency produced by an action potential or by tetanic stimulation (28, 29, 30, 31). MEPP frequency should be a particularly sensitive method for exploring the effect of temperature on the steady-state level of intracellular Ca. However, the results are surprising: in the frog, MEPP frequency is low and almost independent of temperature at 5-15°C, whereas above 16°C the system has a high activation energy (32, 33). Furthermore, various treatments have markedly differential effects above and below this apparently critical (transition) temperature of 16°C. For example, the divalent cation ionophore A23187 (4) and the thiol-oxidizing agent diamide, where the effect is particularly dramatic (34). The sensitivity of MEPP frequency at the mammalian nmj to temperature is more complex with a Q_{10} that is positive from 5-15°C, negative 15-28°C, and positive (with a high activation energy) 28-38°C (35, 36). These results have been discussed in greater detail elsewhere (37) and we may tentatively conclude that

(a) rising temperature causes a phase change, presumably in a hospholipoprotein system at 16°C at the presynaptic terminals of frog, and that there may be two such changes (at about 15 and 28°C) at the mammalian nmj

(b) inspection of the results suggests that rather low concentrations of intracellular Ca are markedly more effective in effecting exocytosis above the critical temperature in frog

(c) in those pharmacological studies where only modest changes in intracellular Ca and MEPP frequency ensue, it is important to study the action of the drug above and below the critical temperature

(d) temperature will modify the various elements of the Ca-control system of the terminals, particularly of the (Mg+Ca)-ATPase, rising temperature serving to reduce intracellular Ca and hence MEPP frequency, so accounting for the negative Q_{10} at 15-28°C at the mammalian nmj (32). However, these effects are secondary to the dominant effect of temperature on the Ca-phospholipoprotein interaction of the exocytosis mechanism (34).

12. EFFECT OF PHARMACOLOGICAL TREATMENTS ON THE MOLECULAR EVENTS OF FUSION-FISSION

It is evident from the foregoing that the majority of drugs that modify vertebrate neuromuscular transmission act via a change in intracellular Ca levels and it is pertinent to ask whether any treatments modify specifically the interaction between Ca, the vesicle and the plasmalemma. Firstly, Sr and Ba can act as substitutes for Ca in triggering fusion-fission when entering the terminals during depolarization; Sr seems to act in an almost identical fashion to Ca, whilst Ba is a poor substitute (38). Mg, Be, Zn and Ni on the other hand, compete with Ca once they have gained entry and so reduce release. Secondly, the Ca-dependence of evoked release at the frog nmj at low quantal contents is known in detail. Below about 4×10^{-5}M, the plot of log extracellular Ca concentration versus m has a slope of 1, whilst at higher concentrations the slope is about 4 (2), showing that the action of Ca is co-operative and possibly suggesting that Ca interacts with a Ca-binding protein in triggering exocytosis.

The following treatments all have very marked effects on transmitter release at the nmj and the results suggest that it is possible that their action may be directly on the process of fusion-fission rather than modifying intracellular Ca:

Diamide (+)
Diazene dicarboxylic acid bis (N'-methylpiperazide) and its salts (+)
? Sodium bisulphite (action via disulphide bonds) (+)
Glutaraldehyde (-)
Stretch (+)
Increased hydrostatic press (-)
Hypertonic (+) and hypotonic (-) solutions
? Black widow spider venom (+), secondary effects
Vinblastine (+)
Vincristine (+)
? Botulinum toxin (second phase of action) (-)
Elevated temperature (+)

The list is a short one at present, much shorter than the catalogue of treatments given above that can block transmission or modify events at the vertebrate nmj via modifications of free cytosolic Ca. Treatments that, it is suggested, modify exocytosis directly have some, but not necessarily all, of the following features in common:

(i) Their action is independent of extracellular Ca

(ii) Their effects on spontaneous and evoked release are in the same direction, be it excitatory or inhibitory.

(iii) The plot of log m versus log extracellular Ca is not only shifted, either to the left in the case of stimulatory treatments (diamide (39); hypertonic solutions (40)) or to the right in inhibitory treatments (botulinum toxin (41), but it is significant that the slope of the line is reduced. The inference is that these treatments alter the degree of co-operativity and hence make a fundamental change in the nature of the interaction between Ca and the fusion process. Since there is, as yet, no evidence that these treatments alter intracellular Ca concentration, the exocytosis mechanism must be changed to become more (diamide, hypertonic solutions) or less (botulinum toxin) sensitive to the level of cytosolic Ca. This finding is in marked contrast with the effects of treatments that are believed to operate via a change in calcium permeability where the line is also shifted either to the left or right but the slope remains exactly the same as that of the controls (usually 3-4), suggesting that the basic response to Ca remains the same. Examples are 4-aminopyridine (42), reduction in extracellular Ca (43), lowered extracellular pH (44), lead (45, 46) and bekanamycin (47). In a number of these latter experiments modified Lineweaver-Burk plots confirm that these treatments are acting competitively with extracellular Ca for entry.

(iv) Diamide, vinblastine and vincristine increase the frequency of multiquantal MEPPs, explicable as the fusion of two or more vesicles (34, 48). Thus, these agents appear to promote both vesicle-plasmalemma and vesicle-vesicle fusion.

(v) Many of these treatments are peculiarly temperature-sensitive. The importance of the effect of temperature and the possibility of a phase change in a phospholipoprotein system has been discussed above and the action of diamide illustrates this point. It raises MEPP frequency only 2- to 5-fold at 13-15°C but 55-60-fold at 17-20°C (34). Botulinum A toxin causes both the typical decrease in MEPP frequency and the block in evoked release in the mouse hemidiaphragm preparation at 37°C, but a reduction in temperature to 20°C causes a remarkable increase in nerve-evoked release, whilst leaving the low MEPP frequency unaffected (49). Finally, Kita et al (50) have concluded that tonicity and temperature appear to affect spontaneous and evoked release in a similar fashion.

(vi) The observations in (v) may be amplified further, because many of these treatments can have complex interacting effects. Black widow spider venom causes a massive increase in transmitter release and at least part of the effect is probably because of a redistribution of membrane proteins that can be inhibited by concanavalin A (which modifies membrane protein mobility (51, 52)) and also by elevated temperature or by hypertonicity of the medium (52). The direct effects of hypertonicity in increasing MEPP frequency and in increasing evoked release are also inhibited by pre-treatment with concanavalin A. The reduction in MEPP frequency seen at the rat nmj after botulinum type A intoxication was also reversed by black widow spider venom (41).

We can draw some tentative conclusions from these studies. The treatments appear to fall into two categories: firstly those that modify proteins, particularly via thiol oxidation (diamide, vinblastine (37), and secondly those that modify the physical organization or fluidity of the membrane on a molecular scale (e.g. temperature, hypertonicity, stretch, hydrostatic pressure, concanavalin A). Any of these changes seem to make phospholipoprotein complexes of the plasmalemma and the vesicle membrane more (or less) sensitive to a given local concentration of cytosolic Ca in effecting membrane fusion when it binds to its appropriate binding site.

13. THE PROPERTIES OF THE Ca-BINDING SITE

The plot of extracellular Ca concentration versus m is sigmoidal, the log-log plot has a slope of about 3-4 (2), the classical picture of allosteric activation via a Ca-binding protein. It is impossible to resist the temptation to speculate that in vivo spontaneous release operates in the non-cooperative range (n=1), where MEPP frequency is governed by Poisson statistics, whilst evoked release operates in the co-operative range and is governed by binomial statistics. Certain drugs modify those properties. For example, changing intracellular Ca may shift the line to the point where it is within the co-operative range right down to the lowest levels of extracellular Ca.

14. Ca-TRIGGERED EVENTS IN EXOCYTOSIS

Ca usually acts as a cellular regulator either by promoting the phosphorylation of a key protein or by producing molecular configurational changes (as with troponin C or calmodulin). The possible role of phosphorylation at the nmj via the agency of cAMP has been discussed elsewhere (37); the effects of dibutyryl cAMP and phosphodiesterase inhibitors are limited and it seems that the action of cAMP is restricted to modulating various elements of the Ca-control system (e.g. the Ca pump, the voltage-dependent Ca channels and the mitochondria). It does not seem likely that vesicle-plasmalemma protein phosphorylation via cAMP is the key event in exocytosis.

A substantial number of publications now testify to the activation of protein kinase C in parallel with the elevation of cytosolic Ca, the two systems operating as synergistic modulators of secretory processes by the phosphorylation of membrane proteins. 12-0-tetradecanylphorbol-13-acetate (TPA) is a potent activator of protein kinase C and its effect on the amphibian nmj has been tested in preliminary experiments. It causes a significant, but not dramatic rise in MEPP frequency both in normal saline and when extracellular K is raised to 10mM (S.J. Publicover, pers. comm.) and it is unlikely that TPA is here mimicking exocytosis although, as with cAMP, phosphorylation might be fine-tuning the key proteins. Rather, it is more probable that the phosphorylation step is at the voltage-dependent Ca-channels, so increasing inward Ca-flux.

Alternatively, therefore we must consider the interaction of Ca with one of the many Ca-binding proteins that have been described in secretory cells, synexin, synapsin, chromobindins, calmodulin, clathrin light chains, 32K and 34K proteins, calelectrins, calrembins. There is some evidence that calmodulin is implicated in chromaffin granule secretion; with trifluoperazine (TFP), a calmodulin inhibitor, the granules accumulate close to the plasma membrane but exocytosis does not occur. Thus, secretion consists of a TFP-insensitive stage when the granules are translocated to the plasma membrane and a second TFP-sensitive stage of granule-plasmalemma fusion (53). However, TFP had only modest effects at the frog nmj, increasing MEPP frequency and quantal content, but inhibiting by 25% the stimulatory effects of depolarization (54). Calmidozolium (R24571) had only limited effects on MEPP frequency (Publicover, this symposium) and there seems to be, surprisingly, no clear evidence implicating the ubiquitous calmodulin in the process of exocytosis at the amphibian nmj (37). Rather, the results are consistent with calmodulin regulating both the voltage dependent Ca-channel and the Ca pump, so that calmodulin inhibitors produce modifications in the free Ca of the cytosol.

We may conclude that proteins of the vesicle and plasma membranes are of particular importance in fusion-fission events of release of quantal packets of transmitter at the vertebrate nmj. The proteins are particularly sensitive to oxidation of their -SH groups and to the physical state of their phospholipid environment. Ca triggers exocytosis by binding in a strongly co-operative manner either to these proteins or to a linked, regulatory protein.

15. REFERENCES

1. Katz B: The release of neural transmitter substances. Liverpool University Press, 1969.
2. Crawford AC: The dependence of evoked transmitter release on external calcium ions at very low mean quantal contents. J. Physiol. 240: 255-278, 1974.
3. Alnaes E, Rahamimoff R: On the role of mitochondria in transmitter release from motor nerve terminals. J. Physiol. 248: 285-306. 1975.
4. Statham HE, Duncan CJ: The action of ionophores at the frog neuromuscular junction. Life Sci. 17: 1401-1406, 1975.
5. Thesleff S, Molgo J: A new type of transmitter release at the neuromuscular junction. Neuroscience 9: 1-8, 1983.
6. Tauc, L: Nonvesicular release of neurotransmitter. Physiol Rev. 62: 857-893, 1982.
7. Blaustein MP, Ratzlaff RW, Kendrick NK: The regulation of intracellular calcium in presynaptic nerve terminals. Ann. N.Y. Acad. Sci. 307: 195-212, 1978.
8. Blaustein MP, Ratzlaff RW, Schweitzer ES: Control of intracellular calcium in presynaptic nerve terminals. Federation Proc. 39: 2790-2795, 1980.
9. Glossmann H: Molecular approach to the calcium channel. XI Congress International Society for Heart Research, London 1983.
10. Ferry DR, Goll A, Glossmann H: Putative calcium channel molecular weight determination by target size analysis. N.S. Arch. Pharmacol. 323: 292-297, 1983.
11. Nicholls DG, Åkerman KEO: Biochemical approaches to the study of cytosolic calcium regulation in nerve endings. Phil. Trans. Roy. Soc. Lond. 296: 115-122, 1981.
12. Duncan CJ: Properties of the Ca^{2+}-ATPase activity of mammalian synaptic membrane preparations. J. Neurochem. 27: 1277-1279, 1976.
13. Åkerman KEO, Nicholls DG: Ca^{2+} transport and the regulation of transmitter release in isolated nerve endings. Trends in Biochem. Sci. 8: 63-4, 1983.
14. Lin SC, Way EL: Calcium-activated ATPases in presynaptic nerve endings. J. Neurochem. 39: 1641-1651, 1982.
15. Ross DH, Cardenas HL: Calmodulin stimulation of Ca^{2+}-dependent ATP hydrolysis and ATP-dependent Ca^{2+} transport in synaptic membranes. J. Neurochem. 41: 161-171, 1983.
16. Campbell AJ: Intracelular calcium. Its universial role as regulator. John Wiley & Sons Ltd, 1983.
17. Blaustein MP, Ratzlaff RW, Kendrick NC, Schweitzer, ES: Calcium buffering in presynaptic nerve terminals. I. Evidence for involvement of a nonmitochondrial ATP-dependent sequestration

mechanism. J. Gen. Physiol. 72: 15-41, 1978.
18. Tillotson DL, Gorman ALF: Localization of neuronal Ca^{2+} buffering near plasma membrane studied with different divalent cations. Cell. Molec. Neurobiol. 3: 297-310, 1983.
19. Lüllmann, H, Peters T, Preuner J: Role of the plasmalemma for calcium homeostasis and for excitation-contraction coupling in cardiac muscle. In: AJ Drake-Holland and MIM Noble (eds) Cardiac Metabolism. John Wiley & Sons Ltd, 1983, p. 1.
20. Blaustein MP, Ratzlaff RW, Schweitzer, ES: Calcium buffering in presynaptic nerve terminals. II. Kinetic properties of the nonmitochondrial Ca sequestration mechanism. J. Gen. Physiol. 72: 43-66, 1978.
21. Somlyo AP: Cellular site of calcium regulation. Nature 309: 516-517, 1984.
22. Molgo J, Thesleff S: Studies on the mode of action of botulinum toxin type A at the frog neuromuscular junction. Brain Res. 297: 309-316, 1984.
23. Rahamimoff H, Abramovitz E: Ca transport and ATPase activity of synaptosomal vesicles from rat brain. FEBS Letts. 92: 163-167, 1978.
24. Israel M, Manaranche R, Marsal J, Meunier FM, Morel N, Frachon P, Lesbats B: ATP-dependent calcium uptake by cholinergic synaptic vesicles isolated from Torpedo electric organ. J. Membr. Biol. 54: 115-126, 1980.
25. Rephaeli A, Parsons SM: Calmodulin stimulation of $^{45}Ca^{2+}$ transport and protein phosphorylation in cholinergic synaptic vesicles. Proc. Nat. Acad. Sci. U.S.A. 79: 5783-5787, 1982.
26. Duncan CJ, Publicover SJ: Inhibitory effects of cholinergic agents on the release of transmitter at the frog neuromuscular junction. J. Physiol. 294: 91-103, 1979.
27. Miyamoto MD: The actions of cholinergic drugs on motor nerve terminals. Pharmacol. Rev. 29: 221-247, 1978.
28. Rahamimoff R, Yaari Y: Delayed release of transmitter at the frog neuromuscular junction. J. Physiol. 228: 241-257, 1973.
29. Balnave RJ, Gage PW: On facilitation of transmitter release at the toad neuromuscular junction. J. Physiol. 239: 657-675, 1974.
30. Datyner NB, Gage PW: Phasic secretion of acetylcholine at a mammalian neuromuscular junction. J. Physiol. 303: 299-314, 1980.
31. Kita K, Narita K, Van der Kloot W: Effects of temperature on the decline in miniature end-plate potential frequency following a tetanus. Brain Res. 190: 435-445, 1980.
32. Duncan CJ, Statham HE: Interacting effects of temperature and extracellular calcium on the spontaneous release of transmitter at the frog neuromuscular junction. J. Physiol. 268: 319-333, 1977.
33. Barrett EF, Barrett JN, Botz D, Chang DB, Mahaffey D: Temperature-sensitive aspects of evoked and spontaneous transmitter release at the frog neuromuscular junction. J. Physiol. 279: 253-273, 1978.
34. Publicover SJ, Duncan CJ: Dantrolene and the effect of temperature on the spontaneous release of transmitter at the frog neuromuscular junction. Experientia 37: 859-860, 1981.
35. Hubbard JI, Jones SF, Landau EM: The effect of temperature change upon transmitter release, facilitation and post-tetanic potentiation. J. Physiol. 216: 591-609, 1971.
36. Ward D, Crowley WJ, Johns TR: Effect of temperature at the neuromuscular junction. Am. J. Physiol. 222: 216-219, 1972.

37. Duncan CJ: Role of calcium in triggering the release of transmitters at the neuromuscular junction. Cell Calcium 4: 171-193, 1983.
38. Erulkar SD: The modulation of neurotransmitter release at synaptic junctions. Rev. Physiol. Biochem. Pharmacol. 98: 64-175, 1983.
39. Carlen PL, Kosower EM, Werman R: The thiol-oxidizing agent diamide increases transmitter release by decreasing calcium requirements for neuromuscular transmission in the frog. Brain Res. 117: 257-276, 1976.
40. Kita H, Van der Kloot W: Action of Co and Ni at the frog neuromuscular junction. Nature, New Biol. 245: 52-53, 1973.
41. Cull-Candy SG, Lundh H, Thesleff S: Effects of botulinum toxin on neuromuscular transmission in the rat. J. Physiol. 260: 177-203, 1976.
42. Molgo J, Lemeignan M, Lechat P: Effects of 4-aminopyridine at the frog neuromuscular junction. J. Pharmacol. Exp. Therap. 203: 653-663, 1977.
43. Colomo F, Rahamimoff R: Interaction between sodium and calcium ions in the process of transmitter release at the neuromuscular junction. J. Physiol. 198: 203-218, 1968.
44. Landau EM, Nachshen DA: The interaction of pH and divalent cations at the neuromuscular junction. J. Physiol. 251: 775-790, 1975.
45. Manalis RS, Cooper GP, Pomeroy SL: Effects of lead on neuromuscular transmission in the frog. Brain Res. 294: 95-109, 1984.
46. Pickett JB, Bornstein JC: Some effects of lead at mammalian neuromuscular junction. Am. J. Physiol. 246: C271-C276, 1984.
47. Uchiyama T, Molgó J, Lemeignan M: Presynaptic effects of bekanamycin at the frog neuromuscular junction. Reversibility by calcium and aminopyridines. Eur. J. Pharmacol. 72: 271-280, 1981.
48. Pecot-Dechavassine M: Action of vinblastine on the spontaneous release of acetylcholine at the frog neuromuscular junction. J. Physiol. 261: 31-48, 1976.
49. Dreyer F, Schmitt A: Transmitter release in tetanus and botulinum A toxin-poisoned mammalian motor endplates and its dependence on nerve stimulation and temperature. Pflugers Arch. 399: 228-234, 1983.
50. Kita H, Narita K, Van der Kloot W: The relation between tonicity and impulse-evoked transmitter release in the frog. J. Physiol. 325: 213-222, 1982.
51. Rubin LL, Gorio A, Mauro A: Effect of concanavalin A on black widow spider venom activity at the neuromuscular junction: implications for mechanisms of venom action. Brain Res. 143: 107-124, 1978.
52. Gorio A, Mauro A: Reversibility and mode of action of black widow spider venom on the vertebrate neuromuscular junction. J. Gen. Physiol. 73: 245-263, 1979.
53. Burgoyne RD, Geisow MJ, Barron J: Dissection of stages in exocytosis in the adrenal chromaffin cell with use of trifluoperazine. Proc. Roy. Soc. Lond. B 216: 111-115, 1982.
54. Publicover SJ: Presynaptic action of trifluoperazine at the frog neruromuscular junction. N.S. Arch. Pharmacol. 322: 83-88, 1983.

DISCUSSION OF THE ARTICLE BY C.J. DUNCAN (U.K.)

R. Rahamimoff (Israel): I am not sure that the experiments of Crawford about changing of the slope at various calcium concentrations can be taken as evidence of allosteric interaction because Ellen Berrett and her colleagues have pointed out quite rightly that in moving from one slope to another, Crawford also changed the rate of stimulation of the preparation very substantially. We know that in a higher rate of stimulation there are also sodium-dependent processes. It may well be that the slope doesn't change so abruptly and the allosteric interpretation may be perhaps still premature.

D.M.J. Quastel (Canada): With regard to the Crawford results, Cooke, Okamoto and I (J. Physiol 228, 459-497, 1973) published similar data the same year. In our hands, we found a sigmoid curve with no break in the slope at all, on a log-log plot.

MODULATION OF PRESYNAPTIC TRANSMITTER RELEASE BY ATP DERIVED FROM POSTSYNAPTIC SOURCES

D.O. SMITH AND C.A. LINDGREN

1. INTRODUCTION

Both ATP and adenosine have long been known to be neurotransmitters released from presynaptic cells. However, ATP may also be released from stimulated postsynaptic cells. This has been observed in frog (1) and human (2) muscle as well as *Torpedo* electric organ (3). Moreover, exogenous ATP has been shown to reduce transmitter release from presynaptic terminals of preganglionic sympathetic neurons (4) and motoneurons (5). Likewise adenosine also depresses synaptic transmission in the central nervous system (6,7). One prime explanation for this phenomenon might be that Ca^{2+} entry into the terminals is somehow altered. However, this seems unlikely, because the depressant effect is independent of extracellular Ca^{2+} concentrations (7). Thus, some other mechanism must be involved, and it remains unknown.

In this study, we have stimulated a presynaptic axon and have detected ATP release from the target postsynaptic cell. Furthermore, this results in increased ATP levels in the presynaptic terminal and an associated depression of transmitter release. These data suggest that ATP may indeed be a localized inhibitory modulator in this system.

2. METHODS

2.1. Experimental preparation

All experiments were performed on the efferent nerve innervating the opener muscle of the first walking leg of the crayfish (*Orconectes virilis*). In each preparation, the closer muscle and the blood vessels were removed, exposing the inner surface of the opener muscle. The nerve, consisting of one excitor and one inhibitor axon, could then be seen clearly as it projected along the surface of the muscle.

After exposing the nerve, the tissue was placed in a 23°C circulating bath of well-oxygenated saline solution (8) for 25 min. The entire preparation, including the opener muscle, its associated exoskeleton, and the nerve, was then quick frozen in liquid nitrogen either immediately (*nonstimulated*) or after repetitive stimulation (*stimulated*) of the nerve for 1 min at 50 Hz. Transfering the preparation from the bath to the liquid nitrogen takes about 5 s, during which time there is a negligible change in ATP levels in crayfish (9).

2.2. Electrophysiology

The nerve was stimulated in the meropodite region using a suction electrode. The efficacy of stimulation was monitored by recording action potentials from the nerve along the proximal surface of the opener muscle using microelectrodes filled with 4M NaCl. Synaptic activity was monitored by recording excitatory postsynaptic potentials intracellularly in the opener muscle using microelectrodes filled with 3M KCl. In addition, activity at single synaptic release sites was recorded using focal extracellular microelectrodes. The criteria for selecting recording sites and specific technical details are presented in Niles and Smith (10).

2.3. Assay procedures

2.3.1. ATP. To determine the ATP levels in a small sample of the nerve, it was necessary to dissect the appropriate tissue without allowing any significant metabolic activity to occur. Lyophilization of the tissue has been shown to prevent hydrolysis of high-energy phosphates, permitting the nerve to be dissected at room temperature without a decrease in ATP levels (11). Thus, the frozen specimens were placed in a lyophilizer and brought to a vacuum before they could thaw. Following this freeze-drying, a small nerve sample was dissected from the terminal region. The dissected nerve was then weighed. The dry weight of the samples was usually less than 1 ug.

The ATP was assayed using the luciferin-luciferase ("Firefly") method (12). After weighing the tissue, it was placed in 50 ul of 0.1N NaOH and vortexed. A 10-ul aliquot was then added to 100 ul of buffer solution containing 50-mM glycylglycine (pH 8.1), 7.5-mM dithiothreitol, 2-mM $MgCl_2$, 2-mM EGTA, and 0.04% bovine serum albumin. 50 ul of luciferin-luciferase (DuPont) was added. After swirling the preparation for 5 s, the reaction cuvette was placed in a photometer (Aminco), and the light intensity at 5 s was recorded. The amount of ATP represented by this peak was then calculated from standard curves according to the method described by Lust et al. (12). With this assay, quantities of ATP as small as 10^{-14} mol could be detected.

In some experiments, ATP levels in the bath solution were

assayed. A 50-ul sample drawn from the bath was added directly to the 50-ul of luciferin-luciferase in these cases.

2.3.2. Total adenylates. Levels of total adenylates (ATP, ADP, and AMP) were measured in a three-stage modification of the basic ATP assay. The lyophilized tissue was placed in 50 ul of 0.1N HCl and vortexed. A 10-ul portion of this tissue extract was placed into 200 ul of imidazol buffer solution containing 100 mM imidazol, 2 mM $MgCl_2$, 75 mM KCl, and 3mM phosphoenolpyruvate. The ATP levels in a 50-ul sample were then measured using the luciferin-luciferase procedure. To measure ATP+ADP, another 10 ul aliquot of the tissue extract was added to 200 ul of the imidazole buffer solution plus pyruvate kinase (25 units/ml of buffer solution), which catalyzes phosphorylation of ADP to ATP. This was allowed to incubate for 30 min at room temperature. The ATP content, representing tissue ATP+ADP in this case, was then assayed in a 50-ul sample. Total adenylates (ATP+ADP+AMP) were measured in a third 10-ul aliquot of tissue extract. Myokinase (7.5 units/ml imidazole buffer solution), which catalyzes formation of ADP from AMP, and pyruvate kinase were added to the incubation medium, and the tissue was allowed to incubate for 60 min at room temperature. ATP levels measured in these samples represent total adenylates in the tissue. The corresponding values for ADP and AMP content were calculated from these data.

2.3.3. Phosphoarginine. The phosphagen, phosphoarginine (PArg), was measured following its conversion to ATP *via* arginine kinase. Specifically, lyophilized tissue was placed in 50 ul of 0.1N NaOH. A 20-ul aliquot was added to 200 ul of an imidazole buffer containing 50-mM imidazole-HCl (pH 7.0), 1-mM $MgCl_2$, 0.6 uM purified ADP, and 0.02% bovine serum albumin. Following a 30-min incubation, tissue ATP levels were then assayed from this preparation. A second 20-ul aliquot of the tissue extract was added to 200 ul of this imidazole buffer plus arginine kinase (20 units/ml of buffer solution). After the 30-min incubation, ATP levels, representing tissue ATP+PArg, were assayed. The PArg values were calculated from these data.

The success of this PArg assay depends upon adequate purification of the ADP added to the reaction mixture. We used the enzymatic procedure specified on p. 153 in Lowry and Passonneau (11) without modification.

3. RESULTS

3.1. Tissue levels of high-energy phosphates

Tissue levels of ATP, ADP, AMP, and PArg obtained in a typical assay are presented in Table 1. The excitor and inhibitor axons could not be separately dissected in the terminal regions, so the assay measured their combined ATP levels. We were able to calculate concentrations, as mol/l of cell water, using the methods described in Smith (9).

Following stimulation of the excitor axon for 1 min at 50 Hz, the ATP content of the nerve increased from 13.4 to 19.0 nmol/mg dry weight, as shown in Table 1. This difference (42%) is statistically significant at the 0.05 level. The ADP and AMP levels also rose (Table 1). These changes represent a significant (0.05 level) increase in total adenylates. In contrast, the calculated energy charge did not differ significantly; the values were 0.87 and 0.86 in the nonstimulated and the stimulated tissue, respectively.

Similar results were observed in four other series of experiments.

Table 1. Levels of high-energy phosphates in single crayfish axons[a]

Compound	Nonstimulated		Stimulated[b]	
	nmol/mg dry wt	mM	nmol/mg dry wt	mM
ATP	13.4±1.5	2.4±0.3	19.0±2.1	3.4±0.4
ADP	4.7±1.1	0.8±0.2	6.4±0.8	1.1±0.1
AMP	0.1±0.1	0.02±0.02	0.3±0.2	0.1±0.04
Total	18.2±1.8	3.2±0.3	25.7±2.4	4.5±0.4
PArg	12.2±0.5	2.2±0.1	14.6±3.9	2.6±0.7

[a]: avg±s.e. (n=8); [b]: 50 Hz for 1 min

The ATP content was assayed in axons which were stimulated at 50 Hz for times ranging from 1 to 10 min. Smaller changes in ATP were observed following these longer stimulation times, but the levels were always higher than in nonstimulated cells. After 10 min of repetitive action potentials, the ATP measurements were not significantly different from those of nonstimulated cells.

An immediate source of the increased ATP might be the phosphagen, PArg. Thus, its levels were also assayed. The results, presented in Table 1, indicated that the PArg content did not change significantly during the 1-min stimulation. Indeed, the lack of any decrease in ADP indicates that the increased ATP is not derived from endogenous neuronal adenylates.

3.2. Control measurements of ATP levels in stimulated mammalian nerve

For comparison, we measured the effects of stimulation on ATP levels of rat phrenic nerve in a region about 1 cm proximal to the diaphragm muscle. Small nerve bundles, with an average (±s.d.)

dry weight of 123 ($\pm$88) ug, were dissected from 4 different animals. The ATP content was then measured in nonstimulated and stimulated tissue; stimulation was delivered at 20 Hz for 1 min. The average ($\pm$s.e.) values were 11.5 ($\pm$2.3) and 6.0 ($\pm$0.5) nmol/mg dry weight, respectively. In contrast to the crayfish results, stimulation resulted in a 48% reduction of the ATP levels in rat nerve. This difference is significant at the 0.05 level.

The observed drop in ATP levels during high-frequency repetitive stimulation of the rat nerve is consistent with data reported by previous investigators (13,14). However, the rise in ATP and total adenylates seen in single crayfish axons is unique, and further experiments were performed to elucidate the underlying mechanisms.

3.3. Dependence on synaptic transmission

Further investigation in crayfish indicated that the increase in ATP associated with nerve stimulation required synaptic transmission. Specifically, postsynaptic activation was necessary. Two series of experiments led to this conclusion.

3.3.1. Calcium dependence. The magnitude of the ATP increase during stimulation became less as $[Ca^{2+}]_e$ was decreased. Furthermore, addition of 10-mM Co^{2+} to the extracellular bath, which effectively blocks Ca^{2+} currents and transmitter release in this system, blocks the stimulation-induced rise in ATP levels. Verapamil (10^{-4}M) and Ruthenium Red (10^{-5}M), which were found in this study to decrease average quantal release by 31% and 35%, respectively, after 25 min also reduced considerably the amount by which ATP rose during stimulation.

Conversely, the increases in ATP levels observed during stimulation could be mimicked by incubation of nonstimulated tissue in saline containing the Ca^{2+} ionophore A23187. Tissue was bathed in normal saline for 20 min, and then A23187 (50 ug/ml) was added for 1 to 5 min. Reduction of the amount of ionophore, the incubation time, or the level of $[Ca^{2+}]_e$ resulted in smaller increases in ATP content.

3.4. <u>Dependence on postsynaptic activation.</u> Collectively, these results with altered Ca^{2+} levels indicate that presynaptic transmitter release is required for the stimulation-induced rise in ATP. This requirement was further characterized in a separate series of experiments in which antagonists of glutamate and γ-amino-butyric acid (GABA), the putative excitatory and inhibitory transmitters, were added to the bath. Picrotoxin (2 mM) was added to the bath, for it effectively blocks the inhibitory responses. It has proved more difficult to block the excitatory junctional potential completely. However, γ-methyl-glutamate (2 to 20 mM) was found to reduce postsynaptic potential amplitudes by about 50% (cf. 15). Thus, it was added to the bath.

Axonal ATP did not increase during stimulation under any of these conditions. This strongly argues that presynaptic ATP levels are associated with postsynaptic activation.

3.5. <u>Postsynaptic release of ATP</u>

By demonstrating that increased presynaptic ATP resulted from postsynaptic activation, these results suggested that ATP might be transfered <u>via</u> the bath. Further experiments tested whether ATP was, indeed, released into the bath solution during repetitive stimulation. In 33 experiments using 12 different animals, 50-ul aliquots of the bath saline were drawn into a pipette tip placed less than 1 mm from the opener muscle surface. The amount of ATP in these samples was measured immediately before and after nerve stimulation.

The average ATP content of these 50-ul bath samples increased by 33% following nerve stimulation. To determine whether the increase was derived from postsynaptic sources, ATP was measured after fresh solution containing 20-mM γ-methyl-glutamate and 2-mM picrotoxin were introduced to the bath. The bath levels of ATP did not increase significantly following nerve stimulation. Thus, postsynaptic sources are the origin of most ATP released into the bath during stimulation. Furthermore, addition of the blocking agents resulted in a 42% decrease in the background ATP levels in the bath. This suggests that resting levels of glutamate or GABA leakage may lead to a sizeable release of ATP in the absence of stimulation.

In separate control experiments, the stability of ATP in the bath was studied. Specifically, measured quantities of ATP (ranging from 10 to 20 pM) were added either to saline alone or saline along with a dissected preparation. Bath ATP levels were then assayed, and recovery was found to be virtually 100% as late as 100 min after addition of the ATP.

3.6. Presynaptic accumulation of ATP in the bath

Studies were then performed to determine whether ATP levels in the bath affect axonal ATP content. ATP, in concentrations ranging from 1 to 10 mM, was added to the bath solution, and the entire preparation was allowed to incubate for 3 min. The tissue was then rinsed in distilled water, and axonal ATP was measured. ATP associated with the axon increased from background levels (ranging from 5 to 8 nmol/mg dry wt in this series of experiments) to average (±s.e.) levels of 14.3±1.5 and 28.3±3.5 nmol/mg dry wt in the presence of 5- and 10-mM ATP, respectively. These differences are statistically significant (0.05 level).

These studies suggested that the observed increase in nerve ATP was quite probably due to transfer from the muscle tissue *via* the extracellular fluid. It was important, then, to determine the apportionment of exogenous ATP between intra- and extracellular spaces of the isolated nerve. Thus, tissue samples from 8 different preparations were incubated for 3 min in bath solutions containing ^{14}C-sucrose (673 mCi/mmol) and ^{3}H-ATP (26 Ci/mmol). The concentrations of sucrose and ATP added to the bath were 0.2 mM and 5 mM, respectively. The tissue ATP and the activities of ^{3}H-ATP and ^{14}C-sucrose in the dissected nerve were then assayed using the luciferin-luciferase method and a scintillation counter, respectively. The sucrose space, the ATP space, and the total ATP within these spaces were calculated from the ^{14}C, the ^{3}H, and the ATP content of the tissue sample and the bath saline.

The results of these experiments indicated that the ATP within the extracellular (sucrose) space could account for 40% of the total ATP associated with the nerve. When this was subtracted from the total nerve ATP, the average net values were still significantly higher than ATP levels observed in the absence of exogenous ATP (5.4 ±0.8 nmol/mg dry wt in this series of experiments). This conclusion was verified by measurements of the ATP space, which was found to be 1.94 times larger than the sucrose space. These data are consistent with the conclusion that exogenously applied ATP becomes associated with the intracellular space of the nerve as well as the extracellular space.

The possibility that exogenous adenosine (1 mM) might also lead to increased axonal ATP levels was tested in similar experiments and ruled out. Indeed, ATP content decreased by 35%.

3.7. Effects of exogenous ATP on transmitter release

To test the effects of elevated levels of extracellular ATP on transmitter release, the average quantal release (m) was measured from single synaptic sites using focal extracellular electrodes. Data were first obtained in normal saline and then, from the same site, 12 to 15 min after addition of ATP to the bath. Sufficient Ca^{2+} was also added to compensate for the chelating effect of the ATP.

In the presence 0.5- and 5-mM ATP, average quantal release was reduced by 82 and 76%, respectively. These reductions were statistically significant (0.05 level). Addition of neither ADP (0.5 mM), AMP (0.5 mM), nor adenosine (1mM) had any significant effect on quantal release. Furthermore, spontaneous quantal release rate was not affected by any of these agents. Thus, ATP uniquely depressed evoked quantal release in these experiments.

4. DISCUSSION

This study demonstrates that ATP levels in the terminal region of a nerve increase during stimulation. The observed adenylate levels in nonstimulated tissue, which generally ranged from 3 to 4 mM, are comparable to those reported in other studies. Following stimulation, the ATP and total adenylate values were increased by 1.0 and 1.3 mM, respectively. Such large increases--about 40%--have not previously been documented. Slightly increased levels of ATP in nervous tissue during electrical stimulation have been reported in other preparations. At moderate stimulation rates ($\leq$30 Hz), ATP content has been seen to rise by 8% and 17% in mixed nerve bundles from rabbit (13) and crayfish (9), respectively. However, these changes were not significant statistically in either case. Furthermore, in the rabbit, stimulation at 50 Hz for 15 s led to a drop in both ATP and phosphocreatine; this is similar to the results observed in rat phrenic nerve in this study.

The results obtained from crayfish indicate that the increased neuronal ATP may well result from activation of the underlying muscle. Experiments showed that the muscle releases ATP into the extracellular space during stimulation of the presynaptic nerve. Moreover, elevation of ATP concentrations in the bath results in increased nerve ATP levels. Specifically, 40% of this increase could be attributed to ATP within the extracellular space surrounding the nerve while the remainder entered the intracellular compartment of the nerve. This conclusion was supported by the observation that the ATP space was nearly twice the volume of the sucrose space.

A potential difficulty with this argument warrants consideration. Namely, 10-mM exogenous ATP was required to increase neuronal ATP by the same amount that was observed following 1 min of 50-Hz stimulation (7.5 pmol). During this period, though, bath ATP only rose to 12 nM. However, the results suggest that local ATP concentrations may, indeed, be much higher. Namely, the 7.5 pmol increase in neuronal ATP is associated with an extracellular water (or sucrose) space of 0.67 nl. If 40% of the total ATP increase is in this space, then the extracellular ATP within this region would rise by 3.0/0.67 pmol/nl, or about 4.5 mM. Thus, the effects of nerve stimulation on neuronal ATP are consistent with a distribution of 4.5 mM increased extracellular ATP in the region of the nerve and an intracellular accumulation of 4.5 pmol/ug dry wt. This is, in fact, quite similar to the observed neuronal accumulation of ATP when the tissue was exposed to 5-mM exogenous ATP.

In summary, the results are consistent with the following mechanism which explains the changes in neuronal ATP during stimulation. During synaptic activation, ATP is released into the extracellular space due to the action of glutamate or GABA on the receptors. This results in a large increase in local ATP concentration within the extracellular space surrounding the muscle, and some of this ATP is taken up by the nerves. This phenomenon may also occur in other preparations, for postsynaptic release of ATP has been reported in frog (1), Torpedo (3), and human (2).

The role of this ATP released from the muscle is not clear. However, It may play a role in the regulation of axon terminal excitability and synaptic transmission, for quantal release is depressed by exogenously applied ATP at concentrations similar to those expected to be released from the muscle into the periaxonal space. Furthermore, ATP and its related nucleoside, adenosine, have been shown to depress presynaptic activity (4-6), and adenosine, in particular, inhibits calcium spikes (16). These congruent observations indicate that ATP emanating from the muscle may decrease axonal excitability and/or synaptic transmission through a negative-feedback mechanism.

5. ACKNOWLEDGMENTS

This work was supported by NIH grant NS13600. Technical assistance of Ms. Curry Gibson and Mr. Michael Chapman are gratefully acknowledged. The Verapamil was kindly provided by Knoll Pharmaceutical Company.

6. REFERENCES

1.Abood LG, Koketsu K, Miyamoto S: Outflux of various phosphates during membrane depolarization of excitable tissues. Am. J. Physiol. 202:469-474, 1962.

2.Forrester T: An estimate of adenosine triphosphate release into the venous effluent from exercising human forearm muscle. J. Physiol. (Lond.) 224:611-628, 1972.

3.Israel M, Lesbats B, Meunier FM, Stinnakre J: Postsynaptic release of adenosine triphosphate induced by single impulse transmitter action. Proc. R. Soc. Lond. B 193:461-468, 1976.

4.Silinsky EM, Ginsborg BL: Inhibition of acetylcholine release from preganglionic frog nerves by ATP but not adenosine. Nature 305:327-328, 1983.

5.Ribeiro JA, Walker J: The effects of adenosine triphosphate and adenosine diphosphate on transmission at the rat and frog neuromuscular junctions. Br. J. Pharmac. 54:213-218, 1975.

6.Schubert P, Lee K, Kreutzberg GW: Neuronal release of adenosine derivatives and modulation of signal processing in the CNS. Prog. Brain Res. 55:225-237, 1982.

7.Dunwiddie TV: Interactions between the effects of adenosine and calcium on synaptic responses in rat hippocampus in vitro. J. Physiol. (Lond.) 350:545-559, 1984.

8.Smith DO: Extracellular potassium levels and axon excitability during repetitive action potentials in crayfish. J. Physiol. (Lond.) 336:143-157, 1983.

9.Smith DO: Levels of high-energy phosphates in crayfish nerve during prolonged repetitive impulse activity.J. Physiol. (Lond.) 301:271-280, 1980.

10.Niles WD, Smith DO: Effects of hypertonic solutions on quantal transmitter release at the crayfish neuromuscular junction. J. Physiol. (Lond.) 329:185-202, 1980.

11.Lowry O, Passonneau JV: A Flexible System of Enzymatic Analysis. Academic Press, New York, 1972, Pp.223-235.

12.Lust WD, Feussner GK, Barbehenn EK, Passonneau JV: The enzymatic measurement of adenine nucleotides and P-creatine in picomole amounts. Anal. Biochem. 110:258-266, 1980.

13.Greengard P, Straub RW: Effect of frequency of electrical stimulation on the concentration of intermediary metabolites in mammalian non-myelinated fibers. J. Physiol. (Lond.) 148:353-361, 1959.

14.Montant P, Chmouliovsky M: Energy-rich metabolites in stimulated mammalian non-myelinated nerve fibers. Experientia 24:782-783, 1968.

15.Lowagie C, Gerschenfeld HM: Glutamate antagonists at a crayfish neuromuscular junction. Nature 248:533-535, 1970.

16.Proctor WR, Dunwiddie TV: Adenosine inhibits calcium spikes in hippocampal pyramidal neurons in vitro. Neurosci. Letters 35:197-201, 1983.

DISCUSSION OF THE ARTICLE BY D.O. SMITH (U.S.)

I. Parnas (Israel): We found that when we add glutamate to the crustacean neuromuscular system, we block release. It occurs to me now that it is possible that glutamate excites release of ATP and does indirectly block release. Did you try to find if you just add glutamate to the bath, does it produce the same uptake or release of ATP?

D.O. Smith (U.S.): When we apply glutamate to the bath, we do see released ATP from the muscle. Furthermore, when we add γ-methyl glutamate to the non stimulated tissue, we see a slight elevation of ATP.

I. Parnas (Israel): So we have a problem now. When we add a material that blocks release, how do we know that it acts directly on the terminal to reduce release or does it release something else from the tissue. It will complicate things now tremendously.

D.O. Smith (U.S.): We have no answer to that at this point.

B. Katz (U.K.): If you want to make the muscle excitable so you can test the effect of direct simulation, probably, a couple of mM of TEA might do the trick. Anyway, it does it in other crustacean muscles. You could add TTX and magnesium to prevent activation of the nerve terminals.

E. Heilbronn (Sweden): Does your system have enzymes breaking down ATP in the synaptic cleft?

D.O. Smith (U.S.): We looked at it indirectly by examining how long the ATP in the bath remains unhydrolyzed. We can add ATP to the bath, in the absence of stimulation, for as long as 5, 10 and sometimes even 30 minutes and then record the ATP in the bath and you find similar levels.

E. Heilbronn (Sweden): So you would not think you have a system similar to what Zimmerman was discussing in Hungary; breaking down to adenosine and having that presynaptically taken up and rephosphorylated?

D.O. Smith (U.S.): No. ATP is quite stable in this preparation over the time periods we have looked at.

E. Heilbronn (Sweden): So how do you get ATP in the presynapse?

D.O. Smith (U.S.): We don't know yet. I might add that the effect of ATP on synaptic transmission is quite dependent on having glucose in the Ringers. If we repeat these experiments looking at quantal release after adding ATP to the bath without glucose present, we don't see the same results. In glucose free medium, the ATP will sometimes increase average quantal release. Our hypothesis is that we have an unhealthy cell under those conditions and we are simply improving the ATP content of the cell.

R. DeLorenzo (U.S.): You have some very interesting results. I think the role of ATP is going to be very intriguing. I just want to point out that the adenosine story is not very clear. There is controversy in the literature from different laboratories as to whether adenosine really does regulate calcium uptake. My point is that your very exciting results may be related to the adenosine effect because adenosine, in some systems, has not been able to block calcium uptake. Therefore, you may have a parallel between ATP and adenosine which acts electrically the same.

D.O. Smith (U.S.): This is a good point.

R. Rahamimoff (Israel): Do you know the chelating constant of ATP for calcium?

D.O. Smith (U.S.): Yes, I do, but I don't have it right here.

R. Rahamimoff (Israel): Maybe with ATP that is released if it is in concentrations of 5-10mM, you are doing local buffering of calcium in the extracellular space and by that affecting transmitter release.

D.O. Smith (U.S.): That is something we thought about. In all our experiments we have corrected the calcium concentrations for this chelating effect. One thing we would like to do is repeat these experiments in which we don't correct the calcium concentrations for the chelation. Another thought would be to repeat your experiments that were reported this morning in which you buffered the extracellular calcium.

REGULATION OF VOLTAGE-SENSITIVE CALCIUM CHANNELS IN BRAIN BY MICROMOLAR AFFINITY BENZODIAZEPINE RECEPTORS

R.J. DeLORENZO, W.C. TAFT, W.T. ANDREWS

1. INTRODUCTION

Ca^{2+} is an important mediator of molecular events in the functioning nerve terminal (1,2). An increase in cytoplasmic Ca^{2+} has been shown to regulate a variety of neuronal cell functions, including neurotransmitter release, stimulation of protein phosphorylation, and synaptic morphological changes (3). It is well-established that these events are directly dependent on the entry of extracellular Ca^{2+} into the presynaptic nerve terminal through specific voltage-gated Ca^{2+} channels (4,6). The central role of Ca^{2+} channels in stimulus-secretion coupling mechanisms has led to the extensive investigation of Ca^{2+} channel function in a wide variety of intact and broken cell preparations (7). Despite extensive electrophysiological characterization of Ca^{2+} channels in brain, the molecular nature and pharmacological characteristics of this major neuronal Ca^{2+} channel remain largely unclear.

1.1 Calcium-Channels In Non-Neuronal Tissue

The properties of Ca^{2+} channels in some peripheral tissues, particularly in the heart and vascular smooth muscle, have been extensively studied (7-10). Ca^{2+} regulates muscle function in both smooth and cardiac muscle. As in brain, the Ca^{2+} channels in these tissues are voltage-sensitive, although channel activity can also be modulated by hormones, neurotransmitters and various drugs (9). These channels are blocked by polyvalent cations such as Mn^{2+}, Co^{2+}, and La^{3+}. In addition, several distinct classes of organic Ca^{2+} channel antagonists have been identified in the cardiovascular systems, including verapamil, diltiazem, and the dihydropyridines, nitrendipine and nimodipine. The ability of the dihydropyridines to act as Ca^{2+} channel antagonists has led to their clinical use in the treatment of cardiac arrythmias, hypertension and other cardiovascular disorders. Further, these drugs have been used as specific pharmacological probes for the study of Ca^{2+} channels in heart and smooth muscle preparations. A high affinity [^{3}H] nitrendipine binding site has been described in heart membrane that has binding properties that correlate well with nitrendipine inhibition of Ca^{2+} flux (11). This observation suggests that the [^{3}H] nitrendipine binding site may represent the voltage-sensitive Ca^{2+} channel in that tissue and represent an initial insight into the molecular identity of this receptor site.

1.2 Organic Calcium-Channel Blockers In Brain

Voltage-sensitive Ca^{2+}-channels in brain are also blocked by polyvalent cations (12). Because of the success of verapamil and dihydropyridines as Ca^{2+} channel probes in cardiac and smooth muscle, these compounds have been employed to study Ca^{2+} channels in brain. In brain preparations, specific high affinity [^{3}H] nitrendipine binding sites have been observed which have similar pharmacological properties to those found in the heart (13). However, the function of these drugs as Ca^{2+} channel blockers in brain seems to be substantially different. Dihydropyridine Ca^{2+} antagonists do not block voltage-sensitive Ca^{2+} uptake in synaptosomes (14) at concentrations where binding to the "nitrendipine receptor" is maximal. Although some studies have appeared which show nitrendipine inhibition of Ca^{2+} uptake in cultured neuronal cell lines, nitrendipine does not block neurotransmitter release in these cells (15) or neurons (16). These observations suggest that the functional characteristics of Ca^{2+} channels in brain are more complex than in heart, and that multiple types of Ca^{2+} channels may exist.

1.3 Benzodiazepine Receptors

Benzodiazepines (BZs) are important therapeutic agents with precise cellular and molecular activities, which have been the object of intense investigation. Specific BZ binding sites have been described that bind BZs in the nanomolar concentration range (17). Nanomolar-affinity BZ receptors (nM BZRs) have been linked on the molecular level with the γ-aminobutyric acid (GABA) receptor and the chloride ion channel. Furthermore, nM BZR binding has been correlated with some of the pharmacological effects of the BZs, including inhibition of pentylenetetrazol-induced seizure activity, muscle relaxant properties, and impairment of mouse rotorod performance. Studies from this laboratory (18) have identified and characterized a distinct class of specific BZ binding sites that bind BZs in the micromolar concentration range, designated micromolar-affinity BZ receptor (uM BZRs). Binding to uM BZRs was readily distinguished from nM BZR binding on the basis of kinetic and pharmacological properties (18). In addition, uM BZR binding potency correlates with BZ inhibition of maximal electric-shock-induced seizures, but not with other effects correlating with nM BZR binding. Thus, it was suggested that uM BZRs may play a role in mediating some of the neuronal stabilizing effects of the BZs (18).

1.4 Benzodiazepines Act Like Calcium Channel Blockers In Brain

Benzodiazepines (BZs) have been shown by several investigators to block Ca^{2+} uptake in nerve terminal preparations (19-22). In addition, BZs inhibit Ca^{2+}-dependent neurotransmitter release in synaptosomes (3), indicating that they regulate functionally significant Ca^{2+} influx. These BZ effects are reversible, dose-dependent, and occur in the micromolar (uM) concentration range. Subsequently a specific class of uM affinity BZ binding sites has been identified and characterized in rat brain membrane preparations (18). These uM BZ binding sites mediate BZ inhibition of synaptosomal Ca^{2+} uptake, and therefore, may serve as important regulators of neuronal cell function. In this report, we will describe our efforts to study BZs as Ca^{2+} channel antagonists in brain, and our pursuit of the relationship between the uM BZ binding site and the voltage-sensitive Ca^{2+} channel.

2. MICROMOLAR AFFINITY BENZODIAZEPINE RECEPTORS

2.1. Comparison of nM and uM BZ Receptors

[^{3}H] Diazepam (DZ) binding in rat brain membrane saturates in both the nM and uM concentration ranges, corresponding to binding to nM affinity and uM affinity binding sites (Fig. 1). Binding data analyzed by the method of Scatchard yields a curvilinear plot consistent with the existence of two distinct populations of specific binding sites. The apparent K_D for DZ binding at the uM site is 45 uM and the receptor density (Bmax) obtained from the plot is 360 pmole/mg. [^{3}H] DZ binding is stereospecific, has defined kinetic parameters, and has the characteristics of a general BZ receptor. A wide variety of BZs displace specific [^{3}H] DZ binding in a concentration-dependent fashion including clonazepam (CNZ), flunitrazepam (FNZ), medazepam (MDZ), and Ro5-4864. The order of potency (K_I) of BZs at uM affinity binding sites is markedly different from the

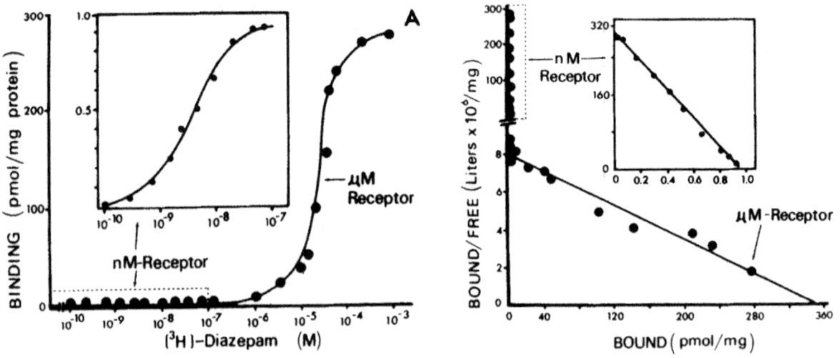

FIGURE 1. A. Saturation curve demonstrating the presence of two distinct classes of saturable benzodiazepine receptors. B. Scatchard transformation of the saturation data showing different binding parameters for the two receptors. Reproduced from reference 21.

potency series at the nM affinity site, indicating that the two receptor populations are pharmacologically distinct.

The apparent K_D for DZ binding at the nM site is 1-2 nM and the receptor density (Bmax) is in the range of 1 pmole/mg. These pharmacological characteristics clearly distinguish the nM BZ receptor from the uM BZ binding site. In addition, the relative potency of DZ and the peripheral BZ, Ro5-4864, varies significantly at the two sites. At the nM binding site DZ is 10,000 fold more potent than Ro5-4864, whereas at the uM binding site the two compounds are essentially equipotent. The two classes of BZ receptors also differ functionally. The nM affinity BZ receptor has been shown to exist in a supramolecular structure with the GABA receptor and the Cl^- ion channel, and its binding is influenced by GABAergic ligands. It has been suggested that a function of the nM affinity receptor is to regulate the Cl^- ion channel. In contrast, concentrations of GABAergic ligands which modulate nM BZ binding and Cl^- fluxes have no effect on uM BZ binding. Thus, the uM BZ binding site does not function in a similar fashion to the activated nM affinity site. Due to the observation that uM BZs block Ca^{2+} uptake in synaptosomes, it is of considerable interest to elucidate the cellular functions associated with uM BZ binding.

2.2. Methodological Considerations.

The investigation of binding of a lipophilic drug in uM concentration ranges presents some interesting technical challenges. We have developed a reproducible binding assay by adjusting reaction conditions to minimize technical difficulties. The principle problems encountered with studies of this type are low aqueous solubility of the drugs, high non-specific binding, and high cost of the radioligand used in uM concentrations. First, to reduce solubility problems, stock solutions of all BZs are prepared in absolute ethanol immediately prior to use. Aliquots of the BZs in ethanol are then added first to reaction tubes and are then diluted to final concentrations with aqueous buffer containing the receptor-membrane preparation. Final ethanol concentration routinely ranges from 5-10%. Increasing final ethanol concentrations dramatically reduces specific BZ binding and should be avoided. Second, high non-specific binding to membrane preparations is expected when dealing with uM radioligand concentrations. We observe that 25-30% of total [^{3}H] DZ or [^{3}H] CNZ binding is specific. As a consequence, it is necessary to improve the reproducibility of the assay by minimizing the variable non-specific binding of the radioligand to filters. Therefore, only glass fiber filters presoaked in aqueous buffer are used, and we routinely dilute (100X with ice-cold aqueous buffer) all radioactive samples immediately prior to filtration. This technique effectively reduces free radioligand concentration at the moment of filtration and lowers non-specific filter binding. Third, to reduce the costs associated with binding assays using uM radioligand concentrations, a displacement assay is employed to measure uM BZ binding. Using this technique, we compare radioligand (5-10 Ci/mmole) binding in a series of samples containing identical concentrations of radioligand (usually 1-10 uM) and increasing concentrations of unlabeled ligand. Thus we routinely measure total [^{3}H] DZ (1 uM) binding and displace the specific component of that binding with 1-1000 uM unlabeled DZ. Maximum displacement can be defined as non-specific binding and is generally 70-75% of total binding. Specific [^{3}H] DZ binding is then calculated by subtracting non-specific binding from each data point and correcting for the dilution of [^{3}H] DZ specific activity. The displacement data can then be directly replotted according to Scatchard for comparison and analysis.

To further increase the reproducibility of this assay, and to improve its value as a molecular probe of the uM BZ binding site, we have explored the use of photoactive BZs and the formation of irreversible bonds. Nitro-containing BZs, such as CNZ and FNZ, form covalent linkages with their binding

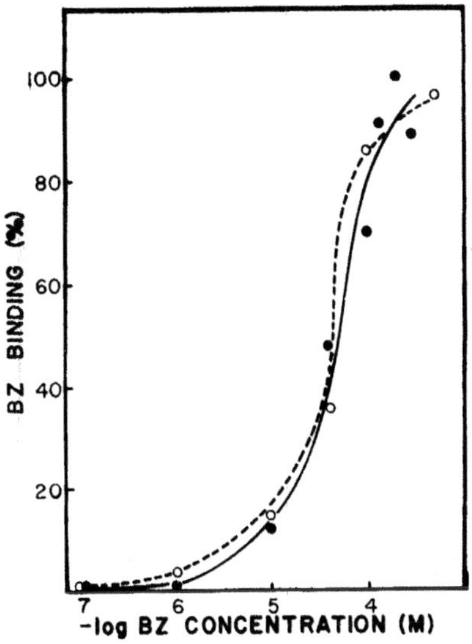

FIGURE 2. Comparison of specific [^{3}H] diazepam membrane binding (○) and irreversible specific [^{3}H] FNZ binding (●). The data give the mean of 6 determinations and represent percent of maximal binding. Reproduced from reference 21.

sites when exposed to short-wave ultraviolet (UV) light (23). Samples are then equilibrated with [^{3}H] CNZ and CNZ under the conditions described above, and exposed uniformly to UV illumination (10 min. at 5 cm) prior to filtration. The formation of irreversible bonds between [^{3}H] CNZ and its binding sites allows increased dilution of samples prior to filtration and more complete washing of the filters without risking dissociation of bound radioligand-receptor complexes. Using this technique, saturable UV-induced uM [^{3}H] CNZ binding is obtained which has similar pharmacological properties to that observed with the standard reversible [^{3}H] DZ binding assay (Fig. 2). Although we are continuing to pursue the pharmacological relationship between reversible BZ binding and irreversible UV-induced BZ binding, the photoaffinity assay shows considerable promise as a technique for reproducible analysis of uM BZ binding sites and investigation of the molecular nature of the receptor.

3. BENZODIAZEPINE INHIBITION OF CALCIUM UPTAKE

3.1. Pharmacological characteristics. To investigate the

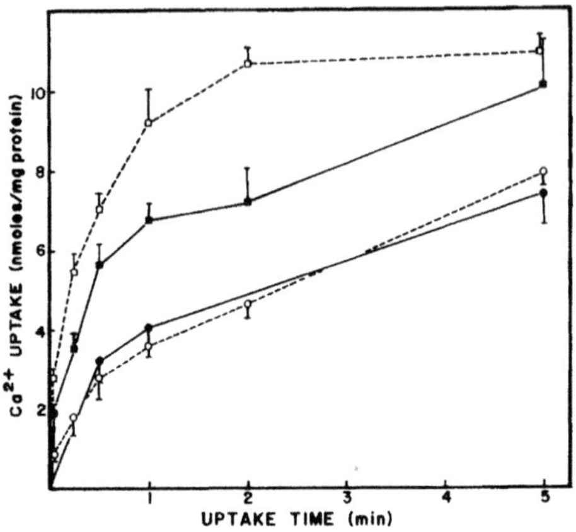

FIGURE 3. Effects of DZ on synaptosomal Ca^{2+} uptake. Data represents $^{45}Ca^{2+}$ uptake into synaptosomes in the presence of low (5 mM) K^+ (○) and high (70 mM) K^+ (□). The effect of DZ (150 uM) on Ca^{2+} uptake in the presence of low K^+ (●) and high K^+ (■) is also illustrated. Values represent the mean of 8 determinations and are representative of 3 experiments (errror bars represent SEM). Reproduced from 21.

effects of BZs on synaptosomal Ca^{2+} uptake, $^{45}Ca^{2+}$ flux was monitored under both depolarized and non-depolarized conditions (21). Calcium uptake is time-dependent and dramatically stimulated by depolarizing concentrations of K^+ (70 mM). Benzodiazepines inhibit the depolarization-dependent component of Ca^{2+} uptake, suggesting that the Ca^{2+} uptake inhibited by BZs enters the nerve terminal via the voltage-sensitive Ca^{2+} channel (Fig. 3). Benzodiazepine inhibition is concentration-dependent in the uM range, and is stereospecific. Kinetic studies show clearly that BZs block only K^+-induced Ca^{2+} influx, and do not affect the efflux of Ca^{2+} under our experimental conditions (21).

Calcium influx due to synaptosomal depolarization occurs predominantly through voltage-sensitive Ca^{2+} channels, but it may also be via TTX-sensitive voltage-dependent Na^+ channels (24). To determine whether BZs inhibit Ca^{2+} uptake by blocking Ca^{2+} and/or Na^+ channels, we studied the effects of DZ on high K^+-induced Ca^{2+} uptake in the

Table 1. Effect of veratridine (V), tetrodotoxin (TTX), and diazepam (DZ) on synaptosomal Ca^{2+} uptake.

Condition	^{45}Ca uptake (% of maximum)
low K^+	41 ± 6
high K^+	100
high K^+ + DZ (200 uM)	52 ± 3
low K^+ + V (50 uM)	91 ± 4
low K^+ + V (50 uM) + DZ (200 uM)	64 ± 7
low K^+ + V (50 uM) + TTX (50 uM)	61 ± 5
high K^+ + TTX (50 uM)	96 ± 4
high K^+ + TTX (50 uM) + DZ (200 uM)	58 ± 6

presence of a concentration of TTX that was shown to block veratridine-induced depolarization. Several lines of evidence confirm that BZ-sensitive Ca^{2+} uptake is mediated by the voltage-sensitive Ca^{2+} channel (Table 1). Veratridine, which induces depolarization by opening Na^+ channels, effectively stimulates Ca^{2+} uptake. Veratridine-induced Ca^{2+} uptake is also blocked by BZs, demonstrating that BZs block depolarization-induced Ca^{2+} flux. The Na^+ channel blocker, tetrodotoxin (TTX), is effective at blocking veratridine-stimulated depolarization, but does not inhibit high K^+-induced Ca^{2+} uptake. Further, DZ (150 uM) is equally effective at inhibition of high K^+-induced Ca^{2+} uptake in the presence or absence of TTX (50 uM). These results indicate that Ca^{2+} uptake induced by high K^+, and its inhibition by BZs, are not mediated by the TTX-sensitive Na^+ channel. We have also investigated the effect of specific voltage-sensitive Ca^{2+} uptake. In additon, the organic Ca^{2+} antagonist verapamil (100 uM) and the dihydropyridine nitrendipine (200uM) had some effect as Ca^{2+} channel inhibitors. In summary, these investigations indicate that BZ inhibition of Ca^{2+} accumulation in synaptosomes is via blockage of Ca^{2+} uptake induced by depolarization and may be mediated by voltage-sensitive Ca^{2+} channels.

3.2 BZs Act as Calcium Channel Antagonists

Voltage-sensitive Ca^{2+} channels in synaptosomes and other neuronal preparations are blocked by Mn^{2+} (12,14). Under our experimental conditions (Table 2), Mn^{2+} characteristically blocked depolarization-induced Ca^{2+} uptake in synaptosomes. Millimolar concentrations of Mn^{2+} inhibited high K^+-induced Ca^{2+} uptake by >90%. Micromolar concentrations of DZ also blocked Ca^{2+} uptake by >95%. In addition, BZ inhibition was additive with that of Mn^{2+} (Table 2). These results indicate that BZs act as Ca^{2+}-channel antagonists in nerve terminal preparations.

Table 2. Comparison of the effects of BZs and Ca^{2+} channel antagonists on K-induced synaptosomal Ca uptake

Agent	K^+-induced Ca^{2+} uptake, % inhibition
DZ (0.2)	56 ± 5.3
DZ (0.6)	95 ± 4.6
Mn^{2+} (1.0)	63 ± 4.1
Mn^{2+} (10)	93 ± 3.3
Mn^{2+} (1.0) and DZ (0.2)	86 ± 4.8

3.3 Irreversible uM BZ Binding Blocks Calcium Uptake

Benzodiazepine inhibition of voltage-sensitive Ca^{2+} uptake in synaptosomes has many of the characteristics of mediation by uM BZ binding sites. Inhibition and binding are both concentration-dependent, saturable in mM concentration ranges, and stereospecific. The potency of Ro5-4864 was similar to that of DZ in both Ca^{2+} uptake and BZ binding studies.

The effect of irreversible BZ binding initiated by UV irradiation on Ca^{2+} flux was also examined. Intact synaptosomes were equilibrated with CNZ or FNZ and then

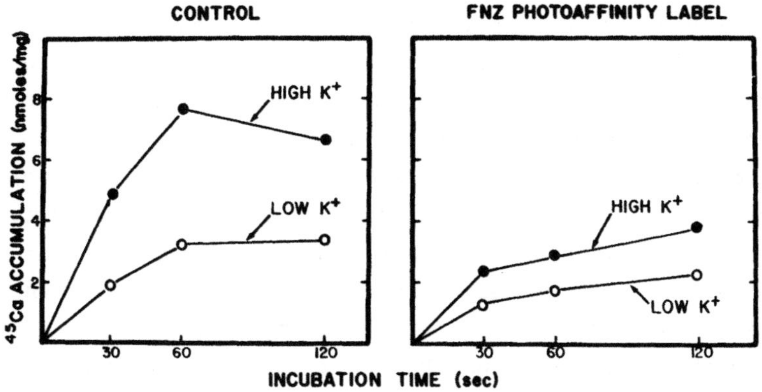

FIGURE 4. Synaptosomal Ca^{2+} uptake in the presence of high and low K^+ in control (left) and irreversibly FNZ labeled (right) synaptosomes. The data represent the means of 4 determinations. Irreversible FNZ labeled synaptosomes showed over 80% inhibition of depolarization dependent Ca^{2+} uptake. Synaptosomes were subjected to FNZ labeling and UV irradiation under identical conditions employed to obtain the FNZ saturation curve (Fig. 2).

exposed uniformly to UV light (21). The photoaffinity labeled synaptosomes were then washed free of unbound BZ and examined for $^{45}Ca^{2+}$ uptake properties. Synaptosomes with UV-induced BZ binding showed markedly reduced depolarization induced $^{45}Ca^{2+}$ uptake in comparison to control preparations (Fig. 4). This effect of membrane-bound BZs on Ca^{2+} influx occurred in the absence of free BZ in the reaction media. This observation indicates that the Ca^{2+} channel antagonist properties of the BZs are mediated by the interaction of these compounds with a population of uM BZ binding sites. In addition, this data strongly suggest that UV-induced BZ binding may serve as an effective probe into the molecular nature of the voltage-sensitive Ca^{2+} channel in brain.

4. MANGANESE INHIBITION OF uM BZ BINDING

These results indicate that uM BZ receptor binding in synaptosome preparations regulates depolarization induced Ca^{2+} uptake in a similar manner to Mn^{2+} . Mn^{2+} is a well established Ca^{2+} channel blocker. Furthermore, Mn^{2+} is believed to directly interact with the Ca^{2+} channel and actually insert and bind to the pore of the channel. Thus, it would be important to determine if Mn^{2+} affects BZ binding to the uM BZ receptor. Recent experiments in this laboratory have demonstrated that 2 mM Mn^{2+} significantly inhibits uM BZ receptor binding. The effect of Mn^{2+} is dose dependent and reversible. These observations suggest that the uM BZ receptor may be closely linked or part of the Ca^{2+} channel. However, further studies are necessary to determine the exact relationship of this uM BZ receptor and the voltage gated Ca^{2+} channel.

5. NITRENDIPINE INHIBITS uM BZ BINDING AT HIGH CONCENTRATIONS

It has been demonstrated that Ca^{2+} channels in synaptosomes and neurons are much less sensitive to verapamil, nitrendipine and other organic calcium antagonists than in heart or smooth muscle cells (15,16). Electrophysiologically measured Ca^{2+} currents in neurons are blocked only by very high concentrations of these organic calcium antagonists or are essentially insensitive to blockade (16). Nitrendipine did inhibit Ca^{2+} uptake in neuronal cell lines, but did not block the Ca^{2+} dependent release of neurotransmitter substances from these cells. These results suggest that the Ca^{2+} channels in brain may be different from those in heart or smooth muscle, or that multiple types of Ca^{2+} channels may exist in a cell.

Since nitrendipine in high concentrations was found in this laboratory to inhibit depolarization dependent Ca^{2+} uptake in synaptosomes, it is possible that like verapamil, this organic Ca^{2+} channel antagonist may be interacting with a distinct site on neuronal membrane in high concentrations that allow it to regulate Ca^{2+} uptake. Since the uM BZ receptor has been implicated as an organic Ca^{2+} channel blocker in brain, we conducted experiments to determine if nitrendipine inhibits uM BZ receptor binding (Fig. 5). The

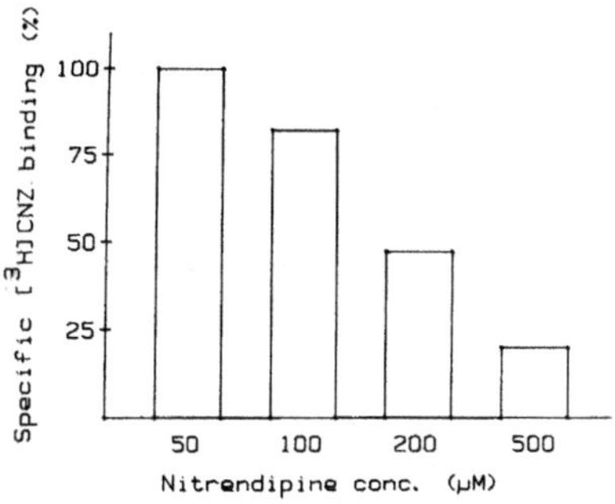

FIGURE 5. Effect of nitrendipine on specific [^{3}H] CNZ binding to brain membrane. Data represent the means of 3 determinations.

data demonstrate nitrendipine at 100-500 uM can significantly displace CNZ binding to the uM BZ receptor. These results suggest that some of the effects at high concentrations of the organic Ca2+ channel antagonists on voltage sensitive Ca2+ channels in neurons may be mediated by their interaction with the uM BZ receptor site.

6. TISSUE DISTRIBUTION OF uM BZ RECEPTORS

Employing the UV- activated BZ binding assay, specific CNZ binding was determined in heart, spleen, kidney, and liver (Fig. 6). Brain had the highest specific binding per mg of membrane. However, all tissues tested manifested specific uM BZ binding. These results suggest that uM BZ receptors are widely distributed in body tissues. Since many cell types contain voltage sensitive Ca^{2+} channels, it is interesting to speculate that these uM BZ binding sites in peripheral tissues may also play a role in regulating Ca2+ currents. Futher investigations of the effects of BZs on Ca^{2+} currents in non-neuronal tissue may provide insights into voltage regulated Ca^{2+} channels.

7. BZs SIMULTANEOUSLY INHIBIT DEPOLARIZATION DEPENDENT CALCIUM UPTAKE AND NEUROTRANSMITTER RELEASE

It is important to demonstrate that inhibition of voltage gated Ca^{2+} uptake in synaptosomes also inhibits voltage

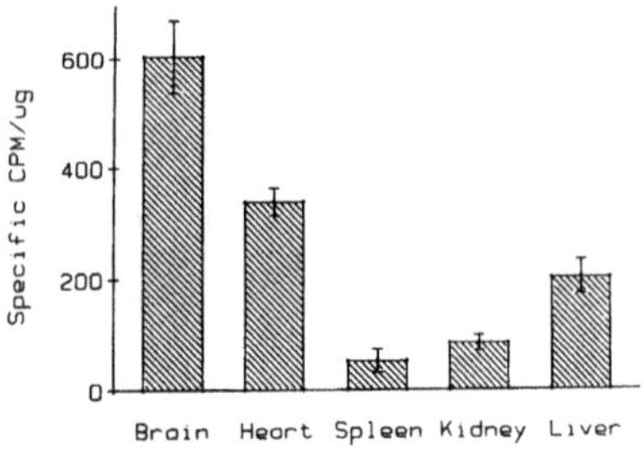

FIGURE 6. Tissue distribution of specific UV-induced [H^3] CNZ binding in membrane preparations. Data give the means of 3 determinations.

sensitive neurotransmitter release. Release of neurotransmitter substances is tightly associated with Ca^{2+} uptake in neurons. The data presented in Table 3 demonstrates that diazepam and clonazepam significantly inhibit norepinephrine and acetylcholine release in intact synaptosomes under conditions that simultaneously inhibit Ca^{2+} uptake (3,19). These results indicate that the BZs act like Mn^{2+} in neurons, since they inhibit both Ca^{2+} uptake and neurotransmitter release.

Table 3. Effects of Ca^{2+} uptake on neurotransmitter release in intact synaptosomes in the presence or absence of Diazepam and Clonazepam.

Condition	$^{45}Ca^{2+}$ Uptake	Neurotransmitter Release: Norepinephrine	Acetylcholine
Control		55	58
Ca^{2+}	42	61	64
Ca^{2+}, K	100	100	100
+ Diazepam	61	59	62
+ Clonazepam	66	64	68

Intact synaptosomes were incubated under various conditions and quantitated for $^{45}Ca^{2+}$-uptake and neurotransmitter release as described previously (19). Concentrations of Diazepam and Clonazepam were 150 uM. The data give the means of 10 determinations and are expressed as percent of the maximally stimulated condition (100%). The largest ± S.E.M. was 5.8.

8. CONCLUSION

Benzodiazepines in micromolar concentrations significantly inhibit depolarization-sensitive Ca^{2+} uptake and neurotransmitter release in intact nerve-terminal preparations. Benzodiazepine inhibition of Ca^{2+} uptake is concentration dependent and stereospecific. Micromolar-affinity benzodiazepine receptors have been identified and characterized in brain membrane and shown to be distinct from nanomolar-affinity benzodiazepine receptors. Evidence is presented that micromolar, and not nanomolar, benzodiazepine binding sites mediate benzodiazepine inhibition of Ca^{2+} uptake. Irreversible binding to micromolar benzodiazepine binding sites also irreversibly blocked depolarization-dependent Ca^{2+} uptake in synaptosomes, indicating that these compounds may represent a useful marker for identifying the molecular components of Ca^{2+} channels in brain. Characterization of benzodiazepine inhibition of Ca^{2+} uptake demonstrates that these drugs function as Ca^{2+} channel antagonists, because benzodiazepines effectively blocked votltage-sensitive Ca^{2+} uptake inhibited by Mn^{2+} and Co^{2+}. These results indicate that micromolar benzodiazepine binding sites regulate voltage-sensitive Ca^{2+} channels in brain membrane and suggest that some of the neuronal stabilizing effects of micromolar benzodiazepine receptors may be mediated by the regulation of Ca^{2+} conductance.

The BZ inhibition of depolarization-dependent Ca^{2+} uptake was shown to be identical to the inhibition of Ca^{2+} channel antagonists. In neuronal and synaptosomal preparations, Ca^{2+} channels have been characterized by their inhibition by Mn^{2+} and Co^{2+} and only by high micromolar concentrations of verapamil, nimodipine, and nitrendipine. Under our experimental conditions, the BZs acted in the same fashion as Mn^{2+} and Co^{2+}. The observation that Mn^{2+}, Co^{2+}, and DZ can individually produce a complete inhibition of voltage-dependent Ca^{2+} uptake also indicates that the BZs are acting as Ca^{2+}-channel antagonists. Furthermore, recent experiments in this laboratory have shown that Mn^{2+} blocks the irreversible binding of BZs to specific uM BZ binding sites. These results demonstrate that the BZs act as Ca^{2+}-channel antagonists and may interact with the same membrane sites as Mn^{2+}. Thus, the BZs may serve as powerful markers for identifying the molecular components of the Ca^{2+} channel. However, further studies are required to determine whether BZs bind to a membrane-bound regulator of the voltage-sensitive Ca^{2+} channel or to the channel itself.

ACKNOWLEDGEMENTS

The helpful discussions and suggestions of A. Kleinhaus were greatly appreciated. This research was supported by Research Career Development Award NSI-EA 1 K04 NS 245 and U.S. Public Health Service Grant NS 13632, NS 06208, and AFOSR-82-0284 to R.J.D. W.C.T. was supported by Aging Training Grant Fellowship Award 5 T32 AG00081.

REFERENCES

1. Rubin RP: The role of calcium in the release of neurotransmitter substances and hormones. Pharmacology Review 22: 389-428, 1972.
2. Rasmussen H, Goodman DBP: Relationships between calcium and cyclic nucleotides in cell activation. Physiology Reviews 57: 421-509, 1977.
3. DeLorenzo RJ: Calmodulin in neurotransmitter release and synaptic function. Fed Proc 41: 2265-2272, 1982.
4. Katz B, Miledi R: Spontaneous and evoked activity of motor nerve endings in calcium Ringer. J Physiol 203: 689-706, 1970.
5. Katz B, Miledi R: Further study of the role of calcium in synaptic transmission. J Physiol 207: 789-801, 1970.
6. Miledi R: Transmitter release induced by injection of calcium ions into nerve terminals. Proceedings of the Royal Society, London 183: 421-425, 1973.
7. Tsien RW: Calcium channels in excitable cell membranes. Ann Rev Physiol 45: 341-358, 1983.

8. Lee KS, Tsien RW: Mechanisms of calcium channel blockade verapamil, D600, diltiazem and nitrendipine in single dialysed heart cells. Nature 302: 790-794, 1983.
9. Reuter H: Calcium channel modulation by neurotransmitters, enzymes, and drugs. Nature 301: 569-574, 1983.
10. Fleckenstein A: Specific pharmacology of calcium in the myocardiam, cardiac pacemaker, and vascular smooth muscle. Ann Rev Pharmacol 17: 149-166, 1977.
11. Triggle D.J., Janis RA: Nitrendipine: Binding sites and mechanisms of action. In: A Scriabine, S Vanov, and K Deck (eds) Nitrendipine. Urban and Schwarzenberg, Baltimore-Munich, 1984, pp. 33-52.
12. Reuter H: Divalent ions as charge carriers in excitable membranes. Prog Biophys Molec Biol 26: 1-43, 1973.
13. Gould RJ, Murphy KMMM, Snyder SH: [^{3}H] Nitrendipine-labeled calcium channels discriminate inorganic calcium agonists and antagonists. Proc Natl Acad Sci USA 79: 3656-3660, 1982.
14. Nachshen DA, Blanstein MP: The effects of some organic calcium antagonists on calcium influx in presynaptic nerve terminals. Mol Pharm 16: 579-586, 1979.
15. Freedman SB, Miller RJ: Effects of nitrendipine on voltage sensitive calcium channels in brain and neuronal cultured cells. In: A Scriabine, S Vanov, and K Deck (eds) Nitrendipine. Urban and Schwarzenberg, Baltimore-Munich, 1984, pp 79-90.
16. Van der Kloot, Kita H: The effects of "calcium-antagonist" verapamil on muscle action potentials in the frog and crayfish and on neuromuscular transmission in the crayfish. Comp Biochem Physiol 50C: 121-125, 1975.
17. Tallman JF, Paul SM, Skolnick P, Gallagher DW: Receptors for the age of anxiety: Pharmacology of the benzodiazepines. Science 207: 274-281, 1980.
18. Bowling AC, DeLorenzo RJ: Micromolar benzodiazepine receptors: identification and characterization in central nervous system. Science 216: 1247-1250, 1982.
19. DeLorenzo RJ: The calmodulin hypothesis of neurotransmission. Cell Calcium 2: 365-385, 1981.
20. Leslie SW, Friedman MB, Coleman RR: Effects of chlordiazepine on depolarization-induced calcium influx into synaptosomes. Biochem Pharmacol 29: 2439-2443, 1980.
21. Taft WC, DeLorenzo RJ: Micromolar-affinity benzodiazepine receptors regulate voltage-sensitive calcium channels in nerve terminal preparations. Proc Natl Acad Sci USA 81: 3118-3122, 1984.
22. Ferrendelli JA, Daniels-McQueen S: Comparative actions of phenytoin and other anticonvulsant drugs on potassium- and veratridine-stimulated calcium uptake in synaptosomes. J Pharma exp Ther 220: 29-34, 1982.
23. Mohler H, Battersby MK, Richards JG: Benzodiazepine receptor protein identified and visualized in brain tissue by a photoaffinity label. Proc Natl Acad Sci USA 77: 1666-1670, 1980.

24. Meyer EM, Cooper JR: Cobalt ions dissociate between calcium uptake through voltage-dependent sodium and calcium channels in synaptosomes. Brain Res 265: 173-176, 1983.

DISCUSSION OF THE ARTICLE BY R. DeLORENZO (U.S.)

M. Lazdunski (France): The stochiometry that you find in binding experiments on synaptosomes is very high; over 300 picomoles per mg protein. This which would suggest that the receptor represents at least 10% of the membrane proteins. Wouldn't that imply that the effect on the calcium channel is indirect?

R. DeLorenzo (U.S.): That is a very good question. When you go to uM concentrations in man, you get sedation. If you give this drug to treat major motor seizures, you reach concentrations of 5-20uM in the blood. At these concentrations, there is a lot of non specific binding. We are now doing exactly what you are eluding to. Looking at what proteins or what membrane constituents the drug is binding to. There are several. There are several major bands, however, that we are trying to characterize and correlate with the calcium channel because we have a direct effect on the calcium permeability without effecting sodium or the resting potential. If we could use this to get a molecular tag on the calcium channel, we would have a powerful tool to study this channel. One of the difficulties, however, is the non specificity that you are eluding to.

M. Lazdunski (France): Did you try the drug on other ionic channels using electrophysiological techniques? Isn't it possible that the drug at these concentrations acts like a local anaesthetic and blocks calcium channels but also sodium channels and calcium-activated potassium channels and that the reason for the sedative action is actually linked to the action on sodium channels?

R. DeLorenzo (U.S.): That is a very good question and is also one of the reasons I wanted to work with Dr. Kleinhause on the leech. She has done those experiments and under conditions where we are blocking the calcium conductances in the leech by approximately 80%, there is no significant effect on the dV/dT of the sodium current. She can conclude that under conditions where we are blocking the Ca^{2+} current, the sodium current is unaffected. I should point out that so far, we showed that this system does not affect the sodium current or the resting potential. However, we have not looked at chloride conductances or potassium conductances and there may be effects on other currents as well.

I. Parnas (Israel): In the leech it is possible to check directly synaptic transmission, for example, from N to L. If you have such a huge spike, I am sure you can get a huge release. Did you check that?

R. DeLorenzo (U.S.): Dr. Kleinhause has done some preliminary studies and you do block this. In synaptosomes, we showed that under conditions where we are blocking calcium uptake, we are also blocking release and that had been one of the problems with the nimodipine and verapamil story in some of the cultured neuronal cells where even though these drugs block calcium uptake somewhat, they are not blocking release. The micromolar receptor binding does block release.

SECTION 4

CALCIUM DEPENDENT MECHANISMS

THE MECHANISM OF BURSTING IN APLYSIA PACEMAKER NEURONS

ROBERT S. ZUCKER AND RICHARD H. KRAMER

1. SUMMARY

Central neurons in the marine mollusc *Aplysia* generate endogenous bursts of action potentials as part of their normal pattern of activity. This pattern is thought to arise from an alternation of depolarizing waves due to calcium influx and hyperpolarizing waves due to potassium efflux. We have found that the hyperpolarization that terminates bursts and prevents firing until the next burst is caused not by a potassium current, but rather by a calcium-dependent inactivation of the calcium current underlying the depolarizing phase. In addition, we found that the depolarizing afterpotential at the end of each burst is due to a calcium-activated nonspecific cation current. These results help to resolve several controversies surrounding the mechanism of bursting in these neurons.

2. INTRODUCTION

The endogenous generation of bursts of action potentials due to oscillations in membrane potential (see Fig. 1A) is widespread in neurons, especially neurosecretory cells (1). The physiological mechanism of bursting has been most thoroughly studied in molluscan central pacemaker neurons (2). These cells show an alternation of depolarizing and hyperpolarizing waves, even in the absence of action potentials. Both phases require calcium in the medium. The depolarization is caused by a persistent inward calcium current that is activated at subthreshold voltages. The hyperpolarization is a calcium-dependent process, caused by the accumulation of intracellular calcium and its effect on membrane channels.

The inward current that triggers a burst has been characterized in *Aplysia* bursting neurons by Eckert & Lux (3) and Gorman *et al.* (4). They found an inward current carried by calcium ions that inactivates little at potentials near rest. Intracellular calcium accumulation during small depolarizations has also been measured using arsenazo microspectrophotometry (5). A depolarizing afterpotential occurring at the end of each burst is thought to reflect another inward current, carried by sodium and calcium ions, which contributes to the depolarization driving bursting (6).

The outward current responsible for burst termination has been attributed to calcium-activated potassium channels. Four kinds of evidence support this belief: 1) the post-burst hyperpolarization, as well as a late outward current following depolarization under voltage clamp, are reduced at hyperpolarized membrane potentials. These hyperpolarizing afterpotentials or late outward currents often fail to reverse at large hyperpolarizations (7), but this could be due to the voltage-dependence of calcium-activated potassium current, which is blocked by hyperpolarization (8). 2) Two types of experiments suggest there is an apparent increase in membrane conductance during the hyperpolarizing phase: current pulses generate a smaller voltage change (9,10), and voltage pulses under voltage clamp lead to smaller net currents (11). However, both procedures measure steady-state conductance, in which a decrease in inward current will have exactly the same effect as an increase in outward current. 3) Injecting calcium into bursting cells leads to an unequivocal increase in potassium current, with a reversal potential near the potassium equilibrium potential (12). 4) The interburst hyperpolarization decays as free intracellular calcium is pumped out (13).

We have investigated two aspects of the control of membrane potential in bursting cells: 1) the current responsible for hyperpolarizing the neuron, terminating the burst and holding the cell at negative potentials until the next burst, and 2) the current underlying the depolarizing afterpotential, which appears to contribute to the depolarizing drive in a burst. We have come to conclusions quite different from the consensus summarized above: 1) Bursts are terminated, and interburst hyperpolarization is maintained, primarily by calcium-dependent inactivation of a resting inward calcium current; and 2) the depolarizing afterpotential is caused by a calcium-activated nonspecific inward currrent. Some of our evidence for these conclusions is presented here. More details may be found in Kramer (14) and in manuscripts now in preparation.

3. PROCEDURE

We have studied bursting in the left upper quadrant bursting cells (L2-L6) in Aplysia abdominal ganglia. In these cells, the pattern of membrane potential changes in a burst (Fig. 1B) is mirrored in a similar pattern of membrane current changes following a depolarizing pulse in voltage-clamped cells (Fig. 1C). The post-burst depolarizing and hyperpolarizing afterpotentials are similar in form and duration to slow inward and outward currents following a pulse. Unlike R15, another bursting pacemaker neuron in the abdominal ganglion, the left upper quadrant bursters usually retain their bursting activity when axotomized near to the soma by the method of Connor (15). They have rather simple large unbranched axons (16); due to the long length constant of Aplysia axons (17), these axotomized neurons are practically isopotential, and are thus especially suited to voltage-clamp studies of membrane currents.

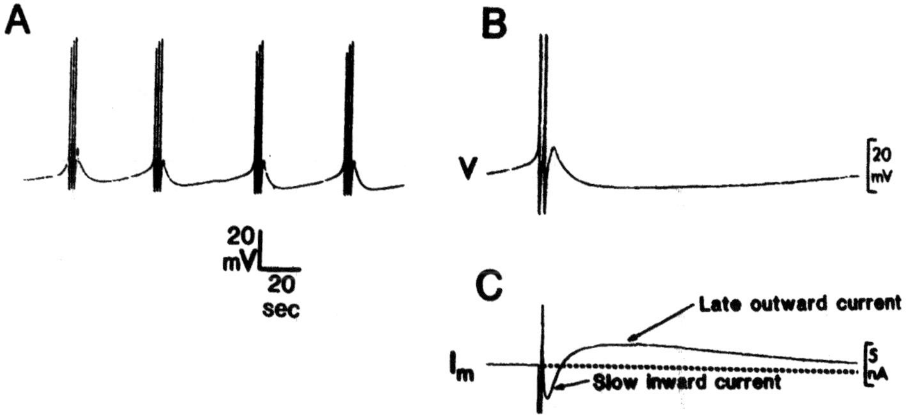

FIGURE 1. A) Endogenous bursting in an upper left quadrant neuron. B) Membrane potential changes (V) in a single burst. C) Membrane current changes (I_m) following a 50 msec depolarization to 0 mV.

4. RESULTS

4.1 Mechanism of burst termination

4.1.1. Calcium-activated potassium current does not terminate bursts. We first began to doubt that calcium-activated potassium current was responsible for burst termination when we observed that the long-lasting hyperpolarization following a burst failed to reverse in cells hyperpolarized by steady current. This occurred even when other potassium-dependent potentials, such as the brief hyperpolarizing afterpotential following each action potential, did reverse. We found that the late outward current generated under voltage clamp also fails to reverse, even when the potassium equilibrium potential is elevated by 35 mV in a high-potassium solution (Fig. 2B). This contrasts sharply with the behavior of calcium-activated potassium current elicited by intracellular calcium injection (Fig. 2A). The latter does reverse at the potassium equilibrium potential, and its reversal potential shifts almost as much as expected when potassium concentration is raised. Moreover, potassium channels opened by calcium injection are fully blocked by 50 mM TEA in

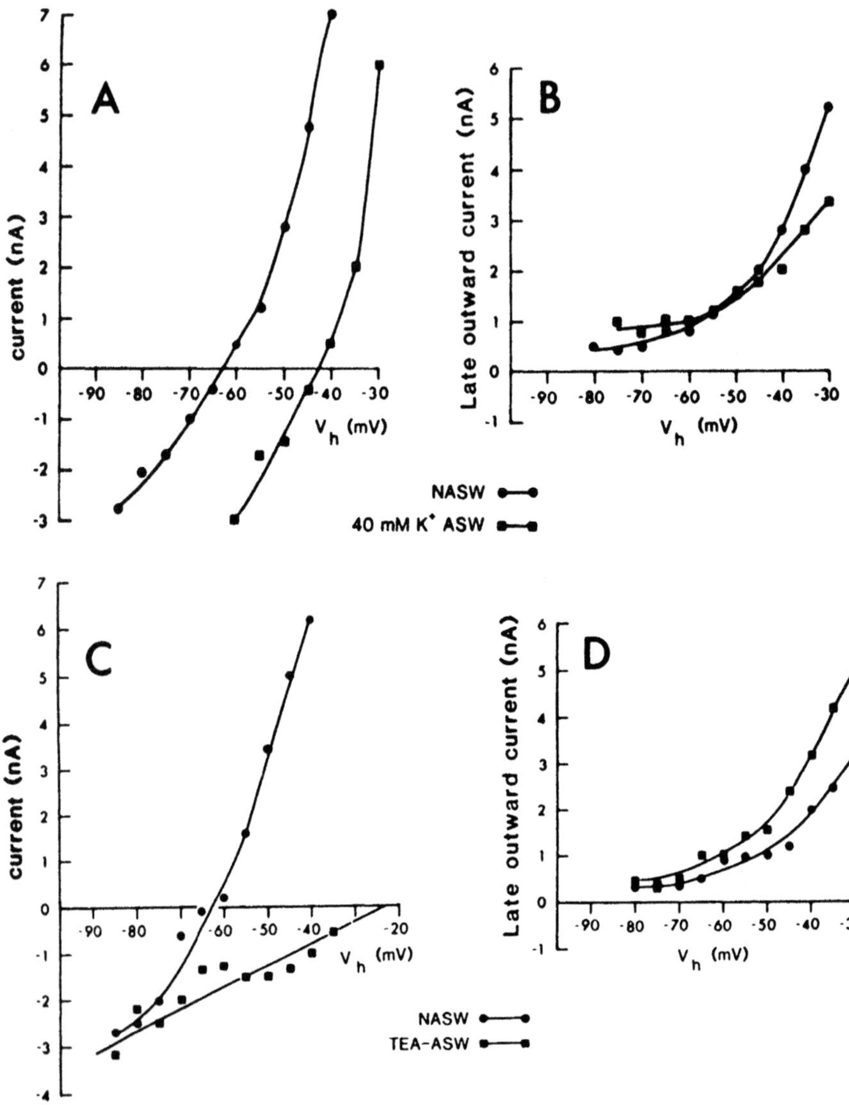

FIGURE 2. A) Currents elicited by calcium injection at different membrane potentials in normal artificial sea water (NASW) and high potassium sea water. B) Potential-dependence of peak late outward current following a 70 msec pulse to 0 mV. C) Effect of 50 mM TEA on calcium-activated current. D) Effect of TEA on late outward current.

the bath (Fig. 2C), as reported by Hermann & Gorman (18). In contrast, TEA has little effect on the late outward current following a depolarization (Fig. 2D). Similar results have recently been obtained in R15 by W. B. Adams and I. B. Levitan (personal communication).

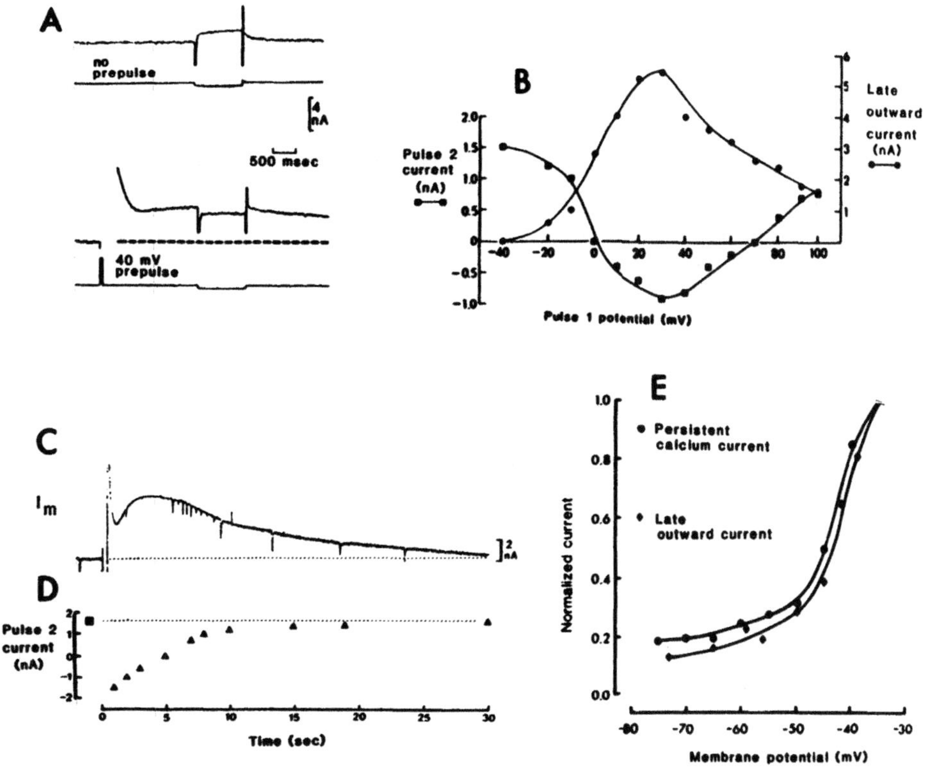

FIGURE 3. A) Effect of a 40 mV prepulse on persistent inward current turned off by a 5 mV test hyperpolarization. B) Dependence of test pulse 2 current and late outward current on size of depolarizing pulse 1. Time course of late outward current (C) and block of persistent inward current turned off by a test hyperpolarization (D) following a 70 msec pulse to +30 mV. E) Voltage-dependence of late outward current and persistent inward current.

4.1.2. Calcium-dependent inactivation of calcium current terminates bursts. Eckert & Lux (3) and Gorman et al. (4) showed that there is a persistent inward current carried by calcium ions that is partially activated at potentials in the pacemaker range (−30 to −50 mV) between the peaks of the depolarizing and hyperpolarizing waves. The steady inward current may be seen in Fig. 3A as a net outward current in voltage-clamped cells during a small hyperpolarization (the leak current is inward during a hyperpolarization). We now believe that the late outward current underlying a post-burst hyperpolarization is due to the block of this persistent inward calcium current by calcium coming in during the burst.

The magnitude of the late outward current following a depolarizing pulse depends on the size of the pulse. As the pulse amplitude grows, the late outward current grows until the pulse reaches a level of about +25 mV; for larger depolarizations, the late outward current is reduced (Fig. 3B). This bell-shaped dependence of late outward current on pulse amplitude is a characteristic of all processes dependent on calcium influx (19,20). It reflects the rate of calcium entry during a pulse, which grows with depolarization as more calcium channels open, then declines with further depolarization as the calcium equilibrium potential is approached. That the late outward current following a depolarization is due to calcium entry is supported by the fact that it is abolished by calcium antagonists (cobalt and manganese) and by EGTA injection to buffer changes in intracellular calcium.

Fig. 3B shows another interesting result: as the late outward current rises and falls after depolarizations of different sizes, so does the persistent inward current detected by a small hyperpolarizing pulse. Fig. 3C gives another indication that the late outward current is caused by a reduction in persistent inward current. The current elicited by a small hyperpolarization switches from inward (due to leak) back to outward (due to deactivation of persistent inward current) as the test pulse is moved later and later following a depolarization. The time it takes for the persistent inward current to reappear is similar to the time it takes for the slow outward current to disappear. Finally, the slow outward current and persistent inward current display similar voltage-dependence (Fig. 3E). The persistent inward current was measured as in Fig. 3A as the outward current (minus leak current) caused by 5 mV hyperpolarizing pulses from the potential on the abscissa. The peak late outward current was measured at various potentials following a depolarizing pulse.

Taken together, these results strongly suggest that bursts are terminated primarily by the block of resting calcium current by intracellular calcium that has accumulated during the burst.

4.1.3. Role of calcium-activated potassium current. Since potassium currents due to calcium entry or injection have been amply demonstrated in bursting neurones (8,12,19,21), it seemed important to estimate the magnitude and time course of this current during bursting. Fig. 4 shows the membrane current during a simulated burst before and after blocking calcium- activated potassium current with 50 m*M*

TEA. The difference current is a measure of the calcium-activated potassium current. It rises to a peak during each spike-like pulse, and decays rapidly afterwards. Clearly, calcium- activated potassium current is involved in spike repolarization and the hyperpolarizing afterpotentials following spikes. However, this current has almost completely decayed by the time a burst usually ends with a depolarizing afterpotential (open triangle), and it has entirely dissipated by the usual time of the peak of the interburst hyperpolarization (filled triangle). Thus it plays little if any significant role in burst termination or determination of the interburst interval.

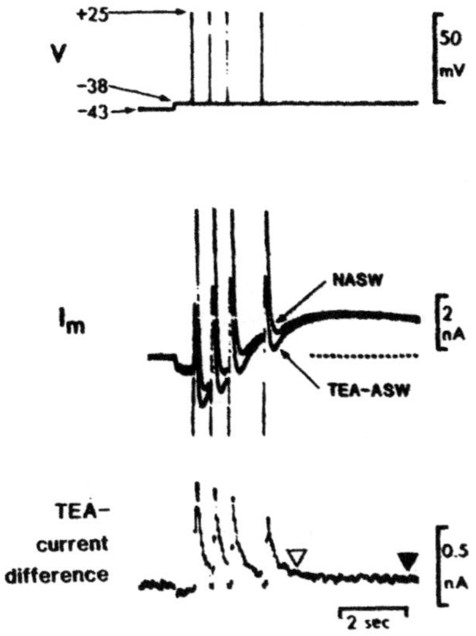

FIGURE 4. Voltage, membrane current in NASW and TEA sea water, and TEA-blocked calcium-activated potassium current during a simulated burst under voltage clamp.

4.2 <u>Basis of depolarizing afterpotential</u>

4.2.1. <u>Calcium-dependence of slow inward current</u>. The depolarizing afterpotential at the end of each burst has been attributed to a slow inward voltage-dependent calcium and sodium current (22). However, our experiments on this current indicate that it is activated by the calcium entering during a depolarization. This current is blocked by EGTA injection or addition of calcium antagonists to the bathing medium (Fig. 5A). Its magnitude shows a bell-shaped dependence on depolarizing pulse amplitude (Fig. 5B), similar to the dependence of the late outward current on pulse potential (Fig. 3B), which is also calcium-dependent (see above).

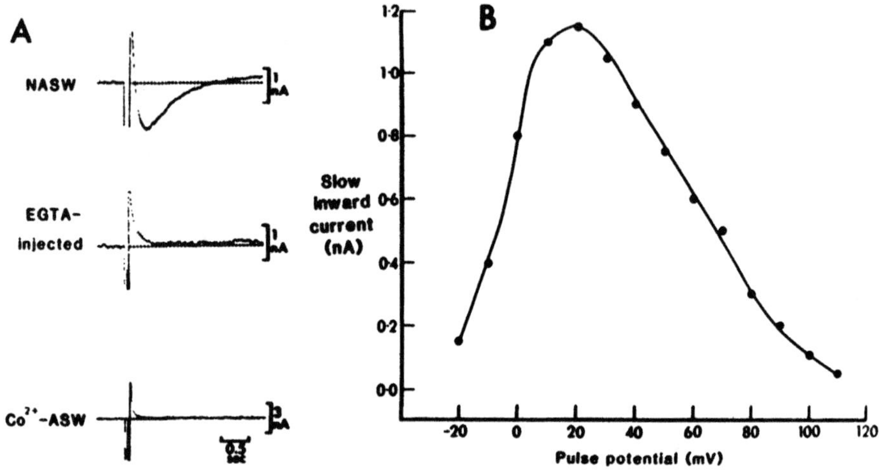

FIGURE 5. A) Slow inward current following a 20 msec pulse to +20 mV in NASW, after EGTA injection, and in sea water with cobalt substituted for calcium. B) Relation between slow inward current and pulse amplitude.

4.2.2. Ions carrying slow inward current. The slow inward current could be a reflection of a modulating effect of intracellular calcium ions on a membrane pump. However, ouabain, which blocks the sodium-potassium pump, and lithium substitution for sodium, which blocks the calcium-sodium pump, have no effect on the slow inward current. The inward current could represent potassium current due to an external accumulation of potassium ions flowing out during a depolarization, shifting the potassium equilibrium potential to more depolarized levels than the resting potential. However, blocking potassium efflux during a pulse with TEA has no effect on the slow inward current.

One ion substitution that dramatically reduces the slow inward current is sodium replacement (Fig. 6A). Unlike the voltage-dependent sodium channels which generate action potentials, this current is not blocked by tetrodotoxin. Calcium removal also blocks slow inward current, but that is because it is a calcium-activated current. Chloride substitution has no effect, and changing potassium has little effect. The effects of sodium removal and EGTA injection distinguish this current from the persistent inward calcium current that flows at rest (see above). The calcium current is enhanced by EGTA injection, and is not affected by sodium removal (4). Thus it appears that the slow inward current is due mainly to a calcium- activated increase in sodium permeability of the membrane.

4.2.3. <u>Calcium injection elicits an inward current</u>. If calcium influx during depolarization activates an inward current, we should be able to activate these channels by calcium injection. The trouble is that calcium will also open potassium channels in these neurons (12). At hyperpolarized membrane potentials close to the potassium equilibrium potential, calcium injection often leads to a biphasic inward-outward current (23), suggesting the existence of two separate calcium-dependent processes. When the calcium-activated potassium current is blocked with TEA, a calcium-activated inward current is left (Fig. 6B). This response is accompanied by a measurable increase in membrane conductance, and depends linearly on membrane potential, with a reversal potential at about -25 mV (Fig. 6C). Sodium removal always reduces the calcium-activated conductance and shifts the reversal potential to more hyperpolarized levels. Calcium removal has a similar but smaller effect. Removing both sodium and calcium shifts the reversal potential about 20 mV to the left and cuts the conductance to about half, so that what remains appears like a potassium current. Changing chloride concentration or pH have no effect on the calcium-activated inward current. Thus the current responsible for depolarizing afterpotentials resembles that elicited by excitatory neurotransmitters at neuromuscular junctions, namely a relatively nonspecific increase in cation permeability.

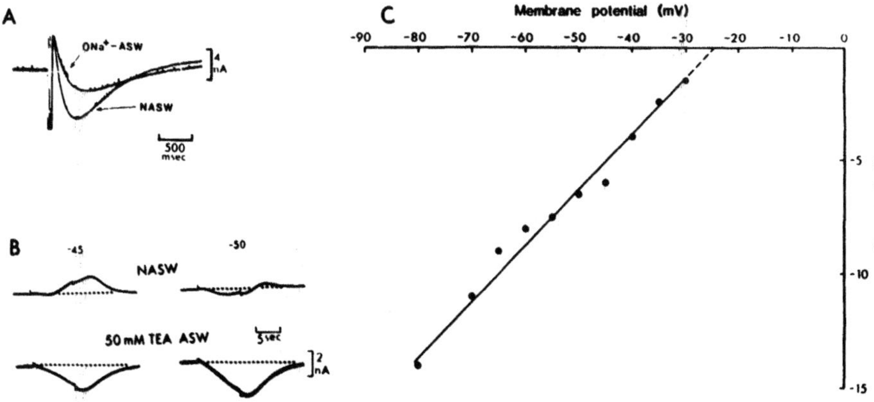

FIGURE 6. A)Effect of tetramethyl ammonium substitution for sodium on slow inward current. B) Calcium-elicited currents in sea water with and without TEA at two membrane potentials. C) Potential-dependence of inward current elicited by calcium injection in TEA sea water.

5. DISCUSSION

5.1. A new model for bursting

We have discovered two new currents in bursting pacemaker neurons -- an outward current due to the calcium-dependent inactivation of resting calcium current, and a nonspecific inward current activated by calcium. Fig. 7 shows how we believe these currents contribute to the pacemaker cycle. A voltage-dependent calcium current (I_{Ca}) causes the depolarizing wave that triggers a burst of spikes. As calcium flows into the cell and accumulates under the membrane ($[Ca]_i$), it activates a nonspecific cation current (called I_B in the figure), which contributes to the depolarizing drive and leads to the depolarizing afterpotential at the end of a burst. Calcium-activated potassium current ($I_{K(Ca)}$) is also activated, which contributes to spike repolarization, and tends to counteract slightly the depolarizing drive. Elevated calcium also blocks the calcium channels which are depolarizing the cell, and eventually this shuts off the depolarizing current, leading to cell hyperpolarization. Since the calcium channels are voltage-dependent, the hyperpolarization is regenerative. Finally, removal of calcium from membrane sites releases the calcium channels from inactivation, and they begin to depolarize the cell again. Bursting therefore arises from the interaction of calcium current, internal calcium accumulation, and three calcium-dependent processes.

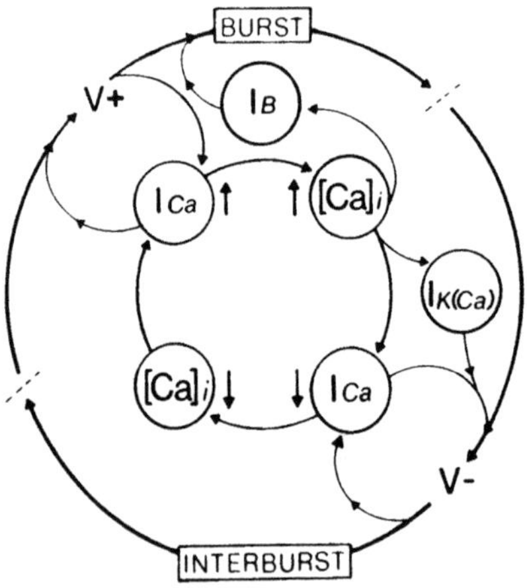

FIGURE 7. Schematic model of the bursting mechanism. Cycles of membrane potential are on the outer circle, and of calcium current and internal calcium levels on the inner circle. Calcium-activated nonspecific and potassium currents are also shown.

How does the cell sort out the three calcium-activated effects? In particular, why does calcium-dependent inactivation of calcium current dominate in the interburst interval, while calcium-activated inward current dominates at the end of the burst and calcium-activated potassium current is prominent only at the end of each spike? We think the three calcium-dependent processes have different calcium sensitivities and stoichiometries. Inactivation of calcium channels is most sensitive. It is present at rest, so that EGTA injection increases calcium currents near rest (4). Thus this effect outlasts the others while submembrane calcium is
dropping after a burst. If calcium-activation of nonspecific cation channels requires only one calcium ion, while calcium-activation of potassium channels requires three calcium ions (24), then the latter current would decay more rapidly as internal calcium declines.

5.2. Some other discrepancies resolved

Our results help to resolve a number of problems in the literature on bursting neurons. For example, estimates of the reversal potential for calcium-activated potassium current have often been more positive than the potassium equilibrium potential (22), while single channel studies of similar channels show a high potassium selectivity (25). Our results indicate that calcium injection activates nonselective as well as potassium channels, so naturally the reversal potential of the mixture is more positive than the potassium equilibrium potential.

Hermann & Gorman (18) reported that low concentrations of external TEA blocked calcium-activated potassium current, but Thompson (26) found that very high concentrations of TEA were needed to block late outward currents attributed to calcium-activated potassium channels. Our results indicate that the late outward current is primarily a calcium-dependent block of calcium current, which is not especially TEA-sensitive.

Eckert & Lux (3) and Gorman et al. (4) report that the inward current leading to depolarization is carried by calcium ions, while others (e.g. 10,27) report that the current is carried by sodium ions. Our discovery of a calcium-activated nonspecific cation channel contributing to the depolarization may be relevant to this controversy. Moreover, we have found that prolonged soaking in sodium-free medium often reduces all calcium-dependent processes. It is possible that block of the sodium-calcium exchange pump leads ultimately to an increase in intracellular free calcium concentration and calcium-dependent inactivation of calcium channels. Then sodium removal will eventually reduce the persistent inward calcium currents underlying bursting (see 3), making it appear like a sodium current.

These controversies illustrate the difficulties that have beset the study of bursting mechanisms in the past. We hope that our results will help to clarify these and other problems, and advance our understanding of the mechanisms responsible for generating endogenous repetitive activity in the many neurons that display it.

6. ACKNOWLEDGMENTS

This work was supported by NIH research grant NS 15114.

7. REFERENCES

1. Rapp, P. E.: An atlas of cellular oscillators. J. Exp. Biol. 81: 281-306, 1979.
2. Meech, R. W.: Membrane potential oscillations in molluscan 'burster' neurones. J. Exp. Biol. 81: 217-279, 1979.
3. Eckert, R., Lux, H. D.: A voltage-sensitive persistent calcium conductance in neuronal somata of Helix. J. Physiol. 254: 129-151, 1976.
4. Gorman, A. L. F., Hermann, A., Thomas, M. V.: Ionic requirements for membrane oscillations and their dependence on the calcium concentration in a molluscan bursting pace-maker neurone. J. Physiol. 327: 185-217, 1982.
5. Gorman, A. L. F., Hermann, A., Thomas, M. V.: Intracellular calcium and the control of neuronal pacemaker activity. Fed. Proc. 40: 2233-2239, 1981.
6. Thompson, S. H., Smith, S. J.: Depolarizing afterpotentials and burst production in molluscan pacemaker neurons. J. Neurophysiol. 39: 153-161, 1976.
7. Johnston, D.: Voltage, temperature, and ionic dependence of the slow outward current in Aplysia burst-firing neurones. J. Physiol. 298: 145-157, 1980.
8. Gorman, A. L. F., Hermann, A.: Internal effects of divalent cations on potassium permeability in molluscan neurones. J. Physiol. 296: 393-410, 1979.
9. Junge, D., Stephens, C. L.: Cyclic variation of potassium conductance in a burst-generating neurone in Aplysia. J. Physiol. 235: 155-181, 1973.
10. Barker, J. L., Gainer, J.: Studies on bursting pacemaker potential activity in molluscan neurones. I. Membrane properties and ionic contributions. Brain Res. 84: 461-477, 1975.
11. Smith, S. J.: The mechanism of bursting pacemaker activity in neurons of the mollusc Tritonia diomedia. Ph.D. Dissertation, Univ. of Washington, Seattle, 1978.
12. Meech, R. W.: Intracellular calcium injection causes increased potassium conductance in Aplysia nerve cells. Comp. Biochem. Physiol. 42A: 493-499, 1972.
13. Gorman, A. L. F., Thomas, M. V.: Changes in the intracellular concentration of free calcium ions in a pace-maker neurone, measured with the metallochromic indicator dye arsenazo III. J. Physiol. 275: 357-376, 1978.
14. Kramer, R. H.: The ionic mechanism of bursting pacemaker activity in Aplysia neurons. Ph.D. Dissertation, Univ. of California, Berkeley, 1984.
15. Connor, J.: Time course separation of two inward currents in molluscan neurons. Brain Res. 119: 487-492, 1977.
16. Winlow, W., Kandel, E. R.: The morphology of identified neurons in the abdominal ganglion of Aplysia californica. Brain Res. 112: 221-249, 1976.

17. Graubard, K.: Voltage attenuation within Aplysia neurons: The effect of branching pattern. Brain Res. 88: 325-332, 1975.

18. Hermann, A., Gorman, A. L. F.: Effects of tetraethylammonium on potassium currents in a molluscan neuron. J. Gen. Physiol. 78: 87-110, 1981.

19. Meech, R. W., Standen, N. B.: Potassium activation in Helix aspersa neurones under voltage clamp: a component mediated by calcium influx. J. Physiol. 249: 211-239, 1975.

20. Eckert, R., Tillotson, D.: Calcium-mediated inactivation of the calcium conductance in caesium-loaded giant neurones of Aplysia californica. J. Physiol. 314: 265-280, 1981.

21. Gorman, A. L. F., Thomas, M. V.: Potassium conductance and internal calcium accumulation in a molluscan neurone. J. Physiol. 308: 287-313, 1980.

22. Adams, D. J., Smith, S. J., Thompson, S. H.: Ionic currents in molluscan soma. Ann. Rev. Neurosci. 3: 141-167, 1980.

23. Hofmeier, G., Lux, H. D.: The time course of intracellular free calcium and related electrical effects after injection of $CaCl_2$ into neurons of the snail, Helix pomatia. Pflügers Arch. 391: 242-251, 1981.

24. Wong, B. S., Lecar, H., Adler, M.: Single calcium-dependent potassium channels in clonal anterior pituitary cells. Biophys. J. 39: 313-317, 1982.

25. Pallotta, B. S., Magleby, K. L., Barrett, J. N.: Single channel recordings of Ca^{2+}-activated K^+ currents in rat muscle cell culture. Nature 293: 471-474, 1981.

26. Thompson, S. H.: Three pharmacologically distinct potassium channels in molluscan neurones. J. Physiol. 265: 465-488, 1977.

27. Smith, T. G., Jr., Barker, J. L., Gainer, H.: Requirements for bursting pacemaker potential activity in molluscan neurones. Nature 253: 450-452, 1975.

DISCUSSION OF THE ARTICLE BY R.S. ZUCKER (U.S.)

I. Parnas (Israel): For the late outward current if it is only a decrease in calcium conductance, you should be able to reverse it at the calcium reversal potential. Your lines were parallel. Is it possible that you have an additional slight increase, for e.g., in K conductance and then the current would never reverse?

R.S. Zucker (U.S.): Seeing a reversal at the calcium equilibrium potential is almost impossible.

I. Parnas (Israel): But the lines do not approach. When you reduce the late outward current, how do the lines go?

R.S. Zucker (U.S.): I am unclear as to what lines you mean, since I have not shown any voltage-current plots that should cross at the equilibrium potential for the late outward current. I am only able to measure this current at potentials of 100mV below the calcium equilibrium potential, where the calcium conductance is highly voltage-dependent. Therevore, I do not expect it to reverse.

I. Parnas (Israel): Let me ask you a different question. How do you eliminate the possiblility that in addition to the decrease in calcium conductance, you have a slight increase in potassium conductance?

R.S. Zucker (U.S.): You may, but it has to be a TEA insensitive one so it would have to be a different channel from the usual calcium-dependent potassium current. It is possible.

B. Minke (Israel): I would like to add that we found a similar phenomenon in insect photoreceptors. When the presentation of intense bright light induces an increase in the intracellular calcium, we also see an afterdepolarization. However, we performed additional measurements. We measured $[Ca^{2+}]$, with a calcium selective microelectrode at the extracellular space, and we saw that this afterdepolarization - which you show as an inward current - is accompanied by large accumulation of calcium in the extracellular space, which is later reversed when you have the hyperpolarization phase. Therefore, we thought that maybe it is involved in sodium-calcium exchange although you eliminated this possibility by using lithium.

R.S. Zucker (U.S.): Yes, we substitute lithium for sodium.

B. Minke (Israel): But you checked only the late phase, the hyperpolarization phase?

R.S. Zucker (U.S.):

No, both phases, and they are not affected.

J. Moore (U.S.): Could you tell me what Gorman and his colleagues missed, was it an interpretation or do you differ in experimental evidence. This was not clear.

R.S. Zucker (U.S.): Many of these experiments have never been done before. There are alot of new experiments being done here. Gorman mainly referred to Junge & Stephens (Junge, D. and Stephens, C.L. (1973). Cyclic variation of potassium conductance in a burst-generating neurone in Aplysia. J. Physiol. 235, 155-181) and he did not himself really study the hyperpolarization. He studied the accumulation of calcium, the removal of calcium, and the fact that when you inject calcium, you do get a potassium current. He quoted Junge & Stephens (1973) as saying that, for example, there is a change in conductance. Junge & Stephens measured the steady-state conductance. They depolarized the cell with current and achieved more or less depolarization. But they did not take into account the fact that there are very sensitive voltage-dependent conductances in these cells which change very rapidly near the resting potential and they did not measure the instantaneous conductance. They measured the steady-state conductance and I have a slide which explains how Jonge & Stephens interpreted their results and how everyone has quoted them. I can repeat their experiments, I can get their results and I can explain them by the voltage-dependent conductances. Junge & Stephens (1973) also claimed that they could reverse the post-hyperpolarization. However, they did not use voltage clamp. When we clamp this current, we get no reversal. When they hyperpolarized a cell after a burst, they thought they saw a reversal of the interburst hyperpolarization. Actually, they saw a relaxation of voltage-dependent calcium conductance. Therefore, I think the evidence is very old, from over one decade ago, and it is quoted by hundreds papers since then while no one has bothered to check it.

D.L. Alkon (U.S.): Just a quick comment. The sensitivity of your conductances to intracellular calcium is quite different from ours and I wonder that maybe the sensitivity of a conductance to inactivation by the elevation of intracellular calcium may have something to do with its state of activation in the first place. In your case, you may have resting calcium conductance in the first place, and perhaps very little resting calcium-dependent K current. In other situations, it may be just the opposite.

(Alkon D.L., Lederhendker I., and Shoukimas J.J. (1982). Primary changes of membrane currents during retention of associative learning. Science 215, 693-695.

Alkon D.L., Acosta-Uriquidi J., Olds J., Kuzma G. and Neary J.T. (1983). Protein kinase injection reduces voltage-dependent potassium currents. Science 219, 303-306.

Acosta-Uriquidi J., Alkon D.L. and Neary J.T. (1984). Ca^{2+}-dependent protein kinase injection in a photoreceptor mimics biophysical effects of associative learning. Science 224, 1254-1257.

Neary J.T., Crow T. and Alkon D.L. (1981). Change in specific phosphoprotein band following associative learning in Hermissenda. Nature 293, 658-660.

Neary J.T. and Alkon D.L. (1983). Protein phosphorylation/dephosphorylation and transient, voltage-dependent potassium conductance in Hermissenda crassicornis. J. Biol. Chem. 258, 8979-8983.)

R.S. Zucker (U.S.): We find that if we look at different cells, the mix of currents is quite different from cell to cell. Other cells, for example, have more calcium-activated K current and less calcium inward current at rest. So there are differences at least in channel density, and possibily also due to differences in the resting calcium concentration.

A PARAMETRIC STUDY OF THE NEUROMUSCULAR JUNCTION DURING ONTOGENESIS AND UNDER DIFFERENT EXTERNAL CONDITIONS

A. Wernig, H. Jans and H. Zucker*
Department of Physiology, University of Bonn, Wilhelmstr. 31
D-5300 Bonn 1, West Germany
*Max-Planck-Institute for Psychiatry, Muenchen

1. SUMMARY

1.1. In a morphometric study on axonal and ChE stained cutaneus pectoris muscles of frog, synaptic and muscle fiber parameters were determined.

1.2. Synaptic length, synaptic complexity (=number of intraterminal branches and number of contributing axons), features of sprouting (sprouts and jumping fibers) and the abandoned gutter length as well as muscle fiber diameter were measured.

1.3. Values for parameters obtained from a single muscle scatter considerably. Coefficients of variation range from 40% (muscle fiber diameter) to 56% (number of intraterminal branches). This diversity is clearly counterbalanced by correlations between muscle fiber and synaptic parameters as well as between different synaptic parameters. Larger muscle fibers bear larger junctions and large junctions have higher numbers of intraterminal branches, contributing axons and have longer abandoned gutters.

1.4. In order to investigate growth and age related changes, muscles from 5 small frogs (body size 4.5 to 6.0 cm) were compared with muscles from 5 large frogs (body size 6.1 to 7.0 cm). In the average, there is an increase in muscle fiber diameter, synaptic length, synaptic complexity and abandoned gutter length in larger frogs.

1.5. To exclude growth-related differences and reveal changes with aging, muscle fibers of identical size were compared. Especially for large muscle fibers, synaptic lengths and abandoned gutter lengths are larger in the older (=larger) animals.

1.6. It appears, therefore, that with aging abandoned gutters on a muscle fiber accumulate. This also indicates that there is formation of new synaptic contacts rather than reoccupation of previously abandoned gutters. Alternatively, or in addition, remodelling of synaptic contacts (Wernig, Pécot-Dechavassine & Stoever, 1980a,b) could be more elaborate in large fibers of older animals.

1.7. In animals kept in the laboratory under certain conditions for 16 weeks there is an increase - as compared to freshly caught frogs - in the number of intraterminal branches and in the features of sprouting while abandoned gutters seem to be reduced.

1.8. This supports previous observations that remodelling, i.e., the extent of nerve sprouting and nerve retraction, is influenced by external factors.

2. INTRODUCTION

Recent morphological and physiological investigations have revealed signs of nerve sprouting and nerve retraction at the neuromuscular junctions of normal adult frogs. Such signs were first observed in the light microscope in axonal- and cholinesterase (ChE)-stained preparations and consequently confirmed in electronmicroscopic studies (Wernig, Pécot-Dechavassine & Stoever, 1980a,b; Wernig, Anzil & Bieser, 1981a; Wernig, Anzil, Bieser & Schwarz, 1981b; Anzil, Bieser & Wernig, 1984). In stained muscles one finds deposits of ChE, within the junctional area, often arranged in palisades, without a nerve present or with a nerve which, however, has apparently lifted up from the muscle fiber (see below). In such regions, the electronmicroscope revealed the presence of secondary folds so one can be confident that these parts are abandoned gutters which previously bore a synaptic contact. Abandoned gutters are by no means rare observations since as many as 80 - 90% of the synapses in a muscle can have one or several abandoned gutters. Often the length of individual abandoned gutters is only a few micrometers but they can be as long as several hundred micrometers and at some junctions amount to 40 - 60% of the total gutter length (Wernig et al, 1980a,b; see below). Furthermore, it is important to notice that the number of abandoned gutters is different during different seasons of the year (Wernig et al, 1980a,b); as will be shown here, their occurrence also differs with the age of the animal. The conclusion which emerges from these observations is that the neuromuscular junction is not a static structure but undergoes a marked continual remodelling. Another necessary component of remodelling would be nerve sprouting with formation of new synaptic contacts or reoccupation of previously abandoned gutters. All these signs have indeed been found. In the light microscope with axonal- and ChE-stained preparations, one frequently finds very small (2 - 5 µm) synaptic contacts besides much larger, more or less continuous contacts all within one junctional area and formed by the same axon collateral. At these small contacts, ChE and receptors for acetylcholine (ACh) are present (Anzil et al, 1984) and in electrophysiological investigations spontaneous discharge of transmitter has been observed (Bieser, Wernig & Zucker, 1984). Some of these contacts have also been investigated with the electronmicroscope and in all cases features of a synaptic contact were found. In a few cases, the subsynaptic apparatus revealed signs of an immature contact (Wernig et al, 1980a,b; Anzil et al, 1984); in particular, secondary folds invariably present at mature junctions, were either missing or were smaller than normal. It is quite likely, therefore, that new synaptic contacts are continually forming in the adult muscles under physiological conditions.

For a quantitative evaluation of nerve sprouting and nerve retraction, one is confined to evaluations in the light microscope of stained muscles. Therefore, it is important to realize the limitations of the staining methods. In particular, we know that among the small contacts, which show signs of sprouting and new synapse formation (see above), there is probably a large fraction with mature secondary folds. Since we do not know the fate of these contacts (they could mature, expand, and fuse with others or remain small as they are), their occurrence has been recently investigated in frogs of different age (Jans & Wernig, in preparation; also, see results below). On the other hand, it is quite likely that sprouting and new synapse formation also occur without formation of small isolated contacts, e.g., by extending existing continuous synaptic gutters. Another occurrence, the extent of which is almost impossible to determine with light microscopy, is the reoccupation of previously abandoned gutters. There are two pieces of evidence which indicate that reoccupation does occur; abandoned gutter length is reduced within a few weeks after animals are brought to laboratory conditions (see below) and similarily, frogs of similar body size (and thus presumably of similar age) have reduced abandoned gutter lengths in winter as compared to summer frogs (Wernig et al, 1980a,b).

In addition to quantifying the number of small contacts, the total length of individual junctions was measured as well as the total number of axonal branches in a junction. In so doing, one has to confront the unique diversity in length and in complexity among individual junctions within a single muscle. Therefore, in order to compare different muscles, somewhat elaborate statistical treatments have to be performed and these are described in detail in the Methods section.

The general ideas of this morphometric investigation were, first, to reveal possible feedback mechanisms between the presynaptic nerve (the axon terminal arborization and its contacts on a single muscle fiber) and the muscle fiber in the course of ontogenesis, and second, to study the influence of external factors (like temperature and reduced motor activity) on the extent of nerve sprouting and nerve retraction. Since frogs grow continually during their life, it was important to demonstrate that the signs of synapse plasticity outlined above are not mere growth-related changes.

3. METHODS

Frogs (R. temporaria) were caught locally and the experimental animals were selected from these on the basis of their body length and weight. For the comparison of small versus large frogs, animals were obtained in August (mean local temperature 20°C) and studied within a few days. 5 frogs with rump-to-nose lengths of 4,5 to 6,0 cm (small frogs) were compared with 5 frogs sized 6,1 to 7,5 cm (large frogs). A group of frogs of similar body size (6 cm) was obtained locally during a high temperature period in October (19°C maximum and 7°C minimum day temperatures); 6 frogs were immediately examined (freshly caught frogs) and 5 frogs were examined after keeping them for 16 weeks at a water temperature of 14°-16°C (laboratory frogs). The latter frogs were kept in a small (60 x 30 cm) water tank with dry places and were force-fed twice a week on meal worms. The laboratory was equipped with timing devices to provide daily light and dark periods.

3.1. Synaptic and muscle fiber parameters

Synapses were evaluated in cutaneus pectoris muscles stained in toto for axon and ChE (Pécot-Dechavassine, Wernig & Stoever, 1979) and mounted on glass slides. Only synapses on superficial muscle fibers in which all or most parts were located on the surface were analysed and the results from the two cutaneus pectoris muscles of any one animal were lumped together.

The staining method is important in that it allows one to distinguish features of nerve sprouting and nerve retraction in the light microscope. Since these have already been described in detail elsewhere (Wernig et al, 1980a,b; Anzil et al, 1984), schematic drawings based on camera lucida drawings will be presented here (Fig 1). Measurements were made in the light microscope using an eyepiece micrometer in a 12,5x occular with 40x or 100x objectives. For each muscle fiber evaluated, the sarcomere spacing was determined and consequently all length measurements were normalized for a sarcomere length of 3 µm.

Synaptic length was determined as the summed nerve length on a muscle fiber associated with ChE including the connecting ChE-free parts. Synaptic complexity was assessed as the number of axonal branches on a muscle fiber plus the number of parent axons (number of branches) and the number of contributing axons separately (number of axons); the latter was somewhat arbitrarily defined as the number of axons crossing the muscle fiber border, regardless of their actual origin. As criteria for sprouting, we considered small axonal branches (up to about 5 µm) with a continuous (Fig 1b) or spot-like contact (Fig 1a) or without apparent contact (Fig 1c) (sprouts) and larger branches with more than one small contact (jumping fibers, Fig 1d,e). The length of jumping fibers was defined as the axon from the last continuous contact or the last branch point, all such distances on a muscle fiber being added.

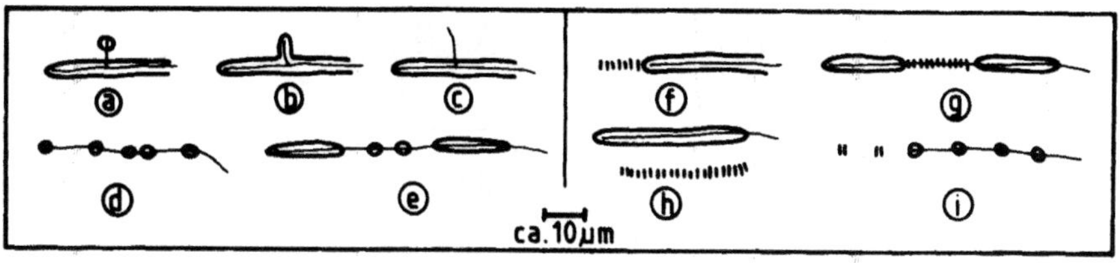

FIGURE 1 a-g. Camera lucida drawings from representative axonal- and ChE-stained muscles. Sprouts (a-c) and jumping fibers (d,e) are considered as signs of sprouting. Abandoned gutters are defined from the presence of ChE in palisades or diffuse bands; axons are usually missing (f,h,i) or have apparently lifted up from the gutter (g).

Abandoned gutters are recognizable from the arrangement of ChE reaction product in palisades or diffuse bands; axons are usually missing or are present and have apparently lifted off the junction to allow the staining solution to enter the widened cleft (Fig 1f-i). Abandoned gutter lengths on a muscle fiber were summed. Total gutter length is defined as synaptic length plus abandoned gutter lengths on a muscle fiber.

3.2. Statistical treatments

The large ranges in values for muscle fiber and synaptic parameters obtained from a single muscle, and also the differences among animals within an experimental group, demanded a novel statistical approach (Wernig, Carmody,

Anzil, Hansert, Marciniak & Zucker, 1984). When comparing two experimental groups, the occurrence of failures, small, mean and large values was separately evaluated. First, the margins for the smallest and largest values were determined somewhat arbitrarily from the pooled data (e.g., the 10% smallest and largest values; failures were considered separately). Consequently, the occurrence of such values in the data from individual animals was determined and expressed as fraction of the total number of observations in this animal. These numbers were used to compare the two experimental groups in a parameter-free rank test (Wilcoxon-Mann-Whitney). The same rank test was used for the mean values.

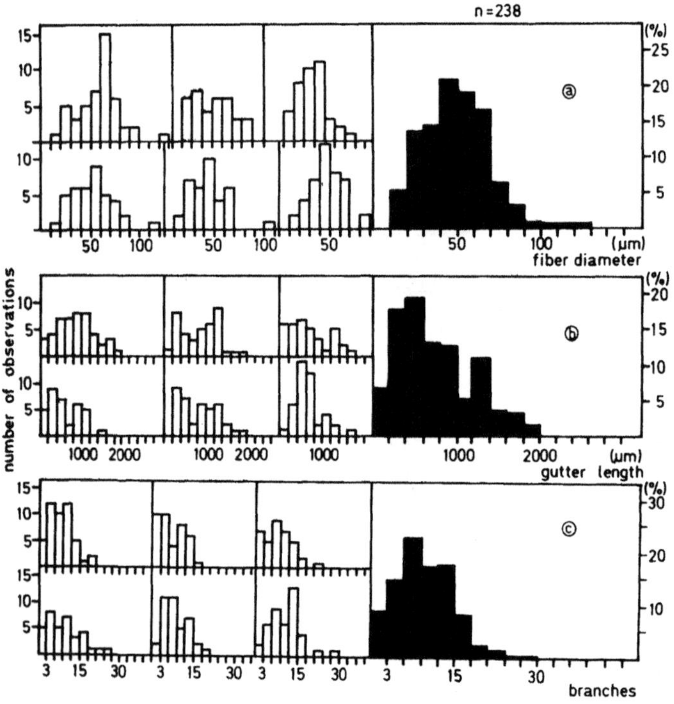

FIGURE 2 a-c. Frequency distributions of fiber diameters (a), gutter lengths (b) and number of branches (c) for 6 freshly caught frogs (October) with 6 cm body length. Superficial junctions on cutaneus pectoris muscles were evaluated and results from the two muscles were lumped together. Left hand side: results from individual animals. Right hand side: cummulated distributions.

4. RESULTS

Every researcher working on frog muscles is aware of the variety in muscle fiber diameters present within a single muscle. It is possible, that a similar diversity is already present in young animals, but it is clear that synapses and muscle fibers both grow considerably during ontogenesis. So the diversity in fiber size found in adult muscles should reflect growth-related differences. In addition, if animals of different body size (small and large frogs) are compared, then age-related differences should also become apparent. The possible relationships between synaptic and muscle fiber parameters on the other hand should reveal feedback mechanisms between pre- and postsynaptic elements, if they exist.

FIGURE 3

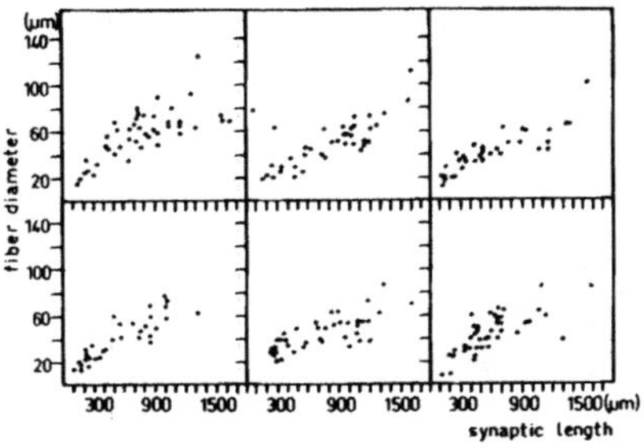

FIGURE 4

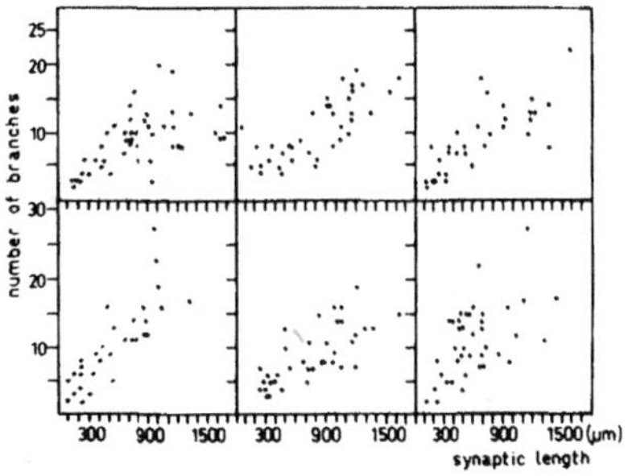

FIGURE 3. Plots of muscle fiber diameter versus synaptic length for 6 frogs sized 6 cm. A total of 238 junctions was evaluated.

FIGURE 4. Plots of number of branches versus synaptic length for muscles from 6 frogs sized 6 cm.

4.1. The diversity in parameters within individual muscles

In frogs of 6 cm body size, muscle fiber diameters in cutaneus pectoris muscles (Fig 2a) range from about 20 to 100 um. It is obvious from the histograms obtained from individual animals (left hand side in Fig 2a) that the large range of values is present within each animal and there is some additional scatter among individual animals. So the coefficient of variation of the cumulated data (right hand side in Fig 2a) is as large as 40%. An even larger scatter is found when considering total gutter length (Fig 2b) and total number of branches contributing to the synaptic contact on a single muscle fiber (Fig 2c). Coefficients of variation amount to 56% for gutter length and 51% for number of branches. This diversity is clearly counterbalanced by correlations between muscle fiber and synaptic parameters as well as between different synaptic parameters (Figs 3,4). For example, values for synaptic length from each of the 6 animals are plotted against muscle fiber diameter (Fig 3) and number of branches (Fig 4). In both comparisons there are positive correlations present with somewhat different slopes and ranges for the individual animals. Thus, larger muscle fibers bear larger junctions and large junctions have a higher number of intraterminal branches than smaller ones. It can be concluded, therefore, that mechanisms exist regulating the size of muscle fiber and the innervating axonterminal either commonly or in a feedback way.

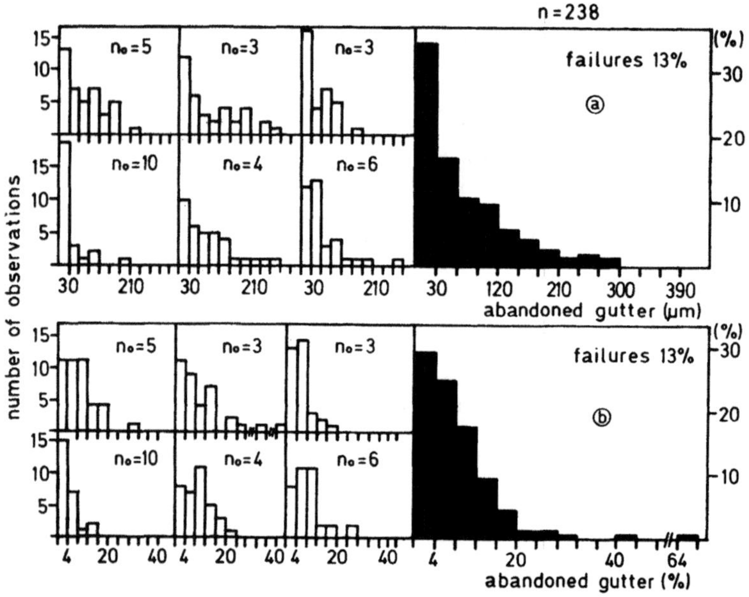

FIGURE 5a. Abandoned gutter lengths in a total of 238 junctions (right hand side). 13% of the junctions did not show abandoned gutters. Values from individual animals on the left hand side. n_o = number of junctions without abandoned gutters.

FIGURE 5b. Abandoned gutter length relative to total gutter length (i.e., synaptic length plus abandoned gutter length).

FIGURE 6

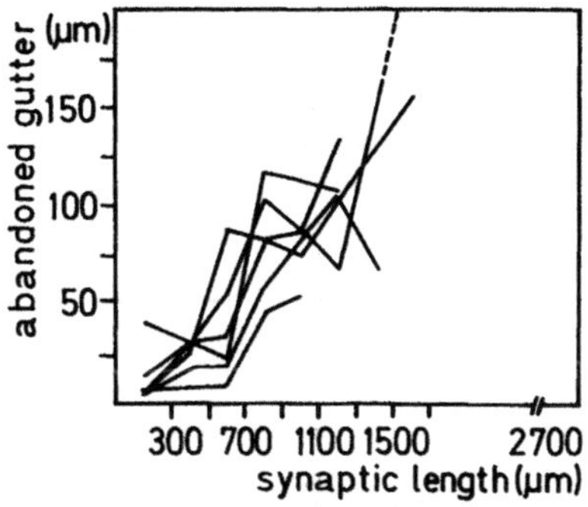

FIGURE 7

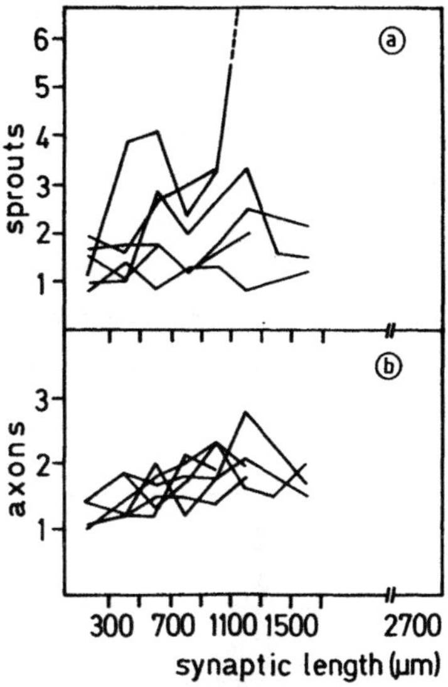

FIGURE 6. Average abandoned gutter length for given ranges of synaptic length for 6 frogs (body size about 6 cm). Lines connect points belonging to individual animals.

FIGURE 7 a-b. Average number of sprouts (a) and axons (b) for given ranges of synaptic lengths. Same frogs as represented in Figs 2 - 6.

Abandoned gutters can be recognized in stained muscles from the presence of ChE reaction product in palisades or as diffuse bands instead of the usual double lines (see Fig 1). The gutter length on a muscle fiber not occupied by nerve or Schwann cell differs considerably from fiber to fiber and was as large as 300 um in the present data (Fig 5a), this amounts to 20 more percent of the occupied gutter length (Fig 5b). 13% of the junctions did not have abandoned gutters. When comparing the abandoned gutter length on a muscle fiber with the nerve-occupied gutter length (synaptic length), it becomes obvious that large junctions are accompanied by considerably larger abandoned gutters than smaller junctions (Fig 6). It appears, therefore, that synapse remodelling is more elaborate in large junctions. On the other hand small contacts, our index for sprouting (see above), are equally frequent in small and large junctions (Fig 7a). There is an increase with synaptic length in the number of parent axons (Fig 7b). Clearly then, while the number of intraterminal branches (Fig 4) and the number of axons innervating a single muscle fiber are larger in large junctions, small contacts do not accumulate. Most likely, therefore, small contacts continually grow and possibly fuse with other small contacts to develop into larger branches. We have also noticed that small contacts had been abandoned again.

4.2. Age-related changes

Since frogs continually increase in body size (Mueller, 1976), size can can be used as a rough measure for their age. Two groups of frogs, small frogs (S group, rump-to-nose length 4,5 to 6,0 cm) and large frogs (L group, 6,1 to 7,5 cm) collected in the local vicinity during August (mean temperature 20°C) were compared.

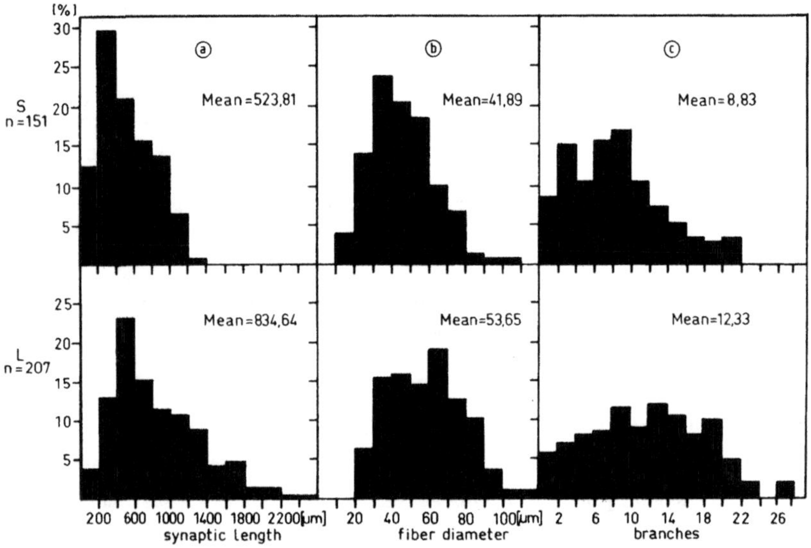

FIGURE 8 a-c. Relative frequencies (%) of synaptic lengths (a), muscle fiber diameters (b) and number of branches (c) for 5 small frogs (S) and 5 large frogs (L).

On the average, fiber diameters, synaptic lengths and number of intraterminal branches are larger in the larger animals. However, the range of values further increases (Fig 8) owing to a shift to larger values, small values remaining comparable. To exclude growth-related differences and reveal changes with aging, muscle fibers of similar size were compared. For given

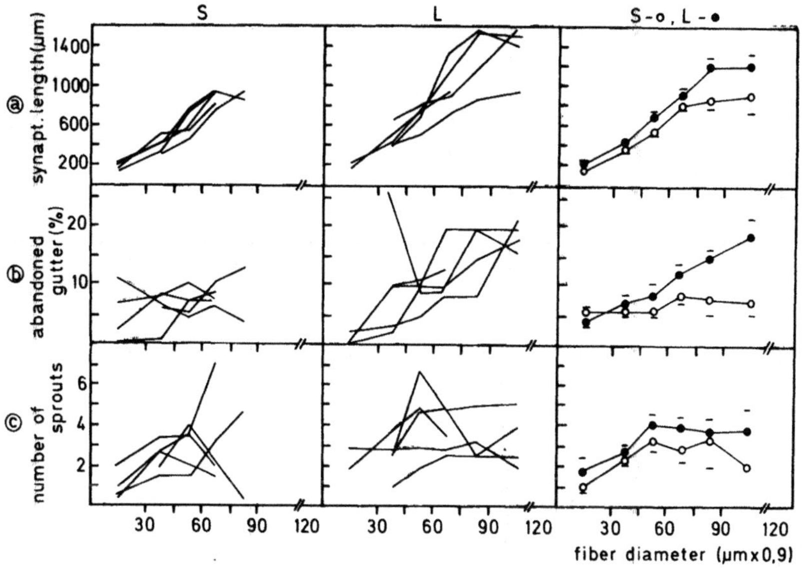

FIGURE 9 a-c. For given ranges of fiber diameters in 5 small (S) and 5 large (L) frogs, average synaptic length (a), average relative length of abandoned gutters and average number of sprouts (c) were determined. Lines connect points obtained from a single animal (left and middle) and the two experimental groups (right). The bars indicate the standard error of the mean (S.E.M.).

ranges of fiber size, mean synaptic lengths are generally larger in the larger frogs, the difference being especially marked for synaptic lengths on large muscle fibers (Fig 9a). In the latter case, the range of synaptic lengths is also large (comp. Fig 3). The number of intraterminal branches is similar for given ranges of synaptic lengths (Fig 10). It was demonstrated before (see Fig 6, above) that abandoned gutter length is positively correlated with synaptic length. We also compared the abandoned gutter length as fraction of synaptic gutter length and found that for equal ranges of muscle fiber diameters, abandoned gutters are considerably larger in the larger animals (Fig 9b). This clear difference indicates that abandoned gutters on a muscle fiber accumulate with age of the animal; since at the same time also synaptic length is larger, it appears that in older animals there is formation of new synaptic contacts rather than reoccupation of previously abandoned gutters.

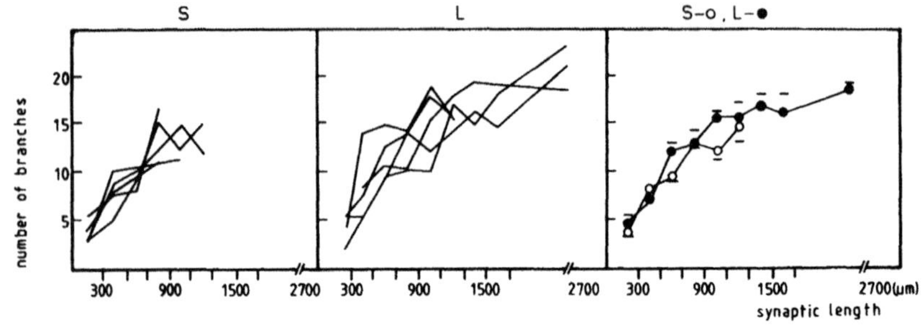

FIGURE 10. Mean number of branches was evaluated for given ranges of synaptic length for 5 small (S) and 5 large (L) frogs. Values from individual animals (left and middle) and the two groups (right). Bars: S.E.M.

The occurrence of small contacts is not correlated with muscle fiber diameter or synaptic length (see Fig 7a, above). For given ranges of muscle fiber diameters (Fig 9c) or synaptic lenghts (not shown) they are marginally more frequent in larger animals. This certainly renders as unlikely the notion, that sprouting is confined to young animals.

4.3. Temperature- and season-dependent changes

It was shown in previous investigations that in equal sized frogs (rump-to-nose length 6 cm) signs of nerve sprouting and new synapse formation are prominent in winter frogs while abandoned gutters are considerably more frequent in summer frogs (Wernig et al, 1980a,b). In those investigations, frogs were taken from nature in August (mean temperature 20°C) and January (mean temperature 0°C). In a more recent investigation from this laboratory (Jans & Wernig, in prep.), 6 cm frogs were collected from nature in October (mean maximum and minimum temperatures 19°C and 7°C) and divided into two groups; one group was immediately studied (freshly caught frogs) while the other was kept in the laboratory (in small tanks in which mobility of the animals was thus restricted) for 16 weeks at 14 - 16°C water temperature (laboratory frogs). The main findings will be summarized here and published in detail elsewhere.

There are no significant differences in synaptic contact lengths or muscle fiber diameters detectable between both groups. However, there are marked differences ($p < 5\%$, Wilcoxon-Mann-Whitney test for mean values and relative frequencies of small and large values, see Methods) for the number of sprouts and the lengths of jumping fibers. Small contacts are more frequent in the laboratory frogs for all ranges of muscle fiber sizes (Fig 11) and for all synaptic lengths (not shown). As a consequence, also the mean total number of branches in a junction is larger for most classes (Fig 11b). Clearly then, frogs have grown a significant number of additional intraterminal branches in a time span of 16 weeks. One could also expect at the same time that there is reoccupation of abandoned gutters. There is indeed a reduction in the length of abandoned gutters visible but this is confined to large muscle fibers (Fig 11c). In the context of previous findings (Wernig et al, 1980a,b; see above), this indicates that reoccupation did occur but its extent was not yet sufficient to cause general and significant reductions in abandoned gutter length.

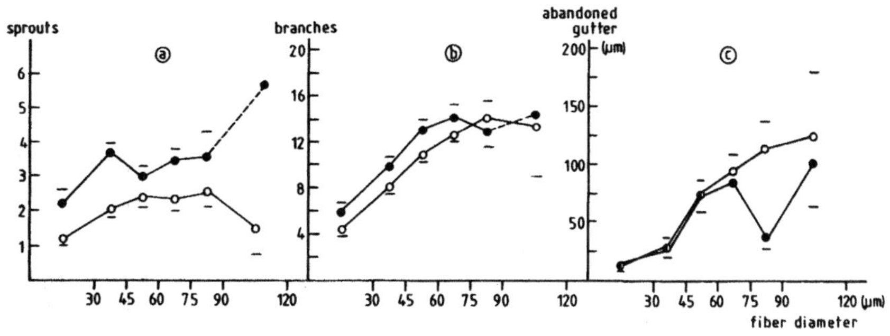

FIGURE 11 a-c. Mean values and S.E.M. for number of sprouts (a), number of branches (b) and abandoned gutter length (c) for given ranges of fiber diameters for two groups of frogs. There were 6 frogs evaluated in the group of freshly caught frogs (o) and 5 frogs in the laboratory group (•). The numbers of junctions evaluated were 238 and 189 respectively.

5. DISCUSSION

The present observation of positive correlations between muscle fiber and nerve terminal parameters allows the rather clear conclusion that there exist growth-regulating interactions. These mechanisms are not known. It is easiest to assume, that increase in muscle fiber diameter somehow causes growth of the presynaptic nerve though the influence could also function into the opposite direction. A third, least likely possibility would be that muscle fiber growth and nerve terminal growth both are caused by some third yet unknown factor. The first notion is also strongly supported by the observation that arrest in muscle fiber growth in young rats also stopped enlargement of the endplates (Slack, Pockett & MacClement, 1983). In the following, several possible mechanisms will be considered, for further discussions see reviews by Grinnell & Herrera (1981) and Grinnell, Nudell, Trussell, Herrera, Pawson & D'Alonzo (1984). Synaptic contacts are not static and there is continual nerve sprouting and nerve retraction (Wernig et al, 1980a,b; see below), but synapse formation is confined to the junctional region and normally does not occur outside. So one has to postulate a circumscribed area which permits synapse formation. Lets assume, sprouting is inherent, then a net enlargement of the synaptic contacts on growing muscle fibers could simply be due to the muscle fiber providing a larger area permitting the formation of synaptic contacts.

The amount of transmitter released from the innervating nerve terminal on the other hand is not trivial for the muscle fiber. The endplate potential (epp) amplitude determines the activity of the muscle fiber in that only epps which trigger action potentials will cause muscle fibers to contract. At low frequencies of nerve stimulation around 1 Hz, epps far above threshold are produced in many fibers in sartorius muscle while other fibers have subthreshold epps (Grinnell & Herrera, 1980). At higher action potential frequencies such as occur under physiological conditions (Sayers & Tonge, 1983), epp amplitudes decline and it is quite likely that epps near threshold occur quite regularily. Growing muscle fibers have increasingly lower input resistances and accordingly larger synaptic conductance changes are necessary to cause suprathreshold epps. In this way, activity of the muscle fiber could function as the regulating factor: One would have to postulate that inactive muscle fibers produce or start to produce more of a nerve growth promoting factor than active fibers. Increase in synaptic contact length most likely allows an increase in the amount of transmitter released per action potential, and in this way muscle fibers would assume a higher activity level which in turn could reduce the growth-stimulating muscle factor.

It is not known at this time whether other features than synaptic length are varied to cause a change in transmitter release. There is accumulating evidence that synaptic contacts are rather inhomogeneous, and it is possible that this diversity reflects plasticity of junctions. The spacing of release sites differs widely and is often larger for large parts of branches (M.

Krause and A. Wernig; in preparation), transmitter quanta are smaller at ends of branches (Bieser et al, 1984) as are transmitter release rates (Bennett & Lavidis, 1982) and average transmitter release per unit length is different in different muscles (Grinnell & Herrera, 1980). These are presently know examples of synaptic diversity to which in the future others most likely will be added.

The increase in synaptic length with increase in muscle fiber diameter is followed by an increase in the number of intraterminal branches. Partial correlation coefficients (Pflanzagel, 1978) indicate dependence of branching on synaptic length and not on muscle fiber diamemter, quite likely, therefore, intraterminal branching and length growth go hand-in-hand. In addition, new axon collaterals are permanently formed and contribute to existing junctions.

The direct evidence for sprouting and new synapse formation obtained previously (Wernig et al, 1980a,b; 1981a; Anzil et al, 1984) could be sufficiently explained by the continual growth of junctions. The mere presence of abandoned gutters, and even more so the increase in length with age, however, can best be explained by a continual remodelling of nerve-muscle junctions. Reoccupation of gutters as well as new synapse formation would be part of this remodelling. Nerve retraction as a physiological process in the neuromuscular junction is not easy to understand. In our investigations (Wernig et al, 1980a,b; freshly caught frogs in the present study) abandoned gutters were markedly more frequent under circumstances in which the animals motor activity presumably was high. A simple hypothesis would be that under elevated synaptic activity, the supply of substrates to peripheral parts of terminal arborizations via axoplasmic transport is reduced and consequently axons shrink and retract; there might be other mechanisms responsible though for the intervals of abandoned gutters within junctions. In addition, during periods of high motor activity signs of sprouting are generally reduced. So remodelling can be thought of as a continual process whereby inherent nerve sprouting is counterbalanced by activity of the neuron which precludes sprouting and causes nerve retraction. In addition, factors originating from the muscle fiber which themselves are modulating during growth and activity might be active.

The apparent accumulation in abandoned gutter length with age indicates that reoccupation is less prominent than one could expect and new synaptic contacts are formed while abandoned gutters remain unoccupied.

The time scale of remodelling is still not known. There are significant differences in the number of sprouts when comparing summer and winter frogs (Wernig et al, 1980a,b), and in the present investigation such differences occured within 16 weeks. However, much faster events are plausible and are being investigated.

6. ACKNOWLEDGEMENTS

This work was supported by a grant from the Land Nordrhein-Westfalen. Brigitte Wernig and Gabi Conrad performed much of the histochemical work. We thank Mathilde Hau for typing the manuscripts and Norbert Haas for the photographic work. We thank also Dr. Brian Freeman for critical reading and improving the manuscript.

REFERENCES

1. Anzil A.P., Bieser A. and Wernig A. (1984). Light and electron microscopic identification of nerve terminal sprouting and retraction in normal adult frog muscle. J. Physiol. 350, 393-399.
2. Bennett M.R. and Lavidis N.A. (1982). Variation in quantal secretion at different release sites along developing and mature motor terminal branches. Developmental Brain Res. 5, 1-9.
3. Bieser A., Wernig A. and Zucker H. (1984). Different quantal responses within single frog neuromuscular junctions. J. Physiol. 350, 401-412.
4. Grinnell A.D. and Herrera A.A. (1980). Physiological regulation of synaptic effectiveness at frog neuromuscular junctions. J. Physiol. 307, 301-317.
5. Grinnell A.D. and Herrera A.A. (1981). Specificity and plasticity of neuromuscular connections: Long-term regulation of motoneuron function. Prog. Neurobiol. 17, 203-282.
6. Grinnell A.D., Nudell B.M., Trussell L.O., Herrera A.A., Pawson P. and D'Alonzo A.G. (1984). Modes of Communication in the Nervous System. In: Comparative Neurobiology. F. Strumwasser and M. Cohen (Eds). John Viley and Sons, New York.
7. Mueller H.K. (1976). The frog as an experimental animal. In: Frog Neurobiology. R. Llinas and W. Precht (Eds). Springer-Verlag, Berlin-Heidelberg-New York, pp. 1023-1039.
8. Pecot-Dechavassine M., Wernig A. and Stoever H. (1979). A combined silver and cholinesterase method for studying exact relations between the pre- and postsynaptic elements at the frog neuromuscular junction. Stain. Technol. 54, 25-28.
9. Pflanzagel J. (1978). Allgemeine Methodenlehre der Statistik, Walter de Gruyter, Berlin.
10. Sayers H. and Tonge D.A. (1983). Persistence of extra-junctional sensitivity to acetylcholine after reinnervation by a foreign nerve in frog skeletal muscle. J. Physiol. 335, 569-575.
11. Slack J.R., Pockett S. and MacClement B.A.E. (1983). Regulation of postnatal growth of motor endplates in rat soleus muscle. Exp. Neurol. 80, 321-328.
12. Wernig A., Pecot-Dechavassine M. and Stoever H. (1980a). Sprouting and regression of the nerve at the frog neuromuscular junction in normal conditions and after prolonged paralysis with curare. J. Neurocytol. 9, 277-303.
13. Wernig A., Pecot-Dechavassine M. and Stoever H. (1980b). Signs of nerve regression and sprouting in the frog neuromuscular synapse. In: Ontogenesis and Functional Mechanisms of Peripheral Synapses. J. Taxi (Ed). Elsevier, Amsterdam, pp. 225-238.
14. Wernig A., Anzil A.P. and Bieser A. (1981a). Light and electronmicroscopic identification of a nerve sprout in muscle of normal adult frog. Neurosci. Lett. 21, 261-266.
15. Wernig A., Anzil A.P., Bieser A. and Schwarz U. (1981b). Abandoned synaptic sites in muscles of normal adult frog. Neurosci. Lett. 23, 105-110.
16. Wernig A., Carmody J.J., Anzil A.P., Hansert E., Marciniak M. and Zucker H. (1984). Continual nerve sprouting with features of synapse remodelling in soleus muscles of adult mice. Neuroscience 11, 241-253.

DISCUSSION OF THE ARTICLE BY A. WERNIG (GERMANY)

S. Thesleff (Sweden): Did you try to make a comparison between physiology and morphology?

A. Wernig (Germany): Yes, we did: Once you have seen this morphological diversity you do not expect any more that all parts of a junction are functionally similar. When monitoring the quantum size at several points within a junction we found typical deviations (Bieser et al. 1984); quantum size is always smaller at the very tips of new branches and larger at the spot-like nerve-muscle contacts which we assume to be newly formed contacts. Deviating quantum amplitudes are not as large as you have shown but they are larger than the average. Therefore, the morphological diversity has a counterpart in the function at least when we look at the quantum size.

S. Thesleff (Sweden): Do you know of any other procedure, despite partial denervation, which would increase sprouting? It would be interesting to use it and to compare it with your results.

A. Wernig (Germany): There are more sprouts in winter frogs (Wernig et al. 1980a,b) and in laboratory frogs in the present investigation but we have not yet done any electrophysiological work along this line. However, we have been stimulating the nerve stem in the living animal now. We immobilized animals, stimulated the nerve to one cutaneus pectoris muscle and then compared the physiology of the two muscles and we have results. The synapse is economizing. Normally, when you stimulate the nerve with 1 Hz or so, you have a very high output (higher than necessary). Following chronic stimulation of the nerve for 2 days, the first impulse releases a smaller amount of transmitter but the facilitation is augmented. But here the morphological investigation has not been performed yet.

PHENYTOIN, TRANSMITTER RELEASE AND CALCIUM FLUX

Jonathan Pincus

INTRODUCTION

Most of the major anticonvulsants including phenytoin, phenobarbital and carbamazepine are known to have presynaptic actions that affect evoked transmitter release (1-6). Most of these substances also have postsynaptic actions. In the case of phenytoin, the presynaptic actions are of special interest because it has long been held that the depression or elimination by phenytoin of post-tetanic potentiation is the experimental analog of its anticonvulsant properties (7) and post tetanic potentiation is very much a result of presynaptic effects (8). We have studied the actions of phenytoin at the neuromuscular synapse and found that its actions there were complex and were dependent on the calcium concentration of the external media (5-7). In normal media containing 1.8mM calcium, phenytoin inhibited evoked release. In low calcium (0.2mM) media, the drug increased evoked release. The drug was found to increase spontaneous release, which is to say it increased the frequency of miniature endplate potentials (mepps), irrespective of the calcium concentration of the media. In normal media it did so and in media containing no added calcium but with EGTA it also increased mepp frequency. To explain these differing effects, it was suggested that phenytoin reduced calcium influx during stimulation and also inhibited some portion of the calcium sequestration or extrusion process.

Acetylcholine release

We have recently completed a series of experiments to determine if phenytoin has a similar effect on transmitter release in rat brain synaptosomes and, if so, if predictions concerning the effect of the drug on calcium movements were correct. One group of experiments was undertaken to determine how phenytoin affects the release of labelled acetylcholine from rat brain synaptosomes which had been preloaded with tritiated methylcholine (9).

METHOD

Rat brain synaptosomes were prepared on a Ficol gradient and were loaded with labelled acetylcholine by pre-incubation for 30 minutes at 37° with tritiated methylcholine in Krebs Ringer bicarbonate. Labelled synaptosomal pellets were resuspended in Krebs Ringer bicarbonate containing eserine with either 1mM Ca or 0 Ca with 1mM EGTA. Some samples contained phenytoin 2 x 10^{-4}M and others were controls. Synaptosomal aliquots were then incubated at 37° in Krebs Ringer bicarbonate containing 5.6mM KCl or in a depolarizing solution of KRB containing 56mM KCl.

After one, five, ten and twenty minutes, reaction was stopped by transferring the synaptosomes to ice for five minutes and centrifuged. The acetylcholine in the supernatant was separated from choline by an enzymatic liquid cation exchange method. Choline was converted to choline phosphate by the action of choline kinase and ATP. The resultant solution was mixed with sodium tetraphenylboron and butronitrile. This was centrifuged to separate the phases. The tritiated acetylcholine was quantitatively dissolved in the organic phase and the choline phosphate in the aqueous phase. Aliquots were removed from the organic phase and the tritiated acetylcholine was counted in a liquid scintillation spectrometer. Acetylcholine release was expressed in counts per minute per mg of protein and the data of each experiment were normalized.

RESULTS

In the absence of calcium, depolarization of synaptosomes in 56mM KCl did not cause or increase acetylcholine release. In media containing $CaCl_2$ (1mM), depolarization lead to a mean 41% increment in acetylcholine release that was significant at each incubation period. There were 21 experiments. This demonstrates the calcium dependence of depolarization-linked acetylcholine release. In calcium containing media, phenytoin (2 x 10^{-4}M) reduced depolarization-linked release nearly to the levels of undepolarized samples (Fig 1). However, phenytoin had a dual effect for in undepolarized samples with 1mM Ca and 5.6mM KCl, phenytoin induced a 26% increment in acetylcholine release which was highly significant. A similar effect of phenytoin was encountered even in calcium free media containing EGTA. Here, undepolarized samples demonstrated a 13% increment in acetylcholine release when exposed to phenytoin (Fig 2). This was also significant ($p < .02$). Phenytoin had no discernable effect on samples that had been depolarized with 56mM KCl in media

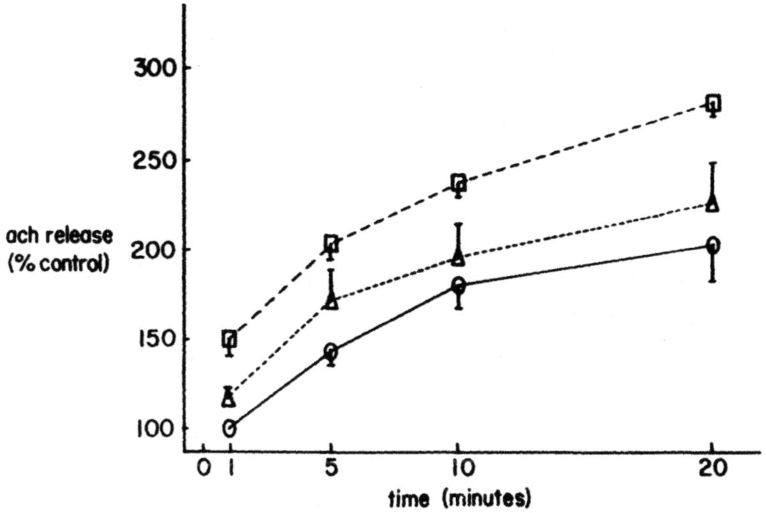

FIGURE 1. ^{3}H Acetylcholine release (normalized) from rat brain synaptosomes. Upper line represents control values from depolarized samples. The middle line shows the same with phenytoin 2 x 10^{-4}M. The bottom line represents control release from undepolarized synaptosomes.

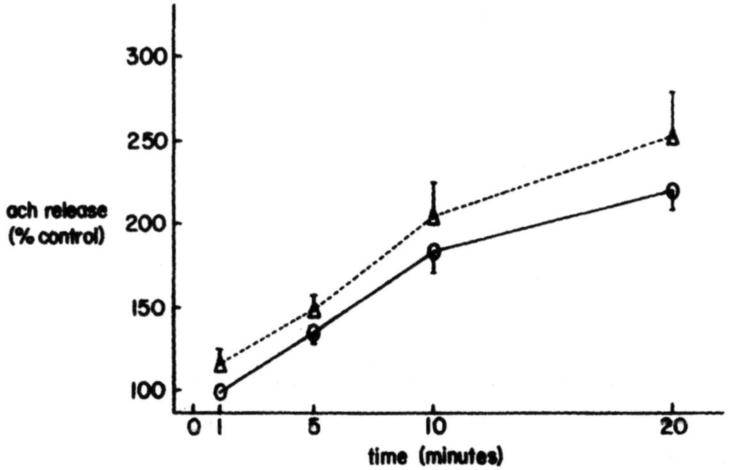

FIGURE 2. ^{3}H Acetylcholine release (normalized) from undepolarized rat brain synaptosomes. The bottom line represents control values. The upper line shows increased release with phenytoin 2 x 10^{-4}M ($p<0.02$).

from which calcium was omitted and 1mM EGTA was added.

These experiments showed that phenytoin at 2×10^{-4}M reduces depolarization-linked acetylcholine release in calcium containing media, but also that it increases basal release irrespective of the external calcium concentration. These results are quite consistent with those obtained at the frog neuromuscular synapse (5-7).

DISCUSSION

The demonstration of an increase in acetylcholine release in 0 calcium media containing EGTA indicates that this effect cannot be ascribed to an increased influx in calcium, but rather to an intrasynaptosomal effect of phenytoin. We would suggest that phenytoin antagonizes both Ca influx into the presynaptic terminal and some portion of the calcium sequestration-extrusion mechanism by means of which intracellular free Ca is kept at low levels (10). These actions of phenytoin would reduce depolarization-linked release under normal circumstances by limiting Ca entry during depolarization. Blocking the sequestration-extrusion mechanism would augment spontaneous release by causing the intracellular accumulation of free calcium.

In these experiments, phenytoin had no effect on acetylcholine release from synaptosomes incubated in 0 calcium with EGTA and depolarized with 56mM KCl. Under these rather special ionic circumstances, the Na-Ca exchange process may be near or below its reversal potential (11) and therefore inoperative. In this case, the gradient for Ca flow across the depolarized membrane would favor outward movement (12). If phenytoin blocked the Na-Ca exchange and Ca influx, phenytoin would not be expected to affect acetylcholine release in depolarized synaptosomes in very low calcium media.

Norepinephrine release

METHODS

We are extending this series of experiments to determine if phenytoin has a similar dual effect on the release of norepinepherine. For these experiments, rat brain synaptosomes were prepared on a Ficol gradient and were loaded by incubation in Krebs Ringer bicarbonate containing tritiated norepinephrine (^{3}HNE) at 37° for 10 minutes. They were then washed at 4° prior to being placed on an 0.45 micron filter (with 0.25mg synaptosomal protein) and perfused at a high, constant rate, 10 drops (about 0.5ml) per minute. With a drop counter, aliquots of 5 drops were collected, transferred to vials containing aquasol and counted in a liquid scintillation spectrometer.

The perfusion fluid was Krebs Ringer bicarbonate containing 1.0mM Ca or 0 Ca with 1mM EGTA with or without phenytoin 2 x 10^{-4}M. After a constant level of counts in the perfusate was established, the synaptosomes were depolarized with 56mM KCl. In some experiments, NaCl was replaced with choline Cl. The basal release of tritiated norepinephrine was determined by incubation of loaded synaptosomes in a vial containing 5.6mM KCl with or without phenytoin 2 x 10^{-4}M. Experiments were performed in media with 1.0mM Ca or in Ca free media with 1mM EGTA.

Depolarization with 56mM KCl lead to a marked increment in ^{3}HNE release in media containing Ca 1.0mM. However, depolarization failed to increase release in calcium free media containing EGTA, thus demonstrating the calcium dependence of depolarization-linked norepinephrine release.

RESULTS

Phenytoin significantly reduced depolarization-linked release in Ringer containing both calcium and NaCl. Phenytoin also reduced depolarization-linked release in media in which choline chloride replaced sodium chloride, but the effect of phenytoin under these conditions was less marked. Depolarization in NaCl resulted in a 204% increase in release over baseline. With phenytoin, depolarization only resulted in a 74% increase, a significant difference. Substitution of choline for NaCl lessened depolarization-linked release from 204 to 180%, but this difference did not reach conventional levels of significance (Table 1).

DISCUSSION

The substitution of choline for NaCl did modify the effect of phenytoin as the drug was significantly more effective in reducing tritiated norepinephrine release in the sodium than in the choline medium. This is consistent with the theory that the drug blocks Ca entry. However, this phenytoin effect is less marked in media containing choline nor in Na media. It would thus appear that the effect of phenytoin on depolarization-linked release is mediated partly by inhibition of Na entry. Rahamimoff & others (12) have shown that Na augments transmitter release and phenytoin is known to reduce Na conductance (13,14).

Table 1.

[3H] Norepinephrine Release from Synaptosomes
(% increase over baseline with depolarization ± SEM. K 56mM, Ca 1.0mM)

	Control	Phenytoin $2x10^{-4}$M	
Na 139	203.7 ±3.6	73.6 ±17.5	(p< .001) n=11
Choline 139	180.0 ±13.3	120.7 ±8.7	(p< .01) n=17
	(p< .2) n=14	(p< .02) n=14	

The basal release of tritiated norepinephrine was increased by phenytoin but the increment attributable to phenytoin was only about 5%. This phenomenon was noted irrespective of the external Ca concentration. Because the effect is so small, however, these experiments are being repeated but the results to date are consistent with those previously obtained with relation to acetylcholine release both with synaptosomes and at the frog neuromuscular junction.

Calcium influx and buffering

We have investigated synaptosomal calcium transport in more detail to assess the extent to which calcium movements at the nerve terminal could be influenced by phenytoin and other drugs.

METHODS AND RESULTS

^{45}Ca uptake by rat brain synaptosomes over a 10 minute period was estimated in the presence or absence of drugs in undepolarized solutions containing 5mM potassium and depolarized by 69mM potassium. Depolarization caused a marked increase in uptake of ^{45}Ca and this uptake was reduced by about a third by phenytoin 2 x 10^{-4}M (15,16) (Fig 3).

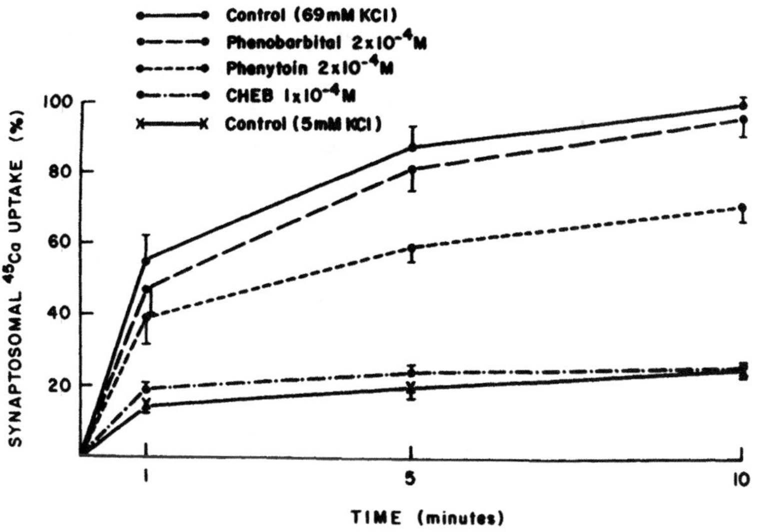

FIGURE 3. ^{45}Ca uptake by rat brain synaptosomes and the effect of various drugs. Upper control and drug containing media were depolarized. The bottom line shows release from undepolarized control samples.

Rat brain mitochondria were prepared and tested for their ability to take up ^{45}Ca over a 10 minute period. Uptake was most rapid in the first minute, the rate diminished thereafter. Of the total uptake, 50-75% occured within the first minute. Mitochondrial inhibitors such as ruthenium red, potassium cyanate and DNP reduced calcium uptake markedly. At 2 x 10^{-4}M, phenytoin caused a small, insignificant reduction (15,16).

ATP dependent calcium uptake could be demonstrated in lysed synaptosomes in which mitochondrial uptake was inhibited by ruthenium red. Of the total uptake in a 5 minute incubation, 40%-50% occured during the first minute. Uptake was 3-4 fold higher in ATP containing samples then in samples to which ATP was not added. Phenytoin did not affect calcium uptake in any samples with or without added ATP (16).

Calcium efflux

METHODS

We determined the effect of phenytoin on ^{45}Ca efflux from synaptosomes. This was determined by the method of Blaustein & Oborn (17). Synaptosomes from a gradient were suspended in 0.5ml of media containing (in mM) NaCl 132, KCl 5, $MgCl_2$ 1.3, KH_2 PO_4 1.2, glucose 10, Tris base 20 and $CaCl_2$ 1.2 (approximately 2mgs of protein per ml). There was a preincubation period of 15 minutes at 30°. The terminals were then loaded with ^{45}Ca by adding 0.5µl ^{45}Ca containing isotonic K saline (137mM KCl substituted for NaCl). Calcium uptake was terminated after a 2 minute incubation at 30° by the rapid addition of 29ml of a Ca free solution containing 1mM EGTA. The EGTA containing solutions used for halting calcium uptake and measuring calcium efflux were calcium free with either 132mM NaCl or 132mM LiCl, 5mM KCl. Drug and non-drug containing solutions were compared.

At timed intervals of 3 minutes, 2ml samples were suction filtered onto 25mm diameter glass fibre filters which were immediately washed 3 times with 12ml ice cold Ca free Li solution. "Zero" time samples were filtered and washed immediately after terminating ^{45}Ca uptake. The filtration and wash procedure took less than 15 seconds to complete. The washed filters were placed in liquid scintillation counting vials containing 1ml of 0.1% triton X-100 stored overnight and counted in liquid scintillation spectrometer.

The ^{45}Ca efflux was calculated by subtracting from the 0 time sample the residual radioactivity on the filters. Results were expressed as umoles Ca per gram of protein and normalized.

RESULTS

Ca^{45} efflux from synaptosomes was clearly influenced by external Na. The Ca efflux decreased significantly when the external NaCl was replaced by LiCl. The rate of Ca efflux at the end of the first minute in Na containing medium was 2.7µmol/gram of protein. These results are similar to those reported by Blaustein & Oborn (17).

Phenytoin 2 x 10^{-4}M, added at 0 time when the ^{45}Ca efflux measurement was initiated, significantly reduced Ca efflux in Na medium at 1,2 and 3 minutes but at 30 seconds, this effect had not yet developed. Phenytoin 1 x 10^{-4}M had no effect. Phenytoin did not affect ^{45}Ca efflux in Li media from which NaCl had been omitted (Fig 4).

CONCLUSION

In conclusion, phenytoin affects depolarization-linked acetylcholine release and ^{3}HNE release from rat brain synaptosomes much as it affects evoked release at the neuromuscular junction. The increase in basal acetylcholine release that was caused by phenytoin even in the calcium free medium may be the result of an intrasynaptosomal effect, possibly inhibition of the Na-Ca exchange. Phenytoin has no effect on Ca uptake by mitochondria or the endoplasmic reticulum and it is therefore possible that the entire increase in basal transmitter release caused by phenytoin is attributable to inhibition by the drug of the Na-Ca exchange.

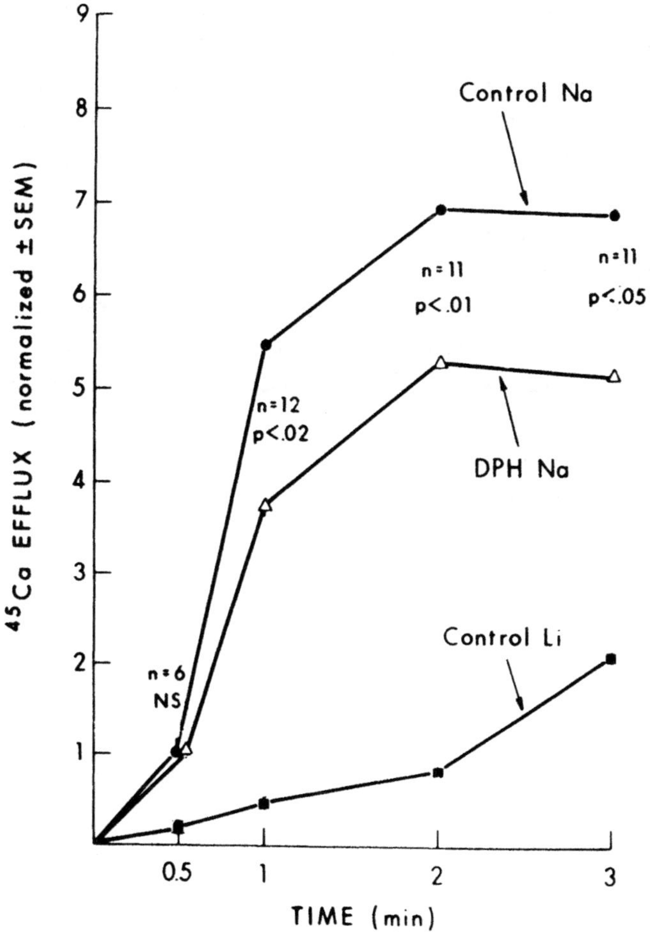

FIGURE 4. ^{45}Ca efflux from rat brain synaptosomes. Upper lines show efflux in the presence of Na and the effect of phenytoin (DPH) 2 x 10^{-4}M. When LiCl is substituted for NaCl, ^{45}Ca efflux is much reduced (bottom line). Phenytoin did not affect this (not shown).

It is not easy to predict whether this action of the drug would be inhibitory, that is anticovulsant, in vivo. Increasing free Ca at nerve terminals facilitates the spontaneous, random, uncoordinated release of transmitter, but this mechanism would also be expected to increase Ca dependent potassium conductance. The latter effect would tend to hyperpolarize the terminals and make them less excitable. Some years ago, Krnjevic proposed that general anaesthetics might activate a calcium-dependent potassium contuctance (18). Carlen et al found that ethanol augments the calcium activated potassium response (19) and the capacity of general anaesthetics to hyperpolarize frog motoneurons and rat hippocampal pyramidal cells correlates with their anaesthetic potency (20). Phenytoin may induce similar changes that are quantitatively less than those encountered with general anaesthetics. Indeed, Ayala reported that phenytoin reduces a slow outward potassium current that is calcium dependent in Aplysia (21).

Thus far, it appears that phenytoin has a similar effect on acetylcholine and norepinephrine release from synaptosomes and these effects are consistent with what we know about the effect of the drug on calcium movement in synaptosomes. Norepinephrine is usually associated with an inhibitory effect in the central nervous system and acetylcholine with an excitatory one. Whether phenytoin has a differential effect on the release of other transmitters such as GABA remains to be determined.

REFERENCES

1. Thompson T.D. and Turkanis S.A. (1973). Barbiturate induced transmitter release at a frog neuromuscular junction. Brit. J. Pharmacol. 48, 48-58.
2. Proctor W.R. and Weakly J.N. (1976). A comparison of the presynaptic and postsynaptic actions of pentobarbitone and phenobarbitone in the neuromuscular junction of the frog. J. Physiol. 258, 257-268.
3. Pincus J.H., Insler N.F. (1981). Barbiturates and transmitter release at the frog neuromuscular junction. Brain Res. 213, 127-137.
4. Yaari Y., Pincus J.H. and Argov Z. (1977). Inhibition of synaptic transmission by diphenylhydantoin. Ann. Neurol. 1, 334-338.
5. Pincus J.H., Yaari Y. and Argov Z. (1980). Phenytoin: Electrophysiological effects at the neuromuscular junction. Adv. Neurol. 363-376.
6. Yaari Y., Pincus J.H. and Argov Z. (1979). Phenytoin and transmitter release at the neuromuscular junction of the frog. Brain Res. 160, 479-487.
7. Esplin D.W. (1957). Effects of diphenylhydantoin on synaptic transmission in cat spinal cord and stellate ganglion. J. Pharm. Exp. Ther. 120, 301-323.
8. Lev-Tov A. and Rahamimoff R. (1980). A study of tetanic and post-tetanic potentiation of miniature endplate potentials at the frog neuromuscular junction. J. Physiol. 309, 247-273.
9. Pincus J.H. and Weinstein H. (1984). Acetylcholine release from synaptosomes and phenytoin action. Brain Res. 296, 313-317.
10. Alnaes E. and Rahamimoff R. (1975). On the role of mitochondria in transmitter release from motor nerve terminals. J. Physiol. 248, 287-306.
11. Mullins L.J. (1979). The generation of electric currents in cardiac fibers by Na/Ca exchange. Am. J. Physiol. 263, C103-C110.
12. Rahamimoff R., Erulkar S.D., Lev-Tov A. and Meiri H. (1978). Intracellular and extracellular calcium ions in transmitter release at the neuromuscular synapse. Ann. N.Y. Acad. Sci. 307, 583-598.
13. Pincus J.H. (1972). Diphenylhydantoin and ion flux in lobster nerves. Arch. Neurol. 26, 4-10.
14. Lipicky R.J., Gilbert D.L. and Stillman I.M. (1972). Diphenylhydantoin inhibition of sodium conductance in squid giant axon. Proc. Natl. Acad. Sci. 69, 1758-1760.
15. Pincus J.H. and Hsiao K. (1981). Phenytoin inhibits both synaptosomal ^{45}Ca uptake and efflux. Exp. Neurol. 74, 293-298.
16. Pincus J.H. and Hsiao K. (1981). Calcium uptake mechanisms affected by some convulsant and anticonvulsant drugs. Brain Res. 217, 119-127.
17. Blaustein M.P. and Oborn C.J. (1975). The influence of sodium on calcium fluxes in pinched-off nerve terminals in vitro. J. Physiol. 247, 657-686.
18. Krnjevic K. (1974). In: Molecular Mechanisms in General Anaesthesia. M.J. Halsey, R. Millar and J.A. Sutton, Eds., Churchill Livingston, N.Y.
19. Carlen P.L., Gurevich N. and Durand D. (1982). Ethanol in low doses augments calcium mediated mechanisms measured intracellularly in hippocampal neurons. Science 215, 306-308.
20. Nicoll R.A. and Madison D.V. (1982). General anaesthetics hyperpolarize neurones in the vertebrate central nervous system. Science 217, 1055-1057.

CURRENTS IN THE SYNAPTIC TERMINALS OF BARNACLE PHOTORECEPTORS

Ann E. Stuart, University of North Carolina, Chapel Hill
Jon H. Hayashi, University of North Carolina, Chapel Hill
John W. Moore, Duke University, Durham, N.C.
Ralph E. Davis, University of North Carolina, Chapel Hill
Part of work carried out at the Marine Biological Lab, Woods Hole, MA

INTRODUCTION

The photoreceptors of the giant barnacle are distinguished anatomically by large axons, visible under the dissecting microscope even to their presynaptic terminal arborizations. Intracellular recordings made from these cells near their presynaptic arbor have shown that the ocellar response to light is conducted with little decrement to the presynaptic terminals, and have demonstrated the presence of voltage-sensitive calcium channels in this region of the cell (Hudspeth & Stuart, 1977; Ross & Stuart, 1978; Edgington & Stuart 1979,1981). Using optical methods, Ross and colleagues have studied the time course of the resulting changes in intracellular calcium concentration (Stockbridge & Ross, 1984).

Our ability to record simultaneously from a photoreceptor and the cell postsynaptic to it (the I-cell, Stuart & Oertel, 1978) has encouraged us to study synaptic transmission between this primary sensory neuron and its second order cell. The receptor makes a chemical, sign-reversing synapse with the I-cell (Oertel & Stuart, 1981), the receptor's transmitter increasing the conductance of the I-cell to potassium. Input/output relations of the synapse reveal that the receptor continuously releases transmitter whether in the light or in the dark. Furthermore, they show that the synapse has the ability to shift its operating range so as to adapt to the receptor's membrane potential, set by the light intensity (Stuart, Hayashi & Moore, 1982).

This paper reports our initial attempts to voltage clamp the presynaptic terminals of these receptors.

METHODS

The preparation consisted of the median photoreceptor of the giant barnacle, *Balanus nubilus*, and the supraesophageal ganglion in which they terminate. The light-sensitive dendrites and the somata of the receptors form a perpheral ocellus, and the cells' axons course in the median ocellar nerve a distance of about 1cm to the supraesophageal ganglion. The ganglion consists of two bilaterally symmetrical hemiganglia joined by a commissure; the receptors enter this commissure, bifurcate, and send a primary branch to each hemiganglion where they make concise presynaptic arborizations about 100um from the branchpoint.

The dissection and methods of recording from the receptors have been published (Hudspeth & Stuart, 1977; Ross & Stuart, 1978). For the present experiments, the preparation was arranged in a two-chambered bath so that the ganglion and no more than 2mm of the median ocellar nerve was in one compartment; the ocellus and the rest of the nerve was in the other compartment. Vaseline sealed the gap between compartments through which the nerve was led.

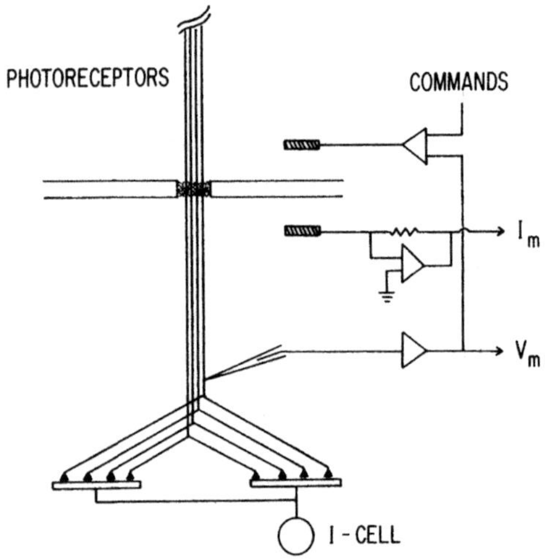

FIGURE 1. Schematic diagram of the nerve and electrode arrangement in the barrier method for voltage clamp of the four photoreceptor synaptic terminals.

To clamp the voltage of the receptor terminals, one of the four receptors in the nerve was impaled with a microelectrode as close as possible to the synaptic arbor (Fig 1). Recording sites ranged from a position in the primary branches less than 100um from the release sites to, at the other extreme, a position in the axon about 1mm from the sites. The receptor's voltage was compared with a command voltage by a voltage clamp circuit and current was passed between bath electrodes through the axon in the nerve to bring the receptor's voltage to the commanded value. Previous control experiments simultaneously comparing the voltage of two receptors, one providing feedback and one not, indicate that the voltage in the other three receptors tracks that in the receptor providing the feedback (Hayashi, Moore & Stuart, 1982). Current was measured by a current-to-voltage converter placed between the ganglion bath and ground.

Another arrangement, designed to eliminate any problems caused by poor tracking of the photoreceptors, was developed. In this alternative method of clamping the terminals, the median ocellar nerve was severed and the axons teased apart. A bundle of fibers containing 1-3 receptor axons was drawn into a tightly-fitting suction electrode. A receptor's terminal was then impaled and the current necessary to hold its voltage at a commanded value was supplied from the suction electrode. Although the light-sensitive portion of the receptor had been cut off, the impaled element could be identified as a receptor if hyperpolarizing pulses delivered to it evoked large impulses in the connectives of the supraesophageal ganglion (Stuart & Oertel, 1978). The opening of the suction electrode was located at about the same position as the Vaseline seal in the first method of clamping, about 2mm from the entry of the nerve into the ganglion. The two methods of clamping gave qualitatively similar results.

The current-passing bath electrodes were silver-silver chloride wires. Intracellular electrodes had resistances of about 50 MOhms. Experiments were carried out at 15°C. The compartment containing the ocellus was continuously perfused with normal saline, while the compartment containing the ganglion and photoreceptor terminals was perfused with a variety of test salines as described in the RESULTS. Normal saline contained 462mM NaCl, 8mM KCl, 20mM $CaCl_2$, 12mM $MgCl_2$ and 10mM Tris (hydroxymethyl) aminomethane-HCl buffer at pH 7.7. Tetraethylammonium chloride (TEA) was substituted for Na; in low Ca solutions, Mg replaced Ca.

RESULTS

Voltage clamp of the terminal region in normal saline

When the receptor terminal region is bathed in normal saline (containing 20mM Ca), depolarizing voltage clamp steps from the dark resting potential

(-60mV) evoke large outward currents (Fig 2). We assume that much of this current is flowing through voltage-sensitive channels, thought to be equivalent to the 'delayed rectifier' potassium conductance (DRGk) of squid axon, which activate rapidly and are blocked by TEA (Ross & Stuart, 1978). In the squid axon, depolarizations turn this conductance on within milliseconds. Because our clamp of the photoreceptor cell is much too slow for us to resolve such a rising phase, this current may be called 'instantaneous' on our time scale of tenths of seconds.

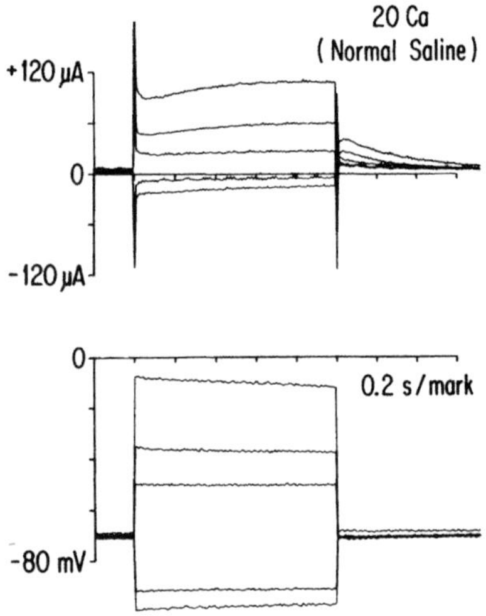

FIGURE 2. Presynaptic currents (upper traces) from photoreceptor terminal regions in response to voltage steps (lower traces) from a holding potential of -70mV in normal barnacle saline (20mM Ca).

A slight early dip in these outward currents hints that an inward current is buried in the net outward current. We interpret this dip, which becomes more pronounced at stronger depolarizations, as due to an inward calcium current. The dip is followed by an increased outward current which we interpret as a calcium-activated potassium conductance (CaGk) (Edington & Stuart, 1979; Bolsover, 1981). The decay of the potential at strong depolarizations shows that the voltage control cannot be maintained with such large current levels.

At the offset of a depolarizing step, a tail of outward current occurs. The tail is thought to consist of currents through both the DRGk and CaGk channels, the former decaying rapidly and the latter much more slowly.

Hyperpolarizing voltage steps give rise to a net inward current that decays slightly during the step. Our interpretation of this decay is that at least part of it reflects the turning off of a small calcium current assumed to be continuously flowing into the cell at the dark resting potential; this idea derives from the observation of a small, steady hyperpolarization of the postsynaptic cell at dark rest, indicating a tonic release of transmitter from the receptor cell (Stuart, Hayashi & Moore, 1982).

Voltage clamp of the terminal region in tetraethylammonium (TEA)

When we added 20mM TEA to the bathing saline to block the DRGk, depolarizing pulses led to an outward current much reduced in amplitude (about 10-fold) from that seen in Fig 2. The current in response to the 500 msec pulse under these conditions was obviously biphasic with a clear inward component preceding the outward component. In this concentration of calcium (20mM), however, voltage uniformity was poor. The inward current, instead of increasing smoothly with successive depolarizations, frequently appeared abruptly and fully-on, indicating that a partial action potential may have been generated elsewhere (most likely at the release sites) and propagated toward the point at which the voltage was controlled. The abrupt current increase seen is that required to suppress this 'spike' at the point of voltage control. In 20mM calcium, neither of our clamp methods yielded consistent voltage control.

Considerably better voltage control could be achieved if, in the presence of 20mM TEA, the external calicum concentration was reduced to 8mM or below. Fig 3A shows the currents obtained under these conditions. The inward currents increase and decrease smoothly in time and with changes in voltage. The inward component of the current reaches a peak in about 30-50 msec and is followed by an outward component that reaches a steady state over the next several hundred milliseconds. The outward current component, thought to be carried mainly by CaGk channels, becomes a slowly declining tail current when the command voltage is returned to the resting level. A brief decrease in this current upon repolarization may be a calcium tail current.

Both the inward and outward components of the current are blocked by removing calcium from the bath and adding 16mM cobalt (Fig 3B). In the cobalt saline, weak to moderate depolarizations evoked a net outward current which decayed during the pulse. We haven not yet identified this current, which was also observed by Bolsover (1981) in voltage clamp records taken from the soma of the Woods Hole barnacle, Balanus eburneus.

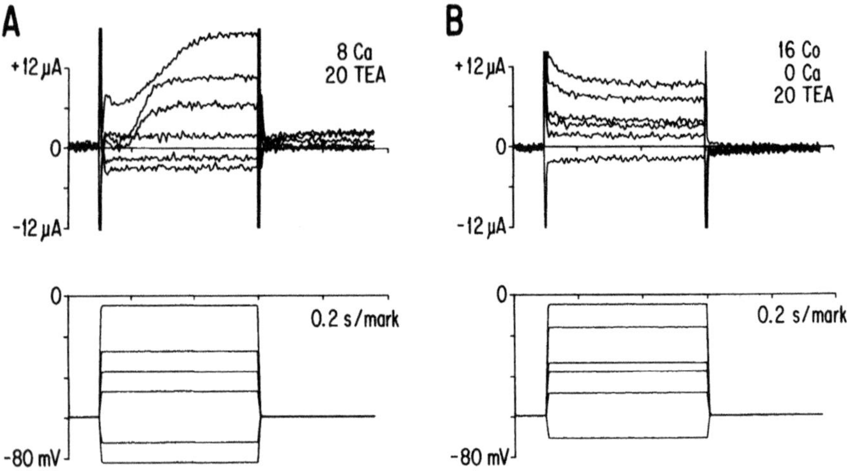

FIGURE 3. Current and voltage records as in Fig 2 when the photoreceptor terminal is bathed in: A) 8mM Ca + 20mM TEA; B) 16mM Co + 20mm TEA.

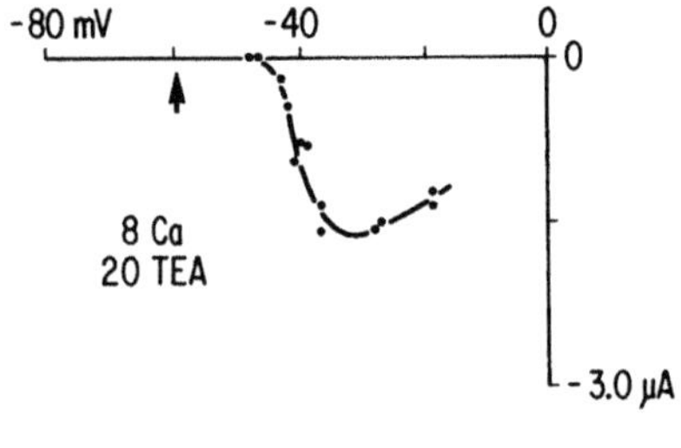

FIGURE 4. Current-voltage relation of the peak inward currents for the experiment shown in Fig 3A where the holding potential was -60mV (arrow), near the resting potential in the dark. Inward currents were measured as described in the text.

In Fig 4, the peak net inward currents observed in the experiment of Fig 3A are plotted against voltage. The currents flowing in reponse to depolarizing steps were measured as the difference between the 'instantaneous' currents (seen as soon as the clamp had settled) and the minimum values of the subsequent current dips. In the preparation of Fig 3A, the activation potential was about -48mV and the point of half-maximum activation about -40mV. While the region of activation varied from preparation to preparation, it lay between -60 and -48mV (for holding potentials at the dark resting values, between -55 and -60mV, n=10). The activation of calcium currents in this range would be expected from the input/output relation of the synapse (Stuart, Hayashi & Moore, 1982) which shows that transmitter is released from the photoreceptor at the dark resting potential. The inward current grows to a maximum value over the next 15-20mV depolarization.

Voltage clamp of the terminal region in barium

If calcium carries the inward current seen in Fig 3A, then barium might also be expected to carry this current and do so in the absence of calcium. Fig 5 shows currents from an experiment carried out in 4mM barium saline free of calcium. Depolarizing pulses led to inward currents as expected, but the currents remained inward for the duration of the pulse. At the offset of the pulse, there was no prolonged tail of outward current as there had been in the calcium salines. These observations suggest that barium flows into the cell through the calcium channel but does not activate the CaGk channels.

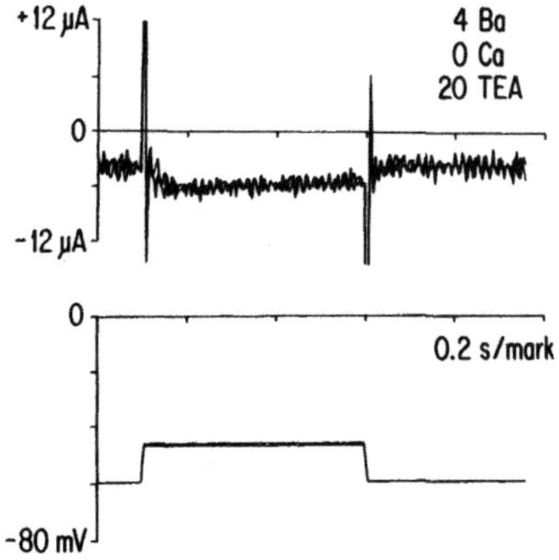

FIGURE 5. Current and voltage records as in Fig 2 when the photoreceptor terminal was bathed in a solution containing 4mM Ba (0mM Ca) and 20mM TEA.

DISCUSSION

The experiments presented here are a first step in the analysis of currents flowing across the synaptic membrane of a photoreceptor cell. The present results are consistent in several respects with the observations of the input/output relation of this synapse (Hayashi, Moore & Stuart, 1982; Hayashi, Moore & Stuart, in preparation): (1) The calcium current activates in the region of the dark resting potential and reaches a maximum at a level 15-20mV depolarized from this value. The postsynaptic response grows over a similar range. (2) In response to hyperpolarizations from the dark resting level, the presynaptic current decays, as if a small inward current is being turned off. The input/output relations show that there is a small tonic release of transmitter from the receptor at dark rest.

The control of voltage in the presynaptic arbor

In general, the attainment and control of a uniform voltage over nerve terminal synaptic contacts (which often extend over process lengths equal to several diameters) has proved difficult, even when the normal DRGk has been reduced by the application of specific blocking agents such as TEA and 4-AP at the squid giant synapse (Llinas, Steinberg & Walton, 1981; Smith, Augustine & Charlton, personal communication). In the barnacle preparation, Stockbridge & Ross (1984) have shown that the major portion of calcium entry is localized to the receptor's concise arborization of stubby secondary and tertiary processes in each hemiganglion (Schnapp & Stuart, 1983). However, it is not known how remote (electrically) these sites of calcium entry are from the point of penetration with microelectrodes in this cell. When measured with hyperpolarizing pulses (Hudspeth, Poo & Stuart, 1977), the membrane resistivity of the receptor is large and its length constant is about forty times the distance from the location of the electrode in the present experiments to the synaptic sites. But when the cell is depolarized, this resistivity decreases by an amount difficult to estimate, especially in the terminal branches where the voltage-sensitive conductances may be concentrated. Consequently, there may well be a voltage drop between the point of measurement and the site of entry of calcium.

We have attempted to minimize the path length between the Vaseline barrier (or opening of the suction electrode) and the point of microelectrode penetration (the 'access resistance', Moore, 1971), in order to achieve the maximum gain and speed of control of the potential at the site of the microelectrode penetration. For best potential control of the terminal region, the electrode was inserted as close as possible to the release sites. Although the clamp had an obvious difficulty in supplying the large currents at strong depolarizations when the preparation was bathed in normal barnacle saline, the control of the measured membrane potential was quite good under the much smaller current demands when the bathing solution contained TEA.

Another measure of the quality (stability) of a clamp (discussed by Cole & Moore, 1960) is how smoothly the inward currents increase in the negative resistance region, where the channel is increasingly activated. In normal saline, there was a smooth increase in the dip in the current records for depolarization of the receptor (Fig 2). When TEA was appplied in sufficient concentration to block the DRGk, the increase in inward currents with increasing depolarization often became abrupt. There records were similar to those seen in squid axon when action potentials were generated outside the region under voltage control and then propagated into it (Cole & Moore, 1960; Taylor, Moore & Cole, 1960). In such cases, extra current had to be supplied by the voltage clamp to absorb the longitudinal currents and thus maintain control at the point of potential measurement.

In 20mM calcium and 20mM TEA, action potentials were observed in the receptor under current clamp (Ross & Stuart, 1978). When we reduced the concentration of the bath calcium from 20mM to 8mM, we saw smaller action potentials under current clamp. Under voltage clamp, there was no evidence of action potentials propagating into the controlled region, suggesting better control of the potential across the axon and terminals in the reduced calcium solution. But voltage gradients may still exist along the arbor's processes because most of the calcium current probably enters at the release sites and travels along the axon to the ocellar bath. Further evaluation by other means is required before we can dismiss voltage gradients as a problem in this preparation.

It is interesting to note the conflicting effects of adding TEA to the saline in these experiments. On the one hand, because TEA blocks the DRGk, it gives better voltage uniformity and, therefore, better control of the axon and terminal (Joyner, Moore & Ramon, 1975). On the other hand, TEA simultaneously causes the control to be poorer because of the very high calcium channel density, which is sufficient to give overshooting action potentials under current clamp.

<u>The nature of currents</u>

Previous work has shown that both the peak and the undershoot of action potentials elicited in the receptor in the presence of TEA are sensitive to the external calcium concentration (Ross & Stuart, 1978; Edgington & Stuart, 1981). Thus, we expected a net inward current carried by calcium ions, generating the spike when the DRGk channels are blocked, and a CaGk to generate the undershoot. Our results in which cobalt and barium were substituted for calcium are consistent with these expectations. Even at moderate depolarizations, a significant entry of calcium occurs and rapidly turns on the potassium conductance. Since this potassium conductance may then mask the peak inward currents, we have not plotted values for calcium currents beyond the region of their activation.

At the offset of large depolarizing steps in 8mM calcium and 20mM TEA, there are clear tail currents which decay very slowly (if at all for the time shown in Fig 3A). This is to be expected because it takes a considerable time for pumps to reduce the intracellular Ca concentration to a point where the potassium flow through the CaGk channels is return to its normal low level.

When calcium was removed from the saline, and barium carried the inward current, presumably through the calcium channel, a 500 msec pulse led only to a net inward current. Apparently barium does not activate the CaGk. It may be possible to use barium to study in greater detail the characteristics of the calcium channel, uncontaminated by the potassium currents.

ACKNOWLEDGEMENTS

The authors thank Edward Harris, Laura Fleck and Karen Litt for steadfast technical assistance. The support of R.E.D. by the Grass Foundation is gratefully acknowledged. This work was supported by U.S.P.H.S. grants EY03347 to A.E.S. and NS0347 to J.W.M.

REFERENCES

1. Bolsover S.R. (1981). Calcium-dependent potassium current in barnacle photoreceptor. J. Gen. Physiol. 78, 617-636.
2. Cole K.S. and Moore J.W. (1960). Ionic current measurements in the squid giant axon membrane. J. Gen. Physiol. 44, 123-167.
3. Edgington D.R. and Stuart A.E. (1979). Calcium channels in the high resistivity axonal membrane of photoreceptors of the giant barnacle. J. Physiol. 294, 433-445.
4. Edgington D.R. and Stuart A.E. (1981). Properties of tetraethylammonium ion-resistant K^+ channels in the photoreceptor membrane of the giant barnacle. J. Gen. Physiol. 77, 629-646.
5. Hayashi J.H., Moore J.W. and Stuart A.E. (1982). Transmitter release as a function of holding potential at the synapse of the barnacle photoreceptor. Neurosci. Abst. 8, 522.
6. Hudspeth A.J., Poo M.M. and Stuart A.E. (1977). Passive signal propagation and membrane properties in median photoreceptors of the giant barnacle. J. Physiol. 272, 25-43.
7. Hudspeth A.J. and Stuart A.E. (1977). Morphology and responses to light of the soma, axon and terminal regions of individual photoreceptors of the giant barnacle. J. Physiol. 272, 1-23.
8. Joyner R.W., Moore J.W. and Ramon F. (1975). Axon voltage clamp simulations III. Postsynaptic region. Biophys. J. 15, 37-54.
9. Llinas R., Steinberg I.Z. and Walton K. (1981). Presynaptic calcium currents in squid giant synapse. Biophys. J. 33, 289-322.
10. Moore J.W. (1971). Voltage Clamp Methods. In: Biophysics and Physiology of Excitable Membranes. W.J. Adelman (Ed.), Van Nostrand Reinhold Co., pp.143-167.
11. Oertel D. and Stuart A.E. (1981). Transformation of signals by interneurones in the barnacle's visual pathway. J. Physiol. 311, 127-146.
12. Ross W.N. and Stuart A.E. (1978). Voltage sensitive calcium channels in the presynaptic terminals of a decrementally-conducting photoreceptor. J. Physiol. 274, 173-191.
13. Schnapp B.J. and Stuart A.E. (1983). Synaptic contacts between physiologicallly identified neurons in the visual system of the barnacle. J. Neurosci. 3, 1100-1115.
14. Stockbridge N. and Ross W.N. (1984). Localized Ca^{2+} and calcium-activated potassium conducances in terminals of a barnacle photoreceptor. Nature 309, 266-268.
15. Stuart A.E., Hayashi J.H. and Moore J.W. (1982). Transmission from the median photoreceptors of second-order cells. Invest. Ophthalmol. Vis. Sci. 22, 275.
16. Stuart A.E. and Oertel D. 91978). Neuronal properties underlying processing of visual information in the barnacle nervous system. Nature 275, 287-290.
17. Taylor R.E., Moore J.W. and Cole K.S. (1960). Analysis of certain errors in squid axon voltage clamp measurements. Biophys. J. 1, 161-202.

DISCUSSION OF THE ARTICLE BY A.E. STUART (U.S.)

A. Mallart (France): Is the action potential sodium, calcium or both?

A.E. Stuart (U.S.): Normally, the cell does not give an action potential at all. If we put the receptor in TEA, we can evoke a calcium action potential. There is no sodium component (Ross W.N. and Stuart E.A. (1978). J. Physiol. 274, 173-191).

A. Mallart (France): Did you use another potassium channel blocker to reveal any other K conductances?

A.E. Stuart (U.S.): We have not experimented yet with various K channel blockers to see, for example whether we can block the outward current that is revealed when we put the preparation in cobalt.

A. Mallart (France): Did you ever observe a transient outward current?

A.E. Stuart (U.S.): Yes, we do in cobalt.

R. Rahamimoff (Israel): Is it possible to inject ions into the cells, something like EGTA, to see whether the release voltage-dependence curve shifts?

A.E. Stuart (U.S.): That is a wonderful experiment which we would like to do. We have in the past successfully injected dyes (procion), HRP, and choline into these terminals.

I. Steinberg (Israel): Is the transmitter known?

A.E. Stuart (U.S.): Les Timpe in my lab has shown that ACh, GABA, 5HT, aspartate, glutamate, tyramine, octopamine, norepinephrine, serine, or adenosine have no effect on the I-cell. We have not yet found a drug that will block the I-cell's response. Consequently, the transmitter is not yet known.

A. Wernig (Germany): Do you know the reversal potential of the I-cell's response?

A.E. Stuart (U.S.): Yes, the postsynaptic response is due to an increased conductance to K. The reversal potential is normally about -80mV (Oertel D. and Stuart A.E. (1981). J. Physiol. 311, 127-146).

G. Falk (U.K.): Have you measured the input-output relation along the steep part of the curve? That is, what is the size of the presynaptic signal compared to the postsynaptic signal.

A.E. Stuart (U.S.): The gain across the synapse appears to be 6-8.

F.J. Sigworth (Germany): Do you think the variability in the calcium activation curves from preparation to preparation is an artifact or real?

A.E. Stuart (U.S.): As you know, there is great difficulty in general when working with calcium currents in getting activation curves that are not variable and we are happy to get an activation curve that we can repeat within an experiment. I think that when we are able to clamp the pre- and postsynaptic cells simultaneously, which is what we are aiming to do, we will know when transmitter is released from the receptor and will better understand the variability.

CALCIUM CHANNELS FORMATION AND MODULATION IN SECRETING BASOPHILS AND MAST CELLS

I. PECHT and R. SAGI-EISENBERG

1. INTRODUCTION

A transient increase in the cytosolic free Ca^{+2} ions concentration has been shown to be a coupling element between extracellular stimuli and the secretion induced from a wide variety of exocytotic cell types (1-3). The direct demonstration of such a transient change in $[Ca^{+2}]$ in several cell types has recently been accomplished through the development of an elegant and specific probe which enables direct monitoring of that parameter (4-6). The details of the machinery involved in permeating the cells plasma membrane to Ca^{+2} ions influx, in modulating it and controlling the cytosolic levels of this ion, are now beginning to emerge.

Mast cells and basophils serve as an instructive system for studies of the coupling mechanism between a stimulus and the ensuing secretion. The common stimulus to secretion of the above cells, is usually by a specific antigen. The respective antibody of the IgE class resides in the cells' membrane onto which it binds with high affinity to a specific Fc_ε receptor (7,8). Antigen binding usually leads to crosslinking of the Fc_ε receptors. This has been suggested to be the actual primary membrane stimulating event. Thus alternative modes of crosslinking Fc_ε receptors, e.g. by antibodies specifically raised against them (9,10), have also been shown to trigger secretion. The membranal and cellular events following the primary crosslinking event and their time course are being actively investigated (11).

The role of transiently increasing Ca^{+2} ions concentration in the cytosole was examined by employing Ca^{+2} ionophores (12), or membrane permeabilization in the presence of Ca^{+2} buffers (13), microinjection of Ca^{+2} (14) and fusion with Ca^{+2} containing liposomes (15). The results established that such a rise in $[Ca^{+2}]$ is a sufficient requirement to evoke secretion. Hence it was suggested that Fc_ε^R aggregation leads to a transient opening of plasma membrane Ca^{+2} channels (16,17).

Our main effort has concentrated on resolving the Ca^{+2} channel forming elements in these cells membranes and their regulation. The two main cell types subjected to detailed biochemical investigation along the above lines are rat peritoneal mast cells and the rat basophilic leukemic line 2H3. The latter cell line has the advantage that the large numbers of cells required for biochemical experimentation and preparations, can be grown in vitro (18). We succeeded in isolating from the RBL-2H3 cells, a membrane component forming the Ca^{+2} channel, responding to Fc_{ε} receptor crosslinking (19). Evidence implicating the Ca^{+2} and phospholipid dependent protein kinase C, recently isolated from these RBL cells (20) in the modulation of cytosolic free Ca^{+2} concentration is now also coming forth.

2. PROCEDURE

2.1 Materials and methods

2.1.1. Reagents. Quin-2 AM (lot 393105) and the ionophore A23187 were purchased from Calbiochem, Luzern Switzerland. Stock solution of 10 mM Quin 2 AM in anhydrous dimethylsulfoxide was stored at -4°C till use. [^{3}H] TPP$^+$ (bromide salt, 2.5 Ci/mmol) was purchased from the Nuclear Research Center, Negev, Beer-Sheva, Israel. γ-^{32}P ATP (triethylammonium salt (>5000 Ci/mmol) and [^{3}H] hydroxytryptamine creatinine sulfate (11 Ci/mmol) were purchased from the Radiochemical Centre. Phospholipids (bovine brain) and other chemicals, unless otherwise specified, were obtained from Sigma. DEAE-cellulose (DE-52) was obtained from Whatman. IgE was a monoclonal anti-DNP antibody generously provided by Dr. Z. Eshhar. The antigen was a conjugate of eight molecules of DNP per one molecule of bovine serum albumin (19).

2.1.2. Cells. Rat basophilic leukemia cells (RBL-2H3) were maintained in 250 ml tissue culture flasks (Nunc. Inter.Med, Denmark) in Eagle's minimal essential medium (Gibco Island, N.Y.) supplemented with 15% heat inactivated fetal calf serum (Biolab, Jerusalem, Israel), 4 mM glutamine and antibiotics in 5% CO_2 at 36°C. The cells were subcultured every 48 hr. Attached cells were suspended by repeated pipettation, washed and transferred to new flasks in 20 ml of medium (about 2×10^6 cells/flask).

2.1.3. Quin 2 loading and fluorescence measurements. The RBL-2H3 cells were incubated in batches ($5x10^7$ cells/ml each) after being sensitized with the monoclonal, DNP specific, IgE (1 hr, 37°C) and washed three times with Tyrode buffer. Cells were incubated in the presence of 100 µM Quin 2-AM for 15 min at 37°C. To the incubation medium an esterase inhibitor was added (BW 284 C51, Wellcome Reagents Ltd) to minimize extracellular hydrolysis of the Quin 2-AM. After incubation, aliquots of $2x10^6$ cells suspensions were centrifuged, supernatants discarded and the pellet resuspended in 2 ml of phosphate-free medium in a 1 cm optical path quartz cuvett. An electrically driven mini stirer was inserted into the cuvette ensuring continuous homogeneity of the suspension without affecting the fluorescence measurement. A thermostated (37°C) Perkin-Elemer MPF-44A spectrofluorometer was employed (Excitation at 339 nm, emission at 492 nm). Uptake of Quin 2-AM and hydrolysis of the acetoxymethyl ester were monitored by following the changes in emission spectra of both the cells and supernatants.

$[^3H]TPP^+$ distribution in membrane potential determination were performed according to our previously described procedure (21).

2.1.4. Degranulation experiments. Cells harvested 48 hr after passage, were suspended, washed and resuspended ($1x10^6$ cells/ml) in Tyrode medium containing 137 mM NaCl, 2.7 mM KCl, 0.4 mM NaH_2PO_4, 2 mM $CaCl_2$, 1 mM $MgCl_2$, 20 mM Hepes and 5.6 mM glucose, pH 7.4. The cells were incubated with a monoclonal, DNP specific IgE antibody and 5µCi/ml ^3H-5-hydroxytryptamine creatinine sulfate (11 Ci/mmol, Radiochemical Centre, Amersham) for 1 hr at 37°C. After washing with Tyrode buffer, degranulation was induced by adding 100 µl cell suspension aliquots of each of a 96-wells flat-bottom microtiter plate (Nunclon) containing the indicated reagents. At the end of 1 hr incubation at 37°C, the cells were spun down and 100 µl aliquots of supernatants counted. The total radiolabeled 5-hydroxytryptamine incorporated into the cells was estimated by dissolving the cells in 50 µl of 2N NaOH and counting the radioactivity of the whole sample. Release is presented as percentage of the total 5-hydroxytryptamine content.

2.1.5. Protein kinase C, isolation and assay procedure. Cells harvested 48 to 72 hr after passage (about $5x10^7$ cells) were washed and resuspended in 2.5 ml of buffer containing 20 mM Tris-HCl, pH 7.5, 50 mM 2-mercaptoethanol and 10 mM EDTA at 0 to 4°C. The cells were sonicated for 30 seconds and the sonicate centrifuged for 5 min at 600 g to sediment unbroken cells. The supernatant was applied to a DE-52 column (3x1 cm) pre-equilibrated with a buffer containing 20 mM Tris HCl pH 7.5, 20% sucrose, 50 mM 2-mercaptoethanol and 2 mM EDTA. The column was then washed with 22.5 ml of this buffer and the protein kinases eluted by a 30 ml linear gradient of NaCl (0-0.4 M) in the same buffer. Fractions of 1.5 ml each were collected.

Protein kinase activities were assayed by measuring the incorporation of the radioactive phosphate of γ-^{32}P ATP into calf thymus histone H1. The complete assay mixture for protein kinase C activity contained (0.25 ml): 20 mM Tris HCl pH 7.5, 10 mM $MgCl_2$, 100 µg H1 histone, 20 µM-^{32}PγATP (0.5 Ci/mmol), 2 mM $CaCl_2$ (unless stated otherwise), 160 µg/ml PS and 50 µl of the fraction to be assayed. After incubation for 10 min at 30°C the reaction was stopped by transferring 100 µl aliquots on separate 2.5 cm squares of phosphocellulose paper. These papers were washed and left overnight in 10% trichloroacetic acid, dried and their radioactivity measured by liquid scintillation spectrometry.

3. RESULTS AND DISCUSSION

Early studies of the increase in calcium influx into triggered mast cells and basophils relied on tracer experiments employing $^{45}Ca^{+2}$. These were however limited in their resolution of both the time course and the precise compartmentation of the Ca^{+2} ions. The recent development of the calcium indicator Quin-2 solved this problem (4-6). This indicator is taken up into the cells as the membrane soluble, acetoxy methyl ester and upon hydrolysis by intracellular esterases is converted into the fluorescent chelator-probe which monitors directly Ca^{+2} concentrations. Earlier attempts to directly monitor changes induced in [Ca^{+2}] upon cell stimulation employed the chromophoric indicator murexide and were limited to a narrow range of Ca^{+2} ions concentrations (22). As illustrated in Fig. 1a,stimulation by antigen of RBL cells causes a graded increase in the intracellular free Ca^{+2} ions concentration as reflected by the Quin 2 emission. This process is dependent on the presence of external Ca^{+2},indicating that Ca^{+2} influx into the cells, has indeed occurred.

Activation of the channel responsible for the observed Ca^{+2} influx could be achieved by a change in transmembranal electrical potential similar to that observed for example in nerve cells. Alternatively,the channels may be activated by a receptor-ligand interaction.

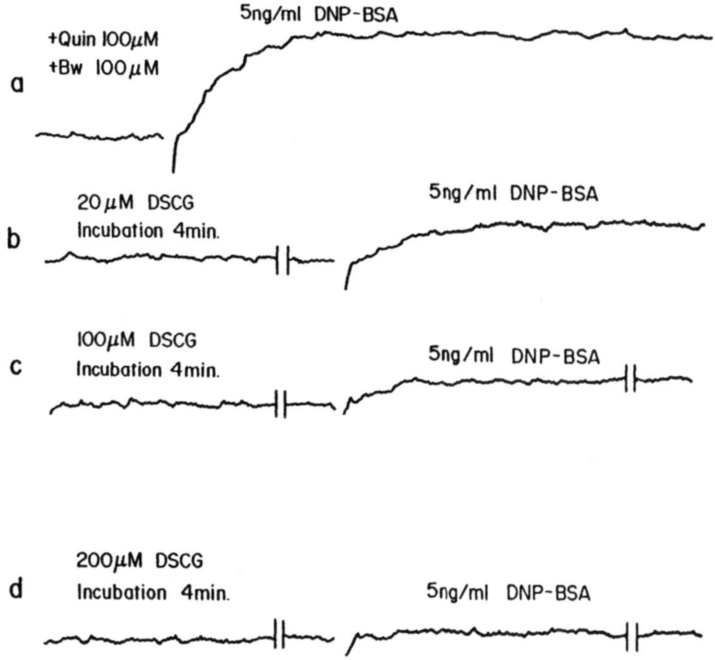

FIGURE 1. Fluorescence emission changes of the Ca^{+2} indicator Quin-2 in RBL-2H3 cells in response to stimulation by antigen and the anti-allergic drug cromolyn. The cell suspension ($1x10^6$/ml) loaded with Quin-2 AM and a DNP specific monoclonal IgE, as described under materials and methods were stimulated by addition of the antigen- a polyvalent conjugate of dinitrophenyl groups on bovine serum: albumin (DNP_8-BSA). Temperature of the constantly stirred suspension 37°C, excitation 339 nm, emission 492 nm.

a) Increase in cytosolic free Ca^{+2} concentration upon addition of 5 ng/ml of DNP_8BSA.
b) Effect of 20 μM cromolyn on the above process.
c) Effect of 100μM cromolyn on the above process.
d) Effect of 200μM cromolyn on the above process.

We have examined membrane potential changes at different functional states of the RBL-2H3 cells by monitoring the transmembranal distribution of the lipid-soluble cation, tetraphenyl-phosphonium ion in its radioactive labeled form (23). The crosslinking of IgE, brought about either by a specific antigen or alternatively by anti-IgE antibodies, caused a marked depolarization of the RBL cells. This process paralleled secretion (21). However, depolarization of the membrane induced by manipulation of the ionic composition of the cells medium, failed to induce degranulation (Figure 2).

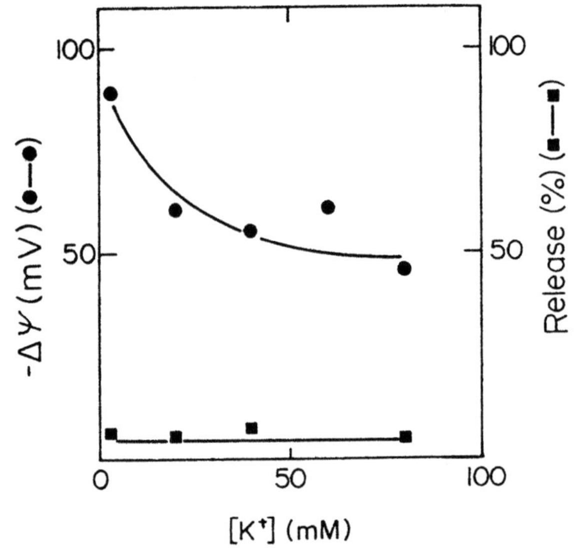

FIGURE 2. The change in overall membrane potential, $-\Delta\Psi$ (●-●) and serotonin release (■—■) caused by increasing the extracellular potassium ions concentration. The membrane potential was monitored by following the [^{3}H] TPP$^+$ uptake by the cells and the release by measuring radioactive serotonin. (For further details see material and methods).

Hence, this and other experiments led us to conclude that the depolarization observed upon triggering the cells to secretion, is a secondary event rather than part of the direct coupling process (24). Furthermore, the decrease in the cells'overall potential is most probably due to depolarization of the mitochondrial membrane caused by its response to the increase in cytosolic Ca^{+2} concentration (24). Thus, the increase in membrane permeability toward Ca^{+2} induced by the immunological stimulus is not mediated through a potential activated membrane channel but rather by a receptor - ligand mediated one.

A major tool in our effort to identify and eventually isolate this ligand activated Ca^{+2} channel has been a drug. This was the anti-allergic asthma prophilactic agent: cromolyn (1,3-bis (2 carboxychromon-5-yloxy)-2-hydroxy propane) (25,26). This drug was shown to specifically bind to the external membranes of rat peritoneal mast cells and the RBL-2H3 cells (27, 28). In later studies we have shown that this binding is related to the inhibitory effect this drug exhibits on the antigen-induced degranulation of these cells (27,28). That the inhibition of secretion is indeed caused by blocking of the antigen induced Ca^{+2} influx is illustrated in Figure 1 (b,c,d). Thus increasing concentration of the drug cromolyn causes a clear decrease in the Quin-2 amplitude induced by a constant dose of antigen.

The cromolyn-binding membrane component of the RBL cells was isolated by two different methods, both based on its affinity to the drug (29). More recently a preparative procedure was designed which allows isolation of this protein in amounts (∿100 μg) providing for the start of its chemical characterization.

The functional role of the cromolyn binding protein (CBP) has and is being pursued in several ways. On the level of the whole cell, we have selected variants of the RBL-2H3 cells which are defective in their capacity to bind cromolyn (30). These variants were shown to be unable of responding to an immunological stimulus by a specific antigen and undergo degranulation. In contrast, however, by using a Ca^{+2} ionophore (A-23187) which directly introduces calcium through the cells' membrane, the degranulation of these cells is induced. This suggested that the defect of these variants is in a membrane element involved in the Ca^{+} channel forming system (30). Further support to this notion came from experiments where the purified CBP has been implanted into the membrane of the variants using vesicles made from Sendai virus envelopes as a vehicle (31). Fusion of the CBP containing vesicles with the variants membranes leads to the restoration of these cells' capacity to degranulate in response to an IgE-mediated stimulus (31). Specific inhibitors have clearly established that this functional restoration is indeed due to the implanted CBP making it the prime candidate for a Ca^{+2} gate element.

On the molecular level, we have investigated the electrical conductance of planar lipid bilayers containing CBP alone or together with other components of the RBL plasma membranes. Results of these studies allow to identify the the Ca^{+2} channel forming membrane component as the CBP (32). In lipid bilayers containing the unfractionated components of the RBL cells membrane, can be induced by crosslinking Fc_{ϵ} receptor through IgE residing in them. Similarly, conductance was also induced in such membranes by aggregation of the CBP with a monoclonal antibody raised against this protein. When the planar lipid bilayer were prepared with CBP only, similar conductances were induced by the same crosslinking procedure with the monoclonal antibody. In the presence of Ca^{+2} ions, 2 pS single channel conductances were recorded. This persisted also in the presence of both Ca^{+2} and Na^{+} ions. However, when only sodium ions were present, single channel conductance of 10 pS were observed. Both time and amplitude characteristics of the open single channels were independent of the applied voltage. Finally, these channels were shown to be blocked by both cromolyn and specific blocking monoclonal antibodies (32). More detailed studies are currently in progress aimed at determining the mechanism of channel formation, the relationship between CBP and other membrane components involved in its function, the ion selectivity of the channel and the stoichiometry of channel formation in terms of CBP.

As it emerges from the above described conductance studies, formation of the open Ca^{+2} channel requires aggregation of two or more CBP molecules (31,32). Naturally further details of the gating mechanism of this system need to be resolved. Apart from the obvious channel closure by dissociation of CBP aggregates, modulation by chemical modification such as phosphorylation may also be operative. A possible link between the Ca^{+2} gating process and protein phosphorylation emerged from the observation that cromolyn induces in mast cells specific phosphorylation of a single protein (33,34). The recently discovered protein kinase C may be the intracellular controlling element (35). Indeed this enzyme has also been demonstrated in RBL-2H3 cells and partially purified and characterized from them (20). Further evidence for the involvement of kinase C in regulation of Ca^{+2} permeability comes from studies employing phorbol esters which are activating ligands of this enzyme (36,37). Protein kinase C has recently been identified as having a central role in control of growth and proliferation of various cell types (36). The enzyme activity has an absolute requirement for Ca^{+2} ions and phospholipid, notably phosphatidylserine (PS). We have also shown it to exhibit a characteristic specificity for exogeneous substrates which is different from that of the cAMP dependent kinase (38). Two further interesting ligands were found to directly activate kinase C; one is diacylglycerol (DAG) which is cellularly produced upon receptor activated hydrolysis of phosphoinositides (39), hence linking kinase C activity to membrane lipid turnover. The second is the synthetic tumor promoting agent: 12-0-tetradecanoylphorobol 13-acetate (TPA). This ligand, like DGA, activates the kinase by increasing its affinity for Ca^{+2} ions by more than three orders of magnitude (20).

In the intact RBL-2H3 cells, TPA by itself fails to change the basal level of secretion. However, at a concentration range of up to 15 nM, it markedly increases the secretion induced by suboptimal doses of antigen (Figure 3).

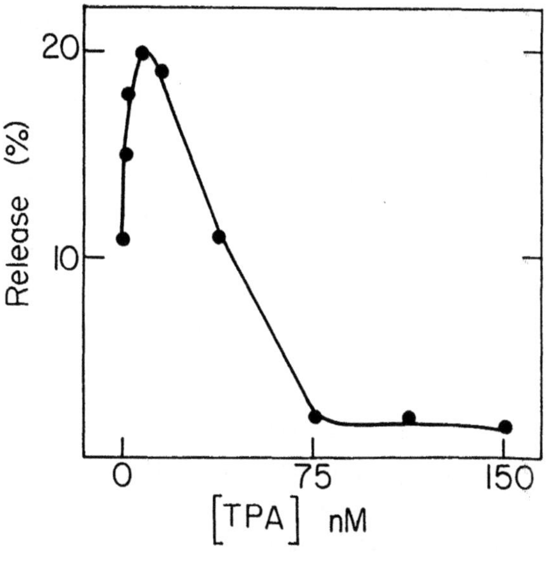

FIGURE 3. Concentration depending of RBL-2H3 degranulation on TPA. Serotonin release was monitored, induced by 1.3 ng/ml of DNP_8-BSA at the indicated concentrations of TPA. Further details under materials and methods.

Further rise in TPA concentration causes a decrease in degranulation and above 75 nM complete inhibition of secretion is observed. In contrast, TPA was found to synergize with the ionophore A-23187 causing a threefold increase in secretion at ~15 nM TPA and a suboptimal dose of 100 ng/ml of ionophore. Furthermore, this synergism levels off and no inhibition of degranulation takes place at higher concentrations of TPA (38). Thus, we observe a selective inhibition of RBL degranulation caused by TPA when the stimulus is via $Fc_{\varepsilon}R$ crosslinking, whereas triggering by allowing direct influx of Ca^{+2} ions by the ionophore, is not subjected to this constraint. These experiments clearly indicate the regulatory role kinase C has in the receptor-dependent Ca^{+2} channeling process. Moreover, this enzyme is involved in terminating the degranulation process as illustrated by the biphasic dose response to TPA (Figure 3). While the Fc_{ε} receptor cross-linking process leads to Ca^{+2} channel opening and a controlled net influx of these ions which is limited by the TPA (Figure 4), the essentially free influx of Ca^{+2} carried by the ionophore, leads to higher levels of secretion which, are not inhibited by excess TPA. Since protein kinase C is the prime cellular target of TPA (37), this enzyme must occupy a key regulatory role (24).

Once again Quin-2 is employed for monitoring the free cytosolyc Ca^{+2} concentrations as affected by TPA (Figure 4). While it did not cause by itself changes in the $[Ca^{+2}]$level, it inhibited, in a time dependent fashion the increase in $[Ca^{+2}]$ induced by antigen (20): If TPA is added prior to antigen, the rise in $[Ca^{+2}]$ is inhibited while if added afterwards, depending of the specific point in time, the Ca^{+2} influx is gradually terminated (Figure 4). When secretion is parallely monitored, one observes that it may linger on after the transient rise in $[Ca^{+2}]$ has decayed. This is not surprising if the signalling role of that intransient is considered. Thus, as soon as the degranulation has been started by the rise in $[Ca^{+2}]$, its decay most probably can not stop it.

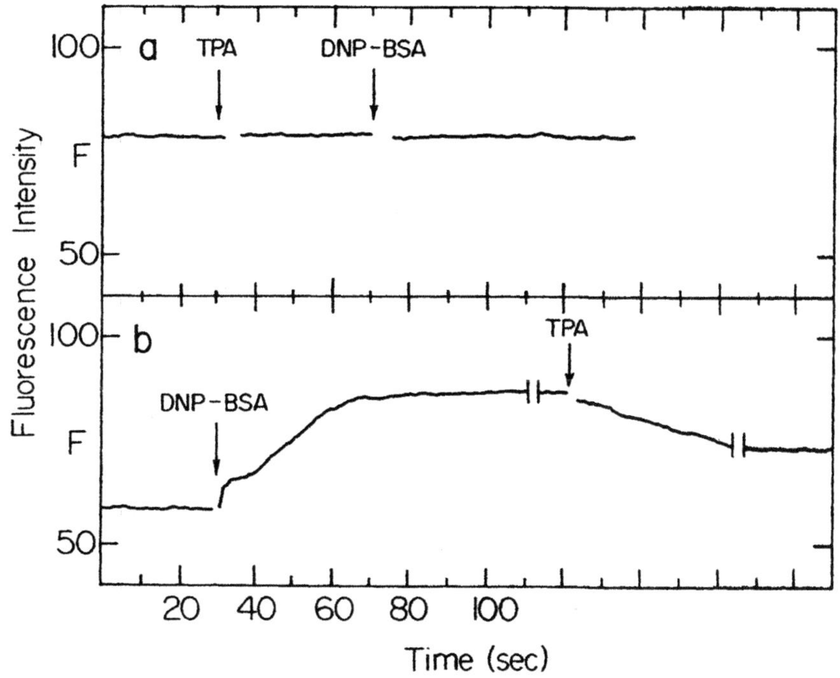

FIGURE 4. Effect of TPA on antigen induced increase in cytosolic free Ca^{+2} concentrations monitored by the emission of Quin-2. a) Exposure of the cells to 1.5 nM TPA for ca. 30 sec and then to 5 ng/ml of DNP_8-BSA. b) Stimulation of the cells by 5 ng/ml of DNP_8-BSA followed by the introduction of 75 nM after 2 min.

In Figure 5 a tentative summarizing scheme is presented. It illustrates the primary events of $Fc_{\varepsilon}R$ aggregation which brings about the associative Ca^{+2} channel formation and the net influx of these ions. The proposed changes in phospholipids which yield ligands like DAG for the kinase C are also assumed to be early events initiated by the $Fc_{\varepsilon}R$ clustering. Hence kinase C is activated already at an early stage and would keep the secretion within limits.

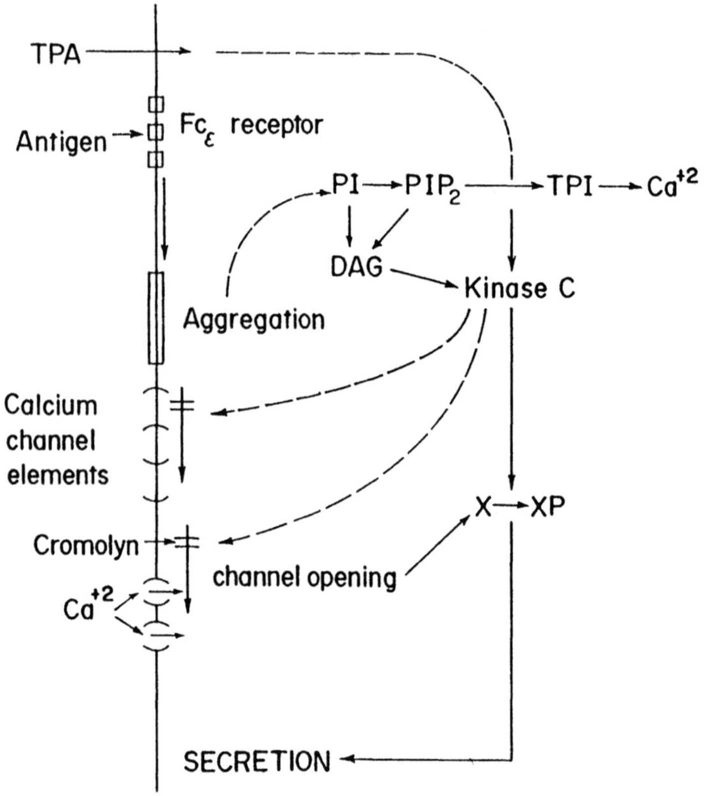

FIGURE 5. Tentative scheme describing several events linking stimulus of the basophils by antigen with secretion.

REFERENCES

1. Douglas WW: Ciba Found. Symp. 54:61-90, 1978.
2. Michell RM, Kirk CJ, Jones LM, Downes CP, Creba JA: Philos. Trans. Ray. Soc. Lond. B 296:123-137, 1981.
3. Foreman JC, Hallet MB, Mongar JL: J. Physiol. (London) 271:193-214, 1977.
4. Tsien RY: Biochem. 19:2396-2404, 1980.
5. Tsien RY: Nature 290:527-528, 1981.
6. Tsien RY, Pozzan T, Rink TJ: J. Cell Biol. 94:325-334, 1982.
7. Froese A: in K. Ishizaka (ed.) Mast cell activation and mediator release. Prog. Allergy vol. 34. S. Karger, Basel pp. 142-187, 1984.

8. Metzger M, Goetze A, Kanellopoulu J, Holowka D, Fewtrell C: Fed. Proc. 41:8-11, 1982.
9. Ishizaka T, Tomioka H, Ishizaka K: J. Immunol. 106:705-710, 1971.
10. Ishizaka T, Ishizaka K: J. Immunol. 120:800-805, 1978.
11. Ishizaka T, Ishizaka K: in K. Ishizaka (eds) Mast cell activation and mediator release. Prog. Allergy, vol 34. S. Karger, Basel pp 188-235, 1984.
12. Foreman JC, Mongar JL, Gomperts BD: Nature 245:249-251, 1973.
13. Bennet JP, Cockfroff S, Gomperts BD: J. Physiol. 317:335-345, 1981.
14. Kanno T, Cochrane DE, Douglas WW: Canad. J. Physiol. Pharmac. 51: 1001-1004, 1973.
15. Theoharides TC, Douglas WW: Science 202:1143-1145, 1978.
16. Foreman JC, Mongar JL: Physiol. 230:493-507, 1973.
17. Gomperts BD. in P. Cuatrecasas and M.F. Greaves (eds), Receptors and recognition. Ser. A vol. 2, Chapman and Hall, London pp 43-102, 1976.
18. Siraganian RP, McGivney A, Barsumian EL, Crews FT, Hirata F, Axelrod J, Fed. Proc. 41:30-34, 1982.
19. Mazurek N, Bashkin P, Pecht I: EMBO J. 1:585-590, 1982.
20. Sagi-Eisenberg R, Pecht I: Immunol. Lett. 1984, in press.
21. Sagi-Eisenberg R, Pecht I: J. Memb. Biol. 75:97-104, 1983.
22. Sagi-Eisenberg, R, Geller-Bernstein C, Ben-Neriah Z, Pecht I: FEBS Lett. 161:37-40, 1983.
23. Lichtshein D, Kaback HR, Blume AJ: Proc. Natl. Acad. Sci. USA, 76:650-654, 1979.
24. Sagi-Eisenberg R, Pecht I: EMBO J. 3:497-500, 1984.
25. Cox JSG, Beach JE, Blair AMJN, Clarke AJ, King J, Lee TB, Loveday DEE, Moss GF, Orr TSC, Ritchie JTA: Adv. Drug Res. 5:115-194, 1979.
26. Altounynan REC: Clin. Allergy 10:481-489, 1980.
27. Mazurek N, Geller-Bernstein C, Pecht I: FEBS Lett. 111:194-196, 1980.
28. Mazurek N, Berger G, Pecht I: Nature 286:722-723, 1980.
29. Mazurek N, Bashkin P, Pecht I: EMBO J. 1:585-590, 1982.
30. Mazurek N, Bashkin P, Petrank A, Pecht I: Nature 303:528-530, 1983.
31. Mazurek N, Bashkin P, Loyter A, Pecht I: Proc. Natl. Acad. Sci. USA 80:6014-6018, 1983.
32. Mazurek N, Schuerholz Th, Schindler H, Pecht I: Proc. Natl. Acad. Sci. USA 81: 1984.
33. Theoharides TC, Sieghart W, Greengard P, Douglas WW: Science 207: 80-82, 1980.
34. Wells E, Mann J: Biochem. Pharmac. 32:837-842, 1983.
35. Sagi-Eisenberg R: Trends in Pharm. Sci. in press.
36. Castagna M, Takai N, Kohuchi K, Sano K, Kikkawa U, Nishizuka Y: J. Biol. Chem. 257:7847-7851, 1982.
37. Parker PJ, Stabel S, Waterfield MD: EMBO J. 3:953-959, 1984.
38. Sagi-Eisenberg R, Ross M, Pecht I: Submitted for publication.
39. Kishimoto A, Takai Y, Mori T, Kakhano Y, Nishizuka Y: J. Biol. Chem. 255:2273-2276, 1980.

DISCUSSION OF THE ARTICLE BY I. PECHT (ISRAEL)

M. Lazdunski (France): How much of the drug binding protein with the molecular weight of 60-70,000 do you find in the membrane? Is it a largely represented protein?

I. Pecht (Israel): It is as prevalent in the parental rat basophilic leukemia cells, from which we are isolating it as the Fc_E receptor. For both of these proteins, the figure is in the range of several hundred thousands molecules per cell.

M. Lazdunski (France): What percent of the membrane protein do they represent?

I. Pecht (Israel): Roughly up to one percent.

M. Lazdunski (France): What knowledge do you have on how the drug acts on the secretion system? Do you have a structure-function relationship for this drug and are the chelating properties of the drug necessary for its action.

I. Pecht (Israel): When the calcium influx into RBL cells is monitored by Quin II, it is clear that the drug blocks it in a very effective way (Sagi-Eisenberg R. and Pecht I. (1984). Immunol. Lett. In press.). As far as the drug's structure-function relationship goes, we did not persue it. It is a drug that is available on the market for more than 15 years (Cox J.S.G., Beach J.E., Blair A.M.J.N., Clarke A.J., King J., Lee T.B., Loveday D.E.E., Moss G.F., Orr T.S.C. and Ritchie J.T.A. (1979). Adv. Drug Res. 5, 115-194; Altounynan R.E.C. (1980). Clin. Allergy 10, 481-489). With reference to the relationship between the calcium chelating property of the drug and its activity, we know one thing: The fact that you don't have antigen induced degranulation without having calcium, is obviously clear. Significantly, however, you do not get binding of the drug to the cell without calcium ions present. This is a very persistent obvservation (Mazurek N., Berger G. and Pecht I. (1980). Nature 286, 722-723). So whether indeed, the particular Ca^{2+} chelate that we observed in organic solvents is the one that operates at the cell membrane, I cannot tell you in a clear cut fashion. But it makes alot of sense that it is.

M. Simonneau (France): How many binding sites are necessary for a single cell to be positive?

I. Pecht (Israel): That is a technical question because it depends on the fluorescent probe that you use. If you use what has become very popular, namely, phycobiliproteins as probes (Glazer V. and Stryer L. (1982). J. Cell Biol. 93, 981), it is claimed that you can see a few hundred molecules. It is naturally a more complex question to answer.

M. Simonneau (France): How many calcium channels do you estimate per unit?

I. Pecht (Israel): The number of cromolyn binding protein molecules per cell (basophil) is in the range of 500,000. That we determined from direct binding of a tritium labeled derivitive of the drug.

INITIAL CHARACTERIZATION OF THE CALCIUM CHANNEL ACTIVATED BY THE CROSSLINKING OF THE RECEPTORS FOR IMMUNOGLOBULIN E*

Baruch I. Kanner[1] and Henry Metzger[2]
[1]Department of Biochemistry, Hebrew University-Hadassah Medical School, POB 1172, Jerusalem
[2]Section on Chemical Immunology, Arthritis and Rheumatism Branch, National Institute of Arthritis, Diabetes, and Digestive and Kidney Diseases, National Institutes of Health, Bethesda, Maryland 20205

1. INTRODUCTION

The plasma membranes of mast cells, basophils and a tumor analog, rat basophilic leukemia (RBL) cells, contain a glycoprotein which binds the Fc portion of monomeric IgE with high affinity. Multivalent antigens combining with the Fab regions of the bound IgE or any other procedures which aggregate this IgE, induce exocytotic, calcium-dependent release of mediators such as histamine and serotonin. The aggregation of the IgE-binding glycoprotein in the plane of the membrane is a sole and sufficient signal to initiate calcium-dependent release (reviewed in 1). Non-cytotoxic exocytosis can also be affected by calcium ionophores (2,3).

When cells are incubated with $^{45}Ca^{2+}$ and the receptors are aggregated, an increase in cell-associated $^{45}Ca^{2+}$ is observed (4-6). These observations led to the proposal, reviewed in (7), that the release process can be divided into two stages: a) A receptor mediated entry of Ca^{2+} into the cell and b) excytosis initiated by the rise in the cytoplasmic Ca^{2+} concentration.

2. RESULTS

2.1. IgE-Mediated release causes depolarization.

Addition of the polyvalent DNP-IgG to cells bearing mouse IgE antibodies to DNP led to substantial release of incorporated tritiated serotonin (Fig 1A). Cells that had been exposed instead to rat IgE lacking anti-DNP activity showed minimal release-equivalent to that obtained with the cells bearing the mouse IgE but incubated without Ca^{2+}. Fig 1A also shows the effect of adding 20μM DNP-lysine. The concentration of the hapten has no effect on IgE-mediated release in general [e.g., that induced by anti-IgE, data (not shown)] but is sufficient to disrupt the binding of the antigen (DNP$\leq$10nM) to the IgE antibody which has an intrinsic binding constant of $7 \times 10^7 M^{-1}$ at 37°C (10). The release of the hydroxytryptamine stopped abruptly, a result previously obtained with this system (4) as well as with an analogous one (3).

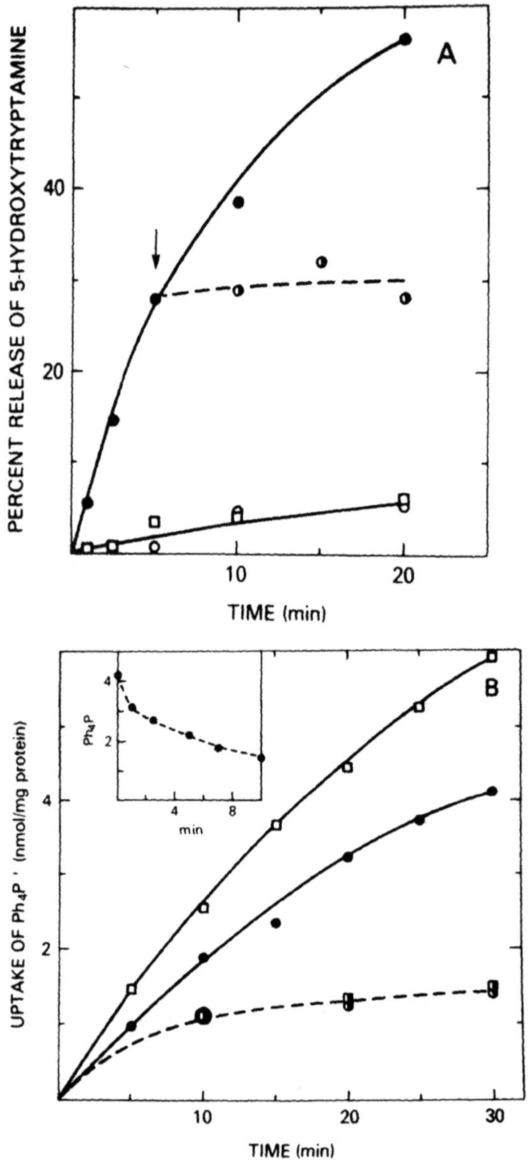

FIGURE 1. Effect of DNP-IgG on cells reacting with mouse IgE anti-DNP and IgE lacking anti-DNP activity. (A) Release of incorporated 5-hydroxy[^{3}H]tryptamine. ●,o,◐, Cells loaded with mouse IgE: ●, control; o, without Ca^{2+}; ◐, addition of DNP-lysine 5 min after addition of DNP IgG; □, cells loaded with rat IgE. (B) Uptake of [^{3}H]Ph_4P^+. ●,◐, Cells loaded with mouse IgE: ●, control; ◐, with CCCP added. □,◧, Cells loaded with rat IgE: □, control; ◧, with CCCP added. (Inset) Control with CCCP (final concentration, 5 μM) added at 30 min.

Fig 1B shows a release experiment carried out under conditions similar to those in Fig 1A except that the tritiated serotonin was omitted and [^{3}H]tetraphenyl phosphonium (Ph_4P^+ - an indicator of the membrane potential) was added at the same time as the antigen. (In control experiments, Ph_4P^+ up to 5 μM had no effect on the release of serotonin.) Rapid uptake of the ion was seen with the cells bearing the non-antibody rat IgE. Under conditions that trigger degranulation (Fig 1A), the cells bearing the mouse IgE showed considerably decreased uptake. In the absence of the DNP-IgG antigen, the cells bearing the alternative IgEs showed equivalent uptake of Ph_4P^+ at 30 min (mean$\pm$SD). Cells with rat IgE, 6.07$\pm$0.01 nmol/mg protein; cells with mouse IgE, 6.04$\pm$0.08 nmol/mg protein. The protein ionophore CCCP, which at 5 uM is expected to depolarize the cells fully, strongly inhibited the uptake of the Ph_4P^+ and to equilavent levels with the cells bearing either type of IgE. When the ionophore was added to cells that had accumulated Ph_4P^+ over 30 min, efflux of the cation was observed (Fig 1B, Inset), and after 10 min, the level was similar to that observed with cells exposed to the ionophore throughout (Fig 1B). A similar low level of bound Ph_4P^+ was observed when the cells were incubated in a medium in which Na^+ was replaced with K^+ - a medium that depolarizes a wide variety of cells. It is likely, therefore, that this residual uptake represents simple binding of Ph_4P^+ rather than voltage-dependent accumulation. In the subsequent experiments, the uptake of Ph_4P^+ in controls incubated with CCCP was subtracted from the experimental values.

The addition of DNP-IgG to cells that bore the rat IgE and had accumulated Ph_4P^+ was without effect. In contrast, those cells bearing the anti-DNP mouse IgE released a considerable amount of the accumulated Ph_4P^+ (Fig 2, upper curves). Release varied between 20% and 40% in different experiments (Table 1).

Studies with other cells (12,13) and with synaptosomes (14,15) indicate that the uptake of Ph_4P^+ reflects both equilibration across the plasma membrane, governed by $\Delta\psi$, and accumulation into mitochondria (which are negative inside compared to the cytoplasm) (14,15). This also seems to be the case with RBL cells: Blocking of the respiratory chain, which generates the mitochondrial $\Delta\psi$ with rotenone and blocking of the proton-translocating ATPase with oligomycin caused a 60% decrease of Ph_4P^+ uptake (Fig 2, zero time point, lower curves). Nevertheless, crosslinking of the receptors for IgE induced depolarization of such cells (Fig 2, lower curves). These results suggest that the depolarization occurs at the level of the plasma membrane. Using the data from this experiment and an estimated volume of 7.5 μl/mg of cell protein, one can calculate that crosslinking of the receptors induced a change in the plasma membrane $\Delta\psi$ from -94mV to -73mV. Upon addition of increasing ammounts of unlabeled Ph_4P^+ (to 100 μM), which preferentially collapses the mitochondrial $\Delta\psi$ (14), the effect of crosslinking the receptors persisted (data not shown). This provides additional evidence that it is the plasma membrane that is depolarized.

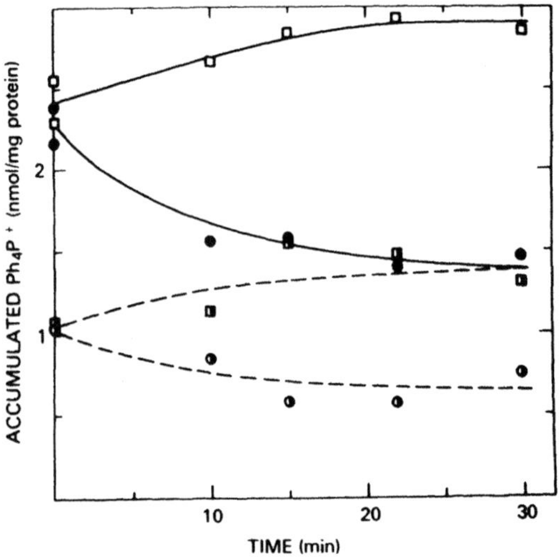

FIGURE 2. Effect of DNP-IgG on that part of the uptake of Ph_4P^+ that is sensitive to uncoupling reagents, in the presence and absence of mitochondrial inhibitors. The cells were incubated with 5 μM Ph_4P^+ (2 μCi per well) for 30 min in the presence or absence of 6 μM oligomycin and 2 μM rotenone, with the appropriate IgE. At zero time, DNP-IgG was added at 1 μg/ml. ●,◐, Cells loaded with mouse anti-DNP IgE: ●, control; ◐, with inhibitors. □ ,◧, Cells loaded with rat nonspecific IgE: □ , control; ◧ , inhibitors.

That the depolarization was dependent upon aggregation of the receptors could also be demonstrated by the use of antibodies directed to the IgE. Addition of the appropriate anti-IgE to cells bearing rat or mouse IgE caused substantial release, whereas with the heterologous combinations there was substantially less secretion. The corresponding data on release of accumulated Ph_4P^+ show the expected correlation of depolarization with receptor-mediated exocytosis. Reaction of antibodies that can agglutinate the tumor cells but do not stimulate exocytosis failed to induce depolarization (data not shown).

2.2. Depolarization is not a consequence of exocytosis.

It was of interest to determine if the observed depolarization preceded or was a consequence of the exocytosis. There is no release of 5-hydroxytryptamine in the absence of Ca^{2+} in the medium (Fig 1A, open circles). Nevertheless, depolarization was observed under these conditions. Our aggretate data (Table 1) show no statistically significant different in the extents of depolarization in the presence and absence of the divalent cation. It should be noted, however, that the absolute magnitude of $\Delta\Psi$ appeared to be reproducibly lower in the absence of Ca^{2+}. We have no explanation for this finding.

Table 1. Effect of Na^+ and Ca^{2+} on receptor-mediated depolarization.

Ion		Depolarization,	Experiments,
Na^+	Ca^{2+}	%	No.
+	+	30.9±10.2	13
+	-	25.0± 7.7	4
-	+	38.5± 4.8	4
-	-	8.2± 2.0	3

*Cells were preincubated with Ph_4P^+ and rat or mouse IgE. The percentage depolarization was calculated by comparing the residual Ph_4P^+ in the cells after addition of DNP-IgG

$$\% \text{ depolarization} = 1 - \frac{Ph_4P^+ \text{ (mouse)}}{Ph_4P^+ \text{ (rat)}} \times 100.$$

In virtually all experiments the comparison was made between the samples taken at several times during the 30 min incubation. Results are shown as means±SD.

2.3. Repolarization when crosslinking of receptors is stopped.

We have already shown the effect on exocytosis when crosslinking of receptors induced by the multivalent DNP-IgG was abrogated by adding monovalent DNP-lysine 5 min later (Fig 1A). The corresponding effects on accumulation Ph_4P^+ are shown in Fig 3. When added to the control cells bearing rat IgG, there was a minimal effect: if anything the hapten caused a small release of Ph_4P^+. However, when added to the cells bearing the mouse IgE the effect was dramatic: instead of the progressive depolarization which reached 40% at 30 min, the cells repolarized and reached control levels within 10 min after addition of the hapten. A similar effect was observed when the hapten was added even 30 min after addition of DNP-IgG (data not shown). These results indicate that the depolarization of the plasma membrane is maintained only so long as there are crosslinked receptors on the membrane. In preliminary experiments, no substantial repolarization was seen when receptor aggregates were not disrupted with hapten, even after 100 min, i.e., after the secretion had plateaued.

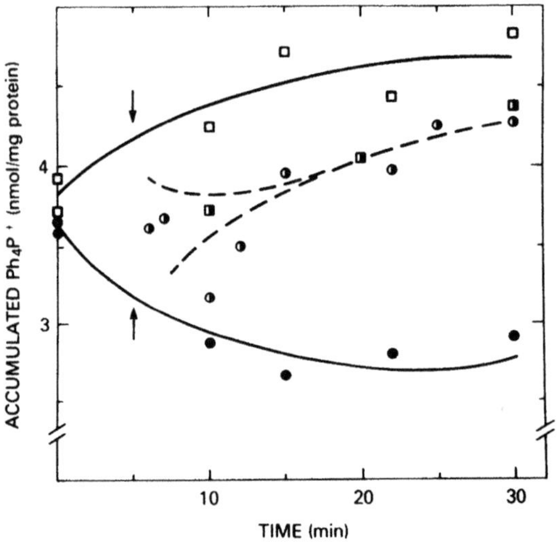

FIGURE 3. Effect of disrupting receptor aggregates on Ph_4P^+ accumulation. Cells were loaded with mouse anti-DNP IgE (circles) or rat nonspecific IgE (squares) and incubated with Ph_4P^+ for 30 min. At time zero, DNP IgG was added. ●,□, Controls; ◐,◧, addition of 20 μM DNP-lysine at 5 min.

2.4. Effect of Na^+ and Ca^{2+} on depolarization.

The depolarization induced by crosslinking of the receptors was observed in the absence of Ca^{2+} (Table 1). This indicates that, if depolarization results from the formation of ion channels, the latter cannot be completely specific for Ca^{2+}. When Ca^{2+} was present but all the Na^+ was replaced by choline, the effect was still observed. However, when both Na^+ and Ca^{2+} were omitted, the depolarization was not seen (Table 1).

An attractive explanation for these findings is that aggregation of the receptors for IgE in the plane of the membrane leads to opening of ion channels that are permeant to calcium, but also to some extent, to sodium. An experimental prediction of this model is that in the absence of calcium, one should be able to measure a sodium flux as a consequence of receptor aggregation and that this flux should be inhibited by calcium.

2.5. Sodium permeability increases upon receptor aggregation.

Addition of polyvalent DNP-IgG to cells bearing mouse IgE antibodies to DNP not only leads to depolarization and calcium-dependent serotonin release, but also causes substantial sodium uptake (Fig 4). This uptake is observed in calcium free medium and containing EDTA, a medium shown to prevent (or abrogate) exocytotic release (1,18) (Fig 1A). Thus, this uptake of sodium is not a consequence of release. The initial rate and the extent of uptake by cells that had been reacted with rat IgE, lacking anti-DNP activity, was much lower (Fig 4). The incubations regularly included 10^{-4}M ouabain in order to inhibit the (Na^+/K^+)-ATPase which would tend to buffer any influx of Na^+. When ouabain was omitted, variable inhibition - ranging from 30 to 80% - of the uptake stimulated by aggregation of receptors observed. This is consistent with the idea that the aggregation induced sodium uptake represents transport across the plasma membrane of the cells. The difference in sodium uptake between cells with aggregated receptors was about 2000 cpm. This represents about 0.05 per cent of the counts added. The internal volume of the cells in each well can be estimated from the value of 7.5 μl of intracellular fluid per mg protein (8). Since in this experiment 320 μg of protein was present per well, the internal volume was 2.4 μl. This is 0.24 per cent of the external volume (1ml). Thus, assuming isotopic equilibration of sodium inside and outside, 20 per cent equilibration was attained at the steady state. Since it is expected that due to the Na^+, K^+-ATPase, there should be an ten-fold sodium concentration gradient (out $>$ in), it appears that sodium uptake at the steady state is of the right order of magnitude.

Two other important features of the sodium uptake are illustrated in Fig 4. First, the process is inhibited by extracellular calcium ions, the inhibition being complete at a concentration of 2mM calcium (Fig 4). Second, addition of 20 μM DNP-lysine also fully inhibits the sodium uptake (Fig 4).

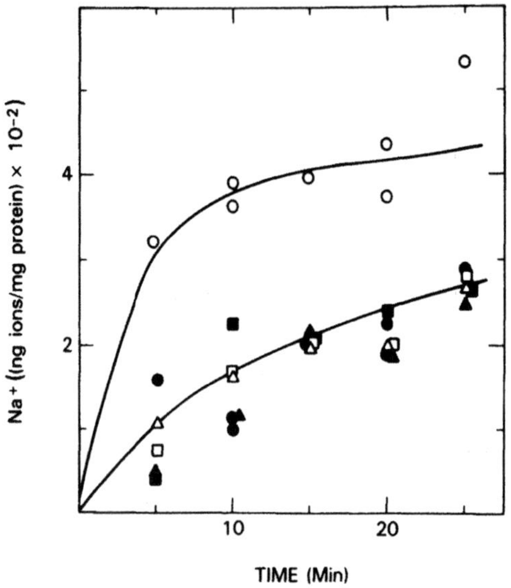

FIGURE 4. Crosslinking of receptor-bound IgE with antigen stimulates the uptake of Na^+ when Ca^{2+} is omitted. The cells were reacted with IgE containing 1mM Ca^{2+} - or Ca^{2+}-free medium containing also 1mM K-EDTA. They were incubated with 1ml of the homologous medium containing in addition 2 µCi [$^{22}Na^+$], 0.1mM ouabain, and 1 µg DNP_{16}-IgG (0.1 µM DNP). As a control, some incubations in addition contained 20 µM DNP-lysine which would inhibit the crosslinking. Open symbols: Cells loaded with mouse anti-DNP IgE and then stimulated with antigen (o) in Ca^{2+}-free medium; (□) in Ca^{2+}-free medium plus DNP-lysine; (Δ) in 1mM Ca^{2+}. Filled symbols: Exactly like above except that the cells had been loaded with non-specific rat IgE.

This concentration of monovalent hapten has no effect of IgE mediated release in general but only on that mediated by DNP-containing antigens (Figs 1 and 3). It also has no direct effect on the membrane potential of the cells (Fig 3). That DNP-lysine at this concentration is not affecting sodium flux per se, is illustrated in Fig 5. In this experiment, aggregation of receptors was initiated by anti-rat IgE antibodies. It can be seen that now the cells containing rat IgE, rather than mouse IgE, display increased sodium uptake and that DNP-lysine now has no effect.

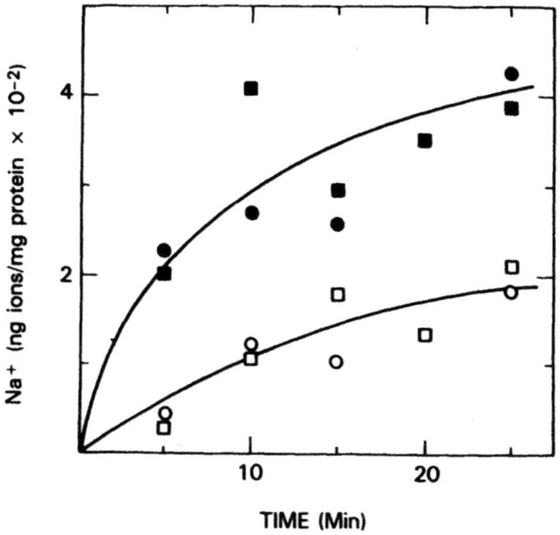

FIGURE 5. Crosslinking of receptor-bound IgE with anti-IgE stimulates the uptake of Na^+ in Ca^{2+}-free medium. The cells were reacted with IgE and then washed with Ca^{2+}-free medium containing 1mM K-EDTA. They were then incubated with 1ml of the same medium containing 2 μCi [$^{22}Na^{2+}$], 0.1mM ouabain, 12 μg anti-rat IgE and with or without 20 μM DNP-lysine. The latter should not inhibit crosslinking in this experiment. Open symbols: Cells loaded with mouse IgE: (o) minus DNP-lysine; (□) plus DNP-lysine. Filled symbols: Cells loaded with rat IgE: (●) minus DNP-lysine; (■) plus DNP-lysine.

2.6. Calcium permeability increases only in the presence of millimolar calcium.

In agreement with previous observations on mast cells (4,5) and RBL cells (6), aggregation of IgE receptors results in an increased calcium uptake (Fig 6). In this experiment, aggregation of the receptors was initiated by DNP-IgG. The cells bearing the mouse (anti-DNP) IgE display markedly increased calcium uptake as compared to cells bearing the rat IgE that lacks anti-DNP acitivty. As expected, the process is inhibited by 20 μM DNP-lysine. The process was also inhibited by valinomycin. This compound, a potassium specific ionophore, is expected to collapse the mitochondrial $\Delta\Psi$ - due to the high cytoplasmic concentration of potassium - and thereby accumulation of calcium into these organelles. The results with valinomycin provide evidence that the calcium taken up from the medium by the cells is transported inwards. It indicates, furthermore, that a substantial portion of the calcium is sequestered into mitochondria or other sites that accumulate this ion by an electrogenic mechanism.

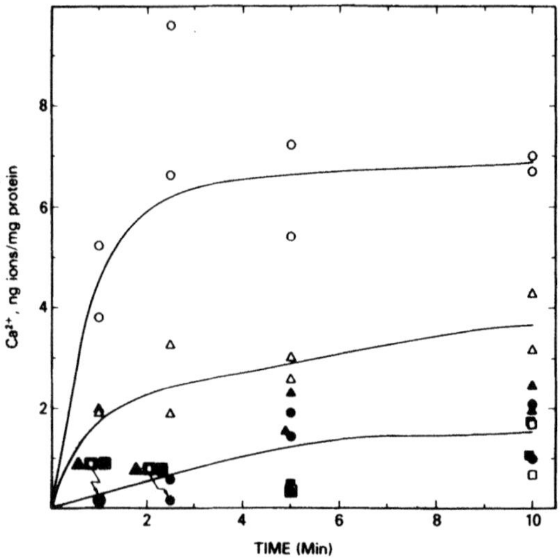

FIGURE 6. Crosslinking of receptor-bound IgE with antigen stimulates the uptake of $^{45}Ca^{2+}$. After loading cells with IgE, they were washed with standard medium (1mM Ca^{2+}) and then studied for uptake of $[^{45}Ca^{2+}]$ as described in Methods, after addition of 1 μg DNP_{16}-IgG (0.1 μM DNP). Open symbols: Cells loaded with mouse anti-DNP IgE and reacted with antigen plus (o) no additions; (□) 20 μM DNP-lysine; (△) valinomycin, 5 μM. Filled symbols: Exactly like above except that the cells had been loaded with rat non-specific IgE.

These experiments were performed with a medium that contained 1mM unlabeled calcium in addition to the 4.5 μM labeled calcium. Calcium flux experiments were also conducted in the absence of added unlabeled calcium. In these instances, no aggregation-dependent uptake of calcium was observed (compare open and filled triangles, Fig 7) in sharp contrast to when 1mM total calcium was present (open and filled circles, Fig 7).

In the absence of millimolar calcium, the uptake was reduced about 20 fold rather than 200 fold. The latter would have been expected if all the uptake sites would have been saturable. Similar findings were made in numerous experiments. Occasionally, cells bearing the rat IgE appeared to accumulate somewhat more calcium than their counterparts bearing mouse IgE when labeled calcium was omitted from the medium.

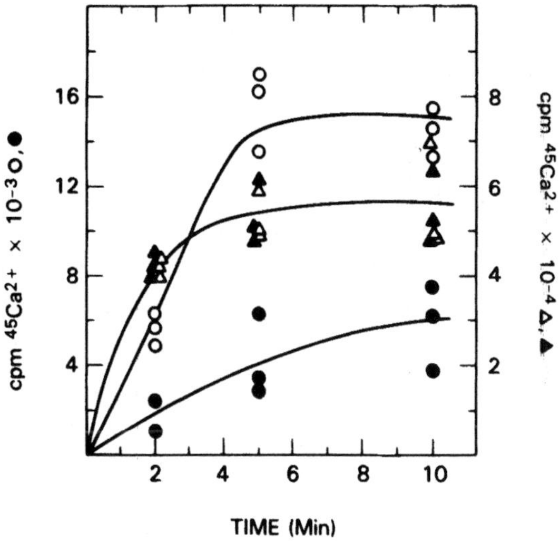

FIGURE 7. The effect of total Ca^{2+} concentration on the uptake of $[^{45}Ca^{2+}]$ after crosslinking receptor-bound IgE with antigen. Cells were loaded with IgE and then washed with standard medium from which Ca^{2+} had been omitted but no EDTA had been added. They were then reacted with 1 μg DNP-IgG and the uptake of $^{45}Ca^{2+}$ measured. Open symbols: Cells loaded with mouse anti-DNP IgE and stimulated with antigen and (Δ) no additions; (o) 1mM Ca^{2+}. Filled symbols: Exactly like above except that the cells had been loaded with nonspecific rat IgE. In this experiment, the cell protein averaged 290 μg/well, and the added $[^{45}Ca^{2+}]$, 8.8 x 10^6 cpm/well.

It has already been noted that the uptake of calcium that is dependent both on the presence of calcium in the medium and on aggregation of the receptors is unlikely to be a consequence of degranulation. As a further check on this as well as to see if the putative channels once opened would remain open after the initial aggregation of the receptors was abrogated, the following protocol was employed. Cells were stimulated with DNP-IgG in the presence of unlabeled calcium for 10 min. The cells were then washed twice with medium containing unlabeled calcium plus 20 μM DNP-lysine. They were then exposed to $[^{45}Ca]$ in the absence of unlabeled calcium and in the presence of 20 μM hapten. No stimulated uptake was observed. These results provide additional evidence that the uptake of calcium is not simply a consequence of secretion.

The mode of aggregation of the receptors is not in itself the critical factor. Thus, when receptors were aggregated by adding species specific anti-rat IgE, the cells bearing rat IgE showed enhanced uptake of calcium compared to cells bearing mouse IgE (data not shown). As expected, the presence of DNP-lysine had no effect in this instance. However, in the absence of adequate calcium, there was no difference between the two types of cells (data not shown).

2.7. The ion channel is not opened by depolarization per se.

Exocytotic release of neurotransmitters from nerves involves voltage-dependent calcium channels. It was of interest, therefore, to determine whether the ion channels studied here could also be opened by changes in potential. In the course of studying aggregation-dependent depolarization, we found that like many other cells, potassium ions play an important role in determining the plasma membrane potential. When all the external sodium was replaced by potassium, a five fold reduction in Ph_4P^+ uptake was observed indicative of a substantial reduction of the membrane potential (Fig 8). This reduction is also seen during aggregation of receptors, i.e., with mouse IgE present (Fig 8).

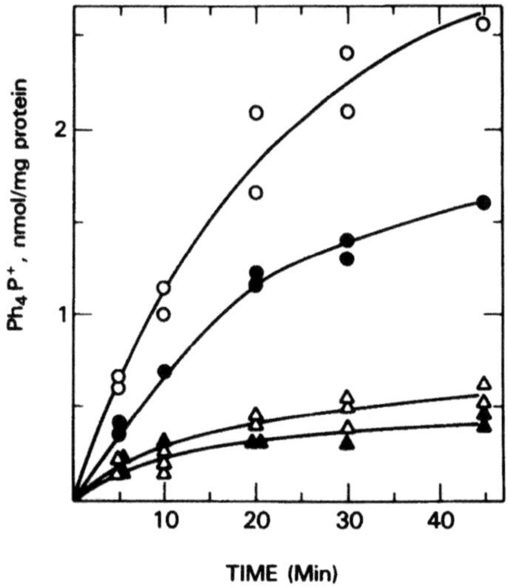

FIGURE 8. High K^+ suppresses the uptake of the potential-sensitive phenylphosphonium probe and modulation of this uptake by aggregating the receptors for IgE. Filled symbols: Cells loaded with mouse anti-DNP IgE and reacted with 1 μg DNP_{16}-IgG in (o) 10mM K^+ or (Δ) 155mM K^+. Open symbols: Exactly like the above except that the cells had been loaded with non-specific rat IgE.

It is clear, however, that unlike with aggregation of receptors, this potassium-induced depolarization by itself is not capable of inducing any release of serotonin (Fig 9). Indeed, in the presence of external potassium, release induced by aggregation of receptors is inhibited (Fig 9). This inhibition is apparently due to the depolarization. That is, with a less negative interior, the driving force for uptake of the positively charged calcium is reduced. Direct evidence in favor of this is the observation that calcium uptake induced by receptor aggregation is strongly inhibited by external potassium (data not shown). Consistent with this explanation we observed that inhibition of aggregation-induced release by potassium was overcome by increasing the concentration of external calcium from 1 to 5mM. Contrariwise in the control, where the receptors were not aggregated, potassium in the presence of 5mM calcium did not induce release (data not shown).

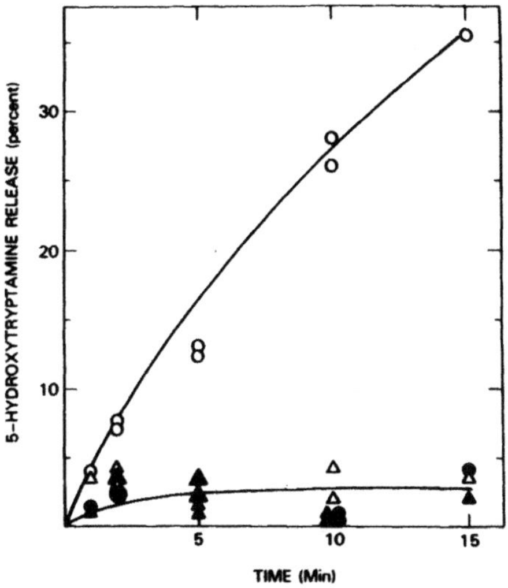

FIGURE 9. High K^+ suppresses the release of serotonin stimulated by the aggregation of the receptors for IgE. Open symbols: Cells loaded with mouse anti-DNP IgE and reacted with 1 μg DNP_{16}-IgC in (o) 10mM K^+ or (Δ) 155mM K^+. Filled symbols: Exactly like the above except that the cells had been loaded with non-specific rat IgE.

3. DISCUSSION

We have observed that RBL cells will accumulate lipophilic Ph_4P^+ in amounts substantially in excess of those seen in the presence of the proton ionophore CCCP. When the mitochondrial respiratory chain and the ATPase that translocates protons were blocked with rotenone and oligomycin, respectively, a residual uptake of Ph_4P^+ was observed. This residual uptake likely reflects the plasma membrane $\Delta\psi$. With reasonable estimates of the cells volume, the calculated value for this potential (-94mV) is similar to that observed with other cells (12).

An average 30% reduction in $\Delta\psi$ was observed when the receptors for IgE were aggregated. This depolarization appears to be a consequence of receptor aggregation per se rather than a consequence of the exocytosis that is thereby induced. Thus, in the absence of Ca^{2+} in the medium - which prevents the exocytosis (Fig 1A) - the depolarization was not substantially diminished (Table 1). The fact that the cells promptly repolarized when receptor aggregation was abrogated early or late in the course of mediator release also suggests that the cells have not simply been made "leaky" during the process of degranulation.

When the cells were depolarized with the ionophore CCCP after having accumulated Ph_4P^+ (Fig 1B, inset), the Ph_4P^+ probe reequilibrated slowly compared to the rate at which the membrane potential is likely to have been discharged. This prevented a precise assessment of the kinetics of depolarization by using this ion. Nevertheless, the evidence is good that receptor aggregation and depolarization are linked kinetically. This was most clearly shown by the experiments in which the aggregation of the receptors induced by the reaction of the multivalent DNP-IgG with the receptor-bound IgE was interrupted by hapten. That secretion was also promptly abrogated indicates that maintenance of aggregation of the receptors is required throughout the secretory process. Interrupting the aggregation caused by Ph_4P^+ probe to reequilibrate at a rate as rapidly as could be anticipated if prompt repolarization had occurred simultaneously with the interruption of receptor aggregation. The increase in the percentage depolarization throughout the course of the secretory reaction, as well as the repolarization experiments, also demonstrates a close stoichiometric linkage between receptor aggregation and the discharge of the membrane potential.

It was observed that the depolarization of the plasma membrane potential induced by aggregation of the receptors was quantitatively similar in the presence and absence of calcium provided sodium was also present in the medium (Table 1). This suggested that ion channels had been generated through which calcium or sodium could pass. These findings also suggested that if these phenomena were caused by formation of a single type of ion channel, that calcium might block the entry of sodium. These predictions have now been confirmed by direct assessment of sodium uptake (Figs 4 and 5).

An unexpected finding was that the increased permeability of calcium induced by aggregation of the receptors for IgE was only observed in the presence of mM calcium (Fig 7). By disaggregating previously associated receptors, we could demonstrate that the flux was not simply a consequence of degranulation. The failure to observe stimulated uptake of calcium at very low concentrations of the ion in the medium was also not simply caused by depletion of free calcium due to binding to newly vacated binding sites for calcium on the surface of the washed cells. This is shown by the fact that only a small proportion of the added calcium became associated with the cells (Fig 7). The results suggest that in the absence of calcium, the presumptive channel has a relatively broad specificity but that mM calcium converts the channel to one that is relatively specific for calcium.

An alternative explanation is that the cell has a pump such as a Ca^{2+}-ATPase that has a relatively low K_m. At μM concentrations the pump may be able to extrude all Ca^{2+} that enters. On the other hand, at mM Ca^{2+} the capacity of the pump may be exceeded when channels are opened because of the aggregation of the receptors. This would produce an apparent influx not observed (but nevertheless present) at the lower concentrations. In any case, this effect of Ca^{2+} concentration is an important phenomenon that deserves further attention.

To what extent are the presumptive channels related to the familiar voltage-dependent channels? Release (16, and Kanner, B.I., unpublished observations) as well as calcium flux induced by aggregation (Kanner, B.I., unpublished observations) are not consistently inhibited by nifedipine - a potent and specific inhibitor of voltage-dependent ion channels (17). Similarly, no consistent results were obtained with verapamil. However, nifedipine blocks only some types of voltage-dependent calcium channels, failing for example to block those that initiate transmitter release (18). Furthermore, high doses of verapamil are needed to block the latter channels (21). Thus, the results with channel blockers on the RBL cells do not themselves provide conclusive evidence against the channels we are studying being of the voltage-dependent type. However, depolarization of the plasma membrane potential by external potassium (Fig 8) did not lead to calcium-dependent release of serotonin (Fig 9), a characteristic in which these channels differ from the more familiar voltage-dependent calcium channels. It is of interest nevertheless, that the latter as well as the channels we have studied transport sodium at low concentrations of calcium (Ref 19, Figs 4 and 5). These considerations render unlikely the possibility that aggregation promotes depolarization which - as in nerve terminals (20,21) - in turn causes voltage-dependent calcium channels to open. On the other hand, the results are consistent with the possibility that only one type of channel is opened by aggregation of receptors. The channel has the interesting feature that it may be modulated by mM calcium to be converted to a calcium specific channel. It is reasonable to postulate that it is these channels that transport the calcium into the cytoplasm initiating the exocytotic pathway.

The receptor for IgE belongs to the class of so-called 'Fc receptors'. Recently another type of Fc receptor, a receptor for IgG on macrophages, has been shown to be linked to an ion channel as judged from studies on membrane vesicles (22,23). However, it is possible that a study similar to ours of this receptor on intact macrophages, might reveal results similar to those presented here. In this case, thus-far uncharacterized calcium modification may have been lost in the membrane vesicle. Alternatively, the channel linked to the IgG receptor may be a different kind of channel. Incorporation of the isolated, purified, IgG receptors into liposomes revealed that the ion channel may be part of the receptor itself. With respect to the receptor for IgE, the ion channel may also be part of the receptor, or it may be (or become) linked directly or indirectly with it. One candidate for such a discrete component is the protein that binds cromoglycate recently described by Pecht and his colleagues (24,25). Reincorporation of the receptor for IgE into liposomes has recently been achieved (26) and represents one obvious approach to studying these questions.

Our conclusions are derived from experiments on intact cells. The complexity of the latter makes it impossible to rule out other interpretations of the specific findings. Therefore, we are currently examining membrane preparations to see if ionic perturbations resulting from aggregation of receptors can be observed with them. If so, these simpler preparations should be helpful in sorting out the events due to changes in the plasma membrane per se. These in turn can then serve as the standard for analyzing even simpler lyposomal preparations. They are entirely consistent with very recently published studies by Beaven et al (27,28) where internal calcium levels were measured with the fluorescent indicator quin 2.

REFERENCES

1. Metzger H. and Ishizaka T. (1982). Transmembrane signalling by receptor aggregation: The mast cell for IgE as a case study. Introduction. Fed. Proc. Am. Soc. Exp. Biol. 41, 7-34.
2. Foreman J.C., Mongar J.L. and Gamperts B.D. (1973). Calcium ionophores and movement of calcium ions following the physiological stimulus to a secretory process. Nature 245, 249-251.
3. Fewtrell C., Lagunoff D. and Metzger H. (1981). Secretion from rat basophillic leukemia cells induced by calcium ionophores. Effect of pH on metabolic inhibition. Biochim. Biophys. Acta 644, 363-368.
4. Foreman J.C., Hallet M.B. and Mongar J.L. (1977). The relationship between histamine secretion and ^{45}Ca uptake by mast cells. J. Physiol. 271, 193-214.
5. Ishizaka T., Foreman J.C., Sterk A.R. and Ishizaka K. (1979). Induction of calcium flux across the rat mast cell by bridging IgE receptors. Proc. Natl. Acad. Sci. USA 76, 5858-5862.
6. Crews F.T., Morita Y., McGivney A., Hirata F., Siraganian R.P. and Axelrod J. (1981). IgE-Mediated histamine release in rat basophillic leukemia cells; receptor activation, phospholipid methylation, Ca^{2+} flux, and release of arachadonic acid. Arch. Biochem. Biophys. 212, 561-571.
7. Pierce F.L. (1982). Calcium and histamine secretion from mast cells. Prog. Med. Chem. 19, 59-109.
8. Kanner B.I. and Metzger H. (1983). Crosslinking of the receptors for immunoglobulin E depolarizes the plasma membrane of rat basophilic leukemia cells. Proc. Natl. Acad. Sci. USA 80, 5744-5748.
9. Kanner B.I. and Metzger H. (1984). Initial characterization of the calcium channel activated by the crosslinking of the receptors for immunoglobulin E. J. Biol. Chem. 259, 10188-10193.
10. Fewtrell C. (1984). Biology of Calcium, In: R.P. Rubin, G. Weiss and J. Putney, Jr. (Eds), Plenum, New York, in press.
11. Kagey-Sobotka A., MacGlashan D.W. and Lichtenstein L.M. (1982). Role of receptor aggregation in triggering IgE-mediated reactions. Fed. Proc. Am. Soc. Exp. Biol. 41, 12-16.
12. Lichstein D., Kaback H.R. and Blume A.J. (1979). Use of lipophillic cation for determination of membrane potential in neuroblastoma-glioma hybrid cell suspensions. Proc. Natl. Acad. Sci. USA 76, 650-654.
13. Kiefer H.R., Blume A.J. and Kaback H.R. (1980). Membrane potential changes during mitogenic stimulation of mouse spleen lymphocytes. Proc. Natl. Acad. Sci. USA 77, 2200-2204.
14. Ramos S., Grollman E.F., Lazo P.S., Dyer S.A., Habig W.H., Hardegree M.C., Kaback H.R. and Kohn L.D. (1979). Effect of tetanus toxin on the accumulation of the permeant lipophillic cation tetraphenyl phosphonium by guinea pig brain synaptosomes. Proc. Natl. Acad. Sci. USA 76, 4783-4787.
15. Nicholls D., Rugolo M., Scott I.G. and Meldolesi J. (1982). -Latrotoxin of black widow spider venom depolarises the plasma membrane, induces massive calcium influx, and stimulates transmitter release in guinea pig brain synaptosomes. Proc. Natl. Acad. Sci. USA 79, 7924-7928.
16. Middelton E., Jr., Drzewiecki G. and Triggle D. (1981). Effects of smooth muscle calcium antagonists on human basophil histamine release. Biochem. Pharmacol. 30, 2867-2869.
17. Bolton T.B. (1979). Mechanism of action of transmitters and other substances on smooth muscle. Physiol. Rev. 59, 606-718.
18. Nachsen D.A. and Blaustein M.P.C. (1979). The effect of some organic

"calcium antagonists" on calcium influx in presynaptic nerve terminals. Mol. Pharmacol. 16, 579-586.

19. Kostyuk P.G., Mironov S.L. and Shuba Ya. M. (1983). Two ion-selecting filters in the calcium channel of the somatic membrane of mollusc neurons. J. Memb. Biol. 76, 83-93.
20. Blaustein M.P., Johnson E.M.J. and Needlemann P. (1972). Calcium-dependent, norepinephrine release from presynaptic nerve endings in vitro. Proc. Natl. Acad. Sci. USA 69, 2237-2240.
21. Blaustein M.P. (1975). Effects of potassium, veratridine and scorpion venom on calcium accumulation and transmitter release by nerve terminals in vitro. J. Physiol. (Lond.) 247, 617-655.
22. Young J.D.-E., Unkeless J.C., Kaback H.R. and Cohn Z.A. (1983). Macrophage membrane potential changes associated with 2b/ 1 Fc receptor-ligand binding. Proc. Natl. Acad. Sci. USA 80, 1357-1361.
23. Young J.D.-E., Unkeless J.C., Kaback H.R. and Cohn Z.A. (1983). Mouse macrophage Fc receptor for IgG 2b/ 1 in artificial and plasma membrane vesicles functions as a ligand-dependent ionophore. Proc. Natl. Acad. Sci. USA 80, 1636-1640.
24. Mazurek N., Bashkin P. and Pecht I. (1982). Isolation of a basophillic membrane protein binding the anti-allergic drug cromolyn. EMBO J. 1, 585-590.
25. Mazurek N., Bashkin P., Loyter A. and Pecht I. (1983). Restoration of Ca^{2+} influx and degranulation capacity of variant RBL-2H3 cells upon implantation of isolated chromolyn binding protein. Proc. Natl. Acad. Sci. USA 80, 6014-6018.
26. Rivnay B., Rossi G., Henkart M. and Metzger H. (1984). Reconstitution of the receptor for immunoglobulin E into liposomes reincorporation of purified receptors. J. Biol. Chem. 259, 1212-1217.
27. Beaven M.A., Rogers J., Moore J.P., Hesketh T.R., Smith G.A. and Metcalfe J.C. (1984). The mechanism of the calcium signal and correlation with histamine release in 2H3 cells. J. Biol. Chem. 259, 7129-7136.
28. Beaven M.A., Moore J.P., Smith G.A., Hesketh T.R. and Metcalfe J.M. (1984). The calcium signal and phosphatidylinositol breakdown in 2H3 cells. J. Biol. Chem. 259, 7137-7142.

DISCUSSION OF THE ARTICLE BY B.I. KANNER (ISRAEL)

M. Lazdunski (France): Is there any electrophysiological experiment on the cell line itself, which would help answer a number of questions that you have raised. Is there a voltage-dependence of the calcium channels? What is the selectivity of the calcium channels? Is there a calcium control of this selectivity?

B. Kanner (Israel): In Germany, there is a group doing these experiments but I do not know the results. The only thing which I may add is that we also added verapamil, nephedipin, and they didn't do anything.

M. Lazdunski (France): The reason why I am asking this question is because a number of us know that TPP experiments have limits. When you do TPP experiments and do electrophysiology side by side, sometimes you get very surprising results. When the potential raises, it raises with both techniques; when it decreases, it decreases with both techniques. However, very often from our experience with several types of cells, the absolute value of potentials are completely different when recorded with two different techniques.

B. Kanner (Israel): I don't make any kind of estimates in terms of absolute value. I consider these TPP experiments as just being qualitative indicators and in order to calculate the membrane potential, that would indeed require a number of assumptions, which I am not prepared to make, so I agree with you.

SEQUENCE OF DEVELOPMENT OF IONIC CHANNELS DURING MOUSE NEURONAL DIFFERENTIATION : A PATCH-CLAMP STUDY

Michel SIMONNEAU

The cell membrane of the vertebrate neuron presents a complex repertoire of ionic channels (1,2). The characteristics of ionic channels changes with development, as examined experimentally in a variety of invertebrate and lower vertebrate preparations (3,4,5).
In this paper, the sequence of development of ionic channels during neuronal differentiation in mouse was analyzed using embryonal teratocarcinoma cells.

Choice of an experimental system

The study of mammalian development has been hampered by the scarcity and the heterogeneity of the embryonic material. Embryonal carcinoma (EC) cells, which resemble uncommitted cells of the early embryo in many aspects, have been widely used as a model to study mammalian cell differentiation (6). These EC cells differentiate into a variety of cell types *in vivo* and *in vitro*. After incorporation into blastocysts and transplantation into host mice, they become part of virtually every tissue of their offspring (7,8). These multipotential embryonal carcinoma cell lines constitute a powerful system for exploring the regulation of gene expression after introduction of exogenous genes into these cell lines (9,10). Moreover, the possibility to obtain large amounts of EC cells at a given state of differentiation makes this system extremely adapt for the isolation of specific mRNA and the identification of the corresponding genes coding for the various developmentally regulated proteins, e.g. laminin or collagen IV (11,12).

These cells lines can be induced to differentiate into nerve cells *in vitro* (13). However, their small size (in the 10 microns range) precluded their study by classical electrophysiological methods.

Fig. 1. Whole-cell recording of an uncommitted embryonal teratocarcinoma cell.
A : Current clamp of a 1003 EC cell. V_h = -50 mV. Responses to current pulses of various intensity. Note the asymmetry of voltage response for large current pulses. Cell capacitance : 20 pF.
B : Voltage clamp of an other 1003 EC cell. V_r = -51 mV. Outward current is always defined as current from the intracellular to the extracellular side of the membrane and is indicated upwards. Membrane holding potentials as indicated on the left of the records. Extracellular solution as indicated in methods.
C : I-V relation of elementary currents in function of the holding potential for the same cell as shown in B.

Materials & Methods :

Cell culture

Experiments were performed on undifferentiated cells of C17-S1 clone 1003 embryonal carcinoma cell line (1003 EC cells). This cell line is multipotential but can be induced to differentiate into nerve cells in defined *in vitro* conditions (13).

Single channel recordings

Membrane currents and membrane potentials were recorded via a glass pipette in the "whole cell" and in the "cell-attached patch" recording configuration (14). Signals were recorded on magnetic tape and analyzed through a Nicolet 1090A Explorer digital oscilloscope and a Hewlett-Packard 9825B calculator. Patch electrodes used in this study were of 5 to 15 MOhms of input impedance, for both cell attached and whole-cell clamp recordings. Experiments were performed at room temperature (20-22°C) in bath solution of the following composition (in mM) : NaCl 140, $CaCl_2$ 1.8, KCl 5, $MgCl_2$ 1, HEPES-NaOH 10 pH 7.3. Pipette contained : KCl 140, $MgCl_2$ 2, $CaCl_2$ .5 , EGTA 5, HEPES-KOH 10, pH 7.2. Records were filtered through a 8-pole Bessel active filter at 1 KHz.
(from Simonneau, Eddé, Nicolas & Jakob, in prep.)

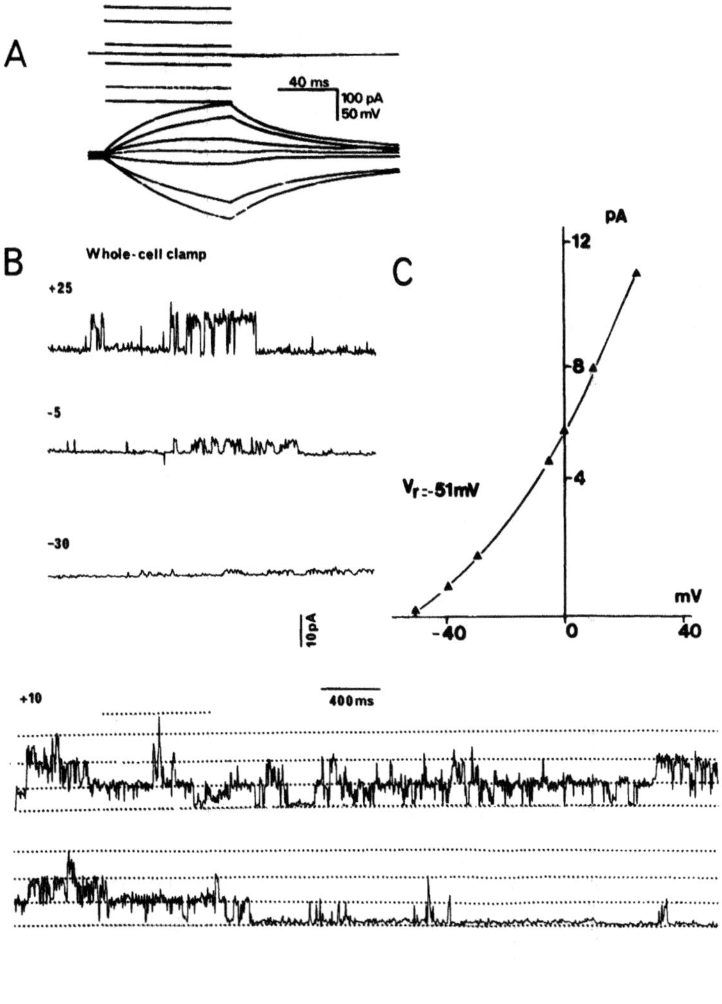
A
40 ms
100 pA
50 mV
B
Whole-cell clamp
+25
-5
-30
10pA
C
pA
12
8
4
V_r:-51mV
mV
-40
0
40
+10
400ms

Fig. 1

Patch clamp recording techniques (14), which do not present this limitation, have permitted the study of these cells during differentiation, using both single channel recording and whole-cell clamp analysis (15,16).

This *in vitro* system appears highly favorable to study the sequence of development of ionic channels during mouse neuronal differentiation. Furthemore, since it permits an experimental manipulation of the sequence of differentiation, by selecting for instance variants presenting a smaller or a larger quantity of a given type of ionic channels, this system makes possible to analyze a possible role of these channels during differentiation.

Electrophysiological characteristics of a mouse embryonic uncommitted cell.

Non-differentiated embryonal carcinoma cells are not excitable. Under whole-cell clamp conditions, with intracellular medium in the patch pipette, they have a resting potential of -42.7 ± 10.6 mV (mean $\pm$ s.d; n=15). They present a limited number and a limited repertoire of channels with mainly three types of ionic channels tentatively attributed to : (1) a Ca^{++} activated potassium channel (unit conductance of 150 pS) (2) a voltage dependent K^{+} channel (unit conductance of 30 pS) and (3) less frequently a voltage dependent chloride channel (unit conductance of 450 pS) found in 3 cell-attached patches (n=60).

The properties of the Ca^{++} activated K^{+} channels are very similar to that described for mammalian fully differentiated cells, for instance the bovine chromaffin cells (17), the rat cultured myotubes (18) or pig pancreas acinar cells (19). They are blocked by quinine and activated by application of the ionophore A 23187. The Ca^{++} activated K^{+} channel presents multiple conductance states with substates of open and close conductances, and with multiple open time constants.

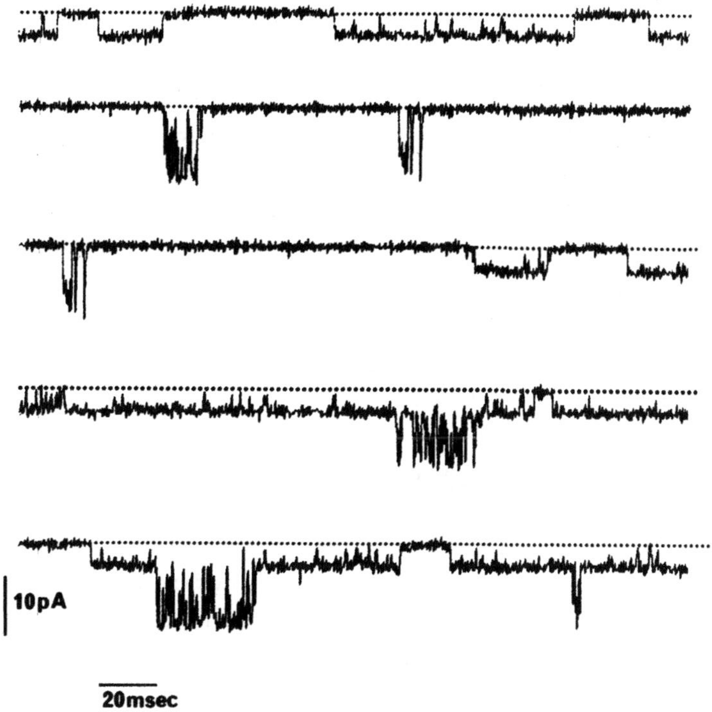

Fig. 2. Cell-attached patch recording from a non-differentiated 1003 EC cell.
Pipette contained the same solution as that described in Fig.1. The holding potential was 120 mV negative to the resting potential. Filter 1 KHz, sampling frequency 4 KHz. (Simonneau & Distasi, unpubl.)

In some cases, it was possible to obtain single channel resolution in whole cell current recordings as illustrated in Fig.1 where at most 4 channels can be open under whole-cell clamp conditions. This suggests that there is a low density of active channels, on the whole surface of the cell.

Fig.2 illustrates the two main types of ionic channels present on a same cell-attached patch of membrane, from a non-differentiated 1003 EC cell.

This repertoire of ionic channels is also present in a variety of non-differentiated embryonal carcinoma cell lines (1003 ,1009, LT_1) and in a cell line derived from the mouse blastocyst inner cell mass (21-4M cell line).

It is also possible to record a similar pattern of single channels in cells directly isolated by immunosurgery (20), from the mouse blastocyst inner cell mass (Simonneau & Condamine, unpublished observations).

For cell-attached patches, with intracellular medium in the pipette, at $V_h = 0$, the probability of opening of the Ca^{++} activated K^+ channels appears to be in the same range of values for both these cell lines and those isolated directly from the blastocyst.

In vitro neuronal differentiation of 1003 EC cells.

We selected to use 1003 EC cells which can be induced to differentiate in neuronal cells when they are deprived of serum and transferred into a defined medium (13). Differentiated cells with neuro-epithelium and neuronal characteristics become the predominant population with a progressive disappearance of EC cells. By a simple observation under a phase-contrast microscope, one can distinguish elongated cells organized in rosettes. They correspond to the neuro-epithelial stage. In these rosettes, after 6 days in culture, it is possible to see process-bearing cells. In culture, these neurite-like processes increase in lenght with time.

Sequence of development of ionic channels

As soon as the cells start to grow small neurites (one cell diameter or less), calcium spikes can be recorded. This calcium conductance appears to be activated only for hyperpolarized values of the resting potential (<-70 mV) and is blocked by millimolar addition of Co^{++} to the bath (Fig.3 & 4). No voltage-dependent Na^+ conductance is found at this stage. Ca^{++} activated K^+ channels and voltage-dependent K^+ channels, with properties similar to those described for uncommitted cells, were recorded in cell-attached patches.

A similar low threshold calcium conductance has already been described in the adult burst-firing mammalian neurones, in the inferior olive and in the thalamus (21, 22 & 23). Similarly, a calcium conductance activated between -75 and -70 mV was found on perfused neurons of rat dorsal root ganglion which presents also an other type of calcium conductance activated in a more depolarized region (-45 to -40 mV) (24).

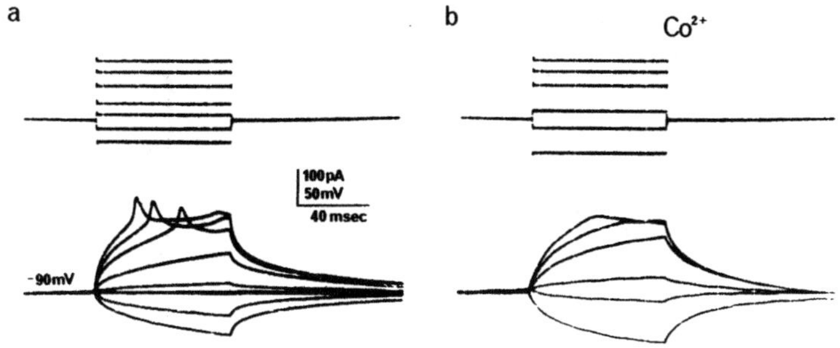

Fig.3. Whole-cell recording under current-clamp of a differentiated 1003 cell presenting short neurites.
The pipette solution was as described in Fig.1. Bath solution (in mM) : NaCl 140, $CaCl_2$ 10, KCl 5, $MgCl_2$ 1, HEPES-NaOH 10 pH 7.3. Holding potential, -90 mV. The calcium spike was blocked by 1mM cobalt bath application (b). (Simonneau & Eddé, unpubl.).

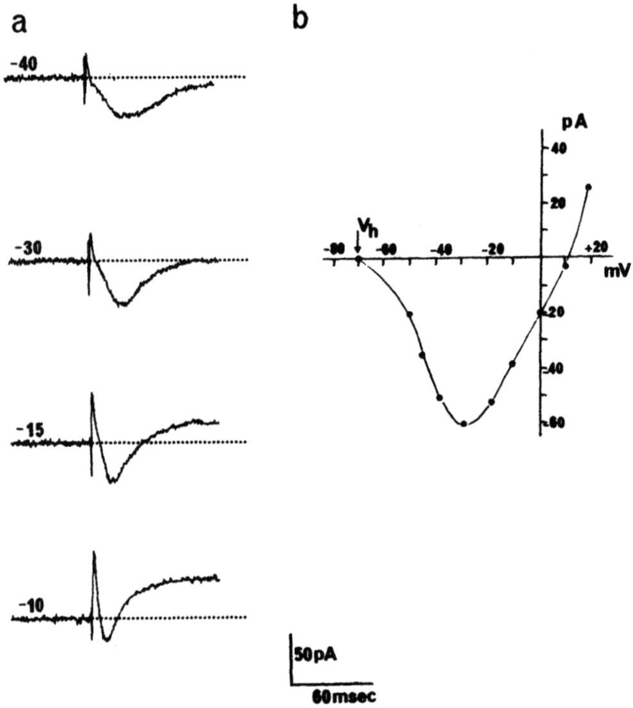

Fig.4. Whole-cell recording under voltage clamp from a differentiated cell with short neurites. The extracellular medium and the pipette solution were as described in Fig.3. The holding potential was -70 mV. Outward K^+ currents were not blocked in this experiment (Simonneau & Eddé, unpubl.).

Fully differentiated cells with long neurites (longer than one cell diameter) have a complete repertoire of ionic channels : voltage dependent Na^{+} and Ca^{++} channels, Ca^{++} activated K^{+} channel and a K^{+} delayed rectifier as described for other mammalian central neurones (1,2,21,22,23 & 24). Fig.5 illustrates an example of currents obtained by whole-cell clamp for a fully differentiated cell.

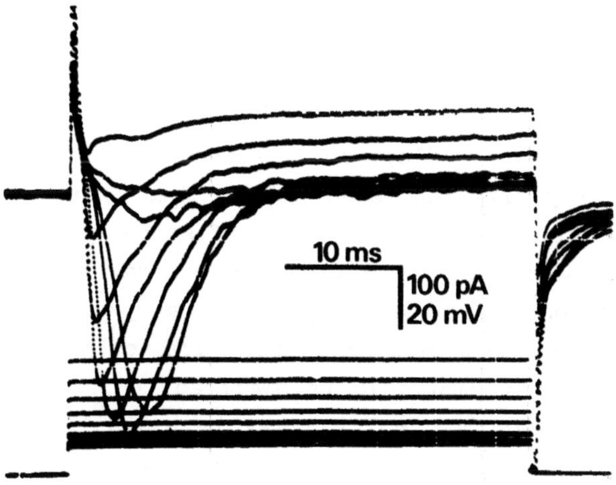

Fig. 5. Whole-cell recording under voltage-clamp for a fully differentiated cell.
Holding potential, -50 mV. The extracellular medium was as described in Fig.1. The pipette solution contained (in mM) NaCl 140, KCl 5, $MgCl_2$ 2, $CaCl_2$ 1.8, HEPES-NaOH, pH 7.3. The bathing solution was the culture medium. Temperature 21°C.

Ca^{++} activated K^{+} channel kinetics at two stages of differentiation : an insight on differences in intracellular calcium regulation during differentiation.

The kinetics of Ca^{++} activated K^{+} channels which reflect the intracellular calcium concentration were examined (i) when the EC cells are not differentiated (ii) for fully differentiated cells.

For V_h = 0 (resting potential), the probability of openings, p, appears to increase of about one order of magnitude between these two stages of differentiation (Fig.6). Furthermore, the open-time distribution is different, with longer bursts in the case of differentiated cells.

These differences reflect an increase in intracellular calcium for fully differentiated cells. This increase may be due to a difference in calcium buffering systems of these two stages of differentiation. Or, it may correspond to the activation of voltage-dependent Ca^{++} channels at rest for differentiated cells and which are not present in the cell membrane of embryonic uncommitted cells. Such an increase in intracellular calcium may be involved in the growth of processes as suggested by various experimental approaches (see, for example, 25).

Fig. 6. Histograms of open intervals of Ca^{++} activated K^{+} channels, for two different stages of differentiation. Semilogarithmic co-ordinates. The histograms were obtained from data corresponding to unitary currents recorded from cell-attached patches, at V_h = 0, with 140 mM KCl in the pipette. The data were recorded on a F.M. Ampex tape recorder and analyzed with a micro-computer PDP 11 (LSI 11-23) system. Note that, in both cases the distribution of all open intervals is described by more than one time-constant. A , histogram for an uncommitted 1003 cell. B, histogram for a fully differentiated cell. (Simonneau, Eddé & Tauc, unpubl.)

A

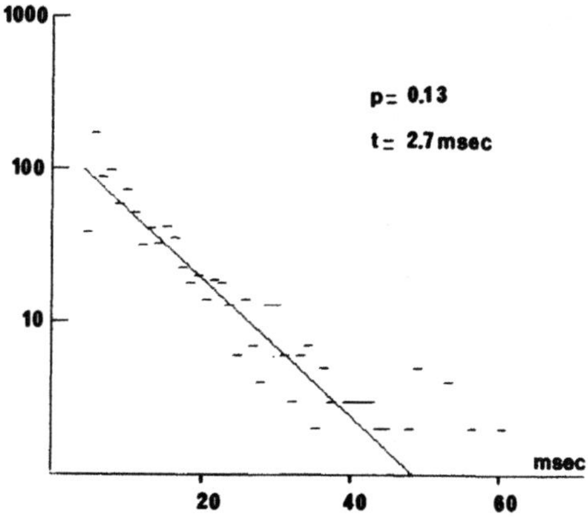
1000
100
10
p= 0.13
t= 2.7 msec
msec
20
40
60

B

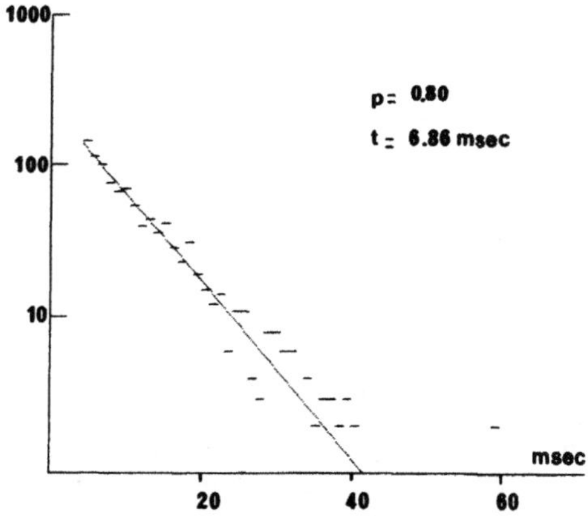
1000
100
10
p= 0.80
t= 6.86 msec
msec
20
40
60

Fig. 6

Conclusion

The sequence of development of ionic channels in these mammalian cells appears different from that described for avian neurones where Ca^{++} channels appear after a Na^{+} conductance (5). However, this sequence is identical to that described in invertebrate (26), in Xenopus neurones (27) and in mammalian neuroblastoma (28).

Furthermore, this system offers an opportunity to test directly a possible role of ionic channels in differentiation by experimental manipulation.

AKNOWLEDGMENTS

The author wishes to thank and aknowledge Drs. H. Condamine, C. Distasi, B. Eddé, H. Jakob, J.F. Nicolas & L. Tauc, for their collaboration in this work and Dr Sheela Vyas, for rewieving the manuscript. This work was supported in part by a Fondation pour la Recherche Médicale Française grant.

REFERENCES

1. Adams, P. 1982. Voltage-dependent conductances of vertebrate neurones. Trends in NeuroSciences 5, 116-119.2.

2. Llinas, R. 1984. Comparative electrobiology of mammalian central neurons. In Brain Slices (Dingledine, R. ed.) pp 7-24, Plenum Press, New York.

3. Spitzer, N.C. 1979. Ion channels in development. Ann. Rev. Neurosci. 2, 363-397.

4 . Hagiwara, S. 1983. Membrane potential dependent ion channels in cell-membranes : phylogenic & developmental approaches. Soc. for Gen. Physiologists. Vol n°3. Raven Press, New York.

5. Bader, C.R., Bertrand, D., Dupin, E. & Kato, A.C. 1983. Development of electric membrane properties in cultured avian neural crest. Nature (London), 305, 808-811.

6. Jacob, F. 1978. Mouse teratocarcinoma and mouse embryo. Proc. Roy. Soc.Lond B 201, 249-270.

7. Brinster, R.L. 1974. The effect of cells transferred into the mouse blastocyst on subsequent development. J. exp. Med. 140, 1049-1056.

8. Mintz, B. & Illmensee, K. 1975. Normal genetically mosaic mice produced from malignant teartocarcinoma cells. Proc. Natl. Acad. Sci. USA 72, 3585-3589.

9. Pellicer, A., Wagner, E.F., El Kareh, A., Dewey, M.J., Reuser A.J., Silverstein, S., Axel, R. & Mintz, B. 1980. Introduction of a viral thymidine kinase gene and the human beta-globin gene into developmentally multipotential mouse teratocarcinoma cells. Proc. Natl. Acad. Sci. USA, 77, 2098-2102.
10. Nicolas, J.F. & Berg, P. 1983. Cold Spring Harb. Conf. on cell proliferation (Teratocarcinoma stem cells), 10, 469-485.
11. Kurkinen, M., Barlow, D.P., Helfman, D.M., Williams, J.G.,& Hogan, B.L.M. 1983. Isolation of cDNA clones for basa lamina components : type IV procollagen. Nucleic Acids Res. 11, 6199-6209.
12. Wang, S.Y. & Gudas, L.J. 1983. Isolation of cDNA clones specific for collagen IV and laminin from mouse teratocarcinoma cells. Proc. Natl. Acad. Sci. USA, 80, 5880-5884.
13. Darmon, M., Bottenstein, J. & Sato, G. 1981. Neural differentiation following culture of embryonal carcinoma cells in a serum-free defined medium. Devl. Biol. 85, 463-473.
14. Hamill, O.P., Marty, A., Neher, E., Sakmann, B. & Sigworth, F.J. 1981. Improved patch-clamp techniques for high-resolution current recording from cells and cell-free membrane patches. Pflugers Arch. 391, 85-100.
15. Eddé, B., Nicolas, J.F. & Simonneau, M. 1984. Evidence for neuronal differentiation of mouse embryonic carcinoma cells : a patch-clamp study. J. Physiol. (London) 346, 82P.
16. Simonneau, M. 1984. Ionic channels in mouse embryonic multipotential carcinoma cells. Biophys. J., 45, 308a.
17. Marty, A. 1981. Ca-dependent K channels with large unitary conductance in chromaffin cell membranes. Nature (London) 291, 497-500.
18. Pallotta, B.S., Magleby, K.L. & Barrett, J.N. 1981. Single channel recordings of Ca^{++} activated K^{+} currents in rat muscle cell culture. Nature (London) 293, 471-474.
19. Maruyama, Y., Gallacher, D.V. & Petersen, O.H.. 1983. Voltage and Ca^{++} activated K^{+} channel in baso-lateral acinar cell membranes of mammalian salivary glands. Nature (London) 302, 827-829.
20. Solter, D., & Knowles, B.B. 1975. Immunosurgery of mouse blastocyst. Proc. Natl. Acad. Sci. USA, 72, 5099-5102.
21. Llinas, R. & Yarom, Y. 1981. Electrophysiology of mammalian olivary neurones in vitro : Different types of voltage-dependent ionic conductances J. Physiol. (London) 315, 569-584.
22. Jahnsen, H. & Llinas, R. 1984. Electrophysiological properties of guinea-pig thalamic neurones in vitro. J. Physiol. (London) 349, 205-226.
23. Jahnsen, H. & Llinas, R. 1984. Ionic basis of the electroresponsiveness and oscillatory properties of guinea-pig thalamic neurones in vitro. J. Physiol. (London) 349, 227-247.

24. Kostyuk, P.G. 1984. Intracellular perfusion of nerve cells and its effects on membrane currents. Physiol. Rev. 64, 435-454.
25. Anglister, L., Farber, I., Shamar, A. & Grinvald, A. 1982. Localization of voltage-sensitive calcium channels along developing neurites : their possible role in regulating neurite elongation. Dev. Biol. 94, 351-365.
26. Goodman, C.S. & Spitzer, N.C. 1979. Embryonic development of identified neurones: differentiation from neuroblast to neurones. Nature (London) 280, 208-213.
27. Spitzer, N.C., & Baccaglini, P.I. 1976. Development of action potential in embryonic amphibian neurons in vivo. Brain. Res. 107, 610-616.
28. Miyake, M. 1978. The development of action potential mechanism in a mouse neuronal cell line in vitro. Brain. Res. 143, 349-354.

DISCUSSION OF THE ARTICLE BY M. SIMONNEAU (FRANCE)

F.J. Sigworth (Germany): What was the membrane potential in which you made the last measurements. I am surprised at the high probability of opening of that channel. Was that at rest?

M. Simonneau (France): In the last slide, it was at rest but in order to increase the size of the single unit of the cell-attached patch, we use a high concentration of K in the pipette (190mM KCl).

FREE COMMUNICATIONS

CALCIUM CURRENTS AND POTASSIUM CONDUCTANCE CHANGES EXPLAIN LOW-DOSE SEDATIVE DRUG ACTIONS IN MAMMALIAN HIPPOCAMPAL NEURONS

P.L. CARLEN, N. GUREVICH, M.F. DAVIES, T.J. BLAXTER, M. O'BEIRNE
Departments of Physiology and Medicine (Neurology),Playfair Neuroscience Unit, Toronto Western Hospital, and Addiction Research Foundation Clinical Institute, University of Toronto, Canada

Summary

It is often assumed that sedative drugs cause neuronal depression by activating GABAergic mechanisms. Using intracellular recordings from in-vitro rat and guinea pig CA_1 and CA_3 neurons, this lab is studying the ionic mechanisms of lower, non-anesthetic doses of ethanol ($\lesssim$20mM), pentobarbital (< 100 uM) and 2 benzodiazepines (midazolam and clonazepam ($\lesssim 10^{-8}$M). All drugs caused a hyperpolarization, usually with moderately increased conductance and decreased spontaneous spiking. Intracellular chloride injection had no effect on these responses. Post-spike long-lasting afterhyperpolarizatons (AHPs), thought to be due to Ca^{++} -mediated K^+ conductance (Ca-gK), were consistently enhanced by these drugs. In the presence of TTX, Ca^{++} spikes were examined. Ethanol and pentobarbital had no effect or tended to diminish whereas midazolam enhanced these Ca^{++} spikes. Zero Ca^{++} perfusate did not prevent ethanol or pentobarbital actions, but blocked midazolam actions. Intracellularly injected cesium, which blocks gK, blocked ethanol effects. Low drug doses had no effect or slightly diminished GABA actions, whereas higher (anesthetic) doses tended to enhance GABA actions.

We conclude that: 1) All these sedative drugs cause neuronal inhibtion at non-anesthetic doses by augmenting Ca-gK and not GABA actions or gCl. 2) Ethanol and pentobarbital, which tend to diminish Ca^{++} spikes, augment Ca-gK possibly by increasing intracellular [Ca^{++}] from intracellular stores. 3) However, benzodiazepine neuronal inhibiton can be explained by increasing inward Ca^{++} flux which then activates gK.

Introduction and Methods

This article summarizes results concerning the neurophysiological effects of low doses of sedative-hypnotic drugs. The data obtained to date suggest that these drugs, when applied in non-anesthetic concentrations to central mammalian neurons in-vitro, augment a calcium (Ca^{++}) mediated potassium (K^{+}) conductance (Ca-gK) and not GABA responses. Part of these data have been presented in detail elsewhere[1,2,3,4,5,6,7,8].

The hypothesis that these drugs augment Ca-gK stems from the findings of Krnjevic and Lisiewicz[9] and Meech[10,11] that intraneuronal injection of Ca^{++} ions causes an increaed K^{+} conductance. Physiologically, Ca-gK is normally triggered by a depolarization-induced influx of Ca^{++} from the surrounding medium. However, in voltage-clamped dorid neurons it was shown that the intraneuronal free $[Ca^{++}]_i$ and not the amount of the Ca^{++} transmembrane current is related to the degree of activation of the Ca-gK [12]. Parodoxically, significantly elevated intracellular $[Ca^{++}]_i$ reduces the depolarization-induced inward Ca^{++} current[13,14,15,16,17]. Therefore, if the Ca-gK were increased by raised $[Ca^{++}]_i$ from intracellular stores, the depolarization-induced inward Ca^{++} current could be reduced. As an indirect monitor of the $[Ca^{++}]_i$ we have used the size and latency of Ca^{++} spikes evoked with relatively large depolarizing currents in the presence of tetrodotoxin (TTX).

These experiments were done using intracellular current clamp recordings from CA_1 or CA_3 neurons in in-vitro rat or guinea pig hippocampal slices. Preliminary data using ethanol have been obtained from rat cerebellar Purkinje cells[1]. The in-vitro slice techniques[18] permit stable intracellular recordings and perfusion or focal application (via the "Picospritzer") of known concentrations of drugs. The slices are placed on a net in a moist atmosphere (33 to 36°C) and subperfused with artificial CSF.

It is important to note that relatively low drug doses were employed. The "legal limit" for ethanol in most jurisdictions is 100 mg% (21.7mM) or less. Lethal blood concentrations range above 100 mM. These experiments used 5 to 20 mM ethanol. Sedative blood pentobarbital concentrations are in the low micromolar range and anesthetic concentrations around 100 µM (O'Beirne et al, unpublished observations). Benzodiazepines are highly protein bound. CNS binding affinities are typically in the low nanomolar range[19]. CSF concentrations are nanomolar[20], whereas serum[21] and brain[22] concentrations are in the micromolar range.

Results

The results from over 250 successful intracellular recordings are summarized in Table 1. The data for each drug will be discussed separately.

Ethanol

Whether bath applied or focally ejected onto the apical dendritic or somatic regions of CA_1 cells[2] or cerebellar Purkinje cells[1], ethanol, in doses of 5 to 20 mM, usually caused a hyperpolarization with a moderate increase in membrane conductance and decreased spontaneous spiking when present. These effects were also elicitable in a Ca^{++} free perfusate with manganese added, or in TTX. If Cl^- ions were injected into the cell, the effects still persisted, suggesting that the ethanol-induced conductance increase and hyperpolarization were not due to increased Cl^- conductance. The other ionic candidate is K^+. Cesium, which is a potent blocker of K^+ conductances, was injected into CA_1 cells, blocking ethanol actions[23] The longlasting post-spike afterhyperpolarization (AHP) is thought to be due to increased Ca-gK[24,25]. This AHP was consistently lengthened and usually deepened by ethanol, even with membrane hyperpolarization. The hyperpolarization (IPSP) following the stratum radiatum evoked EPSP in CA_1 cells is thought to consist of an early GABA mediated Cl^- conductance increase and a later enhanced Ca-gK[26]. This IPSP, particularly the later portion (Carlen et al, unpublished data), was enhanced by ethanol. Focal application of ethanol (5 to 20 mM) to the distal stratum radiatum wherein lie the excitatory synaptic terminals, enhanced EPSPs, whereas focal application to the somatic region tended to enhance the IPSPs. These effects could be on presynaptic transmitter release.

The above data suggested that ethanol increased Ca-gK. Since ethanol actions persisted in zero Ca^{++} perfusate, it was postulated that the enhancement of Ca-gK occurred through an intracellular mechanism. As previously mentioned, Ca^{++} spikes may be an indirect way to monitor $[Ca^{++}]_i$. In the presence of TTX, which blocks voltage-dependent Na^+ channels, Ca^{++} spikes can be elicited in CA_1 cells with relatively high depolarizing currents[27]. Ethanol tended to depress Ca^{++} spikes, suggesting that ethanol might be raising $[Ca^{++}]_i$.

Since ethanol is reputed to act via a GABAergic mechanism[28] this interaction was assessed in the slice. GABA was focally pressure-ejected into the somatic or apical dendritic layer of the slice. Except at very high doses of ethanol (100mM), no augmentation of GABA responses was noted.

Pentobarbital

Pentobarbital, in doses of 1, 10, and 100 µM had almost identical effects on CA_1 or CA_3 cells as those noted for ethanol on CA_1 cells[7,8] (Table 1). At all doses, pentobarbital, either perfused or focally pressure-ejected, caused a hyperpolarization and increased IPSPs and AHPs. These effects were resistant to intracellular Cl^- injection. The hyperpolarization persisted in zero Ca^{++} perfusate. Ca^{++} spikes were diminished as was noted[29] in cultured mice spinal cord neurons. Effects on EPSPS were variable. GABA actions were only enhanced at doses of 100 µM.

Benzodiazepines

Two sedative benzodiazepines have been tested in the laboratory; midazolam[4] and clonazepam[6]. At low namomolar concentrations, both drugs hyperpolarized CA_1 cells and augmented AHPs and IPSPs. However, unlike ethanol and pentobarbital, midazolam was ineffective in zero Ca^{++} perfusate and enhanced rather than depressed Ca^{++} spikes (Table 1). Opposite actions including membrane depolarization and diminished Ca^{++} spikes occurred in the presence of Ro14-7437, a specific benzodiazepine antagonist, even when an agonist (i.e. midazolam) was not in the bath[3].

Discussion

Enhanced gK was proposed as a mechanism of CNS anesthesia in 1972 by Krnjevic[39]. The above data suggest that low doses of these three classes of sedative-hypnotic drugs act through a common mechanism, i.e. by enhancing Ca-gK. The membrane hyperpolarization was resistant to intracellular Cl^- injection suggesting a potassium conductance. The most consistent finding was that all drugs augmented post-spike longlasting AHPs, which are generated by Ca-gK. Also, the orthodromic IPSPs which are in part mediated by Ca-gK were also increased by all three drug types. Finally, against the idea of enhanced gCl is the finding that low doses of all drugs did not enhance GABAergic actions which are presumably mediated mainly by gCl, although recent evidence has demonstrated a GABAergic increased Ca-gK in CA_1 cell dendrites[40].

The data also suggested that these drug effects are mediated through changes in Ca^{++} currents or $[Ca^{++}]_i$, although direct activation of gK cannot be ruled out. Neuronal Ca^{++} metabolism is poorly understood but presently under intensive investigation. It is linked to many intracellular physiological and biochemical processes. Intracellular $[Ca^{++}]_i$ is below μM concentrations and is regulated by several mechanisms. Free intracellular Ca^{++} is buffered rapidly by high affinity cytoplasmic Ca^{++}-binding proteins[30] or more slowly by Ca^{++}- sequestering organelles such as the endoplasmic reticulum[31] or mitochondria[32]. Recently, the physiological role of mitochrondria in buffering Ca^{++} has been questioned [31,33]. Neuronal membrane Ca^{++} pumps may also be important. Human red blood cells have a Ca^{++} pump which is regulated by a Ca^{++}-Mg^{++} dependent ATP[ase] and does not seem to exchange for any other ions[34]. Another Ca^{++} pump exchanges intracellular Ca^{++} for extracellular Na^{+} and this system is powered by energy from the Na^{+} electrochemical gradient[31,35]. Finally, the $[Ca^{++}]_i$ per se has an important influence on inward Ca^{++} currents[13,14,15,16,17].

Midazolam enhanced Ca^{++} spikes and was ineffective in zero Ca^{++} perfusate. These data suggest that this benzodiazepine activates Ca-gK by augmenting transmembrane Ca^{++} currents. On the other hand both ethanol and pentobarbital were effective in zero Ca^{++} perfusate and tended to diminish Ca^{++} spikes, suggesting that these drugs might increase $[Ca^{++}]_i$ from intracellular stores thereby activating Ca-gK. Both ethanol and pentobarbital inhibit intrasynaptosomal sequestration of Ca^{++}[36] and Leslie [37] showed that pentobarbital inhibits synaptosomal Ca^{++} transport. Michaelis[35] recently demonstrated that low dose ethanol inhibits the Na^{+}-Ca^{++} exchange membrane pump. Finally, Seeman et al[38] showed that in erythrocyte ghost membrane, $[Ca^{++}]$ rose by about 6% in the presence of 20 mM ethanol, although pentobarbital had no effect. Certainly further investigations are necessary on the neuronal metabolism of Ca^{++} and its interaction with sedative-hypnotic drugs.

Acknowledgements

Manuscript preparation was done by Ms. Helena Housman. Supported by the Medical Research Council of Canada, Ontario Mental Health Foundation, Alcoholic Beverage Medical Research Foundation, and the Canadian Geriatrics Research Society

Table 1

LOW-DOSE DRUG EFFECTS IN HIPPOCAMPAL CA_1 CELLS

	ETHANOL	PENTOBARBITAL	BENZODIAZEPINES
RMP HYPERPOLARIZATION RESISTANT TO CL^- INJECTION	+	+	+
RMP HYPERPOLARIZATION IN ZERO Ca^{++}	+	+	-
AHPs	↑	↑	↑
IPSPs	↑	↑	↑
Ca^{++} SPIKES	- or ↓	↓	↑
GABA RESPONSES	-	-	-

References

1. Carlen, P.L. and Blaxter, T.J.: Cerebellar Purkinje cells in vitro; membrane properties and the effects of ethanol. Soc. Neurosci. 10, 1984.
2. Carlen, P.L., Gurevich, N. and Durand, D.: Low dose ethanol augments calcium-mediated mechanisms measured intracellularly in hippocampal neurons. Science 215: 306-309, 1982.
3. Carlen, P.L., Gurevich, N. and Polc, P.: The excitatory effects of the specific benzodiapezine antagonist Ro-14-7437, measured intracellularly in hippocampal CA_1 cells. Brain Res. 271: 115-119, 1983.
4. Carlen, P.L., Gurevich, N. and Polc, P.: Low-dose benzodiazepine neuronal inhibition: Enhanced Ca^{++}-mediated K^+ conductance. Brain. Res. 271: 358-364, 1983.
5. Carlen, P.L., Gurevich, N. and O'Beirne, M.: Electrophysiological evidence for increased calcium-mediated potassium conductance by low-dose sedative hypnotic drugs. In: Calcium in Biologicial Systems: Ca^{2+} as a Regulator of Membrane Permeability. Eds. Rubin, R.P., Weiss, G. and Putney, J.W., 1984, In Press.
6. Gurevich, N., Davies, M.F. and Carlen, P.L.: Low dose clonazeam inhibition and CGS-8216 excitation of hippocampal CA_1 neurons in vitro; a potassium mechanism. Soc. Neurosci. 10: and in preparation, 1984.
7. O'Beirne, M., Gurevich, N. and Carlen, P.L.: Pentobarbital inhibiton measured intracellularly in hippocampal CA_3 neurons. Soc. Neurosci. 8: 879, 1982, and in preparation.
8. O'Beirne, M.: The effects of low dosages of sodium pentobarbital on the membrane properties of hippocampal pyramidal cells of the guinea pig: An in-vitro study. M.Sc. Thesis, Department of Physiology, University of Toronto, 1984.
9. Krnjevic, K. and Lisiewicz, A.: Injections of calcium ions into spinal motoneurones. J. Physiol. 225: 363-390, 1972.
10. Meech, R.W.: Intracellular calcium injection causes increased potassium conductance in Aplysia nerve cells. Comp. Biochem. Physiol. 42A: 493-499, 1972.
11. Meech, R.W.: The sensitivity of Helix Aspersa neurones to injected calcium ions. J. Physiol (Lond) 237: 259-277, 1974.
12. Eckert, R., Tillotson, D.: Potassium activation associated with intraneuronal free calcium. Science 200: 437-439, 1978.
13. Eckert, R. and Ewald, D.: Residual calcium ions depress activation of calcium dependent current. Science 216: 730-733, 1982.
14. Eckert, R. and Tilltson, D.L.: Calcium-mediated inactivation of the calcium conductance in caesium-loaded giant neurones of Aplysia California. J. Physiol 314: 265-280, 1981.
15. Hagiwara, S. and Byerly, L.: Calcium channel. Annual Rev. of Neurosci 4: 69-125, 1981.
16. Hofmeier, G. and Lux, H.D.: The time courses of intracellular free calcium and related electrical effects after injection of $CaCl_2$ into neurons of the snail, Helix pomatia. Pflugers Arch. 391: 242-251, 1981.

17. Plant, T.D. and Standen, N.B.: Calcium current inactivation in identified neurones of Helix Aspersa. J. Physiol. 321: 273-285, 1981.
18. Dingeldine, R.: Brain Slices. Ed. R. Dingledine Plenum Press, New York, 1984, pp. 1-442.
19. Braestrup, C. Squires, R.F. Specific benzodiazepine receptors in rat brain characterized by high-affinity (^{3}H)diazepam-binding: Proc. Natl. Acad. Sci. USA 74: 306-309, 1982.
20. Kanto, J., Kangas, L., Siirotola, T.: Cerebrospinal-fluid concentrations of diazepam and its metabolites in man. Acta. Pharmacol. et Toxicol. 36: 328-334, 1975.
21. Garattini, S. Mussini, E. Marucci, F, Guaitani, A.: Metabolic studies on benzodiazepines in various animal species. Garratini, S., Mussini, E., Randall, L.O. Eds. The Benzodiazepines. New York: Raven Press, 1973, 75-97.
22. Klotz, U.: Effect of age on levels of diazepam in plasma and brain of rats. Naunyn-Schmiedeberg's Arch Pharmacol. 307: 167-169, 1979.
23. Carlen, P.L., Puil, E. and Gurevich, N,: Ethanol and K^+ conductance: Enhancement of the afterhyperpolarizations (AHPs) following action potentials or depolarization evoked with amino acids and effects of K^+ and Na^+ channel blockers. Soc. Neurosci. 8: 652, 1982.
24. Gustafsson, B. Wigstrom, H: Evidence for two types of afterhyperpolarizations in CA_1 pyramidal cells in the hippocampus. Brain Res. 206: 462-468, 1981.
25. Hotson, J.R., Prince, D.A.: A calcium-activated hyperpolarization follows repetitive firing in hippocampal neurons. J. Neurophysiol. 43: 409-419, 1980.
26. Nicoll, R.A., Alger, B.E.: Synaptic excitation may activate a calcium dependent potassium conductance in hippocampal pyramidal cells. Science 212: 957-959, 1981.
27. Schwartzkroin, P.A. and Slawsky, M.: Probable calcium spikes in hippocampal neurons. Brain Res. 135: 157-161, 1977.
28. Nestoros, J.N.: Ethanol specifically potentiates GABA-mediated neurotransmission in feline cerebral cortex. Science 209: 708-710, 1980.
29. Heyer, E.J., MacDonald, R.L.: Barbiturate reduction of calcium-dependent action potentials: correlation with anesthetic action. Brain Res. 236: 157-171, 1982.
30. Baker, P.F., Schlaepfer, W.W. Uptake and binding of calcium axoplam isolated from giant axons of LOLIGO and MYXICOLA: J. Physiol. 276: 103-125, 1978.
31. Blaustein, M.P., Nelson, M.T.: Soldium-calcium exchange: Its role in regulation of cell calcium. Carafoli, E., ed. Calcium transport across biological membranes. New York: Academic Press, 217-236, 1982.
32. Lehninger, A.L.: Mitrochondria and calcium ion in transport. Biochem. J. 119: 129-138, 1970.
33. Brinley, F.J. Jr.: Regulation of intracellular calcium in squid ions. Fed. Proc. 39: 2778-2782, 1980.
34. Schatzmann, H.J., Burgin, H.: Calcium in human red blood cells. Ann. N.Y. Acad. Sci. 307: 125-147, 1978.
35. Michaelis, E.: Effects of depressant drugs on sodium-calcium exchange in researled synaptic membranes. In: Calcium in Biological Systems:

Ca^{2+} as a Regulator of Membrane Permeability. Eds. Rubin, R.P., Weiss, G. and Putney, J. W., 1984, In Press.

36. Harris, R.A.: Ethanol and pentobarbital inhibiton of intrasynaptasomal sequestration of calcium. Biochem. Pharmacol. 30: 3209-3215, 1981.
37. Leslie, S.: Sedative-hypnotic drugs and synaptosomal calcium transport. In: Calcium in Biological Systems: Ca^{2+} as a Regulator of Membrane Permeability. Eds. Rubin, R.P., Weiss, G. and Putney, J. W., 1984, In Press.
38. Seeman, P., Chau, M., Goldberg, M., Sauks, T., Sax., L.: The binding of Ca^{2+} to the cell membrane by volatile anesthetics (alcohols, acetone, ether) which induce sensitization of nerve or muscle. Biochimica. et Biophysica. Acta 225: 185-193, 1971.
39. Krnjevic, K.: Exciteable membranes and anesthetics. Fink, B.R. Ed., Cellular Biology and Toxicity of Anesthetics. Baltimore: Williams and Wilkins, 1972, 3-9.
40. Blaxter, T.J., Davies, M.F., Carlen, P.L. and Gurevich, N.: GABA mediates a calcium-dependent potassium conductance in hippocampal CA_1 pyramidal cells. Soc. Neurosci. 10., 1984.

DISCUSSION OF THE ARTICLE BY P.L. CARLEN (CANADA)

Comment: Did you check other psychoactive drugs like phenothiozines or other neuroeleptics?

P.L. Carlen (Canada): None of the drugs we checked are classified as neuroeleptics and we haven't looked at neuroleptics at all.

INTERACTION BETWEEN CORTICOSTEROIDS AND PRESYNAPTIC-ACTING TOXINS FROM SNAKE VENOMS IN RAT PHRENIC NERVE TERMINALS

H.VAN WILGENBURG, R.S. LEEUWIN, K.D. NJIO AND R.D. VELDSEMA-CURRIE,
Dept. of Pharmacology, University of Amsterdam, Polderweg 104,
1093 KP Amsterdam, The Netherlands

1. INTRODUCTION

Corticosteroids and the toxins beta-Bungarotoxin (beta-BuTx) and the angusticeps-type "F7"-toxin from snake venoms all affect neuromuscular transmission at the presynaptic site of the motor-endplate in rodent nerve-muscle preparations (1,2). Prednisolone and dexamethasone (Dex), in low concentrations, increase the amplitude of the miniature endplate potentials (MEPPs) without affecting the rate of transmitter release (3). On the other hand beta-BuTx causes, after an initial decrease followed by an increase of MEPP frequency, failure of transmitter release (4). The polypeptide "F7" one of the components of the venom of the green mamba, *Dendroaspis angusticeps*, augments acetylcholine release (5). These effects of the corticosteroids, beta-BuTx and "F7" toxin on acetylcholine release only occur after a latent period of 30 to 60 min, suggesting that in each case an indirect mechanism is involved.

The aim of the present study is to compare the effects of Dex on acetylcholine release from rat motor-nerve endings with the effects of beta-BuTx and "F7" toxin. The present study also provides data on the interaction of the steroids and the toxins.

2. MATERIAL AND METHODS

Female rats (Cpb:WU(WI), 170-190 g) were decapitated and strips from the left hemidiaphragm were mounted in a 2 ml bath perfused with Krebs-Ringer bicarbonate saline gassed with O_2 + 5 % CO_2. The temperature was kept at 22° C with the aid of a Peltier element. In order to avoid muscle contractions, caused by fasciculations in the presence of the toxins, the muscle fibres were carefully cut across their length, 4 mm on each side of the main nerve branch.

Intracellular recordings were made using glass microcapillary electrodes filled with 3 M KCl and a resistance of 5 to 10 MOhm. The tips of two glass microelectrodes were brought in close proximity and then driven by a single manipulator into the muscle fibre. Voltage clamping was done by feeding the output of the cathode follower to a high gain amplifier (30 V swing) used at a gain of 2000 for passing current through an intracellular electrode. The current produced was assessed by another differential cathode follower as the voltage drop across a resistance of 1 MOhm in series with the current passing electrode.

Miniature endplate potentials (MEPPs) and miniature endplate currents (MEPCs) were measured at intervals of approx. 5 to 10 min from different endplate regions. About 100 MEPPs or MEPCs per recording period could be registered for further analysis. MEPCs were displayed on a Gould digital storage oscilloscope (OS 4020) and plotted from there with a Philips printing recorder (PM 8210). The time constant of both the rising phase and the decay phase were determined from the straight line that could be drawn through the linear portion of a semilog plot that usually was found between 20 and 80 % of the peak.

Results obtained from two or more experiments performed under the same conditions were pooled and analyzed statistically with Student's t test after a confidence test of variance (Bartlett) had first been performed.

"F7"-toxin was a gift from Professor C.Y. Lee, Natl. Taiwan University; beta-Bungarotoxin (purified from *Bungarus multicinctus* venom) was obtained from Boehringer Mannheim, F.G.R.; and dexamethasone sodium phosphate was kindly donated by Merck Sharp and Dohme.

3. RESULTS

3.1. Frequency of spontaneous release. Incubation of the diaphragm in a medium with 200 nM Dex causes a significant change in MEPP frequency during the three hour lasting period of exposure to this drug. Beta-BuTx (22 and 220 nM) has a triphasic effect on the MEPP

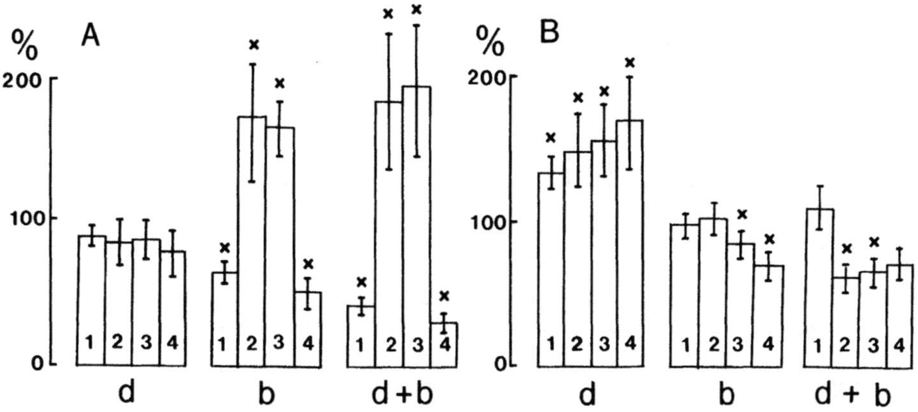

FIGURE 1. Effects of either 200 nM dexamethasone (d), 22 nM beta-BuTx (b), or a combination (d+b) on MEPP frequency (A) and MEPP amplitude (B) relative to control values. Mean values ± S.E.M. Data sampled in the periods 0-30 min (1), 30-60 min (2), 60-90 and 90-120 min (4). * P < 0.05.

frequency. After an initial decrease lasting 20-30 min, a transient increase was followed by a phase dominated by small MEPPs with amplitudes close to the noise level, as well as MEPPs of normal size. The combination of Dex and beta-BuTx did not lead to an obvious change in the effect of the toxin on the frequency (Fig. 1). In the period 30 to 120 min after the onset of perfusion

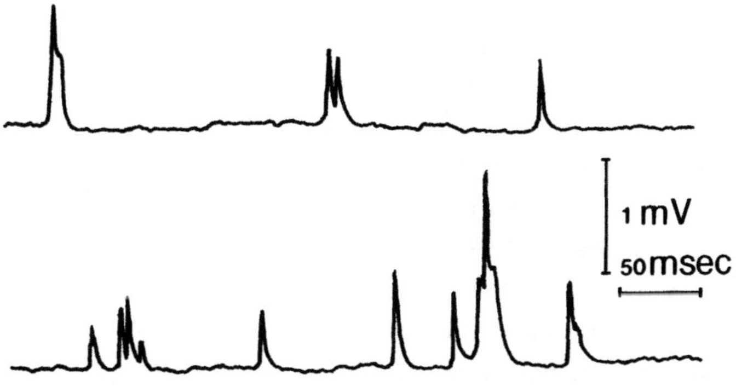

FIGURE 2. Registration of spontaneously occuring MEPPs 110 min after continuous perfusion with 3 µg/ml "F7" toxin. Notice the presence of small bursts of MEPPs and composite MEPPs.

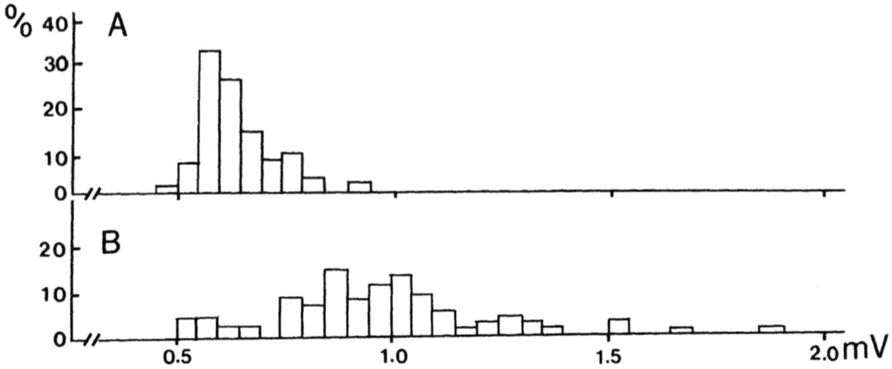

FIGURE 3. MEPP amplitude histograms. A. Contr. B. After 158 min exposure to 3 µg/ml "F7" toxin.

with "F7" toxin (3µg/ml) a transient two to threefold increase of MEPP frequency was seen, accompanied by small bursts of MEPPs. The frequency subsequently decreased and more composite MEPPs appeared (Fig. 2). Dex did not affect the effects of "F7" toxin.

3.2. MEPP amplitudes. After a latent period of about 30 min, Dex (200 nM) increased the mean MEPP amplitude gradually to 150 - 190 % of the control value (Fig. 1). Amplitude histograms changed from pseudo-normal distributions to skewed distributions, often with two or more peaks. Also giant MEPPs with amplitudes several times the amplitudes of the unitary MEPPs were frequently seen. Incubation in beta-BuTx (22 and 220 nM) had no appreciable effect on the amplitudes of the MEPPs, but in a later phase a class of small MEPPs, with amplitudes close to the noise level, remained. In combination with Dex, beta-BuTx abolished completely the increase of the MEPPs and the giant MEPPs caused by Dex (Fig. 1). Perfusion with "F7" toxin (3 µg/ml) leads, after a slow onset of 30 min, to bursts of MEPPs and giant MEPPs, often clearly composed of two or more units (Fig. 2). Spontaneous fasciculations were seen at the same time. The amplitude histograms showed distributions finally dominated by the composite MEPPs (Fig. 3).

3.3. Time constants of MEPCs. The time constants of the rising phase and the decay phase of the MEPCs, recorded at a holding

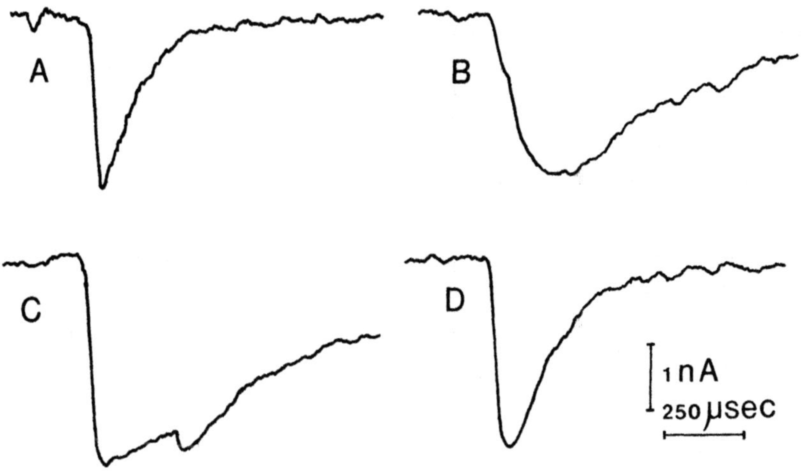

FIGURE 4. MEPCs: A. Contr; B, C and D recorded from the same end-plate 115 min after continuous perfusion with 3 µg/ml "F7" toxin.

potential of - 65 mV, were neither affected by Dex nor by beta-BuTx. On the other hand, either the rising phase of the MEPCs or the decay phase or both were prolonged by "F7" toxin (Fig. 4). The mean rising time increased from 137 ± 6.4 (S.E.M.) µsec (n = 15) to 218 ± 24.5 µsec (n = 38) after 2 h incubation in "F7" toxin (3 µg/ml) and the mean decay time from 1425 ± 48 µsec (n = 25) to 3125 ± 395 µsec (n = 38), $P < 0.01$.

4. DISCUSSION

The experiments described here provide further evidence that glucocorticoids as well as the toxins, beta-BuTx and "F7" from snake venoms, interact with the acetylcholine release mechanism at the neuromuscular junction. In agreement with results from previous experiments, using prednisolone both *in vitro* and *in vivo* (6), we have found that Dex increases the MEPP amplitudes and produces giant MEPPs. The glucocorticoids have a benificial effect in the treatment of the muscle disease, myasthenia gravis (7). They have been shown to antagonize the blocking effect of hemicholinium-3 in nerve muscle preparations (8) and to increase the accumulation of radioactive choline and the turnover and output of acetylcholine in a choline rich medium (9). Furthermore morphometric studies have revealed that synaptic vesicle size is increased and this is correlated with the physiological effects (3). In guinea pig lung, in vitro, perfusion with Dex induces the biosynthesis of a phospholipase A2 inhibitor (10). The slow onset of the effects of Dex on MEPP amplitude and distribution suggets that the corticosteroids here may perhaps also be involved in indirect processes affecting neurotransmitter release.

The effects of Dex are abolished by beta-BuTx. Beta-BuTx is believed to have only presynaptic effects (11). In agreement with this conclusion we could not find any effect neither of beta-BuTx, nor of Dex, nor of the combination of these agents on the time constants of the MEPCs. Beta-BuTx possesses a calcium dependent phospholipase A2 activity which is thought to produce the transient increase and the final failure of transmitter release. It is also believed that beta-BuTx increases the level of free calcium in the cytosol of nerve terminals (4).Glucocorticoids promote changes in intracellular calcium, although little is known about the mechanism by which this occurs (12). Dex induced increase in MEPP amplitude, but in combination with beta-BuTx, like in beta-BuTx alone (13), only a class of MEPPs with amlpitudes smaller than the unitary MEPPs seemed to persist.

In contrast to many other presynaptic-acting snake venoms, it is believed that Dendroaspis venom has no detectable phospholipase A2 activity (14). In mice in vivo "F7" induces fasciculations (15). This has also been seen here with unstimulated rat diaphragm preparations and is correlated with the increased MEPP frequency, small bursts of MEPPs and giant MEPPs. Preliminary experiments, using a combination of "F7" toxin and beta-BuTx, showed that the giant MEPPs were not abolished by beta-BuTx. Under the influence of "F7" toxin both the rising- and decay time of MEPCs were often prolonged and showed small fluctuations caused by the release of subunits. Prolongation of the MEPCs might also be due to the anti-cholinesterase activity of "F7" toxin (16), to postsynaptic myo-necrosis causing an increased diffusion pathway (15), or to direct postsynaptic effects. Ultrastructural investigation, done after finishing the electrophysiological experiments, did not show any sign of myonecrosis. Beside MEPCs with prolonged rising and decay times, MEPCs were also seen in the registrations from the same endplates with practically normal time constants. We would suggest therefore, that under the influence of "F7" toxin one of the effects is that the synchronization of the spontaneous release of quanta of acetylcholine is altered.

Corticosteroids as well as "F7" toxin seem to facilitate acetylcholine release, but the mechanisms by which they affect the neurotransmitter release are probably different, as shown by the interactions with beta-BuTx. If it is supposed that the glucocorticoids and the presynaptic acting snake toxins act after one another in the chain of events underlying the process of neurotransmitter release, then these agents may become useful tools in the further elucidation of this mechanism.

REFERENCES

1. Veldsema-Currie RD, Van Wilgenburg H, Labruyere WT, Langemeijer MWE: Presynaptic, facilitatory effects of the corticosteroid dexamethasone in rat diaphragm: Modulation by beta-Bungarotoxin. Brain Research 294: 315-325, 1984.
2. Lee CY, Tsaur ML, Tsai MC, Carlsson FHH, Joubert FJ: Facilitation of acetylcholine release by Angusticeps-type toxins from Mamba venoms. Neuroscience Letters, Suppl. 14: S215, 1983.
3. Van Wilgenburg H, Njio KD, Belling GAC, Van Den Hoven S: Effects of corticosteroids on the myoneural junction. A morphometric and electrophysiological study. Europ. J. Pharmacol. 84: 129-137, 1982.
4. Howard BD, Gundersen CB: Effects and mechanisms of polypeptide neurotoxins that act presynaptically. Ann. Rev. Pharm. Toxicol. 20: 307-366, 1980.
5. Viljoen CC, Botes DP: Snake venom toxins. The purification and amino acid sequence of toxin F VII from Dendroaspis angusticeps venom. J. Biol. Chem. 248: 4915-4919, 1973.
6. Van Wilgenburg H: The effect of prednisolone on neuromuscular transmission in the rat diaphragm. Europ. J. Pharmacol. 55: 355-361, 1979.
7. Warmolts JR, Engel WK, Whitaker JN: Alternate day prednisolone in a patient with myasthenia gravis. Lancet 2: 1198-1199, 1970.
8. Wolters ECMJ, Leeuwin RS: The effect of corticosteroids on the nerve-diaphragm preparation treated with hemicholinium-3, a possible model for myasthenia gravis. Neurology 26: 574-578, 1976.
9. Veldsema-Currie RD, Wolters ECMJ, Leeuwin RS: The effect of corticosteroids and hemicholinium-3 on choline uptake and incorporation into acetylcholine in rat diaphragm. Europ. J. Pharmacol. 35: 399-402, 1976.

10. Flower RJ, Blackwell GJ: Anti-inflammatory steroids induce biosynthesis of a phospholipase A2 inhibitor which prevents prostaglandin generation. Nature (Lond.) 278: 456-459, 1979.
11. Chang CC, Lee CY: Isolation of neurotoxins from the venoms of Bungarus multicinctus and their modes of neuromuscular blocking action. Arch. Int. Pharmacodyn. 144: 241-257, 1963.
12. Baxter JD, Rousseau GG: Glucocorticoid hormone action. Springer Berlin, 1979, pp 1-24.
13. Llados F, Matteson DR, Kriebel ME: Beta-Bungarotoxin preferentially blocks one class of miniature endplate potentials. Brain Research 192: 598-602, 1980.
14. Harvey AL, Karlsson E: Protease inhibitor homologues from mamba venoms: facilitation of acetylcholine release and interactions with prejunctional blocking toxins. Brit. J. Pharmacol. 77: 153-161, 1982.
15. Veldsema-Currie RD, Njio KD, Van Wilgenburg H, Leeuwin RS, Lee CY: Alterations in ultrastructure and activity of motor nerve terminals caused by "F7" toxin from the eastern green Mamba snake (Dendroaspis angusticeps). In preparation.
16. Lee CY, Lin WW, Carlsson FHH, Joubert FJ: Anticholinesterase activity of Angusticeps-type toxins and protease inhibitor homologues from Mamba venoms. Chin. J. Physiol. 27: 20A, 1984.

DISCUSSION OF THE ARTICLE BY H. VAN WILGENBURG (NETHERLANDS)

Comment: Have you tried any other phospholypase A2 inhibitors?

H. van Wilgenburg (Netherlands):

We intend to check it soon.

D.O. Smith (U.S.): Your extracellular recordings are greatly dependent on electrode position and I noticed that you mentioned that you recorded up to two hours. Did you have any tests which would prove that your electrode had not moved, which may explain some of these results?

H. van Wilgenburg (Netherlands): It was difficult to keep the electrode in place for a long time so what I have done mainly was to record from different endplates and then to use statistical analysis.

FLUCTUATION ANALYSIS APPLIED TO NONSTATIONARY BIOLOGICAL PROCESSES

Riccardo Fesce
Rockefeller University. New York. NY
(Permanent address: Dept. Pharmacology, University of Milano. Italy)

According to the standard theory of the statistics of stationary 'shot noise' (ie. a fluctuating signal produced by the linear summation of elementary events occurring randomly at a constant mean rate), the mean value, variance and skew of the noise are related to the mean rate of the events, $\langle r\rangle$, their amplitude, h, and their time course, w(t), by the following equations: mean=$\langle r\rangle h\int w(t)dt$, variance=$\langle r\rangle h^2\int w^2(t)dt$, skew=$\langle r\rangle h^3\int w^3(t)dt$ (Rice, 1944).

The fluctuating signal recorded from endplates during vigorous quantal secretion of ACh is due to the superimposition of miniature endplate potentials (mepps) and is amenable to such analysis: Segal et al. (1984) demonstrated that: (1) the mepp waveform is well represented by $w(t)=e^{-t/\theta_1}-e^{-t/\theta_2}$; (2) θ_1 and θ_2 can be deduced either by averaging many individual mepps when they do not overlap, or by curve fitting the power spectrum of endplate recordings during intense activity; (3) mepp rate and amplitude are reliably measured from the variance and skew of endplate records, better results being obtained after appropriate high-pass filtering has removed the contributions to the moments due to slow changes in membrane potential or mepp rate.

The power spectra obtained from endplates under certain experimental conditions (e.g. in hypertonic solutions or in the presence of black widow spider venom), do not show white behavior at low frequencies (such white behavior is expected for the spectrum of shot noise with brief events; Schick, 1974). This could be due either to nonstationarity in mepp rate (e.g. bursting secretion) or to membrane noise unrelated to mepps. Previous theoretical analyses of nonstationary shot noise (Rice, 1944; Schick, 1974) indicate that the contribution of nonstationarity to the variance appears as added components in the power spectrum that are limited to the frequency bandwidth of the time dependent rate of occurrence, r(t). In principle, they can be filtered out. I have derived the frequency composition of the skew: the contribution of nonstationarity to the skew is also limited to the bandwidth of r(t), and, in principle, can also be filtered out.

Let's indicate by $\tilde{v}(n)$ the finite Fourier transform of the signal V(t), over a period T, for a frequency $f_n=n/T$: $\tilde{v}(n)=\int_0^T V(t)e^{-2\pi nti/T}$. According to the theory of Fourier series, the departure of the signal from its mean value, $v(t)=V(t)-\langle V\rangle$, is given by:

$$v(t) = \frac{1}{T}\sum_{n} \tilde{v}(n)e^{2\pi nti/T} \qquad \{ n=[-\infty,\infty],\ n\neq 0 \}.$$

It follows quite directly that the variance of the signal, $\langle v^2\rangle$, is:

$$\langle v^2\rangle = \frac{1}{T^2}\sum_{n} \tilde{v}(n)\tilde{v}(-n) = \frac{1}{T^2}\sum_{n} \tilde{v}_2(n)$$

This expression gives the frequency composition of the variance: $\tilde{v}_2(n)/T$ is the power density per unit bandwidth ($2\tilde{v}_2(n)/T$ if only positive frequencies are considered). Similarly, the skew of the signal, $\langle v^3\rangle$, is:

$$\langle v^3\rangle = \frac{1}{T^3}\sum_{n,m} \tilde{v}(n)\tilde{v}(m)\tilde{v}(-n-m) = \frac{1}{T^3}\sum_{n,m} \tilde{v}_3(n,m)$$

This expression gives the frequency composition of the skew: its bispectral density is $\tilde{v}_3(n,m)/T$, or $6\text{real}[\tilde{v}_3(n,m)]/T$ if only positive frequencies are considered.

Now let V(t) be generated by the linear summation of mepps. If many records are available, all with the same time course of the probability of occurrence of the mepps, the expected value of V(t), E[V(t)], is determined by the expected rate of occurrence, r(t): $E[V(t)] = \int_{-\infty}^{\infty} r(t')w(t-t')dt'$.

The expected values of $\langle V\rangle$, $\tilde{V}(n)$, $\tilde{v}_2(n)$ and $\tilde{v}_3(n,m)$ are easily computed; neglecting the errors at the edges of the record (see below), they are given by:

$E[\langle V\rangle] = \langle r\rangle h\int w(t)dt$, $\quad E[\tilde{V}(n)] = h\tilde{w}(n)\tilde{r}(n)$, $\quad E[\tilde{V}_2(n)] = h^2\tilde{w}_2(n)[\tilde{r}_2(n)+\langle r\rangle T]$,

and $\quad E[\tilde{V}_3(n,m)] = h^3\tilde{w}_3(n,m)[\tilde{r}_3(n,m)+\tilde{r}_2(n)+\tilde{r}_2(m)+\tilde{r}_2(n+m)+\langle r\rangle T]$,

where $\tilde{w}_i$ and $\tilde{r}_i$ are defined as $\tilde{v}_i$.

Therefore, the mean is insensitive to nonstationarity, while all contributions from r(t) to the variance and skew of v(t) are limited to the bandwidth of r(t). The latter is recognized as the part of the power spectrum which deviates from the analytical shape of the spectrum of w(t). If the signal is high-pass filtered, so that the whole bandwidth of r(t) is cut, the variance and skew of the remaining signal become linearly related to $\langle r\rangle$, as it is the case under stationary conditions. After the signal has been appropriately filtered, then, the usual procedures used to analyze stationary processes will yield valid estimates of amplitude and mean rate of the events. Alternatively, the estimates can be obtained from unfiltered mean and filtered variance. The contributions of nonstationarity to the random errors in the estimates have the same bandwidth of r(t), as well, and are also reduced by filtering. Filtering also removes extra-components in the variance or skew due to membrane noise of limited bandwidth, unrelated to the mepps.

The errors at the edges arise because all the events occurring in the period t=[0,T] are treated as if they were complete, and the tails of events occurring just before time 0 are ignored. These errors will be small if T is much longer than the duration of a single event, and if r(t) changes without fast progressive trends. When the time course of the rate of occurrence obeys these restrictions, the random errors in the estimates of variance and skew will decrease as the observation time increases, and the amplitude and rate of occurrence of the events can be computed on single, sufficiently long records.

The procedure was tested by analyzing simulated nonstationary endplate records (simulated superimposed volleys of mepps, with random occurrence and duration); it yielded valid estimates of rate and amplitude, with bias and random errors in the order of 10%. Random errors were further reduced when several subsequent records were averaged.

REFERENCES

1. Rice S.O. 1944. Mathematical analysis of random noise. Bell Tel. System J. 23:282-332. Reprinted in: Selected papers on noise and stochastic processes. N. Wax ed. Dover Publications, New York, pp.133-294, 1954.
2. Schick, K.L. 1974. Power spectra of pulse sequences and implications for membrane fluctuations. Acta Biotheor. 23(I):1-17.
3. Segal, J.R., B. Ceccarelli, R. Fesce and W.P. Hurlbut. 1984. Miniature endplate potential frequency and amplitude determined by an extension of Campbell's theorem. Biophys. J. In press.

I gratefully aknowledge the constant help and support of Dr. W.Paul Hurlbut, and thank Dr. Daniel Tranchina and Dr. John R. Segal for helpful discussion.

Supported by PHS grant NS18354 (W.P. Hurlbut)

DISCUSSION OF THE ARTICLE BY R. FESCE AND W.P. HURLBUT (U.S.)

I. Steinberg (Israel): It wasn't obvious how you overcame the problem of non-linear summation.

W.P. Hurlbut (U.S.): The problems of a drifting membrane potential and non-linear summation are both overcome simply by the fact that the analysis is based entirely on the fluctuations about the mean, and these fluctuations are much smaller in amplitude than the drift of the mean. The values obtained from the analysis are those that are appropriate to the average level of membrane potential that existed when you collected the data sample.

CALCIUM INDEPENDENT DEPLETION OF QUANTA AND SYNAPTIC VESICLES INDUCED BY OUABAIN

Haimann, C., Torri-Tarelli, F., Fesce, R., and Ceccarelli, B.
CNR Center of Cytopharmacology and Center of Peripheral Neuropathies, Via Vanvitelli 32, 20129 Milano, Italy

A novel extension of the Campbell's theorem was used to estimate mepps parameters during the intense quantal secretion of ACh induced by ouabain. The variance, skew and power spectra of the fluctuating postsynaptic membrane potential were measured over successive 10 sec records (Segal et al., Biophys. J., submitted). The high frequency of mepps is often not stationary over the 10 sec of a single record, therefore the signal was filtered with a high-pass filter RC of 2 msec to reduce the errors introduced by non-stationarity (Fesce et al., this Symposium). Ouabain (0.1 and 0.05 mM) was applied in frog cutaneous pectoris nerve muscle preparation bathed in modified Ringer's solution containing either Ca (1.8 mM Ca and 4 mM Mg) or no added Ca (4mM Mg and 1 mM ETGA). When Ca was present (6 muscles) after a lag of 1.5-2 hours mepps frequency rose esponentially and reached an average peak rate of 1288 ± 492/sec in about 30 min, then suddenly subsided to low level (10/sec). When ouabain was applied in Ca-free solution (5 muscles) after a shorter lag (1-1.5 hours) the mepps frequency rose to peak rate of 677 ± 399/sec in about 20 min and then gradually subsided (10/sec). In spite of the different time course of secretion in the two experimental conditions the cumulative quantal release was not significantly different ($7.4 \pm 1.3 \times 10^5$ in Ca containing and $8.9 \pm 2.7 \times 10^5$ in Ca-free solutions). 60 min after the peak of secretion some muscles were fixed for the EM observation. The results of morphometric analysis on micrographs of neuromuscular junctions revealed in both cases an almost complete depletion of synaptic vesicles and deep infoldings of presynaptic membrane. In addition, the use of HRP suggests that ouabain impairs the recycling process that tends to conserve the vesicle population in face of intense secretion of neurotransmitter. The relevance of all these results on the vesicle hypothesis of transmitter release will be discussed. Partially supported by M.D.A. Grant (B.C.).

DISCUSSION OF THE ARTICLE BY C. HAIMANN (ITALY)

R.J. DeLorenzo (U.S.): Is it possible that the ouabain treatment causes calcium release from inside the terminal? Even though calcium in the media is not required, ouabain may actually release it. Possibly, with quin II, you will be able to see it.

C. Haimann (Italy): Yes, it is possible and it has been suggested that transmitter release is induced by ouabain through such a mechanism (Baker, P.F. and Crawford, A.C. (1975). A note on the mechanism by which inhibitors of the sodium pump accelerate spontaneous release of transmitter from frog nerve terminals. J. Physiol. 247, 209-226). We haven't tried to measure intracellular calcium with quin II.

R. Rahamimoff (Israel): Is there any way which you can distinguish between accumulation of intracellular sodium or inhibition of the ATPase? Which of the two inhibits recycling?

C. Haimann (Italy): We do not have a way to distinguish between these two possibilities. There is one evidence in the literature (Larkin, J.M., Brown, M.S., Goldstein, J.L. and Anderson, R.G.W. (1983). Depletion of intracellular potassium arrests coated pit formation and receptor mediated endocytosis in fibroblasts. Cell 33, 273-285). In human fibroblasts, the deprivation of intracellular potassium, more than the accumulation of intracellular sodium, causes the impairement of the endocytosis.

CALMODULIN INHIBITORS AND TRANSMITTER RELEASE AT THE FROG NEUROMUSCULAR JUNCTION

S.J. PUBLICOVER AND Z.Y. SAHAF

1. INTRODUCTION

The calcium binding protein calmodulin (CaM) appears to mediate a number of Ca^{2+} activated cellular processes(1), included among which are the secretary activities of a variety of cell types (2). Of particular interest is the fact that CaM has been implicated in the release of transmitters from synaptosomes. Using this preparation De Lorenzo and his group have shown that CaM is present in the presynaptic cytoplasm and in synaptic vesicle preparations and that it has a variety of effects on synaptosomal componenents that are suggestive of an involvement of CaM in exocytotic transmitter release (3). These workers have also shown that transmitter release from synaptosomes is inhibited by CaM-interacting drugs, their hypothesis being that in the intact synaptosome interaction of Ca^{2+} with CaM activates a kinase resulting in phosphorylation of specific proteins (possibly tubulin) and that this phosphorylation is an important step in excitation secretion coupling (3).

Though differences in structure may exist between the presynaptic terminals of central and peripheral synapses, there are clearly great similarities in their functioning, including the role of Ca^{2+} in excitation-secretion coupling. It would therefore seem reasonable to hypothesise that the release of transmitter at the vertebrate neuromuscular junction is similarly controlled by a Ca^{2+}-CaM system. However, studies of the type that have been carried out using synaptosomes cannot be repeated on neuromuscular preparations, and so one of the few methods which might be employed to test such an hypothesis is the use of Ca^{2+}-antagonistic drugs, such as the phenothiazines. Though results from such studies must be treated with caution (4), they do at least provide a way of probing for the possible

---

Statistical analysis: All limits to means given in text and tables are ± 1 S.E.M. All significance levels were determined using Student's t-test.

involvement of CaM in these synapses, and have already provided valuable results in investigations on other secretory systems.

In this communication the results of studies on the frog cutaneous-pectoris neuromuscular preparation using two 'CaM-antagonists' Trifluoperazine (TFP, a phenothiazine) and Calmidazolium (R24571, a non-phenothiazine inhibitor of CaM), are described.

2. TRIFLUOPERAZINE

2.1. Spontaneous transmitter release

The effect of TFP on M.E.P.P. frequency was tested by treating preparations with the drug for 15 - 20 min (a time period that has been shown to be adequate to produce an effect on other cell types) and calculating the ratio of frequency after treatment, to frequency before treatment (normalised frequency). The results are summarised in Table 1. The mean normalised M.E.P.P. frequency exceeded that of the controls under all of the experimental conditions tested (individual normalised frequency varied between 1.05 and 5), but in normal frog saline (1.8 mM Ca^{2+}) the effect was less marked then in low Ca^{2+} saline and the means were not significantly different from that of the controls.

Table 1. Effect of TFP on spontaneous transmitter release. Results are expressed as mean normalised frequency after treatment. D.W. indicates distilled water, the vehicle for TFP.

Saline	TFP	(n)	Mean normalised frequency	P
1.8 mM Ca^{2+}	25 μM	10	1.51 ± 0.32	N.S.
1.8 mM Ca^{2+}	10 μM	6	1.80 ± 0.46	N.S.
0 Ca^{2+}, 1 mM Mg^{2+} 1 mM EGTA	25 μM	5	2.10 ± 0.27	<0.02
1.8 mM Ca^{2+}	(1% D.W.)	6	0.85 ± 0.06	-

2.2. Evoked transmitter release

At fairly high quantal contents (30-40), TFP has no significant effect on evoked release even at concentrations of 50 μM (5) and so a number of experiments were carried out to see if any action of the drug could be detected at lower levels of release (quantal content of 1-2) using a saline with reduced Ca^{2+}-content (0.3 mM Ca^{2+}, 1 mM Mg^{2+}). Under these conditions a rise in quantal content was seen in all of 3 preparations exposed to 25 μM TFP, (normalised quantal content after treatment being 1.51 ± 0.46

($P<0.01$)) and in both of 2 preparations exposed to 2.5 μM TFP (normalised quantal contents of 1.53 and 1.18).

2.3. K^+-induced transmitter release

TFP causes a marked reduction in depolarisation-induced transmitter release from synaptosomes (maximum inhibition at 5-15 μM, Ref. 3) and so a number of parallel experiments were carried out using neuromuscular preparations. Under these conditions a clear inhibitory effect of the drug was observed. Elevation of Ca^{2+} from 2.5 mM to 10 mM caused a rise in M.E.P.P. frequency of approximately 900% but after pretreatment with TFP (25 μM) this response was reduced to only 500% ($P<0.01$). At a concentration of 10 μM TFP appeared to have an inhibitory effect but the difference from controls was not statistically significant. When the experiments were repeated in modified saline (1 mM Ca^{2+}, 2 mM Mg^{2+}, such that elevation of [K^+] to 10 mM only doubled M.E.P.P. frequency) no effect of TFP pretreatment could be discerned.

2.4. Tetanic enhancement of M.E.P.P. frequency

It is possible to increase the frequency of spontaneous release by tetanic stimulation of the nerve in salines containing low Ca^{2+} levels or even in the presence of a reversed electrochemical gradient for Ca^{2+}, probably because the tetanus causes the release of Ca^{2+} from intracellular storage sites (6). The effect of TFP on tetanic enhancement of M.E.P.P. frequency was tested, enhancement being defined as

$$E(t) = (Ft/F)-1$$

Where $E(t)$ = enhancement at time t, Ft = M.E.P.P. frequency at time t and F = pretetanic M.E.P.P. frequency.

2.4.1. Inward electrochemical gradient. Preparations were stimulated at 50 Hz for 50 s in saline containing 50 μM Ca^{2+}. Stimulation induced a steady rise in M.E.P.P. frequency, the mean enhancement for the final 10 s of tetanus under control conditions being 38.69 ± 11.43 (6 preparations). In each experiment the procedure was then repeated after exposing the preparation to 25 μM TFP for 20-25 min. In every case the response was markedly reduced (an example is shown in Fig. 1). The effect of TFP was assessed by comparing the enhancement over the final 10 s of tetanus under control conditions with that recorded after exposure of the preparations to TFP. The mean ratio (enhancement before TFP/enhancement after

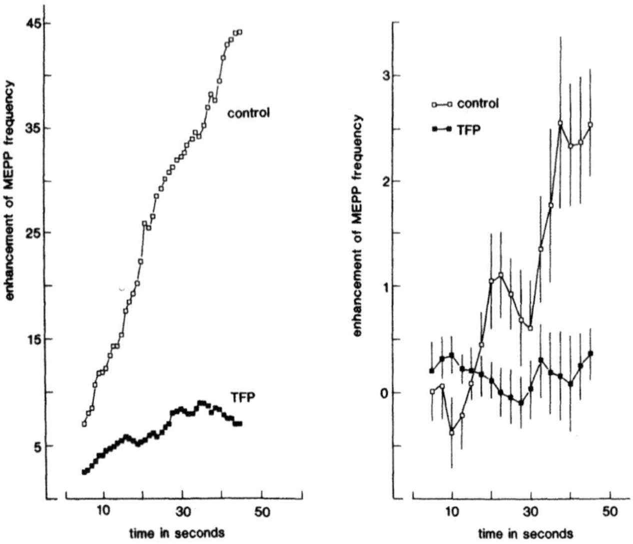

FIGURE 1. Effect of TFP (25 μM) on tetanic enhancement of M.E.P.P. frequency. Preparations were stimulated at 50 Hz for 50 s and data were analysed by the moving bin method (see Ref. 6). LEFT: Inward electrochemical gradient for Ca^{2+}, (□) control response (■) after TFP. Bin size = 10 s, Δ bin = 1 s. RIGHT: Reversed electrochemical gradient for Ca^{2+}, mean of 5 different preparations (± S.E.M.), (□) control response (■) after TFP. Bin size = 10 s, Δ bin = 5 s. After Ref. 2.

TFP) was 4.27 ± 1.10 ($P<0.02$).

2.4.2. Reversed electrochemical gradient. The electrochemical gradient for Ca^{2+} can apparently be reversed by using saline containing 1 mM EGTA, 1 mM Mg^{2+} and no added Ca^{2+} (see Ref. 6). Under these conditions TFP still induced a discernable reduction in the response to tetanus in all of the 5 preparations tested. The data of all 5 preparations were meaned in order to get a clear idea of the progressive rise in M.E.P.P. frequency during the tetanus. The results are plotted in Fig. 1. The mean ratio (enhancement before TFP/enhancement after TFP) was 7.54 ± 2.17 ($P<0.05$).

3. CALMIDAZOLIUM

3.1. Spontaneous transmitter release

R24571 was found to exert its full effect on M.E.P.P. frequency within 10-15 min of treatment. Experiments were therefore carried out similarly to those with TFP. Control M.E.P.P. frequency was recorded (F) after which the preparation was exposed to the drug for 15-20 min, followed by a second

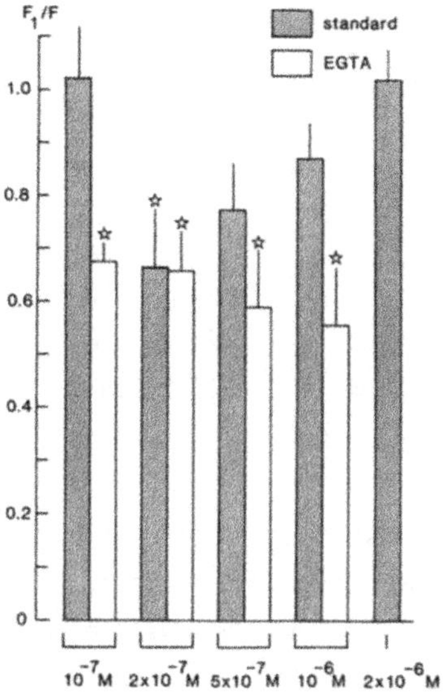

FIGURE 2. Effect of R24571 on spontaneous transmitter release. M.E.P.P. frequency was determined before (F) and after 15-20 min (F_1) of drug treatment. 'Standard' - 1.8 mM Ca^{2+}; 'EGTA' - $0Ca^{2+}$, 1 mM, Mg^{2+}, 1 mM EGTA. (☆) indicates significantly different from control value of 0.98 ± 0.06 ($P<0.05$).

experimental' recording (F_1). The results are summarised in Fig. 2.

In normal frog saline (1.8 mM Ca^{2+}) the effect of R24571 was complex. At 10^{-7}M no effect could be discerned, at 2 x 10^{-7}M the spontaneous release rate was significantly reduced ($P<0.05$), but further increases in concentration resulted in a reduction of this suppressive effect (Fig. 2). In contrast, when the experiments were repeated in Ca^{2+}-free saline (1 mM EGTA, 1 mM Mg^{2+}, 0 Ca^{2+}) a significant inhibitory effect was seen at all the concentrations tested and this action appeared to be more marked at higher concentrations (Fig. 2).

3.2. K^+-induced transmitter release

M.E.P.P. frequency was measured first under control conditions (F), then after 25 min of R24571 treatment (F_1), and finally after exposure to 10 mM K^+ and R24571 (F_2). After treatment with R24571 the response to 10 mM K^+ became extremely variable, ranging from 800 to 3,000%. The mean response to K^+ was higher at concentrations of 10^{-7}M and 5 x 10^{-7}M (Table 2), but was not significantly different from the controls. At 2 x 10^{-6}M the mean response to K^+-saline was slightly lower than that in the controls, but again, this difference was not significant.

Table 2. Effect of R24571 on K^+-induced release of transmitter. See text for details

Conc.	n	F_2/F	F_2/F_1
10^{-7}M	4	16.03 ± 5.09	15.82 ± 4.75
5 x 10^{-7}M	6	16.71 ± 3.67	12.38 ± 2.76
2 x 10^{-6}M	5	8.21 ± 1.38	8.30 ± 1.22
control	4	9.98 ± 0.50	9.78 ± 0.46

4. DISCUSSION

TFP clearly has at least two distinct actions on the process of transmitter release at the neuromuscular junction. The stimulatory effects of TFP on M.E.P.P. frequency and on low levels of evoked release of transmitter may well reflect the same action of the drug, and are probably analagous to the TFP-induced rise in basal release rate that has been observed in other systems (7,8). Since this stimulatory action appears to be enhanced by reduction of $[Ca^{2+}]_o$ (Table 1) it is clearly not a result of increased Ca^{2+} -influx across the plamalemma. Though an effect on CaM cannot be ruled out (inhibition of membrane Ca^{2+}-pumping for instance), it may well be that this stimulatory action of the drug merely results from a less specific effect on membrane characteristics or on intracellular Ca^{2+}-storage sites.

An inhibitory action of the drug was also clearly demonstrated (see 2.3, 2.4). Though an action on the Ca^{2+}-channel of the plasma membrane might produce such effects, this is considered unlikely since TFP inhibits tetanic enhancement of release even when the electrochemical gradient for Ca^{2+} is reversed. Under such conditions, the effect of blockage of Ca^{2+} channels would be reversed, i.e. an increase in tetanic enhancement (see Ref. 6). It is therefore possible, though clearly not proven, that these inhibitory effects of TFP do reflect an action on a Ca^{2+}-CaM system in the presynaptic terminal.

R24571 also appears to have two opposing actions on transmitter release. This is well demonstrated in the action of the drug on spontaneous transmitter release. In Ca^{2+}-free saline M.E.P.P. frequency was reduced by R24571 at all the concentrations used (a finding that is in accord with the high potency of R24571 as a CaM-inhibitor, 9), whereas in normal saline

(1.8 mM Ca^{2+}) R24571 produced a significant reduction in spontaneous release only at 2 x 10^{-7}M, and at higher concentrations the normalised M.E.P.P. frequency rose towards (and marginally above) 1.0 (Fig. 2). It therefore seems likely that a second stimulatory effect of the drug is exerted at the higher concentrations, but that this effect is only observed when normal Ca^{2+}-levels are present in the saline. Initial experiments with higher concentrations of R24571 (10^{-5}M) suggest that in some preparations the drug can produce a rise in M.E.P.P. frequency (3-4 fold) even in Ca^{2+}-free saline. Therefore, though the Ca^{2+} dependence of this effect might indicate that R24571 stimulates spontaneous release by inducing Ca^{2+}-influx, it may well be that release of Ca^{2+} from an intracellular site is involved. It has already been shown that exposure of preparations to EGTA-containing salines can markedly reduce the effect of intracellularly released Ca^{2+} on M.E.P.P. frequency (10). The inhibitory effect of R24571 on spontaneous release was seen in both types of saline (Fig. 2) and therefore is almost certainly exerted at an intracellular site. As in the case of the inhibitory effect of TFP, an action through CaM is a possibility, but is not proven.

The failure of R24571 to inhibit K^+-induced release is somewhat disappointing, and emphasises the lack of any clearly common effect of these two compounds that might be attributable to an action on CaM. Both TFP and R24571 inhibit K^+-stimulated transmitter release from rat brain synaptosomes (8) and it may thus be that in the neuromuscular preparation the stimulatory action of R24571 is sufficiently enhanced by depolarisation of the terminals to mask any inhibitory effect. It is noteworthy that in these experiments the response to K^+ was extremely variable (see 3.2) an observation that may well be indicative of a fine balance between two opposing effects (see Ref. 10).

In general, the results obtained with CaM-inhibitors do not as yet serve to confirm the hypothesis for a role of CaM in neuromuscular transmitter release. Inhibitory effects are clearly discernable in the actions of both drugs, but they are not seen under all experimental conditions (notably evoked release on which preliminary experiments with R24571 have failed to reveal any) and do not parallel each other particularly closely. The only recorded inhibitory effect of a 'CaM-inhibitor' on evoked release is the action of chlorpromazine (11), an effect that was attributed to inhibition of Ca^{2+}-influx by the authors. It is therefore clear that further work is necessary in order to clarify the presynaptic actions of these drugs, and it is in fact conceivable that the inconsistency of their inhibitory action on peripheral synapses as compared with their action on synaptosomes (3,8) reflects not a difference of these synapses from those in the CNS, but a difference in the triggering action of Ca^{2+} in synchronous and asynchronous transmitter release.

REFERENCES

1. Cheung WY: Calmodulin plays a pivotal role in cellular regulation. Science 207: 19-27, 1980.
2. Publicover SJ: Presynaptic action of trifluoperazine at the frog neuromuscular junction. Naunyn-Schmiedeberg's Arch Pharmacol 322: 83-88, 1983.
3. De Lorenzo RJ: Calmodulin in neurotransmitter release and synaptic function. Fed Proc 41: 2265-2272, 1982.
4. Landry Y, Amellal M, Ruckstuhl M: Can calmodulin inhibitors be used to probe calmodulin effects? Biochem Pharmacol 30: 2031-2032, 1981.
5. Cheng K-C, Lambert JJ, Henderson EG, Smilowitz H, Epstein P: Postsynaptic inhibition of neuromuscular transmission by trifluoperazine. J Pharm Exp Ther 217: 44-50, 1981.
6. Lev-Tov A, Rahamimoff R: A study of tetanic and post-tetanic potentiation of miniature end-plate potentials at the frog neuromuscular junction. J Physiol Lond 309: 247-273, 1980.
7. Douglas WW, Nemeth EF: On the calcium receptor activating exocytosis: inhibitory effect of calmodulin-interacting drugs on rat mast cells. J Physiol Lond 323: 229-244, 1982.
8. Leung MTK, Collard KJ: The effect of trifluoperazine and R24571 on the K^+-evoked release of 5-hydroxytryptamine from super-fused synaptosomes. Neuropharmacol 22: 1095-1099, 1983.
9. Van Belle H: R24571: a potent inhibitor of calmodulin activated enzymes. Cell Calcium 2: 483-494 (1981).
10. Statham HE, Duncan CJ, Publicover SJ: Dual effect of 2, 4-dinitrophinol on the spontaneous release of transmitter at the frog neuromuscular junction. Biochem Pharmacol 27: 2199-2022. 1978.
11. Argov Z, Yaari Y: The action of chlorpromazine at an isolated cholinergic synapse. Brain Res 164: 227-236, 1979.

n.b. Results obtained from experiments completed after submission of this manuscript suggest that the data presented in Fig. 2 may be partially incorrect. It appears that the effect of R24571 on spontaneous release from preparations bathed in EGTA-containing saline is essentially similar to that on preparations in standard frog ringer. MEPP frequency is reduced by 20-40% in the range 10^{-7} - 5×10^{-7}M, but at higher concentrations of the drug the inhibitory effect is no longer discernable and a marked stimulatory effect is sometimes observed (see 'Discussion'). It would therefore appear that the stimulatory effect of high concentrations of R24571 is not suppressed in EGTA-containing salines.

DISCUSSION OF THE ARTICLE BY S.J. PUBLICOVER (U.K.)

J.W. Moore (U.S.): Are you familiar with the recent work of Clapham and Neher on the chromaffin cell? It was just published in the August 1984 issue of the Journal of Physiology, 353, 541-564.

S.J. Publicover (U.K.): No.

J.W. Moore (U.S.): I became interested in this issue some years back and had a paper ready for publication in Nature (by Stockbridge and myself) which showed that TFP has presynaptic action because of a strong effect on evoked release in the absence of a change in mepp amplitude. When I saw a paper by a group from the University of Connecticut (Cheng et al., J. Pharmacol. Exp. Ther. 271, 44-50), saying that the effect of TFP was postsynaptic (reducing mepp amplitude), we retracted our paper and hoped to resolve the issue. We repeated their experiment and found that TFP reduced the mepp amplitude only when the nerve was stimulated. Because stimulation is necessary to evoke a release, we have not yet been able to separate the pre and postsynaptic effects of TFP.

Chromaffin cells show both presynaptic and postsynaptic characteristics and allow such a resolution. Clapham and Neher have examined chromaffin cells with the patch clamp technique and have shown that TFP blocks exocytosis, as measured capacitance changes. Although it also reduces the calcium current, it is not reduced enough to account for the full reduction in the exocytosis. Furthermore, they found that TFP desensitizes the ACh channel. Thus, I think that they have a pretty clear picture of TFP effects, both pre and postsynaptic.

S. Silberberg (Israel): We found that in the concentrations of TFP used by you, there was a pronounced postsynaptic effect. Did you observe a postsynaptic effect?

S.J. Publicover (U.K.): Yes, there was quite a pronounced inhibitory postsynaptic effect. However, I did not find any inhibitory effect postsynaptically with calmodazolium. In fact, in some of the preparations, I got the impression that the mepps were bigger after treatment although I haven't quantified it yet.

STUDIES ON THE MECHANISM OF THE INHIBITORY ACTION OF VERAPAMIL ON Ca^{2+} TRANSPORT

Anat Erdreich and Hannah Rahamimoff

1. INTRODUCTION

Verapamil is a synthetic papaverine derivative from the group of Ca^{2+} channel blockers. Its main therapeutic use is in the treatment of angina pectoris and atrial and supraventricular tachy/arrhythmias (1). On the cellular level, it is mainly known for its inhibitory effect on the slow Ca^{2+} channel (2) but it has also been shown to affect other ion fluxes (3,4) and ion flux dependent processes in many systems (5,6,7,8). In order to study the mechanism of action of verapamil at the molecular level, we used synaptic plasma membrane (SPM) vesicles derived from osmotically shocked rat brain synaptosomes (9). As also in other membrane vesicles, ionic gradients across the SPM can be easily created by choice and used to drive transport processes of specific carriers.

2. METHODS

Detailed procedures for preparation of SPM vesicles and their characterization has been described (9). Na^+ gradient (or other gradients) were created by preloading the vesicles in the appropriate solution (see legends to Figures) and diluting them rapidly into an external solution containing the ion to be transported ($^{45}Ca^{2+}$ or $[^3H]$-TPP^+). Reactions were stopped by filtration through nitrocellulose (for $^{45}Ca^{2+}$) or cellulose acetate (for TPP^+) filters, followed by washing with KCl solution, drying and counting (10). Liposomes were prepared according to (11).

3. RESULTS

Figure 1 demonstrates Ca^{2+} uptake into SPM vesicles. In the absence of other driving forces such as ATP, Na-phosphate loaded vesicles take up Ca^{2+} in a Na^+ gradient dependent manner (Figure 1A and 1B, open squares and open triangles). These vesicles take up Ca^{2+} also in the absence of a Na^+ gradient provided a negative inside membrane potential developes across the membrane (Figure 1B, closed triangles). The membrane potential in these vesicles can be measured from the distribution of the lipophilic cation TPP^+ across the vesicles' membrane (10). Calculating this membrane potential yields a value of about -100mV negative inside. The amount of TPP^+ solubilized in the membrane has been measured under conditions of zero membrane potential (vesicles preloaded in a K^+ containing solution diluted into identical medium containing valinomycin) and subtracted from the measured values. This constitutes an additional driving force for Ca^{2+} uptake beside the Na^+ gradient.

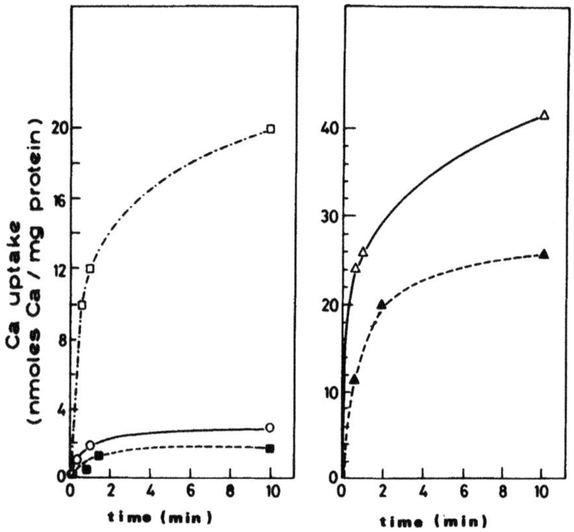

FIGURE 1. Na^+-dependent Ca^{2+} uptake into SPM vesicles from brain tissue. A) 3 µl (about 30 µg protein) of 0.15M Na-phosphate buffer pH 7.4 preloaded SPM vesicles were diluted into 250 µl of either 0.15M KCl, 10mM tris buffer pH 7.4 and 50 µM $^{45}CaCl_2$ (about 0.1 µCi) (□) or 0.15M NaCl, 10mM tris buffer pH 7.4 and 50 µM $^{45}CaCl_2$ (0.1 µCi) (o). The effect of preloading the vesicles with 0.15M K-phosphate buffer pH 7.4 and diluting them into 250 µl 0.15M NaCl, 10mM tris buffer pH 7.4 and 50 µM $^{45}CaCl_2$ (about 0.1 µCi) (■) is also shown. B) 3 µl (about 30 µg protein) of 0.15M Na-phosphate buffer pH 7.4 preloaded SPM vesicles (△) or 3 µl of K-phosphate pH 7.4 preloaded SPM vesicles (▲) were diluted into 250 µl of 0.3M sucrose, 10mM tris buffer pH 7.4 and 50 µM $^{45}CaCl_2$ (0.1 µCi).

Figure 2A shows the decrease in TPP^+ uptake with increase in external KCl concentration. Depolarization increases with KCl concentrations. Figure 2B shows the effect of depolarization by external KCl on the Ca^{2+} uptake. Both types of Ca^{2+} uptake - Na^+ gradient dependent (Figure 2B, open circles) and negative inside membrane potential dependent (Figure 2B, open triangles) are decreased with depolarization.

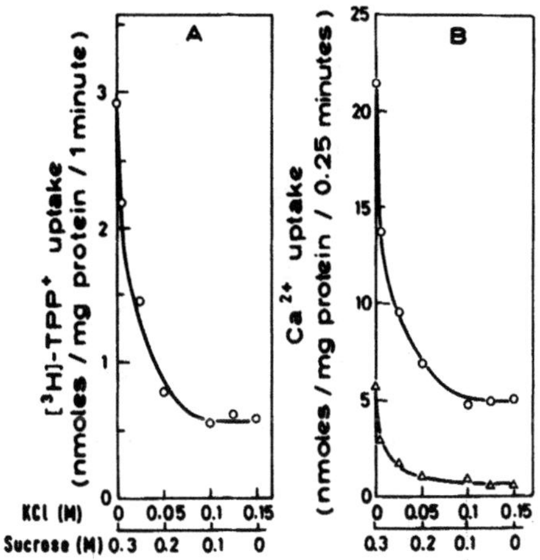

FIGURE 2. A) 3 µl SPM vesicles (30 µg protein) preloaded with 0.15M Na-phosphate buffer or 0.15M K-phosphate buffer pH 7.4 (o-o-o) were diluted into 250 µl of external solution containing 0-0.15M KCl and appropriate concentration of sucrose to maintain osmolarity, 10mM tris HCl pH 7.4, 50 µM $CaCl_2$, 5.5mM $[^3H]TPP^+$ (4 nCi). $[^3H]TPP^+$ solubilized in the membrane was measured and subtracted from the results. B) 3 µl SPM vesicles (30 µg protein) preloaded with 0.15M Na-phosphate buffer pH 7.4 (o-o-o) or 0.15M K-phosphate buffer pH 7.4 (Δ-Δ-Δ) were diluted into 250 µl external solution containing 0-0.15M KCl and sucrose as in A, 10mM tris HCl pH 7.4, 50 µM $^{45}CaCl_2$ (0.1 µCi).

Addition of increasing concentrations of verapamil to the external uptake solution (Figure 3) also decreases the Ca^{2+} taken up by the vesicles. Both types of Ca^{2+} uptake, the Na^+ gradient dependent (Figure 3A, closed circles) and the voltage dependent Ca^{2+} uptake (Figure 3A, open circles), are inhibited by verapamil. The effect is dose dependent, starting at a concentration of 20 µM verapamil. Figure 3B shows that addition of verapamil also decreases TPP^+ uptake by the vesicles in a dose dependent manner in both K-phosphate and Na-phosphate preloaded SPM vesicles. Verapamil did not change the amount of TPP^+ solubilized in the membrane.

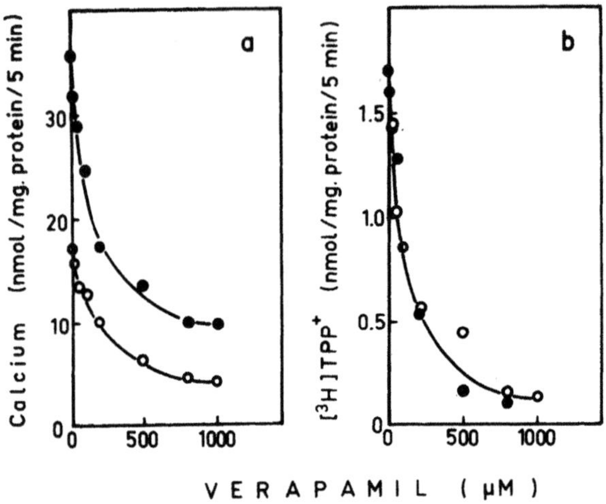

FIGURE 3. A) 3 μl SPM vesicles (30 μg protein) preloaded with 0.15M Na-phosphate buffer pH 7.4 (●-●-●) or 0.15M K-phosphate buffer pH 7.4 (o-o-o) were diluted into 250 μl external solution containing 0.3M sucrose, 10mM tris HCl pH 7.4, 50 μM $^{45}CaCl_2$ (0.1 μCi) and 0-1000 μM verapamil. B) 3 μl SPM vesicles (30 μg protein) preloaded as in A, were diluted into 250 μl external solution containing 0.3M sucrose, 10mM tris HCl pH 7.4, 50 μM $CaCl_2$, 4.48 μM $[^3H]TPP^+$ (4 nCi) and 0-1000 μM verapamil. $[^3H]TPP^+$ solubilized in the membrane was measured and subtracted from the results.

Addition of verapamil to Ca^{2+} preloaded vesicles (Figure 4A, open circles) led to partial Ca^{2+} efflux (Figure 4A, closed circles). In a similar way, verapamil also causes efflux of TPP^+ from TPP^+ containing vesicles (Figure 4B, open and closed circles).

In order to rule out that verapamil acts as a Ca^{2+} ionophore, its effects were compared to A23187 (Table 1). Under conditions of Na^+ gradient dependent and voltage dependent Ca^{2+} uptake (Na-phosphate^{in}/sucrose^{out}), verapamil reduced Ca^{2+} uptake whereas A23187 enhances it. From the measurements of the membrane potential it is clear that Ca^{2+} equilibrium is not reached in the time span of the experiment. This explains the increased Ca^{2+} influx in the presence of A23187. A similar effect is observed when only the voltage dependent component of Ca^{2+} uptake is measured (K-phosphate^{in}/sucrose^{out}). When only the Na^+ gradient dependent component of the Ca^{2+} uptake is measured (Na-phosphate^{in}/KCl^{out}) a different situation occurs: A23187 as well as verapamil decrease Ca^{2+} entry into the vesicles. Their effects are additive, and when both are employed together the inhibition is even greater.

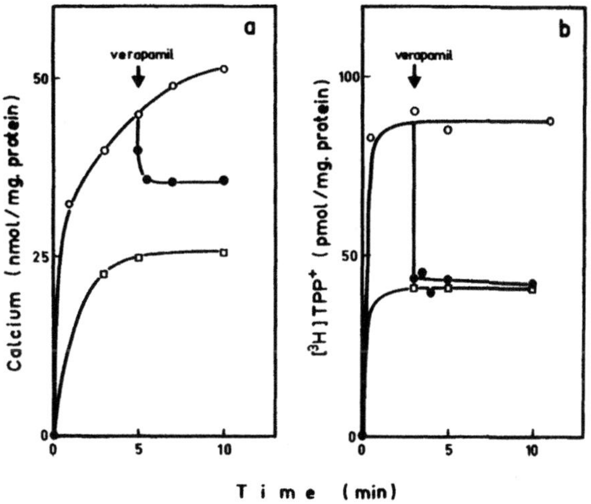

FIGURE 4. A) 3 µl SPM vesicles (30 µg protein) preloaded with 0.15M Na-phosphate buffer pH 7.4 were diluted into 250 µl of external solution containing 0.3M sucrose, 10mM tris HCl pH 7.4, 50 µM $^{45}CaCl_2$ (0.1 µCi) (o-o-o). 200 µM verapamil was added to the above vesicles either at the time of dilution (□-□-□) or at T = 5 min (•-•-•). B) SPM vesicles preloaded as in A were diluted into 250 µl external solution containing 0.3M sucrose, 10mM tris HCl pH 7.4, 50 µM $CaCl_2$ and 0.176 µM $[^3H]TPP^+$ (4 nCi) (o-o-o). 200 µM verapamil was added to the vesicles either at the time of dilution (□-□-□) or at T = 3 min (•-•-•). $[^3H]TPP^+$ solubilized in the membrane was measured and subtracted from the results.

It seems that verapamil does not cause permanent damage to the vesicles' membrane. Its inhibitory effects can be partially reversed, either by dilution of the verapamil inhibited vesicles into a verapamil free aqueous solution (not shown) or by treating them with liposomes to which the lipophilic verapamil molecules probably adhere (Table 2).

Table 1. The effect of verapamil and the Ca^{2+} ionophore A23187 on Ca^{2+} flux in SPM vesicles.

	Added at (T = min)	Medium $(NaPi)^{in}/(suc)^{out}$	$(NaPi)^{in}/(KCl)^{out}$	$(KPi)^{in}/(suc)^{out}$
Control	-	49.9	21.1	35.6
Verapamil	0	17.6	13.3	13.9
A23187	0	64.8	10.9	56.8
Verapamil	5	32.4	16.1	21.4
A23187	5	59.6	9.7	48.3
Verapamil + A23187	5	46.2	5.6	34.1

nMol Calcium/mg protein/10 min.

3 µl SPM vesicles (30 µg protein) preloaded with 0.15M Na-phosphate or K-phosphate buffers pH 7.4 were diluted into 250 µl external solution containing 10mM tris HCl pH 7.4. 50 µM $^{45}CaCl_2$ (0.1 µCi) and 0.3M sucrose or 0.15M KCl. 200 µM verapamil and/or 1 µM A23187 were added to the reaction solution either with the dilution (T = 0 min) or at T = 5 min.

Table 2. Restoration of Ca^{2+} uptake activity to verapamil inhibited SPM vesicles by "liposome washes".

	Ca^{2+} uptake nMoles/mg protein/10 min	Inhibition %
Control		
SPM vesicles	88.6	--
SPM vesicles + verapamil	50.8	43
Following liposome treatment		
SPM vesicles	87.5	--
SPM vesicles + verapamil before treatment	65.0	26
SMP vesicles + verapamil before and after liposome treatment	26.4	70

3 µl SPM vesicles (30 µg protein) preloaded with 0.15M Na-phosphate buffer pH 7.4 were diluted into 250 µl external uptake solution containing 0.3M sucrose, 10mM tris KCl pH 7.4, 50 µM $^{45}CaCl_2$ (0.1 µCi) in the presence of 200 µM verapamil or without it for 10 minutes (control). 100 µl (8 mg lipid) diluted into a verapamil free external solution as above was added to a group of reaction tubes after the initial 10 minutes of Ca^{2+} uptake as above, incubated at room temperature for 30 min and centrifuged for 20 min at 27,000 g at 4°C. The supernatant (containing the liposomes and verapamil) was discarded and 250 µl external uptake solution as above (with or without 200 µM verapamil) was added to the pellet, for another 10 minutes.

4. DISCUSSION

SPM vesicles take up Ca^{2+} in two manners: 1) In response to Na^+ gradient; 2) In response to a negative inside membrane potential. Verapamil ihibits both forms of Ca^{2+} uptake in a dose dependent manner. The drug also causes a dose dependent depolarization of the membrane as manifested by the decrease in TPP^+ uptake. The effects of verapamil could not be attributed to a Ca^{2+} ionophore like action since comparing the effect of A23187 and verapamil had opposing effects on the membrane potential driven Ca^{2+} influx.

It is of interest to note that the effect of verapamil can be partially reversed by either adding excess of phosphatidylcholine liposomes to verapamil ihibited SPM vesicles or diluting them with excess medium. Therefore, it is feasible that verapamil acts also on the lipid phase of the membrane and not only on the transporting molecules.

It is proposed here that in addition to the well known direct inhibition of the slow Ca^{2+} current by verapamil, the drug also causes an indirect effect on the membrane potential and thus may affect a large number of apparently unrelated electrogenic processes.

A.E. is a Foulkes Fellow. Part of this work will be submitted to the Hebrew University Medical School as a partial requirement for a Ph.D. thesis by A.E.

Verapamil was a generous gift from Knoll AG Ludwigshafen, Federal Republic of Germany.

REFERENCES

1. Singh BN, Ellrodt G, Peter CT: Verapamil: A review of its pharmacological properties and therapeutic use. Drugs 15: 169-197, 1978.
2. Fleckenstein A: Specific pharmacology of calcium in myocardium, cardiac pacemakers, and vascular smooth muscle. Ann. Rev. Pharmacol. Toxicol. 17: 149-166, 1977.
3. Galper JB, Catterall WA: Inhibition of sodium channels by D600. Molec. Pharmacol. 15: 174-178, 1978.
4. Ichida S, Okada K, Terao M: Effect of verapamil on ^{45}Ca uptake by synaptosomes. Japan J. Pharmacol. 30: 207-211, 1980.
5. Cohen J, Gutman Y: Effects of verapamil dantrolene and lanthanum on catecholamine release from rat adrenal medulla. Br. J. Pharmacol. 65: 641-645, 1979.
6. Pinto JEB, Trifaro JM: The different effects of D600 (methoxyverapamil) on the release of adrenal catecholamines induced by acetylcholine, high potassium or sodium deprivation. Br. J. Pharmacol. 57: 127-132, 1976.
7. Robinson ICAF, Russel JT, Thorn NA: Calcium and stimulus secretion coupling in the neurohypophysis. Acta Endocrinologica 83: 36-49, 1976.
8. McGee R, Schneider JE: Inhibition of high affinity synaptosomal uptake systems by verapamil. Molec. Pharmacol. 16: 877-855, 1979.
9. Erdreich A, Spanier R, Rahamimoff H: The inhibition of Na-dependent Ca uptake by verapamil in synaptic plasma membrane vesicles. Eur. J. Pharmacol. 90: 193-202, 1983.
10. Erdreich A, Rahamimoff H: The inhibition of Ca uptake in cardiac membrane vesicles by verapamil. Biochem. Pharmacol. 33: 2315-2323, 1984.
11. Barenholtz Y, Gibbs D, Litman BJ, Goll J, Thompson TE, Carlson FD: A simple method for the preparation of homogeneous phospholipid vesicles. Biochemistry 16: 2806-2810, 1977.

DISCUSSION OF THE ARTICLE BY A. ERDREICH (ISRAEL)

I. Friedberg (Israel): Do you think that verapamil functions as an ionophore?

A. Erdreich (Israel): I have shown a slide which indicates that verapamil does not function as an ionophore for calcium. As for other ions, I do not know.

MANGANESE AND ASYNCHRONOUS TRANSMITTER RELEASE AT THE FROG NEUROMUSCULAR JUNCTION

Y. Yaari

1. INTRODUCTION

Manganese ions block nerve membrane Ca channels (Baker, Meves and Ridgeway, 1973) and thus interfere with the Ca_o-dependent synchronous (Meiri and Rahamimoff, 1972) and asynchronous (Kajimoto and Kirpekar, 1972) quantal acetylcholine (ACh) release from motor nerve terminals. However, if Ca_o is removed, adding Mn to the perfusing solution enhances the asynchronous liberation of transmitter quanta from K-depolarized or tetanically activated presynaptic terminals (Kita, Narita and Van der Kloot, 1981).

These observations suggest that Mn may exert a dual action on the mechanism of quantal ACh release: a) A depressant action on the Ca_o-dependent release; b) A facilitatory action on the Ca_o-independent release. I investigated this possibility further by examining how Mn alters the release of transmitter quanta by tetanic nerve stimulation at low concentrations of Ca_o.

2. METHODS

Experiments were performed at room temperature on the sartorius nerve-muscle preparation of the frog (Rana ridibunda). The standard Ringer solution contained (in mM) 115.6 NaCl, 2 KCl, 1.8 $CaCl_2$ and 2 Trisma (titrated to pH 7.4). $CaCl_2$ was lowered and $MnCl_2$ was added by isotonic substitution for NaCl. In Ca-"free" solutions [Ca] was approximately 10^{-5} M. The nerve was stimulated by tetanic trains of square pulses (0.02 msec) at various frequencies (10-50 Hz) and durations (30-60 sec) in intervals of 15 min. Muscle end-plate activity was recorded by convenional methods of intracellular record'ng and displayed on moving paper using a Gould pen-recorder.

The asynchronous release of transmitter quanta occuring spontaneously or evoked by nerve stimulation was measured by counting the number of miniature end-plate potentials (MEPPs) appearing in a unit of time. At rest, the rate of spontaneous release is relatively low (less than 1 per sec). However, tetanic nerve stimulation leads to a progressive rise in MEPP frequency (tetanic potentiation, TP), which remains elevated after cessation of stimulation (post-tetanic potentiation, PTP) and decays to base line (spontaneous MEPP frequency) within several seconds or minutes (del Castillo and Katz, 1954; Erulkar and Rahamimoff, 1978). The magnitude of both TP and PTP increases with the frequency of nerve stimulation, the number of applied stimuli and the $[Ca]_o$ (Lev-Tov and Rahamimoff, 1980). In spite of the strong Ca_o-dependence of potentiation, some TP and PTP persist in the virtual absence of Ca_o (Miledi and Thies, 1971; Hurlbut, Longnecker and Mauro, 1971; Erulkar and Rahamimoff, 1978).

3. RESULTS

In the first group of experiments (n=10), I tested whether Mn affects TP and PTP in preparations perfused with Ca-"free" solutions. As expected in these conditions (Erulkar and Rahamimoff, 1978), the potentiation of MEPP frequency by tetanic nerve stimulation was small or absent. In the experiment illustrated in Fig 1, TP and PTP became apparant only at a stimulation frequency of 50 Hz.

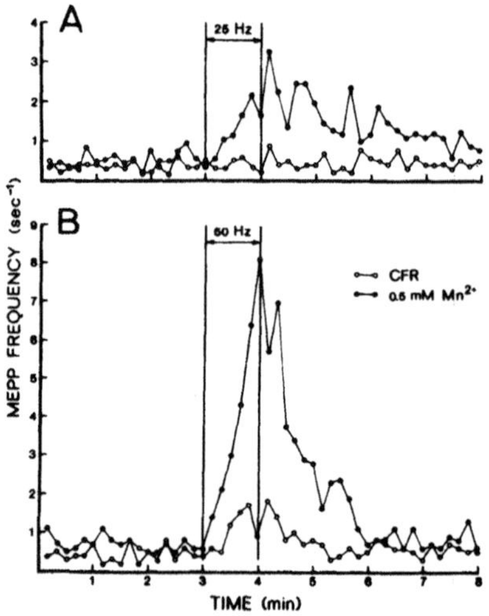

FIGURE 1. Stimulatory effects of Mn on potentiation of asynchronous ACh release in Ca-"free" solution. Each plotted point represents average MEPP frequency for a time bin of 10 sec. Vertical lines mark 1 min of tetanic nerve stimulation at 25 Hz (A) and 50 Hz (B). Open circles, MEPP frequency in control solution; closed circles, in 0.5 mM Mn (15-30 min of exposure).

Following the addition of 0.5 mM Mn, both TP and PTP were markedly enhanced. This effect of Mn was not necessarily accompanied by a change in spontaneous MEPP frequency. However, with prolonged (over 1 hour) exposures to Mn, the MEPP frequency at rest also increased. This was associated with a further rise in potentiation. Substantially bigger effects were observed when [Mn] was increased to 1 or 2 mM, whereas 0.1 mM Mn had negligible or no effects. All effects of Mn were reversible upon its washout.

In the second group of experiments (n=7) I examined how Mn affects potentiation in the presence of a higher $[Ca]_o$ (usually 0.1 mM). PT and PTP in these conditions were much larger than in the previous group (Fig 2A; cf. Erulkar and Rahamimoff, 1978; Lev-Tov and Rahamimoff, 1980). The action of Mn in these conditions was concentration-dependent. At a low [Mn] (0.1 mM), TP was markedly suppressed, whereas PTP and the spontaneous MEPP frequency were either unaffected or slightly enhanced (Fig 2B). At a ten-fold higher concentration (1.0 mM) Mn enhanced TP, PTP and spontaneous release (Fig 2C).

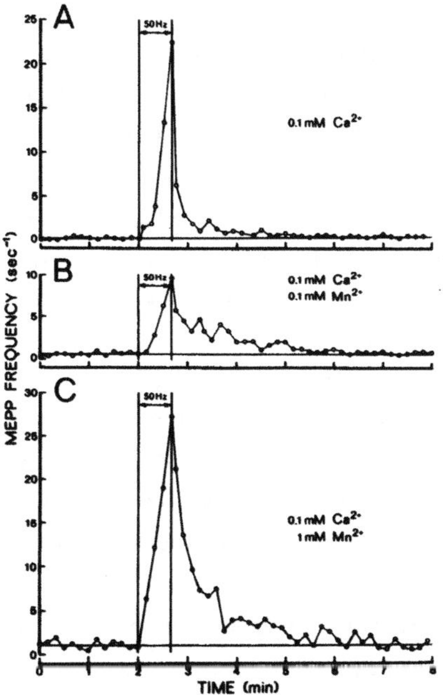

FIGURE 2. Dual effect of Mn on potentiation of asynchronous ACh release in 0.1 mM Ca solution. Each plotted point represents average MEPP frequency for a time bin of 10 sec. Vertical lines mark 45 sec of 50 Hz tetanic nerve stimulation. Horizontal line marks base line MEPP frequency (mean frequency for 5 min before the tetanus). A. Control. B. after 30 min in 0.1 mM Mn. C. after additional 30 min in 1 mM Mn.

4. DISCUSSION

The data thus show that Mn exerts a dual action on the asynchronous release of ACh quanta from tetanically activated frog motor nerve terminals. It suppresses the Ca_o-dependent TP, but facilitates the Ca_o-independent TP. The first action predominates in conditions in which a substantial fraction of TP is mediated by Ca_o. By contrast, when this fraction is minimized (e.g., by Ca_o removal or by the addition of low $[Mn]_o$), increasing $[Mn]_o$ augments TP. Interestingly, Mg exerts a similar dual action on TP, though at a concentration about ten-fold higher than that of Mn (Lev-Tov and Rahamimoff, 1980).

The suppression by Mn of the Ca_o-dependent TP is most probably due to the blockage of presynaptic Ca channels by Mn (Meiri and Rahamimoff, 1972). It is less clear how the Ca_o-independent TP is enhanced by Mn. This effect seems to reflect a general increase in quantal release probability since it requires a higher [Mn] and is accompanied by enhanced PTP and spontaneous release. It has been suggested, that potentiation independent of $[Ca]_o$ involves Ca liberation from intraterminal binding sites (e.g., mitochondria, Alnaes and Rahamimoff, 1975), possibly consequent to a rise in $[Na]_i$ during the period of tetanic nerve stimulation (Erulkar and Rahamimoff, 1978; Lev-Tov and Rahamimoff, 1980). Possibly, Mn may interfere with the buffering mechanisms of intraterminal Ca, thus leading to a bigger rise in $[Ca]_i$ during the tetanus. Indeed, Mn was found to inhibit the synaptosomal Na/Ca exchange system (Blaustein and Ector, 1976; Schellenberg and Swanson, 1981; Trosper and Philipson, 1983), which may play a major role in Ca_i homeostasis also at motor nerve terminals (Meiri, Zelingher and Rahamimoff, 1985). An alternative possibility is that Mn can permeate the terminals during their spike depolarization and trigger the release of transmitter quanta, either by liberating intracellularly bound Ca or by exerting a Ca-like action on the mechanism of exocytosis (cf. Drapeau and Nachshen, 1984).

Supported by the Fund for Basic Research administered by the Israeli National Academy of Sciences and Humanities. I thank M. Silber for secretarial assistance.

REFERENCES

1. Alnaes E and Rahamimoff R: On the role of mitochondria in transmitter release from motor nerve terminals. J. Physiol. 248: 285-306, 1975.
2. Baker PF, Meves H and Ridgway EB: Effects of manganese and other agents on the calcium uptake that follows depolarization of squid axons. J. Physiol. 231: 511-526, 1973.
3. Blaustein MP and Ector AC: Carrier-mediated sodium-dependent and calcium-dependent calcium efflux from pinched-off presynaptic nerve terminals (synaptosomes) in vitro. Biochim. Biophys. Acta 419: 295-308, 1976.
4. del Castillo J and Katz B: Statistical factors involved in neuromuscular facilitation and depression. J. Physiol. 124: 574-585, 1954.
5. Drapeau P and Nachshen DA: Manganese fluxes and manganese-dependent neurotransmitters release in presynaptic nerve endings isolated from rat brain. J. Physiol. 348: 493-520, 1984.
6. Erulkar SD and Rahamimoff R: The role of calcium ions in tetanic and post-tetanic increase of miniature endplate potential frequency. J. Physiol. 278: 501-511, 1978.
7. Hurlbut WP, Longnecker HB, Jr, and Mauro A: Effects of calcium and magnesium on the frequency of miniature endplate potentials during prolonged tetanization. J. Physiol. 219: 17-38, 1971.
8. Kajimoto N and Kirpekar SM: Effects of manganese and lanthanum on spontaneous release of acetylcholine at frog motor nerve terminals. Nature 235: 29-30, 1972.
9. Kita H, Narita K and Van der Kloot W: Tetanic stimulation increases the frequency of miniature endplate potentials at the frog neuromuscular junction in Mn^{2+}-, Co^{2+}-, and Ni^{2+}-saline solutions. Brain Res. 205: 111-121, 1981.
10. Lev-Tov A and Rahamimoff R: A study of tetanic and post-tetanic potentiation of miniature endplate potentials at the frog neuromuscular junction. J. Physiol. 309: 247-273, 1980.
11. Meiri H, Zelingher J and Rahamimoff R: A possible involvement of the Na-Ca exchanger in regulation of transmitter release at the frog neuromuscular junction. In: R Rahamimoff and B Katz (eds) Calcium, Neuronal Function and Transmitter Release. Martinus Nijhoff Publishing, Boston, 1985.
12. Meiri U and Rahamimoff R: Neuromuscular transmission: inhibition by manganese ions. Science 176: 308-309, 1972.
13. Miledi R and Thies R: Tetanic and post-tetanic rise in frequency of miniature endplate potentials in low-calcium solutions. J. Physiol. 212: 245-257, 1971.
14. Schellenberg GD and Swanson PD: Sodium-dependent and calcium-dependent calcium transport by rat brain microsomes. Biochim. Biophys. Acta 648: 13-27, 1981.
15. Trosper TL and Philipson KD: Effects of divalent and trivalent cations on Na^{+}-Ca^{2+} exchange in cardiac sarcolemmal vesicles. BBA 731: 63-68, 1983.

DISCUSSION OF THE ARTICLE BY Y. YAARI (ISRAEL)

Y. Shimoni (Israel): Are you claiming that normally the calcium is shortcircuited by outward K currents, so that there is no calcium influx? If the calcium influx is triggering the afterdischarges, the calcium should still give rise to the oscillation even though the K currents prevent you from measuring the hump.

Y. Yaari (Israel): By blocking action potential repolarization with 4-AP, one prolongs the action potential and thereby enhances the inward current. In the presence of Ca, this current is probably carried by Ca and in Ca absence, it may be carried by other ions such as Na. In both cases, this current possibly triggers the afterdischarges.

Y. Shimoni (Israel): This is very similar to cardiac muscle after oscillations that you see in digitalis which is blocked also by other local anaesthetics (Eisner, Lederer and Sheu (1983). J. Physiol. 340, 239-257).

Y. Yaari (Israel): This prolonged terminal depolarization seen in the absence of calcium may be important for physiological phenomena such as facilitation and tetanic potentiation.

O.F. Hutter (U.K.): What other ion could be responsible for this current? Presumably, it is sodium and you can't take it away.

Y. Yaari (Israel): There is no way to do it because reducing sodium would affect the orthodromic volley and it would then be impossible to interpret the results.

A. Mallart (France): Did you try to block it with cobalt this so called calcium hump you observed in the absence of extracellular calcium?

Y. Yaari (Israel): We have tried unsuccessfully to block it with manganese.

A. Mallart (France): In the presence of 4-AP and 0 Calcium?

Y. Yaari (Israel): Yes. In addition, replacing calcium with barium and then adding the manganese resulted in a partial block. Possibly, 5mM manganese is not sufficient but we haven't used cobalt because it elicited some strange "humps".

SEVERAL POTASSIUM CONDUCTANCES MODULATE THE EXCITABILITY OF FROG MOTOR NERVE TERMINALS

G. David and Y. Yaari

1. INTRODUCTION

A nerve impulse propagating along a motor axon normally initiates a single action potential at points where the unmyelinated terminals branch off from their parent axon. In some vertebrate nerve-muscle preparations, this action potential continues to propagate actively along some portion of the nerve terminals (17), whereas in others it most probably depolarizes the terminals by only a passive spread of current (7). In either case, the resultant depolarization of the nerve terminals is sufficient to trigger the synchronous release of the transmitter, acetylcholine (ACh), into the junctional cleft.

Certain conditions, however, result in the generation of multiple action potentials within the terminal arborization of motor fibres following its invasion by a single orthodromic nerve impulse. These afterdischarges are observed following repetitive nerve activation (32) and can be induced by exposing the nerve muscle preparation to neuromuscular "facilitatory" drugs and various convulsants. Notable examples of these drugs are nicotinic agonists, anticholinesterase agents, quarternary ammonium compounds (27), penicillin (24) and aminopyridines (8,16).

The mechanisms underlying afterdischarge electrogenesis are not clear. Most probably this involves a re-excitation of the distal nodes of Ranvier by local current spread from the more depolarized terminals (32,33). Accordingly, afterdischarges arise following a normally propagating action potential when the spike depolarization of the terminals exceeds that of their parent axon, and reflects back to excite it at the nearest non-refractory node. In contrast to the nodal membrane, the nerve terminal membrane lacks a substantial leak conductance (23). Consequently, spike duration at motor nerve terminals is largely determined by the activation of repolarizing membrane K^+ conductances (g_Ks). This suggests the latter may be involved in afterdischarge electrogenesis, as their blockade would prolong spike depolarization preferentially at the terminals. In the present study we explore this notion and report that frog motor nerve terminals may possess several distinct g_Ks, which contribute to the maintenance of a low terminal excitability. A blockade of these g_Ks may be a common mechanism underlying the generation of afterdischarges at the terminal arborization of motor fibres.

2. METHODS

Experiments were performed at room temperature on sartorius nerve-muscle preparations isolated from small frogs (Rana ridibunda). The preparations were perfused throughout the experiment with a Ringer solution containing d-tubocurarine in a concentration (10 mg/L) sufficient to block all postsynaptic responses (confirmed by intracellular recordings from muscle end-plates). Potential changes at the nerve terminals were recorded by 1M NaCl filled microelectrodes (5-10 Mohm) positioned under the perineural sheath (Sheath of the Henle) of superficial small nerve bundles, the indifferent reference electrode being an AgCl pellet in the bath. As originally described by Gundersen, Katz & Miledi (13), the signals recorded subperineurally are generated by local currents flowing across the "seal" resistance (ca. 0.1-1 Mohm) of the perineural sheath between the terminal and preterminal regions of the axons in the bundle (Fig 1). Thus, when the terminals are more depolarized than the axons, the microelectrode records a positive "field potential", whereas a reverse signal appears when the terminals become relatively hyperpolarized. Signal sizes ranged between 2-5 mV in different preparations. Conventional electrophysiological techniques were used for signal amplification, storage and display. A more detailed description of the methods is given in a recent publication (8).

3. RESULTS

3.1. Induction of afterdischarges by blockers of delayed rectification

At frog motor nerve terminals, spike repolarization is largely mediated by an outward current through a delayed rectifying g_K ($g_{K,V}$), which is blocked by aminopyridines (e.g., 2,3-diaminopyridine, DAP) and tetraethylammonium (TEA) (5,18,21). Exposure of the preparations to a normal Ringer solution containing either DAP (10-100 μM) or TEA (5 mM) or both, induced

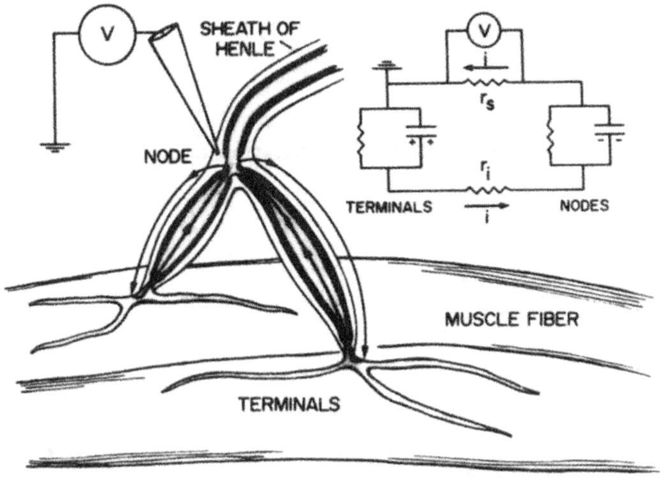

FIGURE 1. Experimental setup for subperineural recording. Microelectrode is inserted underneath the perineural sheath of a small nerve bundle containing several motor axons (only one is shown in the figure). Extracellular currents flowing between the unsheathed nerve terminals and their parent axons traverse in part the perineural sheath and are recorded across its 0.1-1 Mohm resistance. When the terminals are more depolarized than the preterminal axons, local currents flow from the terminals to the nodes through the axoplasmic resistance (r_i) and return through extracellular current pathways which include the sheath resistance (r_s). This is described by the simple analogue electric circuit diagram. The outward currents through the nodes thus produce a positive "field potential" at the site of microelectrode emplacement.

afterdischarges following nerve activation by a single supramaximal stimulus (8). Fig 2 demonstrates the appearance of this hyperexcitable state, as recorded subperineurally from a fine nerve bundle. The normal response of the nerve fibres to the stimulus consists of an orthodromically propagating volley which appears as a mostly monophasic negative wave. The addition of 100 μM DAP to the Ringer solution induced the gradual appearance of a brief (ca. 2-3 msec) positive-going "afterpotential" following the primary response. Concomittently, afterdischarges appeared in some of the stimulated fibres. In different nerve bundles exposed to DAP the post-stimulus latencies of thes action potentials ranged from 2 to 25 msec and often had a constant temporal relationship to the orthodromic volley (Fig 2). In TEA-treated preparations the decay of the positive afterpotential was usually slower and afterdischarges with longer latencies (up to 100 msec) were more commonly observed (see below, Fig 4a).

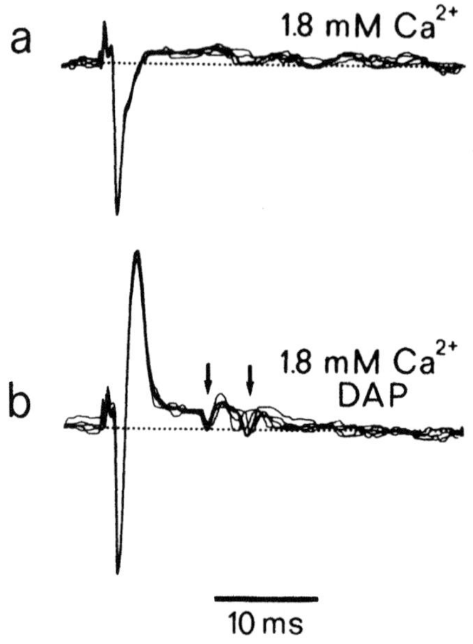

FIGURE 2. Effects of $g_{K,V}$ blockade by DAP on spike potentials of motor nerve terminals. Subperineural recording from a small nerve bundle in control solution (a) and 30 min after the addition of DAP (100 µM) to the perfusing solution (b). Note the appearance of afterdischarges (arrows) in some fibres exposed to DAP. Each record is composed of 5 consecutive superimposed traces.

3.2. Origin of afterdischarges in motor nerve terminals

We considered the possibility that the subperineurally recorded potentials may originate in part from sensory muscle fibres in the "penetrated" bundle. Therefore, in some experiments, the sartorius nerve and muscle were dissected out with the ipsilateral ventral and dorsal roots of the lumbar enlargment (spinal segments VIII, IX). Stimulation of the ventral, but not the dorsal roots, mimicked whole nerve stimulation in eliciting a positive afterpotential and afterdischarges, thus confirming their origin from motor axons. The latter notion was also supported by the demonstration of multiple motor nerve terminal potentials and muscle end-plate potentials evoked by a single nerve stimulus in DAP-treated preparations (8).

When the nerve bundle was cut distal to the recording site, the orthodromic volley was not altered significantly, but the afterdischarges and the positive afterpotential were completely abolished. Similarly, no afterdischarges could be recorded from the main nerve trunk after severing all its peripheral branches. These findings indicate that the afterdischarges are initiated at or near the terminal arborizations and propagate both orthodromically and antidromically. They also corroborate the notion that the positive afterpotential recorded extracellularly from preterminal axons is generated by local current spread from the more depolarized terminals (8,13).

3.3. Ca^{2+}-activated K^+ conductance at motor nerve terminals

The results discussed thus far suggest that $g_{K,V}$ blockers prolong selectively terminal spike depolarization. Indeed, this finding is not unexpected in view of the aforementioned major role of $g_{K,V}$ in spike repolarization at the unmyelinated nerve terminals, as compared with its lesser contribution to the repolarization of the nodal membrane (23). This effect and the decrease in delayed rectification also at subthreshold depolarizations, would tend to increase the effective space constant of the terminal membrane, and promote local current spread between the terminals and their parent axons. An alternative hypothesis for the origin of these currents has been proposed by Gundersen et al. (13). They suggested that the blockade of $g_{K,V}$ and the resultant spike prolongation markedly enhance inward Ca^{2+} current at the terminals, which then appears as an outward current at the preterminal axons. This, most likely, was not the case in our experimental conditions, since the addition of several known blockers of Ca channels in nerve membranes, namely, Mn^{2+} (5 mM), Cd^{2+} (2 mM) and La^{3+} (0.1 mM), did not reduce significantly the amplitude of the positive afterpotential recorded from the preterminal axons. Conversely, and indeed quite surprisingly, these ions caused a prolongation of this potential and often enhanced afterdischarge generation (Fig 3a). This effect was probably consequent to the blockade of some Ca^{2+} entry into the terminals, since it was mimicked by lowering $[Ca^{2+}]_o$ (Fig 3b). Indeed, in Ca^{2+}-free solutions (containing no added $CaCl_2$ and 1 mM $MgCl_2$), the positive afterpotential usually lasted 20-50 msec and was accompanied by intense afterdischarge electrogenesis.

The apparent further prolongation of the terminal spike depolarization by Ca^{2+} blockers and low $[Ca^{2+}]_o$ could be explained by the presence of a Ca^{2+}-activated g_K ($g_{K,Ca}$) at the motor nerve terminals. According to this hypothesis (22,29), Ca^{2+} entering the terminals during the DAP- or

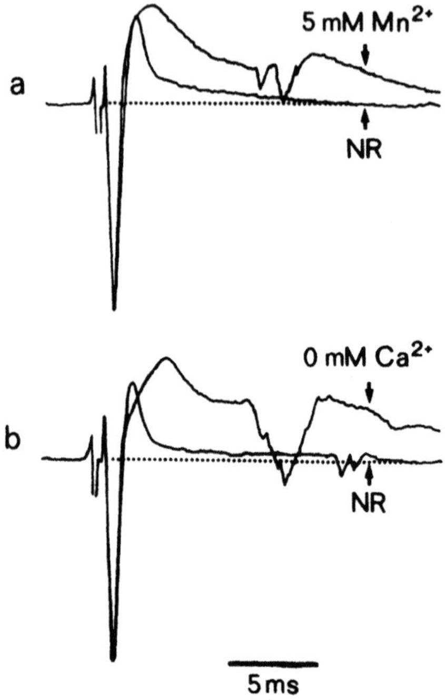

FIGURE 3. Effects of a combined blockade of $g_{K,V}$ and $g_{K,Ca}$ on spike potentials of motor nerve terminals. Subperineural recording from a small nerve bundle in a preparation perfused with normal Ringer (NR) solution containing DAP (100 µM). The addition of 5 mM Mn^{2+} produced a marked prolongation of the positive afterpotential and enhanced afterdischarge generation (a). This effect was reversible upon Mn^{2+} washout and could be mimicked by perfusing the preparation with Ca^{2+}-free solution (b).

TEA-enhanced spike depolarization activates a g_K, which in absence of delayed rectification mediates the repolarization of the terminal membrane. Blocking this $g_{K,Ca}$ further decreases membrane rectification and increases its effective space constant, thereby enhancing and prolonging even more local current spread between the depolarized terminals and their parent axons. Also consistent with this hypothesis we found that the addition of Ba^{2+} to the perfusing solution markedly prolonged the positive afterpotential (Fig 4). This effect was observed even in Ca^{2+}-free solutions. It has been shown that Ba^{2+} ions readily traverse Ca^{2+} channels but block $g_{K,Ca}$ (e.g., ref 12). Not unexpectedly, Ba^{2+} markedly enhanced and prolonged afterdishcarge generation (Fig 4).

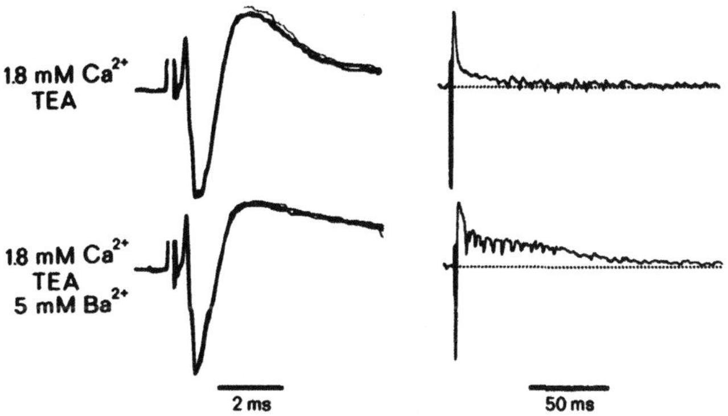

FIGURE 4. Effects of Ba^{2+} on spike potentials of TEA-treated motor nerve terminals. Subperineural recording from a small nerve bundle in a preparation perfused with normal Ringer solution containing 5 mM TEA. Upper traces, control response (at two different time scales); Lower traces, responses 10 minutes following the inclusion of 5 mM Ba^{2+} in the perfusing solution. Left records are composed of 5 superimposed traces.

The effects of g_{Ca} blockers, low $[Ca^{2+}]_o$ and Ba^{2+} on nerve terminal spike potentials were observed only in the presence of TEA or DAP. This suggests that $g_{K,Ca}$ does not participate in normal spike repolarization. Possibly a single action potential is too brief for a substantial $g_{K,Ca}$ activation (3,15).

Raising $[Ca^{2+}]_o$ from its normal concentration (1.8 mM) to 10 mM did not affect significantly the DAP- or TEA-induced positive afterpotential but suppressed afterdischarge generation. The latter effect is consistent with the known action of divalent ions to decrease membrane excitability by screening negative fixed charges (9). It can also explain why lowering $[Ca^{2+}]_o$ was more effective than adding Mn^{2+} in inducing afterdischarges, even though both procedures apparently prolonged the terminal spike depolarization.

3.4. The action of acetylcholine

ACh has been shown to block some g_Ks in several neuronal preparations and thereby to enhance their excitability (e.g., refs 1,6). Since ACh and related drugs cause afterdischarges at mammalian motor nerve terminals, though not at amphibian (27), we examined how ACh affects the spike potentials at the terminals. In DAP-treated preparations, ACh (3mM) prolonged the positive afterpotential recorded from the preterminal axons and enhanced the appearance of afterdischarges. The application of ACh alone also induced the appearance of a small positive afterpotential (Fig 5) but this action was insufficient to evoke afterdischarges. A possible interpretation of these results is that some K^+ channels at the nerve terminal membrane are blocked by the transmitter, ACh.

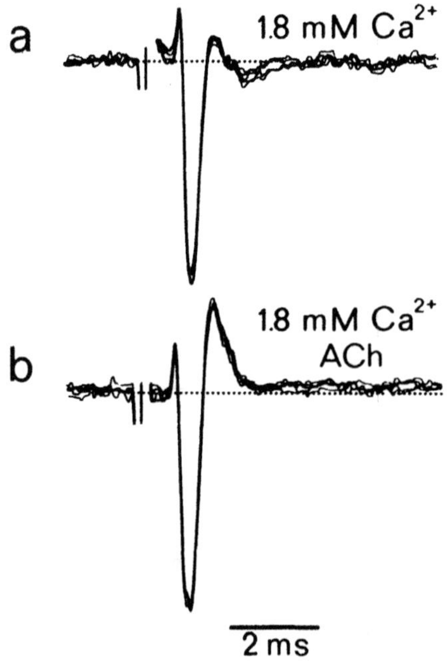

FIGURE 5. Effect of ACh on nerve terminal spike potentials. Subperineural recording from a small nerve bundle in normal Ringer solution before (a) and 5 min after the inclusion of 3 mM ACh in the perfusing solution (b). ACh produced a positive afterpotential which was reversible within 7 min of ACh washout.

4. DISCUSSION

The data show that g_K blockers prolong spike depolarization at motor nerve terminals and promote local current spread between the terminals and the preterminals axons. This hyperexcitable state is associated with afterdischarge electrogenesis at or near the terminal arborization. Presumabely, the prolonged terminal depolarization in these conditions acts similarly to a generator potential at sensory nerve endings in triggering action potentials at the last nodes of Ranvier. The data also suggests the existance of at least two pharmacologically distinct g_Ks at the terminals: a $g_{K,V}$ and a $g_{K,Ca}$. Whether the transmitter interacts with any one of these, or possibly with a distinct ACh-sensitive g_K, remains to be clarified.

As motor nerve terminals are not amenable for intracellular recording, a more direct study of the voltage dependence and kinetics of the terminal g_Ks is not yet possible. However, it is interesting to note that the soma-dendritic membrane of frog spinal motoneurones also contains at least two g_Ks: a fast TEA-sensitive $g_{K,V}$, which is responsible for spike repolarization, and a slow $g_{K,Ca}$, which is sensitive to Ca^{2+} channel blockers and mediates a prolonged afterhyperpolarization (3). Whereas the $g_{K,V}$ is maximally activated by brief intense depolarizations (as those associated with a normal action potential), the $g_{K,Ca}$ requires moderate but prolonged or repetitive depolarizations for its full activation. Hence, $g_{K,Ca}$ is not much activated by a single action potential unless its repolarization is prolonged by TEA (3). Although indirect our findings are compatible with the existance of comparable g_Ks at the nerve terminals. It is therefore tempting to speculate that the two unmyelinated portions of frog motoneurones are similar in their g_Ks, as they are in respect to the existance in both of a voltage-dependent Ca^{2+} conductance (2,3,18,19). Several distinct g_Ks, including a $g_{K,Ca}$, have been found also in the soma-dendritic membrane of mammalian spinal motoneurones (4,20), and data presented by Mallart in this volume suggest the existance of $g_{K,Ca}$ in mammalian motor nerve terminals as well.

The existance of different g_Ks in the soma-dendritic membrane of a variety of other vertebrate neurons is now well established. These include g_Ks which are voltage-dependent, Ca^{2+}-activated or ACh-sensitive (e.g., refs 1,4,11,20,31). The outward membrane currents carried across these g_Ks are intrinsically inhibitory in nature, as their blockade leads to the evolvement of convulsive phenomena (e.g., refs 10,28,34). Possibly these g_Ks control also the excitability of the presynaptic nerve terminals and thus prevent excessive, disorganized excitation of post-synaptic neurones. Indeed, the induction by convulsants of ectopic action potentials in cortical nerve terminals has been suggested to play an important role in epileptogenesis (14,25,30). Since many convulsants produce afterdischarges at amphibian and/or mammalian motor nerve terminals (8,16,24,26,33), it is possible that the blockade of nerve terminal g_Ks is a common mechanism of convulsant action.

Supported by the United States-Israel Binational Science Foundation (BSF) and the Fund for Basic Research Administered by the Israeli National Academy for Sciences and Humanities. G.D. is a Foulkes Fellow.

REFERENCES

1. Adams P.R., Brown D.A. and Constanti A. (1982). M-currents and other potassium currents in bullfrog sympathetic neurones. J. Physiol. 330, 537-572.
2. Alvarez-Leefmans F.J. and Miledi R. (1980). Voltage sensitive calcium entry in frog notoneurones. J. Physiol. 308, 241-257.
3. Barrett E.F. and Barrett J.N. (1976). Separation of two voltage-sensitive potassium currents, and demonstration of a tetrodotoxin-resistant calcium current in frog motoneurones. J. Physiol. 255, 737-774.
4. Barrett E.F., Barrett J.N. and Crill W.E. (1980). Voltage-sensitive outward currents in cat motoneurones. J. Physiol. 304, 251-276.
5. Benoit P.R. and Mambrini J. (1970). Modification of transmitter release by ions which prolong the presynaptic action potential. J. Physiol. (Lond.) 210, 681-695.
6. Bernardo L.S. and Prince D.A. (1982). Ionic mechanisms of cholinergic excitation in mammalian hippocampal pyramidal cells. Brain Res. 249, 333-344.
7. Brigant J.L. and Mallart A. (1982). Presynaptic currents in mouse motor endings. J. Physiol. (Lond.) 333, 619-636.
8. David G., Selzer M.E. and Yaari Y. (1984). Suppression by phenytoin of convulsant-induced afterdischarges at presynaptic nerve terminals. Brain Res. (in the press).
9. Frankenhaeuser B. and Hodgkin A.L. (1957). The action of calcium on the electrical properties of squid axons. J. Physiol. 137, 218-244.
10. Galvan M, Grafe P. and ten Bruggencate G. (1982). Convulsant actions of 4-aminopyridine on the guinea-pig olfactory cortex slice. Brain Res. 241, 75-86.
11. Galvan M. and Sedlmeir C. (1984). Outward currents in voltage-clamped rat sympathetic neurones. J. Physiol. 356, 115-133.
12. Gorman A. and Herman A.L.F. (1979). Internal effects of divalent cations on potassium permeability in molluscan neurones. J. Physiol. 296, 393-410.
13. Gundersen C.B., Katz B. and Miledi R. (1982). The antagonism between botulinum toxin and calcium in motor nerve terminals. Proc. R. Soc. Lond. B. 216, 369-376.
14. Gutnick M.J. and Prince D.A. (1972). Thalamocortical relay neurons: antidromic invasion of spikes from a cortical epileptogenic focus. Science 176, 424-426.
15. Hotson J.R. and Prince D.A. (1980). A calcium-activated hyperpolarization follows repetitive firing in hippocampal neurones. J. Neurophysiol. 43, 409-419.
16. Illes P. and Thesleff S. (1978). 4-Aminopyridine and evoked transmitter release from motor nerve endings. Br. J. Pharmacol. 64, 623-629.
17. Katz B. and Miledi R. (1965). Propagation of electric activity in motor nerve terminals. Proc. R. Soc. B. 161, 453-482.
18. Katz B. and Miledi R. (1967). A study of synaptic transmission in the absence of nerve impulses. J. Physiol. 192, 407-436.
19. Katz B. and Miledi R. (1969). Tetrodotoxin-resistant electric activity in presynaptic terminals. J. Physiol. 203, 459-487.
20. Krnjevic K., Puil E. and Werman R. (1978). EGTA and motoneuron afterpotentials. J. Physiol. 275, 199-223.
21. Mallart A. (1984). Presynaptic currents in frog motor endings. Pflugers Arch. 400, 8-13.
22. Meech R.W. (1978). Calcium-dependent potassium activation in nervous

tissue. Ann. Rev. Biophys. Bioeng. 7, 1-18.

23. Molgo J. and Thesleff S. (1982). Electrotonic properties of motor nerve terminals. Acta Physiol. Scand. 114, 271-275.
24. Noebels. J.L. and Prince D.A. (1977). Presynaptic origin of penicillin afterdischarges at mammalian nerve terminals. Brain Res. 138, 59-74.
25. Noebels. J.L. and Prince D.A. (1978). The development of focal seizures in cerebral cortex: role of axon terminal bursting. J. Neurophysiol. 41, 1267-1281.
26. Randic M. and Straughan D.W. (1964). Antidromic activity at the rat phrenic nerve-diaphragm preparation. J. Physiol. 173, 130-148.
27. Riker W.F. and Okamoto M. (1969). Pharmacology of motor nerve terminals. Ann. Rev. Pharmacol. 9, 173-208.
28. Schwartzkroin P.A. and Prince D.A. (1980). Effects of TEA on hippocampal neurones. Brain Res. 185, 169-181.
29. Schwarz E. and Passow H. (1983). Ca-activated K channels in erythrocytes and excitable cells. Ann. Rev. Physiol. 45, 359-374.
30. Scobey R.P. and Gabor A.J. (1975). Ectopic action-potential generation in epileptogenic cortex. J. Neurophysiol. 38, 383-394.
31. Segal M. and Barker J.L. (1984). Rat hippocampal neurons in culture: Potassium conductances. J. Neurophysiol. 51, 1409-1433.
32. Standaert F.G. (1963). Post-tetanic repetitive activity in cat soleus nerve. Its origin, cause and mechanism of generation. J. Gen. Physiol. 47, 53-70.
33. Werner G. (1960). Neuromuscular facilitation and antidromic discharges in motor nerves: their relation to activity in motor nerve terminals. J. Neurophysiol. 23, 171-187.
34. Yaari Y., Konnerth A. and Heinemann U. (1983). Spontaneous epileptiform activity of CA1 hippocampal neurones in low extracellular calcium solutions. Exp. Brain Res. 51, 153-156.

ELECTROPHYSIOLOGICAL EXAMINATION OF GLUCOSE TRANSPORT INTO MOTOR NERVE TERMINALS IN MATURE ADULT AND AGED RATS

JULIE L. ROSENHEIMER

1. INTRODUCTION

The rate of glucose uptake is different into nerve terminals innervating muscles composed of different fiber types (9). For example, glucose enters terminals innervating glycolytic muscles more rapidly than it enters those innervating oxidative muscles. Furthermore, there is a strong relationship between the efficiency of α-motoneuron energy metabolism, for which glucose is a major substrate, and the type of muscle fiber that the motor neuron innervates (12). Specifically, nerve terminals innervating oxidative muscles exhibit a greater capacity for oxidative enzyme activity than those innervating glycolytic muscles. Therefore, glucose transport into peripheral nerve terminals may be linked to its rate of utilization. Kinetic analyses of glucose uptake into synaptosomes and brain slices support this (3, 13).

Oxidative enzyme activity in muscles changes with age (1, 2, 4). Likewise, the dynamics of end-plate remodeling, that is nerve terminal growth and degeneration, change with age (10). It is unknown, however, whether these changes are accompanied by corresponding age-related changes in glucose uptake by nerve terminals. Therefore, the present study was conducted to examine this question.

Glucose transport into α-motoneuron terminals can be

monitored indirectly at the single-cell level of resolution (9, 11). Namely, addition of glucose to the bathing medium results in a transient increase in spontaneous transmitter release due to hypertonic conditions. This is manifest postsynaptically as an increase in the rate of miniature end-plate potentials (MEPP's) (cf. 5, 6, 8). The rate at which glucose enters into the nerve terminals, thus reestablishing isotonicity, can then be measured electrophysiologically by recording the subsequent decline in the MEPP rate.

The results of this study add support to the suggestion that motor neurons which exhibit different metabolic capabilities also differ in the rate at which their nerve terminals take up glucose (cf.9). Furthermore, they demonstrate a tendency towards decreased rates of glucose uptake with age into these terminals.

2. METHODS

Glucose uptake into motor nerve terminals innervating three muscles composed predominantly of either fast-twitch glycolytic, fast-twitch oxidative/glycolytic, or slow-twitch oxidative fibers, namely the extensor digitorum longus (EDL), diaphragm and soleus, respectively, was measured indirectly in barrier-reared male Fischer 344 rats aged 10 (mature adult) and 28 (aged) months. Muscles were dissected from the rats while under chloral hydrate anesthetic (about 0.5 mg/g body wt). Each preparation was then placed immediately into a bath of physiological saline containing 137 mM NaCl, 5 mM KCl, 2 mM $CaCl_2$, 1 mM $MgCl_2$, 3 mM HEPES, and 11 mM D-glucose (pH 7.3; 285.0 $\pm$ 2.3 mosm). The bath temperature was maintained at 37°C. A mixture of 95% O_2 - 5% CO_2 was bubbled continuously into the saline, which circulated around the preparation at a rate of 10 ml/min. The O_2 tension of the bath was maintained at approximately 80% saturation.

Intracellular records of MEPP's were obtained with 15- to 20-megohm glass microelectrodes filled with 3 M KCL. Recordings from fibers on the muscles' surface exhibiting and maintaining resting membrane potentials of at least -70 mV were used for subsequent data analysis. When MEPP's were located whose rise times were less than 1 ms and amplitudes were approximately 1 mV, they were filmed. The osmolarity of the saline was then increased by approximately 16% with the addition of 30 mM D-glucose, and 30-s film records were obtained at 2.5- to 5.0-min intervals for 25 to 30 min. Experiments were repeated utilizing 30 mM 2-deoxy-D-glucose, a nonmetabolized analogue of D-glucose which shares a common carrier-mediated process (3, 7) or 30 mM L-glucose, the stereoisomer of D-glucose which purportedly enters the nerve terminal through passive diffusion only (7). Data were compared using routine two-tailed statistical tests.

3. RESULTS

Increasing the osmolarity of the circulating saline solution with D-glucose resulted in an initial sharp increase in MEPP frequency which was followed by a steady decline. An example of the change in MEPP frequency after addition of 30 mM D-glucose relative to that observed at rest is illustrated in Fig. 1. Similar results were obtained using 2-deoxy-D-glucose. In contrast, when L-glucose was added to the bath, the initial increase in MEPP frequency was sustained for the duration of the experiment. Using the methods of least squares, exponential curves were fit to the falling phase of the MEPP frequency data (e.g. Fig. 1) which was obtained after the addition of 30 mM D-glucose. The time constants of decay, which estimate glucose uptake by the nerve terminals as a new osmotic equilibrium is established, were calculated from these data and are presented in Table 1.

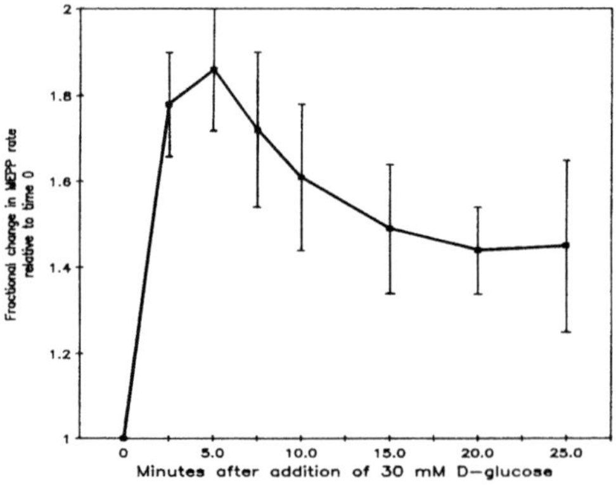

Figure 1. The fractional change in MEPP rate relative to the resting MEPP rate after additionof 30-mM D-glucose. Values represent the average(±S.E.) change observed at the neuromuscular junctions of 10-month EDL muscles. (n = 7).

These results indicate that D-glucose entered nerve terminals innervating the EDL muscle more rapidly than the diaphragm or soleus. This difference was statistically significant (0.05 level) between the EDL and soleus only. Furthermore, the rates of glucose uptake declined somewhat with age at all neuromuscular junctions.

Table1. Time constants for MEPP decay in the presence of D-glucose.

Muscle	Age(mos)	Time Constant (min)
EDL	10 (n=7)	56.50 ± 5.75[a]
	28 (n=5)	68.49 ± 7.04[b]
diaphragm	10 (n=12)	76.34± 17.48
	28 (n=5)	99.01 ± 31.37
soleus	10 (n=4)	125.00 ± 39.06[a]
	28 (n=4)	200.00 ± 72.00[b]

Time constant values are averages ± S.E. All time constants are significantly different from 0. a and b indicate that the corresponding values are significantly different at the 0.01 and 0.005 levels, respectively.

4. DISCUSSION

The main conclusion that can be drawn from these data is that glucose uptake into nerve terminals exhibiting different capacities for oxidative enzyme activities appears to occur at different rates and that these rates decline with age. That is, glucose entry is slowest into cells which exhibit the greatest oxidative potential, and therefore whose energy requirements are derived from the highest ratio of respiration to glycolysis (cf. 12). This could be the case in the nerve terminals innervating the predominantly oxidative soleus or to a lesser extent, the oxidative/glycolytic diaphragm. In contrast, a cell which engages predominantly in bursts of phasic activity may be somewhat less dependent on oxidative metabolism and may rely more heavily on glycolysis for a quick, however less efficient source of energy. Therefore it may take up glucose at a higher rate.

This category would include nerve terminals innervating the fast-twitch EDL.

In each of the three nerves, there appears to be a decrease in glucose uptake associated with age. These changes were of marginal statistical significance. However, they were sizeable (21 to 60%), and consistent. If indeed this age-related decrease is real, it could indicate that nerve terminals become more oxidative with age. Although age-related changes observed in oxidative enzyme activity within the accompanying muscles are variable, they generally indicate that most muscles become more oxidative with age (1, 2, 4).

It should be noted that this technique for monitoring glucose transport into nerve terminals requires high concentrations of glucose to obtain significant changes in MEPP frequencies. Although the concentrations used were less than the K_m of the low-affinity glucose transport system (75 mM), Diamond & Fishman (1973) have also reported a high-affinity uptake system (0.24 mM) in synaptosomes. Substantial age-related changes may occur within the domain of a high-affinity system which cannot be assayed electrophysiologically.

5. ACKNOWLEDGEMENTS

This work was supported by NIH grant AG01572 and an NIH training grant to the Neurosciences Training Program at the University of Wisconsin.

6. REFERENCES

1. Bass, A., Gutmann, E. & Hanzlikova, V. 1975. Biochemical and histochemical changes in energy supply-enzyme pattern of muscles of the rat during old age. Gerontologia 21, 31-45.

2. Caccia, M.R., Harris, J.B. & Johnson, M.A. 1979. Morphology and physiology of skeletal muscle in aging rodents. Muscle & Nerve 2, 202-212.

3. Diamond, I. & Fishman, R.A. 1973. High-affinity transport and phosphorylation of 2-deoxy-D-glucose in synaptosomes. J. Neurochem. 20, 1533-1542.

4. Eddinger, T.J., Moss, R.L. & Cassens, R.G. 1984. The mechanical properties and fiber type composition of skeletal muscles from young and senescent rats. Fed. Proc. 43, 531.

5. Fatt, P. & Katz, B. 1952. Spontaneous subthreshold activity at motor nerve endings. J. Physiol. 117, 109-128.

6. Furshpan, E.J. 1956. The effects of osmotic pressure changes on the spontaneous activity at motor nerve endings. J. Physiol. 134, 689-697.

7. Heaton, G.M. & Bachelard, H.S. 1973. The kinetic properties of hexose transport into synaptosomes from guinea pig cerebral cortex. J. Neurochem. 21, 1099-1108.

8. Niles, W.D. and Smith, D.O. 1982. Effects of hypertonic solutions on quantal transmitter release at the crayfish neuromuscular junction. J. Physiol. 329, 185-202.

9. Pickett, J.B. 1980. Nerve terminals are as metabolically different as the muscle fibers they innervate. Science 210, 927-928.

10. Rosenheimer, J.L. & Smith, D.O. 1985. Differential changes in the end-plate architecture of functionally diverse muscles during aging. (in press).

11. Shimoni, Y. & Rahamimoff, R. 1983. Stereospecific glucose transport across motor nerve terminal membrane: an electrophysiological study. Am. J. Physiol. 245, C308-C315.

12. Sickles, D.W. & Oblak, T.G. 1984. Metabolic variation among α-motoneurons innervating different muscle-fiber types. I. Oxidative enzyme activity. J. Neurophys. 51, 529-537.

13. Siesjö, B.K. 1978. Brain Energy Metabolism chap. 4. New York: Wiley.

DISCUSSION OF THE ARTICLE BY J. ROSENHEIMER (U.S.)

O.F. Hutter (U.K.): Are you suggesting that the property of the muscle in some way determines what is going on in the nerve endings? Or, is it that the property of the neuron determines what goes on both in the nerve endings and in the muscle?

J. Rosenheimer (U.S.): I am simply saying that there may be a correlation. This was proposed by Pickett a number of years ago (Pickett, J.B. (1980). Nerve terminals are as metabolically different as the muscle fibers they innervate. Science 210, 927-928). There may be a correlation between the metabolism of the muscle and the metabolism of the nerve. What I have shown is that there is a difference in uptake into nerves innervating metabolically different muscles for whatever reasons.

O.F. Hutter (U.K.): I am sure you are right that there is a correlation. However, may I put it to you, as a hypothesis, that it is the neuron which determines all. Those neurons which innervate the fast muscles and which make it a fast muscle by discharging rapidly, have the higher metabolism and can take up the glucose more rapidly. They determine the metabolic pattern of the muscle. In a sense, what goes on in the muscle is secondary to what goes on in the nerve. Finally, may I put to you a possibly very challenging experiment: have you thought about doing similar experiments also with intrafusal muscle fibres? The latter show very considerable differences in metabolic activity and innervation and you may find comparable results there. Further, does the rate of uptake depend on the level of activity of the nerve? If you stimulate more, is there more uptake?

J. Rosenheimer (U.S.): I would definitely assume you would. I have not stimulated in my work.

Comment: Can you correlate the uptake with the size of the neurons and their surface area?

J. Rosenheimer (U.S.): I don't think you can. I have done morphology on the endplates of all three muscles and, in fact, the terminal arborization as well as the endplate area in the EDL are significantly smaller than a diaphragm or the soleus. The soleus, in fact, is the largest in area but does not have as many nerve terminals as the diaphragm, for example. The soleus takes up glucose much more slowly so I don't think you can correlate that at all.

THE ASYMMETRY OF GLUCOSE TRANSPORT IN NERVE TERMINALS; POSSIBLE ROLE IN THE ETIOLOGY OF DIABETIC NEUROPATHY

Shlomo Constantini and Rami Rahamimoff

1. SUMMARY

Hyperosmotic neurosecretion (HN), a calcium dependent electrophysiological phenomenon, has been used to measure indirectly glucose transport at the rat neuromuscular junction. The motor nerve terminal membrane possesses a transport system for sugars which is stereospecific, insulin sensitive and asymmetrical. Glucose 'uptake factor', an indirect index for glucose entry efficiency, was 61%±18SD while the 'efflux factor' was 21%±6SD. 3-0-methyl glucose, a non-phosphorylable, non-metabolisable, glucose analogue, is transported asymmetrically as well. Its average 'uptake factor' was 57%±12SD and the average 'efflux factor' was 14%±7SD. This indicates that the apparent asymmetry in transport is not a result of glucose phosphorylation, but presumably resides either in the transport system itself or in its regulators.

We suggest that this asymmetry, combined with fluctuating glucose levels in the extracellular fluid, can cause trapping of glucose in nerve cells, osmotically induced volume changes and thus such asymmetry can contribute to a further damage followed by which might lead to a diabetic neuropathy.

2. INTRODUCTION

Diabetic neuropathy is the most frequent late complication of diabetes mellitus (1) and is a common category of peripheral neuropathy (2). The severity of the neuropathy is linked to the degree of hyperglycemia to which the patient or animal is exposed (3,4). However, the pathophysiology of diabetic neuropathy is still not fully understood.

A primary step towards understanding this pathophysiology might be to investigate the mechanisms which control glucose transport across nerve membranes. Here, we employ an electrophysiological method to further examine glucose transport mechanisms in a very crucial segment of the peripheral nervous system: the motor nerve terminal at the neuromuscular synapse. This subcellular element is small and difficult to handle using conventional biochemical techniques. Therefore, an electrophysiological method was developed (5,6), which enables the study of glucose transport across a single nerve terminal membrane, while the synapse is intact and functional.

3. MATERIALS AND METHODS

As shown before (5,6), hyperosmotic neurosecretion (HN), probably a calcium-dependent phenomenon (7) first described by Fatt & Katz (8) and by Furshpan (9), can serve to estimate glucose entry into a single functional nerve terminal. Fig 1 demonstrates the normal response of miniature endplate potentials (mepp) frequency to a hyperosmotic challenge. When adding a non-penetrating substance, e.g., sucrose (Fig 1A), to the normal Ringer perfusing solution, there is a sustained increase in mepp frequency which declines only when the perfusing solution reverts back to normal Ringer. When D-glucose is the hyperosmotic agent (Fig 1B), following a similar initial increase in frequency, there is a gradual decline towards control levels. This presumably indicates that glucose enters into the terminal and thus dissipates the osmotic gradient. At the second arrow, when the perfusing solution reverts back to normal Ringer solution, there is a decrease in mepp frequency below the control levels. This 'undershoot' phenomenon presumably happens because at this point the terminal is loaded with glucose and the outside normal Ringer solution is relatively hypoosmotic. Following this undershoot, there is a gradual increase in frequency towards control levels, presumably indicating glucose efflux from the terminal.

The rate of decline of mepp frequency can serve as an indirect index for the rate of glucose entry. A 'glucose uptake' factor was defined as the percentage of the initial jump in frequency that was dissipated within 30 minutes. (Decline of mepp frequency in 30 minutes/initial increase in mepp frequency x 100.)

Experiments described in this paper were all done on the Sabra rat neuromuscular junction. Electrophysiological data was collected using conventional intracellular recording techniques, taped on an HP (3964) taperecorder and then sampled by a computer (11/23 DIGITAL MINC). Analysis of the data was done by programs written especially for this purpose.

Normal Ringer solution contained in mM: 138 NaCl, 5 KCl, 2 $CaCl_2$, 1 $MgCl_2$, 11 D-glucose, 15 $NaHCO_3$, 1 NaH_2PO_4, titrated to pH 7.1-7.4 and bubbled with 95% O_2 - 5% CO_2. Hyperosmotic solutions were done by adding the specific sugar examined to the normal Ringer solution.

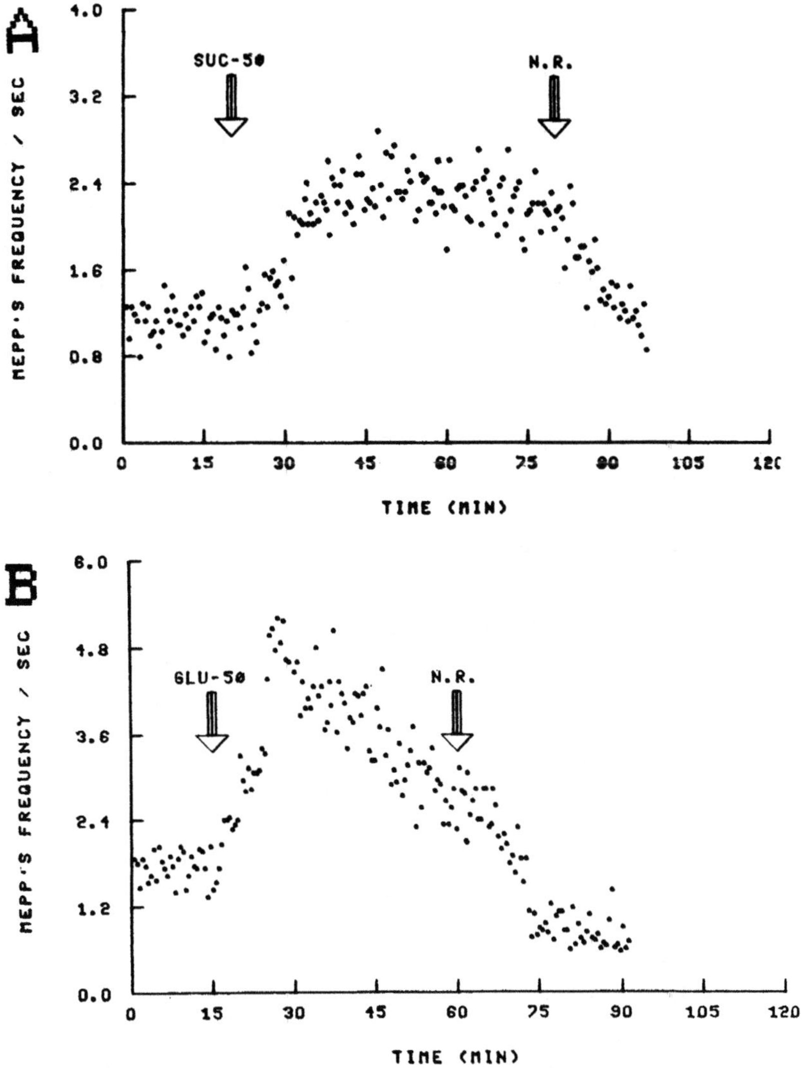

FIGURE 1. The hyperosmotic effect of non-penetrating versus penetrating sugars. a) When adding 50mM sucrose to the perfusing normal Ringer solution, there is an increase in mepp frequency which is sustained until the solution reverts back to normal Ringer. b) With 50mM D-glucose: following the initial increase in frequency there is a gradual decline towards control levels. Every point represents the average mepp frequency in an interval of 30 seconds.

4. RESULTS

4.1. <u>The effect of different sugars on HN</u>

Using the HN technique, we have investigated the transport of six different sugars into the motor nerve terminal. Fig 2 summmarizes the 'initial jump' in frequency and the 'uptake factor' of these sugars. Fig 2B shows that the initial jump in frequency is similar for the six sugars examined (approximately 2.2 fold), indicating that the initial HN response is the result of the osmolarity of the sugar added to the perfusing solution and not its conformation or its metabolic properties. However, the rate of entry into the terminal, expressed by the 'uptake factor' (in parenthesis) is different for various sugars. 2-deoxy glucose (52%±15SD) and 3-0-methyl glucose (57%±12SD), non-metabolisable analogues of glucose, have similar uptake factors to D-glucose (61%±18SD), thus ruling out the possibility that the decline of mepp frequency is a metabolic effect of D-glucose and not the result of glucose entry into the terminal. L-glucose (10%±5SD) has the lowest 'uptake factor', much lower than D-glucose (61%±18SD), indicating that the sugar transport system in the motor nerve terminal is stereospecific, with a preference to the D isomer of glucose.

Using the same technique, it was shown that the motor nerve terminal sugar transport system is temperature sensitive and is significantly facilitated by insulin (6).

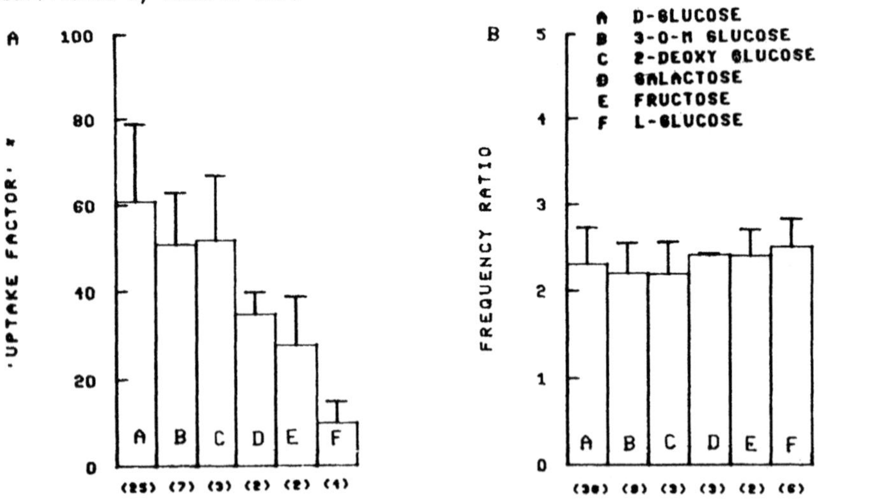

FIGURE 2. Initial increase in frequency and 'uptake factors' of six sugars: while the initial increase in frequency is approximately the same, the uptake factor indicating glucose entry efficiency is different for the different sugars. Number of experiments is indicated in parenthesis.

4.2. The asymmetry of the glucose transport system

Following the loading of the terminal with glucose, perfusion with normal Ringer solution causes the frequency of the mepps to be at sub-control level, demonstrating the effect of relatively hypoosmotic solution (Fig 1B). Afterwards, there is a gradual increase which probably reflects the efflux of glucose from the terminal.

Experiments similar to the one described in Fig 3 raise the possibility that glucose transport is not symmetrical. Following the first challenge with 50mM D-glucose, there is a 'normal' HN effect; an initial increase in frequency followed by a gradual decline towards control level. (Note the compressed time axis.) If one waits long enough, mepp frequency stabilizes at control levels, presumably indicating full dissipation of the osmotic gradient and sugar equilibration between the intracellular and the extracellular fluid. At this point, perfusion with normal Ringer solution shows the effect of a 50mM hypoosmotic solution. The sharp 'initial decrease' in frequency is probably the reverse of the 'initial increase' obtained at the first challenge. However, the gradual increase towards control levels indicates that the efflux of glucose, is much slower than the entry of glucose.

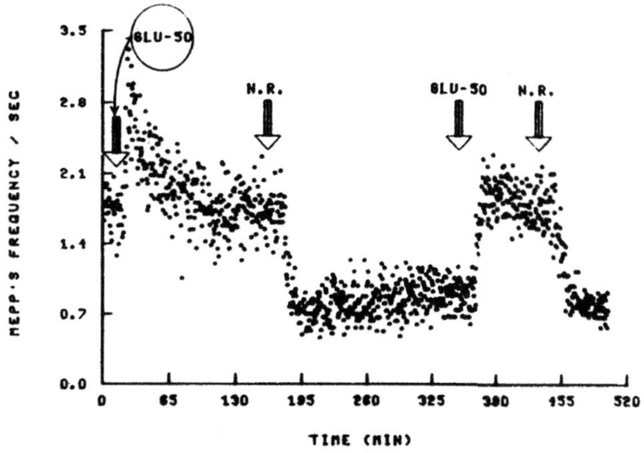

FIGURE 3. The asymmetry in glucose transport. Following a 'normal' HN response to 50mM glucose, mepp frequency returns to control levels indicating an equilibration between intracellular and extracellular fluid. At 160 minute, perfusion with normal Ringer was started and the initial hypoosmotic effect was followed by a slow increase in frequency towards control levels. This slow increase (137-370 minutes) reflects presumably a slow efflux of glucose from the terminal.
Every point is the average mepp frequency of a 30 second period.

The experiments described in Fig 3 require long intracellular recording and are difficult to perform; hence, experiments were done somewhat differently to shorten the sampling time. After dissection, the whole preparation was put in a 50mM hyperosmotic solution for 5-6 hours, leading to glucose equilibration. At this stage, mepp frequency was sampled, followed by a change to normal Ringer solution. The 'efflux factor' thus calculated had a value of 21%$\pm$6SD (3 experiments) compared to the 61%$\pm$18SD found for the 'uptake factor' of glucose (25 experiments).

The asymmetry between entry and efflux of glucose might be due to trapping of the phosphorylated product of glucose, rather than the result of asymmetry of the transporter itself or its regulators. To check this point, 5 experiments were done in the 'shortened' approach with the non-metabolisable, non-phosporylatable analogue of D-glucose: 3-0-methyl glucose. The 'uptake factor' of 30MG was in 7 experiments 57%$\pm$12SD, similar to that of D-glucose. The average 'efflux factor' was found to be 14%$\pm$7SD. One of the experiments with 30MG is shown in Fig 4. Following incubation for 5 hours in a normal Ringer solution enriched with 50mM 30MG, control frequency was sampled. With the change to normal Ringer solution, there was a decrease of 2.53 in frequency, after which there is a gradual very slow increase towards control levels (in this particular experiment, 17% of the return to control was achieved in 30 minutes). Since the results with 30MG are similar to those with D-glucose, it is reasonable to assume that this asymmetry, approximately 3 folds faster entry than efflux, happens in the level of the transporter itself and is not a result of the glucose metabolism inside the terminal.

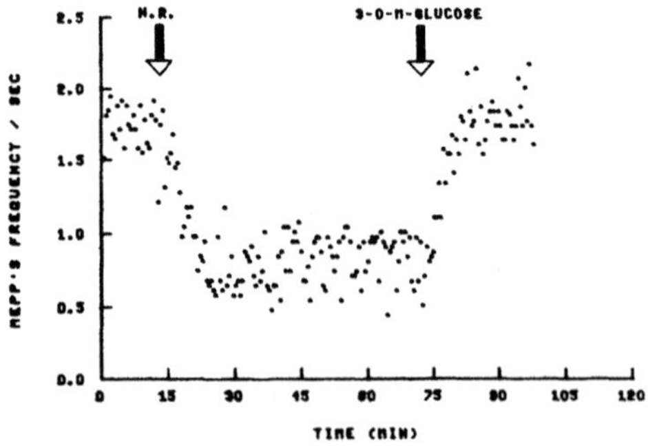

FIGURE 4. Asymmetry in 30MG transport. Preceeding the sampling shown above, the preparation was incubated for 5 hours in a solution of normal Ringer plus 50mM 30MG. The hypoosmotic effect caused by perfusion with normal Ringer solution is followed by a slow increase in frequency towards control levels. 'Efflux factor' of 17% in 30 minutes.

5. DISCUSSION AND CONCLUSIONS

In this work, we use hyperosmotic neurosecretion to measure glucose transport at the nerve terminal. Previous work (10) and some of the data presented here indicate that the rat, as well as the frog (6), possess a facilitated diffusion transport system for glucose in their motor nerve terminal membrane. This system is stereospecific, effected by temperature and sensitive to insulin. In the current work, we bring data to show that the transport of glucose and 3OMG into the terminal is 3-4 times faster than their efflux. We speculate that this asymmetry might be related to the pathogenesis of the diabetic neuropathy.

Asymmetry in glucose transport was first reported in the red blood cell membrane by Wilbrant (11) and reviewed by Widdas (12). It is known, as well, that sodium serves as a co-factor in glucose absorption in the gut (13,14) and in the kidney (15,16). In the nervous system, asymmetry was reported in the Squid giant axon by Baker & Carruthers (17,18), but with the efflux faster than the entry.

We propose that if this asymmetry exists also in human nerve terminals, it may contribute to the development of diabetic neuropathy; patients with severly uncontrolled diabetes (such as Brittle diabetes) can have blood glucose levels fluctuating between 30mM and 4mM. While glucose levels are high in the extracellular fluid, it can enter the terminal and equilibrate with the lateral medium. However, when blood glucose levels go down, glucose is trapped for a certain period of time inside the nerve since the efflux is relatively inefficient. This trapping can cause an osmotic swelling and thus cause a further damage to the nerve cell. Hence, it might be important not only to reduce mean glucose levels in diabetic patients but also to avoid gross fluctuation in blood glucose levels.

ACKNOWLEDGEMENTS

We thank Ms. Miriam Silber for unfailing help in the preparation of this manuscript.

This work was supported by the National Institutes ot Health, the Juvenile Diabetes Foundation, the Muscular Dystrophy Association and the U.S.-Israel Binational Science Foundation. Equipment was purchased with the aid of the Bay Foundation.

During this work, Shlomo Constantini was a Foulkes Foundation Fellow.

REFERENCES

1. Clements R.S., Jr., (1979). Diabetic neuropathy; new concepts of its etiology. Diabetes 28, 604-611.
2. Goto Y., Horiuchi A. and Kongure K. (1982). Diabetic neuropathy. Proc. Internat. Symposium.
3. Ellenberg J. (1976). Diabetic neuropathy; clinical aspects. Metabolism 25, 1627-1655.
4. Pirart J. (1978). Diabetic care. 1, 168-252.
5. Rahamimoff R. and Shimoni Y. (1977). Glucose transport by motor nerve endings. J. Physiol. 271, 18-19P.
6. Shimoni Y. and Rahamimoff R. (1983). Stereospecific glucose transport across motor nerve terminals: an electrophysiological study. Am. J. Physiol. Nov. 1983, C308-C314.
7. Shimoni Y., Alnaes E. and Rahamimoff R.d (1977). Is hyperosmotic neurosecretion from motor nerve tendings a calcium dependent process. Nature 267, 170-172.
8. Fatt P. and Katz B. (1952). Spontaneous subthreshold activity at motor nerve endings. J. Physiol. 117, 109-128.
9. Furshpan E.J. (1956). The effects of osmotic pressure changes on the spontaneous activity at motor nerve endings. J. Physiol. 134, 689-697.
10. Shimoni Y., Constantini S. and Rahamimoff R. (1983). Glucose transport in functional nerve endings. In: The Proceedings of the 1983 Jerusalem International Symposium on Diabetes.
11. Wilbrandt W. (1954). Secretion and transport of non-electrolytes. Symp. Soc. Exp. Biol. 8, 136-162.
12. Widdas W.F. (1980). The asymmetry of the hexose transfer system in the human red cell membrane. Current Topics in Membranes and Transport. 14, 165-223.
13. Crane R.K., Forstner G. and Eicholtz A. (1965). Studies on the mechanism of the intestinal absorption of sugars; an effect of Na^+ concentration in vitro. Biochim. Biophys. Acta 109, 467-477.
14. Kimmich G. and Randles J. (1979). Energetics of sugar transport by isolated intestinal epithelial cells: effects of cytochalasin B. Am. J. Physiol. 237, C56-C63.
15. Ullrich K.J. (1976). Renal tubular mechanisms of organic solute transport. Kidney International 9, 134-148.
16. Heinz E. (1972). Na linked transport of organic solutes. Springer-Verlag, Berlin, New York.
17. Baker P.F. and Carruthers A. (1981a). 3-O-Methyl glucose transport in internally dialyzed giant axon of Loligo. 316, 503-525.
18. Baker P.F. and Carruthers A. (1981b). Sugar transport in giant axon of Loligo. J. Physiol. 316, 481-502.

DISCUSSION OF THE ARTICLE BY S. CONSTANTINI (ISRAEL)

Comment: Does the asymmetry imply that the transport is active?

S. Constantini (Israel): No. It is probably a facilitated diffusion transport mechanism - a carrier system - since it is stereospecific. We haven't checked the effect of uncouplers of oxidative phosphorylation but we think it is a passive transport system.

Comment: I wonder because a passive transport would mean that you have in and out at the same rates, unless it is coupled by something else.

S. Constantini (Israel): There are several possibilities. One possibility is that this transport system is linked to another ion, e.g., sodium or calcium. Other possibilities exist, as have been demonstrated in the red blood cell (RBC) (Wilbrant W. (1954). Secretion and transport of non electrolytes. Sym. Soc. Biol. 8, 136-162).

R.J. DeLorenzo (U.S.): Do you intend to extend this work to include the peripheral nerve, not only the terminal, since a lot of the neuropathies that you see are actually in the axon?

S. Constantini (Israel): Hyperosmotic neurosecretion can serve for measuring glucose uptake only in nerve terminals. For studying glucose uptake in the nerve axons, one would have to use conventional biochemical techniques. There are no good reports comparing entry and efflux in mammalian nerve trunks.

Comment: Are your results in line with what is found in other tissues? In many other tissues, people have studied glucose flux in experimental diabetes and have then found asymmetry?

S. Constantini (Israel): Since we worked with high osmotic gradients, we probably worked in a saturating state of the transport system. Therefore, we cannot calculate Km values. However, we have calculated Vmax which is approximately of Vmax found by others in nerve tissue. Asymmetry is well known in the RBC. It was found by Wilbrant (1954) and further examined by Widdas (Widdas W.F. (1980). The asymmetry of the hexose transfer system in the human red cell membrane. Currents Topics in Membrane and Transport 14, 165-223). The asymmetry was well studied and found to be a result of the transporter asymmetry. It was found that when glucose binds to the transporter from the inside, it binds from the C1 end of the glucose while when it binds from the outside, it binds from the C5 end. We haven't found the mechanisms of the asymmetry in this preparation. I know of only one work in nervous tissue dealing with efflux. This is the work by Baker and Carruthers in the squid giant axon which describes asymmetry in the reverse direction to ours (Baker P.F. and Carruthers A. (1981). 3-O-Methyl glucose transport in the giant axon of Loligo. J. Physiol. 316, 481-502).

A. Erdreich (Israel): Is it possible that the trapping of glucose that you see is due to it's utilization by the myoinositol pathway which is implicated in endothelial cells in diabetes?

S. Constantini (Israel): We have shown that the asymmetry occurs with 3-O-methyl glucose which is not metabolised at all. We therefore assume that it is in the tranport system and not due to metabolic processes.

THE ROLE OF PEPTIDES RELEASED BY SMALL CELL CARCINOMA OF THE LUNG ON NEUROMUSCULAR TRANSMISSION: CALCITONIN

Muhammad A. Abdul-Ghani[1], Shai D. Silberberg[1], Michael F. Johnston[2], Edward A. Kravitz[2], Halina Meiri[1] and Rami Rahamimoff[1]
[1]Department of Physiology, Hebrew University-Hadassah Medical School, Jerusalem, Israel
[2]Department of Neurobiology, Harvard Medical School, Boston, Massachusetts

INTRODUCTION

The action of peptides on neuromuscular transmission is of a potential interest from two different viewpoints. First, under physiological conditions peptides may serve as potential modulators of synaptic transmission. For example, it has been shown that, the vasoactive intestinal peptide (VIP), has augmenting effect on the number of neurotransmitter quanta released by nerve stimulation (Gold, 1982) The adrenocorticotropic hormone (ACTH) also increases transmitter release, but unlike VIP, it has a long-lasting enhancing action (Johnston et al, 1983). Peptides differ not only in the duration of their action, but also in the characteristics of their effect. The 11 amino acid peptide, Substance P has a biphasic effect on transmitter release, (Steinackre, 1977), while enkephalin reduces the quantal content of release at the frog neuromuscular junction (Bixby and Spitzer, 1983).

Secondly, peptides released from malignant tumours may be responsible for some of the neurological complications observed in patients with neoplastic diseases. These paraneoplastic syndromes may cause polyradiculoneuropathy and myopathy (Wilkinson et al, 1967; Posner, 1971; Henson, 1974; Stolinsky, 1980). Eaton and Lambert, (1957) described and characterized a syndrome of a decrease in neuromuscular transmission associated with small-cell carcinoma of the lung. In this myasthenic syndrome, the main pathophysiological defect is at the presynaptic nerve terminals where the nerve impulse causes diminished quantal release of acetylcholine (Lambert and Elmqvist, 1971).

Recently, small cell carcinoma of the lung tumor cells were grown as cell lines in a number of laboratories (Gazdar et al, 1980; Pettengill et al,

1980) and were found to release a large variety of peptides and other substances, which are listed in Table 1.

TABLE I

HORMONES AND OTHER MOLECULES PRODUCED BY CULTURES OF SMALL CELL CARCINOMA

SUBSTANCE	REPORTED BY
ACTH (adrenocorticotropin)	Bertagna et al 1978, Sorenson et al 1981
BOMBESIN	Moody et al 1981
CALCITONIN	Ellison et al 1975, Sorenson et al 1981
β- ENDORPHIN	Bertagna et al 1978, Sorenson et al 1981
ESTRADIOL 17 β	Brinck-Johnsen et at 1979, Sorenson et al 1981
GLUCAGON	Sorenson et al 1981
GLYCOPEPTIDE (31,000)	Bertagna et al 1978
GROWTH HORMONE	Sorenson et al 1981
HUMAN CHORIONIC GONADOTROPIN	Rabson et al 1973, Sorenson et al 1981
LH (luteinizing hormone)	Sorenson et al 1981
LIPOTROPIN	Bertagna et al 1978, Sorenson et al 1981
NEUROPHYSIN	Sorenson et al 1981
PROLACTIN	Sorenson et al 1981
PTH (parathyroid hormone)	Sorenson et al 1981
SOMATOSTATIN	Sorenson et al 1981
VASOPRESSIN	Pettengill et al 1977

(For references see: Rabson et al, 1973; Ellison et al, 1975; Pettengill et al, 1977; Bertagna et al, 1978; Bertagna et al, 1978; Brinck-Johnsen et al, 1979; Moody et al, 1981; Sorenson et al, 1981). These findings led to the hypothesis that one or more of the peptides released by the cancerous cells are transferred via the humoral system to the periphery where they act on the nerve terminals to alter synaptic transmission. Hence, a systematic approach of examining the effects of these peptides on neuromuscular transmission may be worthwhile The augmenting effect of calcitonin on acetylcholine release from the motor nerve endings is reported in the present study.

METHODS

The effect of thyrocalcitonin on the cutaneous pectoris and the sartorious neuromuscular junctions of the frog was investigated.
The isolated preparation was initially perfused continuously with control Ringer solution containing 116 mM NaCl, 2m KCl, 1mM $MgCl_2$, 0.3 $CaCl_2$ buffered at pH 7 with 5 mM Hepes. Later on, the calcitonin (6.3 uM) was introduced into the experimental bath via the perfusion system.
Conventional nerve stimulation and intracellular microelectrode recording techniques were used to record evoked and spontaneous synaptic potentials.
The synaptic potentials were recorded on tape and analyzed with the aid of programs written for the PDP 11/23 MINC microcomputer.

RESULTS

Exposure of the neuromuscular junction to thyrocalcitonin augmented the amount of neurotransmitter liberated from the motor nerve endings in response to nerve stimulation. The amplitudes of successive end plate potentials (epps) increased during the exposure of the preparation to 6.3 uM thyrocalcitonin as demonstrated in figure 1.

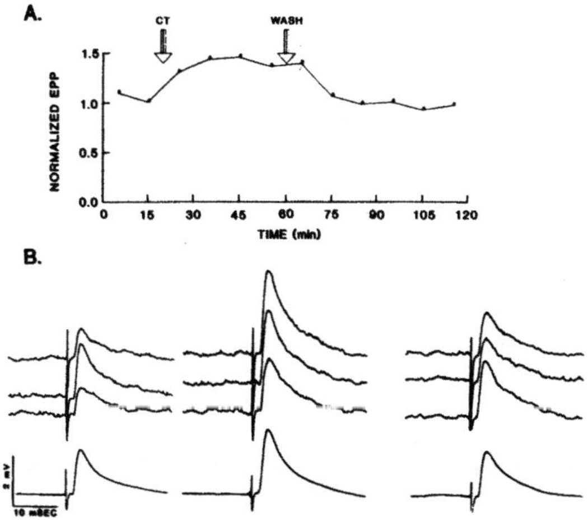

Fig. 1: Effect of calcitonin on epp amplitude

A. Normalized epp amplitude before, during, and after exposure to calcitonin (6.3uM). During the exposure there is a great increase in epp amplitude and complete recovery from the effect is seen after washing with control Ringer (0.3 mM $CaCl_2$, 1 mM $MgCl_2$). Every point represents the average of 50 epps.

B. Individual epps (upper traces) and average epps (lower traces), before (left), during (middle), and after (right) the exposure.

The extent of increase in epp amplitude varied between 30 to 210% in different experiments, depending on the concentration of peptide.

Complete recovery from this effect was observed in 11 out of 15 experiments upon washing the preparation with the control Ringer Solution. In the other 4 experiments only a partial decrease of release was noticed following removal of the calcitonin.

Such an increase in the epp amplitude could reflect either augmented release of the neurotransmitter or an elevated post synaptic sensitivity to acetylcholine or both. In order to distinguish between the various possibilities, the spontaneously appearing miniature end plate potentials (mepps) were recorded throughout the experiment and the quantal content of evoked release was calculated (del Castello and Katz, 1954).The unaltered amplitudes of mepps (fig. 2) indicate no significant change in the post synaptic sensitivity of the muscle to aceltylcholine upon exposure to 6.3 uM calcitonin. In all the experiments an increase in the epp amplitude was accompanied by an augmented quantal content.

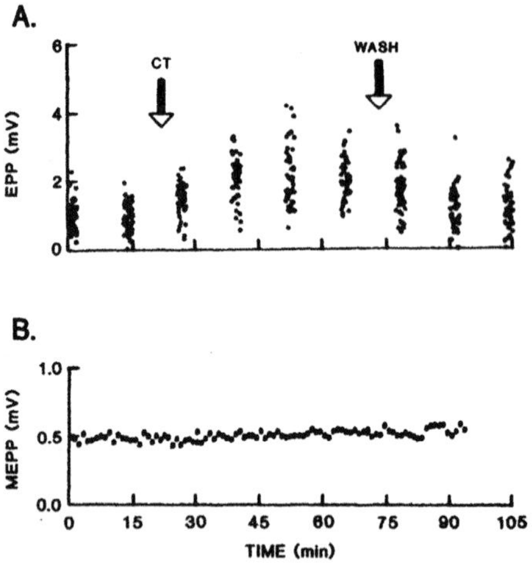

Fig. 2: Effect of 6.3 uM calcitonin on epps and mepps.

A. There is an increase in epp amplitude and a complete recovery from the effect after washing with control Ringer (0.3 mM $CaCl_2$, 1 mM $MgCl_2$).

B. No effect was observed on mepps amplitude recorded simultaneously from the same end plate.

DISCUSSION

The results in this article demonstrated that calcitonin affects synaptic transmission at the neuromuscular junction by increasing the number of quanta released by the nerve impulse.

The action of calcitonin is probably mediated by different subcellular mechanisms than that of ACTH, since the action of the latter exceeds by hours the time of exposure. We do not know whether this action of calcitonin is of a physiological significance. However, it should be noted that calcitonin binding sites were shown in the rat brain (Oligiati et al, 1983).

Our working hypothesis is that a substance, presumably a peptide, released from the small cell carcinoma tumour is acting humorally on the neuromuscular junction to produce muscle weakness, by reducing the amount of transmitter released by the nerve terminal. It is obvious that the observed short term action of calcitonin of increasing the quantal content, does not fit in a straight forward fashion to the working hypothesis. A possibility still exists that the long term exposure to calcitonin may have an inhibitory effect. Such speculation however needs to be examined either in vivo or in organ culture.

Acknowledgments

We thank Ms. Shoshana Wineberg for her help in the preparation of the manuscript. This work was supported by the Muscular Dystrophy Association, the A.L.S. Foundation, the U.S.-Israel Binationial Science Foundation and the Council for Tobacco Research.
Equipment was purchased with aid from the Bay Foundation.

REFERENCES

1. Bertagna, X.Y., Nicholson, W.E., Pettengill, O.S., Sorenson, G.D., Mount, C.D. and Orth, D.N. (1978). Ectopic production of high molecular weight calcitonin and corticotropin by human small-cell carcinoma cells in tissue culture: Evidence for separate precursors. J. Clin. Endocrinol. Metabol. 47: 1390-1393.

2. Bertagna, X.Y., Nicholson, W.E., Sorenson, G.D., Pettengill, O.S., Mount, C.D. and Orth, D.N. (1978). Corticotropin, lipotropin and -endorphin production by a human nonpituitary tumour in culture: Evidence for a common precursor. Proc. Natl. Acad. Sci. USA 75: 5160-5164.

3. Bixby, J.L. and Spitzer, N.C. (1983). Enkephalin reduces quantal content at the frog neuromuscular junction. Nature, 301: 431-432.

4. Brinck-Johnsen, T., Pettengill, O.S., Brinck-Johnsen, K., Sorenson, G.D. and Maurer, L.H. (1979). Estrogen production by an established cell line from pulmonary small-cell anaplastic carcinoma. J. Steroid. Biochem. 10: 339-340.

5. Eaton, L.M. and Lambert, E.H. (1957). Electromyography and electric stimulation of nerves in diseases of motor unit. Am. J. Physiol. 163: 1117-1124.

6. Ellison, M., Woodhouse, D., Hillyard, C., Dowsett, M., Coombes, R.S., Gilby, E.D., Greenberg, P.B. and Neville, A.M. (1975). Immunoreactive calcitonin production by human lung carcinoma cells in culture. Br. J. Cancer 32: 373-379.

7. Gazdar, A.F., Carney, D.N., Russell, E.K., Sims, H.L., Baylin, S.B., Bunn, P.A., Guccion, J.G. and Minna, J.D. (1980). Establishment of continuous clonable cultures of small-cell carcinoma of the lung which have amine precursor uptake and decarboxylation cell properties. Cancer Res. 40: 3502-3507.

8. Gold, M.R., (1982). The effect of Vasoactive Intestinal Peptides on neuromuscular transmission in the frog. J. Physiol. 327: 325-335.

9. Henson, R.A. (1974). Neuromuscular disorders associated with malignant disease. In: Disorders of Voluntary Muscle. J.N. Walton. Ed., Churchill-Livingstone, pp.766-769.

10. Johnston, M.F., Kravitz, E.A., Meiri, H. and Rahamimoff, R. (1983). ACTH causes long-lasting potentiation of transmitter release from frog motor

nerve terminals. Science 222: 1071-1072.

11. Lambert, E.H. and Elmqvist, D. (1971). Quantal components of endplate potentials in myasthenic syndrome. Ann. N.Y. Acad. Sci. 183: 183-199.

12. Lambert, E.H., Rooke, E.D., Eaton, L.M. and Hodgson, C.H. (1961). Myasthenic syndrome occassionally associated with bronchial neoplasm: Neurophyisological studies. In: Myasthenia Gravis, H.R. Viets (Ed.) 2nd International Symposium Proceedings. Springfield, Ill., Charles C. Thomas, Publisher, pp.362-410.

13. Moody, T.W., Pert, C.B., Gazdar, A.F., Carney, D.N. and Minna, J.D. (1981). High levels of intracellular bombesin characterize human small-cell lung carcinoma. Science 214: 1246-1248.

14. Olgiati, V.R., Guidobono, F., Netti, C. and Pecile, A. (1983). Localization of calcitonin binding sites in rat central nervous system: evidence of its neuroactivity. Brain Res. 265: 209-215.

15. Pettengill, O.S., Faulkner, C.S., Wurster-Hill, D.H., Maurer, L.H., Sorenson, G.D., Robinson, A.G. and Zimmerman, E.A. (1977). Isolation and characterization of a hormone-producing cell line from human small-cell anaplastic carcinoma of the lung. J. Natl. Cancer Inst. 58: 511-516.

16. Pettengill, O.S., Sorenson, G.D., Wurster-Hill, D.H., Curphey, T.J., Noll, W.W., Cate, C.C. and Maurer, L.H. (1980). Isolation and growth characteristics of continuous cell lines from small-cell carcinoma of the lung. Cancer 45: 906-918.

17. Posner, J.B. (1971). Neurological complications of systemic cancer. Med. Clin. North American, 55: 625-646.

18. Rabson, A.S., Rosen, S.W., Tashjian, A.H., Weintraub, B.D. (1973). Production of human chorionic gonadotropin invitro by a cell line derived from carcinoma of the lung. J. Natl. Cancer Inst. 50: 669-674.

19. Sorenson, G.D., Pettengill, O.S., Brinck-Johnsen, T., Cate, C.C. and Maurer, L.H. (1981). Hormone production by cultures of small-cell carcinoma of the lung. Cancer 47: 1289-1296.

20. Steinacker, A. (1977). Calcium dependent presynaptic action of Sub-P at the frog neuromuscular junction. Nature, 267: 268-270.

21. Stolinsky, D.C. (1980). Paraneoplastic syndromes. West J. Med. 132: 189-208.

22. Wilkinson, M., Croft, P.B. and Urich, H. (1967). The remote effects of cancer on the nervous system. Proc. R. Soc. Med. 60: 683-692.

DISCUSSION OF THE ARTICLE BY M. ABDUL-GHANI (ISRAEL)

C.J. Duncan (U.K.): Does the mepp frequency change with calcitonin?

M. Abdul-Ghani (Israel): Mepp frequency does not change in the presence of 2.0mM KCl in the medium.

D.M.J. Quastel (Canada): Does calcitonin change mepp frequency in raised potassium?

M. Abdul-Ghani (Israel): Under these conditions, calcitonin highly augments the frequency of mepps bringing it to a level where we could not count them. The frequency was brought back to normal upon removal of calcitonin.

D.M.J. Quastel (Canada): The next question is obvious: Does calcium play a role in it?

M. Abdul-Ghani (Israel): When you remove the extracellular calcium and add EGTA, calcitonin still induces an increase in the mepps frequency. This result was seen in the only experiment done under these conditions.

BIPHASIC EFFECT OF CALCIUM ON TETANIC POTENTIATION AT THE FROG NEUROMUSCULAR JUNCTION

R. Fink[1], S. Ginsburg[1,2] and R. Rahamimoff[1]
[1]Department of Physiology, Hebrew University - Hadassah Medical School, POB 1172, Jerusalem, Israel
[2]Everyman's University, POB 39328, Tel-Aviv, Israel

1. INTRODUCTION

High frequency stimulation of the motor nerve produces an increase in transmitter release, termed tetanic potentation. Previous work showed that calcium ions in the extracellular medium are an important factor in generating tetanic potentiation (Rosenthal, 1969; Weinreich, 1971). Using spontaneous mepps as an indicator of transmitter release, it was suggested that accumulation of calcium ions inside the nerve terminal is largely responsible for this phenomenon (Erulkar, Rahamimoff & Rotshenker 1978; Erulkar & Rahamimoff, 1978; Lev-Tov & Rahamimoff 1980; Rahamimoff, Lev-Tov & Meiri, 1980). The increase in intracellular calcium is presumably caused by calcium entering the terminal via the surface membrane and by calcium released from intracellular stores. Regardless of its origin, it is expected that such elevated intracellular calcium may produce a change in the relationship between transmitter release and extracellular calcium.

Previous work on the frog neuromuscular junction showed that at low frequencies of stimulation, the evoked release vs [Ca]out relation is sigmoidal in nature, the power dependence being about 4 (Jenkinson, 1957; Dodge & Rahamimoff, 1967). According to the residual calcium hypothesis (Katz & Miledi, 1968; Rahamimoff, 1968; Smith & Zucker, 1980) the power relation should decrease when calcium accumulates near sites critical for release.

In the present work tetanic potentiation was examined as a function of [Ca]out. We show that during tetanic potentiation the power relation between transmitter release and extracellular calcium concentration changes in a complicated fashion, consistent with some of the current views on the role of intracellular calcium in regulation of transmitter release.

2. METHODS

The experiments were performed on the sartorius neuromuscular junction of the frog Rana ridibunda. Endplate potentials (epps) were recorded intracellularly from the muscle fiber using conventional techniques. The preparation was bathed in frog Ringer containing 1.5mM Mg and varying [Ca]out (0.2-0.4mM). The bathing chamber was perfused continously to avoid changes in [Ca]out due to leakage of calcium from the muscle. The low quantal content under these conditions prevented both contraction of the muscle and non-linear summation of the postsynaptic response.

The nerve was stimulated by a suction electrode in a fixed pattern controlled by a computer. The pattern consisted of a 8-10 minutes pretetanus control period followed by a 10-seconds tetanus at stimulation frequencies of 0.5 and 33 Hz, respectively. In some experiments, the second phase was followed by an additional 5 min of stimulation at 1.0 Hz to estimate the time required for relaxation of the potentiation back to the control level. Once such a series was delivered at a given [Ca]out, the perfusing solution was changed to a higher concentration in increments of 0.04mM. The time interval between consecutive tetani was 30 min. The epps were recorded on a tape and fed to the computer for averaging and analyzing.

Figure 1 illustrates the synaptic responses to one series of stimulation, delivered at 0.36mM Ca.

3. RESULTS

The three main results presented below refer to the extracellular calcium concentration range (0.2-0.4mM) which could feasibly be examined. Higher concentrations caused muscle contractions during tetanic stimulation, while lower ones produced prohibitive sampling errors.

1. Increase in [Ca]out enhances tetanic potentiation

As previously reported (Jenkinson, 1957; Dodge and Rahamimoff, 1967) we observed an increase in transmitter release when [Ca]out was elevated during low-frequency stimulation. Substracting the low-frequency evoked release from that obtained during tetanic stimulation, we found that the resultant tetanic potentiation was enhanced by increasing [Ca]out.

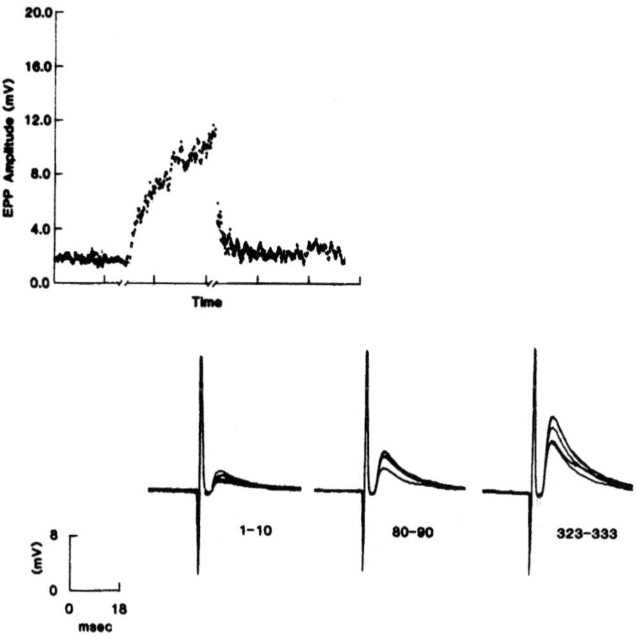

<u>Figure 1:</u> <u>Synaptic responses to one series of stimulation</u>
Top - synaptic responses to a single stimulation series consisting of three periods: 8-min pretetanic control, 10-sec tetanus and 4-min post-tetanus with stimulation frequencies 0.5, 33 and 1 Hz, respectively. Note different time scales. The data within each period was averaged by the moving-bin method (bin=10, d-bin=1).
Bottom - individual end plate potentials recorded at various times (indicated by the stimuli serial numbers) during the tetanus.
The tetanus was delivered at 0.36mM Ca.

2. <u>The relation between release and [Ca]out changes during the tetanus</u>

At low-frequency stimulation, the relation between release and [Ca]out is sigmoidal. If calcium accumulates in the terminal during tetanic stimulation, and if [Ca]in controls transmitter release, it is expected that the power relation should decline during tetanic potentiation. The results confirm this expectation in two ways. First, the slope of the log(epp)-log[Ca]out relation at the end of the tetanus is smaller than the slope obtained at low-frequency stimulation (Figure 2B, right). Secondly, the consecutive slopes of the log-log relation, along the tetanus, decrease progressively (12 experiments performed at 0.3-0.4mM [Ca]out).

3. The slope of the log-log relation is not a monotonic function of [Ca]out

The results presented above are easily compatible with the notion that tetanic potentiation is due to intracellular calcium accumulation. To our surprise, at the lower end of the extracellular calcium concentrations examined (0.2-0.3mM), the slope of the log-log relation (Figure 2A) did not decrease but rather increased during the tetanus (7 experiments). An attempt to explain these apparently paradoxical results is presented below, in the light of some of the current views on calcium-activated transmitter release.

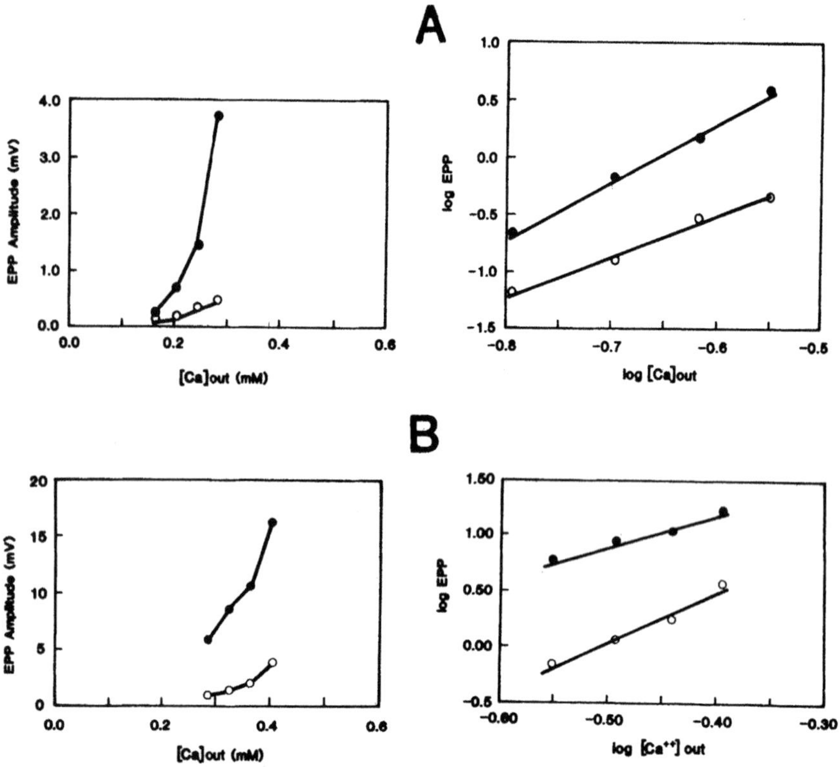

Figure 2: Changes in the relation between release and [Ca]out during tetanic stimulation
A: Epp amplitudes as a function of [Ca]out at the end of the tetanus (left) replotted on double-logarithmic coordinates (right). Top curves refer to the relations at the end of tetani, while bottom curves were obtained at low-frequency stimulations from the same muscle cell. The calcium concentration range was 0.16-0.28mM.
B: The same relations as described in A, obtained at a higher [Ca]out range of 0.28-0.40mM.

4. DISCUSSION

The sigmoidal dependence of transmitter liberation on extracellular [Ca]out led Dodge & Rahamimoff (1967) to suggest a cooperative model of release. According to this model, at least 4 active sites, which bind independently calcium ions, cooperate to release a fixed amount of transmitter. Therefore it is expected, that a fixed slope for the log(epp)-log[Ca]out relation at low extracelular calcium levels and a monotonic decline in the value of this slope at higher concentrations due to saturation of the active sites by calcium, will be found. Further work suggested that the location of the active sites is inside the nerve terminal (Katz & Miledi, 1969; Miledi, 1973; Rahamimoff & Alnaes 1973; Alnaes & Rahamimoff, 1975; Llinas & Nicholson, 1975). Thus the question arises whether the intracellular resting calcium level influences the shape of the relation between extracellular [Ca] and transmitter release after stimulation.

The decrease in the measured slope during a tetanus (Figure 2B) can be interpreted as accumulation of calcium inside the terminal due to the high-frequency stimulation and an approach to saturation. Our observation that at very low [Ca]out the slope of the log-log relation increases during a tetanus (Figure 2A) cannot be explained on the basis of the original set of equations which do not predict such a behaviour. However, a small modification of the original cooperative model, allowing for a possible influence of the resting intracellular calcium level on release, does give rise to predictions that are compatible with the whole set of experimental observations. We shall name the modified version the "generalized cooperative model" and attempt to show that it may serve as one possible explanation for the results. The possibility that intracellular resting calcium may influence transmitter release was suggested by Crawford (1974) and was recently further developed by Parnas & Segel (1981) and Barton, Cohen & Van der Kloot (1983).

Figure 3 describes the generalized cooperative model and Figure 4 illustrates the predicted dependence of the slope (S) of the log-log relation on [Ca]out.

(1) $$m \propto \left(\frac{[Ca]_{in}}{[Ca]_{in} + K} \right)^n$$

(2) $$[Ca]_{in} = [Ca]_R + A$$

(3) $$A = \frac{W[Ca]_{out}}{P + [Ca]_{out}}$$

(4) $$A = \frac{W[Ca]_{out}\,(1+t)}{P + [Ca]_{out}\,(1+t)}$$

(5) $$m \propto \left(\frac{[Ca]_R + A}{[Ca]_R + K + A} \right)^n$$

(6) $$S \equiv \frac{d\log m}{d\log [Ca]_{out}} = \frac{nP}{W} \times \frac{A^2}{[Ca]_{out}(1+t)} \times \left(\frac{1}{[Ca]_R + A} - \frac{1}{[Ca]_R + K + A} \right)$$

Figure 3: The generalized cooperative model
(1): n sites cooperate to produce release, m, after binding independently, with a dissociation constant K to intracellular calcium ions.
(2): The intracellular calcium concentration, [Ca]in, is determined by the intracellular resting level, [Ca]R and an amount, A, contributed by the stimulation. In most conditions A represents predominantly the entry of calcium from the external medium.
(3): It is assumed that A increases in a saturable manner as [Ca]out is raised; P and W are constants which determine the saturable nature and the magnitude of calcium entry, respectively.
(4): The factor, (1+t), a dimensionless function of time, is introduced to account for the influence of repetitive stimulation; to illustrate the main point, we assume a very simple time dependence; in reality, this dependence is probably much more complicated. Note that for t=0, A represents the amount of calcium influx following a single stimulus; different values of t will represent tetanic stimulations (the number of stimuli corresponding to a given value of t may be chosen arbitrarily).
(5): Release, m, based on equations (1) and (2) in conjunction with the more hypothetical assumptions described by (3) and (4).
(6): The slope, S, of the log(m) vs log[Ca]out relation, derived on the basis of (5). Note that the dependence of S on [Ca]out is complex, as it is implicit also in A.
Unless otherwise stated, the following values for the various parameters were chosen in figures 4-6 generated by these equations: n=8; $P=1.5\times10^{-3}$M; t=0; $W=10^{-3}$M; $[Ca]_R=10^{-7}$M; $K=5\times10^{-6}$M.

Two differences between the predictions of the original cooperative model and the generalized one, relate to the slope S (Figure 4):

a. In the original version, the initial slope, at a vanishing [Ca]out, can be taken as a measure of n, the number of critical sites cooperating to produce release; in the generalized case, the value of the slope is always lower than the degree of cooperativity, n.

b. In the original model, the slope declines monotonically as [Ca]out is raised; In the generalized one, the dependence of the slope on [Ca]out is bell-shaped, growing from zero to a maximum and then declining, as [Ca]out is increased.

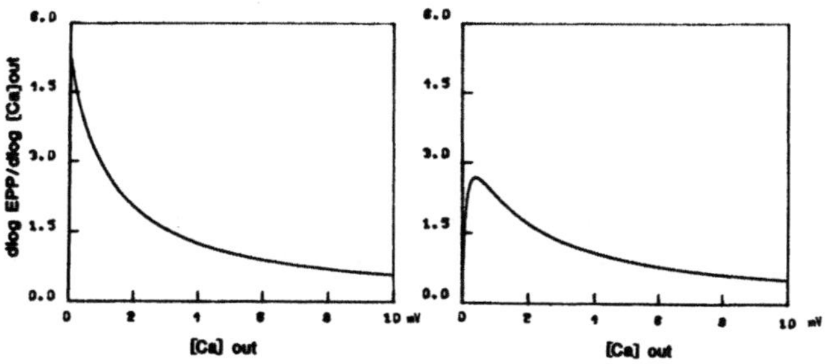

Figure 4: The predicted slope of the log(epp)-log[Ca]out relation as a function of [Ca]out

The slope as a function of [Ca]out declines monotonically when $[Ca]_R=0$ (left) and is bell-shaped when $[Ca]_R=10^{-7}$ M (right). Note that the curve on the left is similar to curves generated by the original cooperative model, while that on the right illustrates the behaviour according to the generalized one.

A third feature of the generalized cooperative model relating to the slope, S, is illustrated in figure 5: The rates of increase and decrease in the slope as a function of [Ca]out, as well as its maximal value, are strongly dependent on the level of intracellular calcium at rest; at very low resting [Ca]in, changes in [Ca]out have a more profound influence on the value of the slope than at higher resting levels.

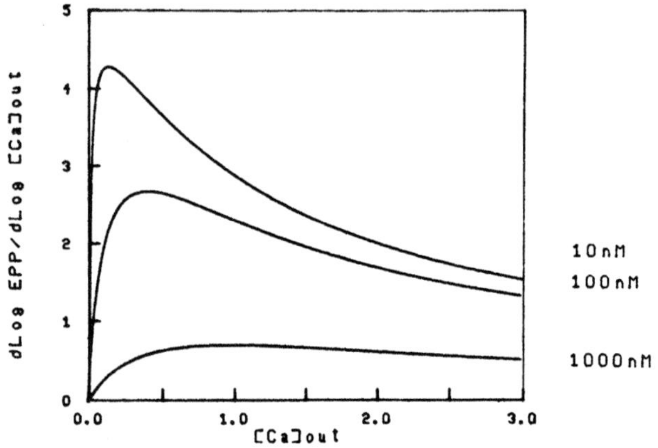

Figure 5: The dependence of S on [Ca]out for a series of $[Ca]_R$. The three curves, from top to bottom, refer to resting intracellular calcium concentrations of 10^{-6}, 10^{-7} and 10^{-8}M.

Figure 6 demonstrates a non-monotonic influence of calcium accumulation during tetani on the slope of the log-log relation at different [Ca]out ranges, as predicted by the generalized model. The two bell-shaped curves in Fig. 6 may correspond to two different "times" during the tetanus (t=0 and t=1) and hence, also, to different amounts of accumulation. Since the two curves cross each other, there will be a low range of [Ca]out for which $S_{t=1} > S_{t=0}$ and a higher range of [Ca]out for which the opposite will hold ($S_{t=1} < S_{t=0}$), in aggreement with the experimental findings.

The generalized cooperative model may thus serve as an explanation for The results presented above. Other models can yield similar predictions. One such model assumes that two conformations of sites, each containing n subunits, act through an allosteric mechanism to produce release. In this case an S function is generated, resembling qualitatively the one derived from the generalized cooperative model, even without introducing the influence of resting [Ca]in. A description of the allosteric model will be presented elsewhere.

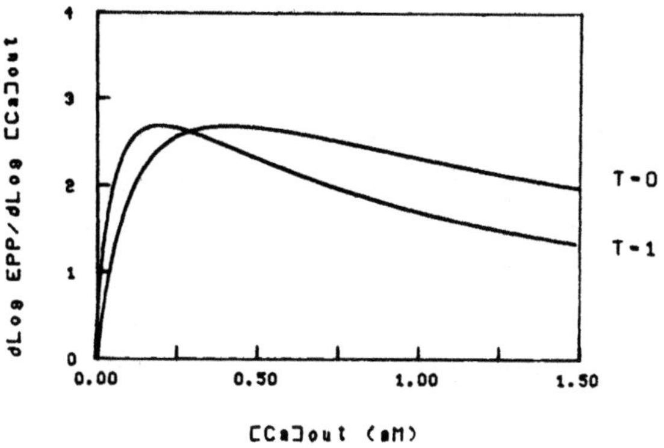

Figure 6: The dependence of S on [Ca]out at different "times" (as it may occur during repetitive stimulations). The two curves were generated for t=0 and t=1. (The point at which they intersect depends on the choice of parameters: The values were chosen arbitrarily as described in the legend to Fig. 3).

It is still too early to state which set of equations describes best the molecular mechanisms of transmitter release. In any event, it is clear that the resting intracellular free calcium concentration affects not only release itself but also certain characteristics of the release process, such as its [Ca]out dependence.

5. ACKNOWLEDGEMENTS

This work was supported by the Fund for Basic Research administered by the Israel Academy of Sciences and Humanities, the Muscular Dystrophy Association and the U.S.-Israel Binational Foundation. Equipment was purchased with the aid of the Bay Foundation.

REFERENCES

1. Alnaes, E. and Rahamimoff, R. (1975) On the role of mitochondria in transmitter release from motor nerve terminals. J. Physiol. 248: 285-306.
2. Barton, S.B., Cohen, I. & Van der Kloot, W. (1983). The calcium dependence of spontaneous and evoked release at the frog neuromuscular junction. J. Physiol. 337: 735-752.
3. Crawford, A.C. (1974) The dependence of evoked transmitter release on external calcium ions at very low mean quantal contents. J. Physiol. 240: 255-278.
4. Dodge, F.A. and Rahamimoff, R. (1967) Co-operative action of calcium ions in transmitter release at the neuromuscular junction. J. Physiol. 294: 91-103.
5. Erulkar, S.D. and Rahamimoff, R. (1978) The role of calcium ions in tetanic and post-tetanic increase of miniature end plate potentials frequency. J. Physiol. 278: 501-511.
6. Erulkar, S.D., Rahamimoff, R. and Rotshenker, S. (1978). Quelling of spontaneous transmitter release by nerve impulses in low extracellular calcium solutions. J. Physiol. 278: 491-500.
7. Jenkinson, D.H. (1957). The nature of the antagonism between calcium and magnesium ions at the neuromuscular junction. J. Physiol. 138: 434-444.
8. Katz, B. and Miledi, R. (1968) The role of calcium in neuromuscular facilitation. J. Physiol. 195: 481-492.
9. Katz, B. and Miledi, R. (1969) Tetrodotoxin resistant activity in presynaptic terminals. J. Physiol. 203: 459-487.
10. Lev-Tov, A. and Rahamimoff, R. (1980) A study of tetanic and post-tetanic potentiation of miniature endplate potentials at the frog neuromuscular junction. J. Physiol. 309: 247-273.
11. Llinas, R. and Nicholson, C. (1975) Calcium in depolarization secretion coupling: an aequorin study in squid giant synapse. Proc. Nat. Acad. Sci., U.S.A. 72: 187-190.
12. Miledi, R. (1973) Transmitter release induced by injection of calcium ions into nerve terminals. Proc. R. Soc. B. 183: 421-425.
13. Parnas, H. and Segel, L.A. (1981) A theoretical study of calcium entry in nerve terminals, with application to neurotransmitter release. Theor. Biol. 91: 125-169.

14. Rahamimoff, R. (1968) A dual effect of calcium ions on neuromuscular facilitation. J. Physiol. 195: 471-480.
15. Rahamimoff, R. and Alnaes, E. (1973) Inhibitory action of Ruthenium red on neuromuscular transmission. Proc. Nat. Acad. Sci. U.S.A. 70: 3613.
16. Rahamimoff, R., Lev-Tov, A. and Meiri, H. (1980). Primary and secondary regulation of quantal transmitter release: Calcium and Sodium. J. Exp. Biol. 89: 5-18.
17. Rosenthal, J. (1969) Post-tetanic potentiation at the neuromuscular junction of the frog. J. Physiol. 203: 121-133.
18. Smith, S.J. and Zucker, R.S. (1980) Aequorin response facilitation and intracellular calcium accumulation in molluscan neurones. J. Physiol. 300: 167-196.
19. Weinreich, D. (1971) Ionic mechanism of post-tetanic potentiation at the neuromuscular junction of the frog. J. Physiol. 212: 431-446.

DISCUSSION OF THE ARTICLE BY R. FINK (ISRAEL)

C.J. Duncan (U.K.): I fully agree with your interpretation of your results but I cannot quite see why at higher levels of calcium you don't find leveling off?

R. Fink (Israel): During the tetanus, you load the terminal with calcium and then the small differences between the additional amounts aren't very effective. Therefore, when you plot it against external calcium concentration, you get a slope smaller than 4.